The Applied Genetics of Plants, Animals, Humans and Fungi

B. C. Lamb

Reader in Genetics
Department of Biology
Imperial College of Science, Technology and Medicine
London

Imperial College Press

Published by

Imperial College Press
57 Shelton Street
Covent Garden
London WC2H 9HE

Distributed by

World Scientific Publishing Co. Pte. Ltd.
P O Box 128, Farrer Road, Singapore 912805
USA office: Suite 1B, 1060 Main Street, River Edge, NJ 07661
UK office: 57 Shelton Street, Covent Garden, London WC2H 9HE

British Library Cataloguing-in-Publication Data
A catalogue record for this book is available from the British Library.

THE APPLIED GENETICS OF PLANTS, ANIMALS, HUMANS AND FUNGI

ISBN 1-86094-179-6

Printed in Singapore by FuIsland Offset Printing

ACKNOWLEGEMENTS, AND COPYRIGHT OF PHOTOS

I am particularly grateful to Mrs Jacqueline Lamb and Mrs Brenda Lamb for help with proofreading, to Dr Simon Zwolinski, Senior Clinical Cytogeneticist at the Royal Victoria Infirmary, Newcastle Upon Tyne, for the human chromosome photographs (Plates 14, 30, 31, 32, 33, 34, 36) and to Professor Mary Seller of the Division of Medical and Molecular Genetics, Guy's Hospital, for invaluable advice on human genetics. I am very grateful to Maxine Whitton and the Vitiligo Society for Plate 26, to Lorraine Kirby of Capespan International for permission to reproduce their poster, Plate 51, and at the Imperial College Biology Department, to Professor Ken Buck for reading the genetic engineering section, to Dr Andrew Goldsworthy for reading the plant physiology section, and to Dr David Dickinson for Plate 4. Copyright of those photographs remains with those people; all other photos were taken by the author and are his copyright.

I shall always be indebted to Denis Bloodworth who taught me biology, to the late Dr Lewis Frost who taught me genetics, to the late Professor Miles Fleming who taught me economics, and to those whose textbooks have often helped with my own teaching, especially those of Professor Monroe Strickberger.

Other people whom I wish to thank for many kinds of help include Dr Rupa Wickramaratne, formerly chief tea and chief coconut breeder in Sri Lanka; Professor Magdon Jayasuria of the Plant Genetic Resource Centre, Sri Lanka; Dr David Ockendon, formerly of the National Vegetable Research Station, Wellesbourne; Dr Martin Howell of the Wessex Human Genetics Institute, University of Southampton; the late Professor Helmut Becker of the Wine School, Geisenheim; various correspondents from the *Daily Telegraph* and *Sunday Telegraph*; Eric Clarke; Doug Hodkinson; Dr Jim Hardie; Ajith Ratnayaka; the late Frank Strong; The Royal Botanical Gardens, Kew; The British Society of Plant Breeders; Dr G. Simm of Scottish Agricultural Colleges; The Monopolies and Mergers Commission and the Office of Fair Trading; the Seven Sisters Sheep Centre, Sussex; members of various breed societies who exhibit at the South of England Show; Catherine Robinson and Hannah; Adefolarin Majekodumi; Tania Joyce; and staff of the International Agricultural Centre, Wageningen, The Netherlands, where I sat in on part of the 1976 International Course on Plant Breeding.

It is a pleasure to acknowledge the help and constructive comments of Imperial College Biology Department students, including Candy Lian, Suhad Moustapha, Steven Cook, Katherine Hammond, Amy Kennell, Robert Elbourn, Richard Frost, Felicity Imoh, Rosalind Carney, Katherine Parsons and Ioanna Stavridou.

DEDICATION

This book is dedicated to my parents, for all they have done for me.

WHY A BOOK ON APPLIED GENETICS?

Plant breeding, animal breeding, human and medical genetics, and the genetics of fungi are usually considered and taught separately, and often only one of these topics is taught to a particular group of students. All these topics are linked, however, by very strong central concepts concerning the generation, control and fate of genetic variation at the levels of genes, chromosomes, genomes, individuals and populations. Mutation, recombination, selection, population genetics and karyotype changes are involved, together with breeding systems and human intervention.

It is therefore possible to teach an integrated course in Applied Genetics based around these central concepts, and which is suitable for those interested in working with plants, animals, humans or fungi. There is so much to learn in modern biology, agriculture, biomedical sciences and microbiology that it is very useful to have some courses which are suitable for people with interests in different types of organism but who can learn the central concepts and then see how they apply in different groups of organisms.

This book is based on - but expanded from - a five-week, full-time course in Applied Genetics taught by the author for many years at Imperial College to final year undergraduates in Biology, Microbiology, Plant Sciences, Zoology, Ecology and other biological subjects. Many students have gone on to research and careers in various areas of applied genetics, including plant breeding, animal genetics, human and medical genetics, molecular genetics and fungal genetics. The students take a first year course in cell biology and genetics equivalent to one-quarter of a year full-time. Chapter 1 of this book includes revision of the basic concepts and terminology of genetics.

In a work of this length, it is not possible to deal with all topics in depth, especially complex statistical concepts and the vast amount of molecular information which has recently become available. Bacteria and viruses have been largely omitted. Readers are referred to other works where necessary. An introduction is given to the economics of agricultural products and breeding programmes, to help students understand commercial realities such as price theory, gluts and shortages, and government interventions. The book is suitable for students in any year of a degree, once they have covered some basic genetics and cytology, such as Mendel's laws and meiosis.

CONTENTS

CHAPTER 1

INTRODUCTION; AIMS OF APPLIED GENETICS; REVISION OF BASIC GENETIC CONCEPTS AND TERMINOLOGY

1.1. Introduction

Genetics is one of the most exciting subjects today. Almost every week there are newspaper headlines or TV programmes about the latest advances in medical genetics or plant or animal breeding. Applied genetics is a very stimulating and topical science, and one whose achievements do an enormous amount to feed and clothe us, and to improve our health. **Applied genetics** is the practical application of genetics to plant, animal and microbial breeding, and to medicine.

As the **primary food producers**, plants are basic to agriculture. Only 20 different plant species provide 90% of our food, with wheat, maize and rice providing more than half of that. See Table 1.1.1 for world production of the top 15 crops. There is an estimated worldwide average of 0.3 ha of arable land per person. Especially in the second half of the 20th century, the work of plant, animal and microbial breeders has been essential in raising standards of health and nutrition throughout the world's six billion people, through providing better crops, meat and microbial products, including antibiotics. In the "Green Revolution", the introduction of high yielding semi-dwarf wheat and rice, with appropriate fertilisers, pesticides and agricultural practices, raised world agricultural output in the period 1972 to 1982 by 25%, with a 33% increase in developing countries and 18% in industrialised countries.

As human populations have increased, starvation has often been avoided by the work of plant and animal breeders, whether working for private firms or for large institutes such as the former Plant Breeding Institute, Cambridge, England, the Centro Internacional de Mejoramiento de Maiz y Trigo in Mexico (Centre for Improvement of Maize and Wheat), and the International Rice Research Institute in the Philippines. On the medical side, there has been a revolution in our understanding of many diseases caused by single genes or by chromosomal abnormalities, in the treatment of some but not all diseases, and in our ability to detect disorders by prenatal testing, for example, by using gene-specific DNA probes. Genetic counselling

1

Table 1.1.1. World production of the top 15 crops, which make up 85% of agricultural production. Figures from ASSINSEL (undated, but 1997).

Crop	Millions of metric tonnes a year
Maize	370
Rice	350
Wheat	320
Potatoes	265
Sugar beets	235
Barley	165
Cassava	150
Sugar cane	130
Sweet potatoes	120
Oats	110
Tomatoes	75
Sorghum	60
Oranges	55
Chickpeas	50
Millet	48

is reducing the suffering and lethality caused by certain inherited conditions, such as thalassaemia and Tay-Sachs disease (see Chapter 13). The massive gene-sequencing effort in the **human genome project** should lead to further advances in understanding.

This book is mainly for undergraduates in any biological, agricultural or medical area, but much of it will be understandable by 16–18-year-old biologists, by farmers, medical practitioners, or the layman with interests in biology. It assumes a knowledge of elementary biology and genetics, including meiosis and Mendel's laws, but this chapter revises many of the **basic concepts** and much of the **terminology**. It is essential to include equations and statistics, but I have concentrated on basic principles rather than on mathematical elaboration. Those needing more equations should consult the quoted literature, especially Falconer and Mackay (1996) and Cameron (1997) on quantitative genetics. **Full references** are given at the end of each chapter, with some works quoted in more than one chapter. Not all the works given in the **Suggested reading** sections are mentioned in the text.

While modern **general textbooks** on genetics often have a lot on applied genetics, it is much easier for the student of applied genetics to find all the information collected into one book, instead of having to search many hundreds of pages for the relevant information. The **examples used** here are usually from organisms used in agriculture, horticulture, industry or medicine, although classic genetic organisms such as *Drosophila*, *Neurospora* and *Escherichia coli* are referred to if they provide the best examples of a phenomenon.

This book is based on a **course** which I have taught for over 20 years at Imperial College of Science, Technology and Medicine, London, to final-year undergraduates in a range of disciplines, including biology, applied biology, microbiology, plant sciences, animal sciences, and ecology. The course lasts five weeks full-time, with about 27 lectures, plus practicals, tutorials, problem sessions and slide shows. The book contains much additional information. It retains the key approach of teaching the **general principles of applied genetics** which apply to plants, animals, microbes and humans, with relevant examples from any of those groups, rather than teaching say plant breeding separately from animal breeding. Specialist aspects of each of those groups are also included, such as reproductive physiology (Chapter 17); the microbial area usually excludes bacteria and viruses, as Prokaryotes have such basic differences from Eukaryotes. The section on applied fungal genetics is partly included because the application of ideas from fungal genetics helped so much with human genetics, especially for cell fusion and parasexual techniques, and because many of the techniques are applicable to other organisms.

The text includes **worked examples of problems** because students find that they greatly improve their understanding of the theory, as well as giving them experience of data analysis and answering practical problems. Readers are encouraged **to try doing the problems** but they should first cover up the answers, which follow immediately.

Readers who are very familiar with genetics can skim over the revision section in Chapter 1.3, checking whether they know all the terms in **bold print**. Understanding the terminology and symbols is essential. This book has extensive cross-referencing between sections and some minor repetitions to help those following their own sequence of topics when reading the book, and because some topics, such as selection, occur in many parts. **Bold print** is used to draw attention to key terms.

When using **genetic symbols**, it is normal to leave no gap between alleles at a locus if the symbols are of one letter only, e.g., *aa*, but to leave a gap when the symbols are of more than one letter, e.g., *vg vg*. In this book, I have left a gap between the two alleles at one locus, whether the symbol is of one or more than one letter, and have put a comma between symbols for different loci, e.g., *a a, B b, C'*

C, w^e w^a. This makes for greater clarity, especially when superscripts and primes are being used. It also helps to distinguish letters used as allele symbols, when spaces are used, and letters used for whole genomes or chromosome regions, when no space is used between them.

It is very important that students and users of applied genetics should have the subject placed firmly in a context of the realities of **economics and government regulations**. These are stressed throughout the book, as well as being covered in Chapter 19. Although the prices and laws mentioned will become out of date, they should still be useful for comparison. Because the laws in the USA may differ from state to state, most of the references to laws are to ones in Britain or the EU. The examples of applied genetics are taken from many countries, with a natural bias towards those best known to the author.

1.2. The aims of applied genetics in plants, animals, humans and fungi; the use of genetic variation; improving the harvest index

With animals, plants and fungi, the **main aims of applied genetics** are to increase the yield and quality of the required product or products, to reduce the costs of production and/or purification, and to reduce vulnerability to disease and to climatic variation. With dairy cows one wants to increase milk yield and milk nutritional value (protein, fats, vitamins), decrease the cost of feed, shelter, transport, land, labour - especially of feeding and milking - and veterinary expenses, and to minimise yield decreases from unfavourable weather. With humans, the aims are humanitarian and medical rather than economic - to understand and find ways of detecting, curing, treating or preventing disorders, especially hereditary ones, including chromosomal and genetic disorders. Because medical expertise, facilities and many drugs are expensive, economic factors cannot be ignored in human applied genetics, e.g., in deciding whether mass screening for a genetic disease is cost-effective. Chapter 19 deals with basic economics and its application to agriculture.

Depending on the organism's biology, genetics and economics, the applied geneticist may use **selection** for better phenotypes and genotypes using **pre-existing genetic variation**, or may use **induced mutations** to increase variation, or use **hybridisation**, or **changes in ploidy** or in **individual chromosomes**. The breeder may have **one main product** to consider, such as the drug penicillin from the fungus *Penicillium*, young leaves from tea plants, milk from dairy cattle, meat from beef cattle, or more than one main product. In sheep the main product is usually meat but wool is important in many breeds, and so are milk and cheese in some breeds. The coconut tree has many products, including coir fibre from the husk, charcoal for filters from the shell, copra (the flesh, used for eating, oil and desiccated coconut), coconut milk from grated flesh plus water, coconut water as a drink, leaves for thatching and wood for building. In coconut plantations there are also valuable **undercrops** including bananas, coffee, pineapples and limes which could be affected by changes in coconut palm canopy density. With **multiple products**, the breeder has to decide which products to try to improve most, taking into account their relative values and whether there are positive or negative correlations between their yields. For example, meat production in sheep might decrease if selection increased the amount of energy devoted to wool production, a negative correlation.

Take **beet** as a plant example, with **sugar** as its product. The total yield depends on the amount of root produced and on the sugar content of the root. The sugar content in 1747 was less than 2% but selection increased this to 6% by the 1820's, and to 9% by the 1830's. By 1858, crosses of *Beta vulgaris* x *B. maritima* gave up to 14% sugar, and modern beets yield over 20% sugar. Triploid varieties are produced by crossing diploids with tetraploids, giving high yields of roots and of sugar. They often have less ash and nitrogen, which is helpful for processing. Yield is more certain as genetic resistance has been incorporated to various diseases, such as *Cercospora* and some viruses. Mechanised production has been simplified by the breeding of plants with "monogerm" seedballs, with only one seed, instead of several seeds per fruit, and which therefore do not require thinning after germination.

With a **human disorder**, one would try to find out whether it was caused by infection, by diet or other environmental factors, by heredity from one to many genes, by chromosome number abnormalities or chromosome structural aberrations, and how best to treat the condition and to prevent its occurrence in future births. Diagnosis today often includes karyotyping, DNA analysis and biochemical testing, as well as physical examination. One can now treat, but not cure, many hereditary disorders by surgery, physiotherapy, diet or drugs, with gene therapy now being tried (Chapters 13 and 14).

Applied genetics in non-human organisms usually involves the manipulation of inherited genetic variation to achieve the aims set out at the beginning of this section. That variation may pre-exist in a population from previous mutations, chromosome aberrations, recombination, or from immigration. See Plate 1 for variation in *Sorghum* panicles. Further variation may be introduced by the breeder by mutation, recombination, genetic manipulation, or by bringing in new variants from other populations and even from other species. Where there are heterozygotes, they can be considered as storing genetic variation, with **segregation** from single heterozygotes, say *A a* giving *A A* and *a a* types as well as more *A a*. With double heterozygotes, say *A a, B b*, **recombination** releases variation, giving *A A, B B*; *A*

A, B b; *A A, b b*; *A a, B B*; *A a, b b*; *a a, B B*; *a a, B b* and *a a, b b*, as well as more of *A a, B b*.

Over long periods of evolution, species develop genes which work well with each other: **co-adapted gene complexes**. Where different species usually grow together, such as clovers and grasses in pasture, one may even get genetic co-adaptation between species. Breaking up well-adapted gene complexes by recombination, mutation or genetic engineering may have adverse effects, and improving the yield of one species in a mixture may have adverse effects on other species. The breeder needs a general biological awareness, as well as specialised knowledge.

The applied geneticist often works in **multidisciplinary teams**, with people specialising in cytology, molecular biology, biochemistry, soils, fertilisers, plant and animal pests and diseases, agronomy, agricultural engineering, etc. Economic forecasting is often crucial to the planning of long-term breeding projects: what will the consumer want by the time a project has been completed? Public and media opinion strongly influences future markets, and so do government actions. The current fashion for vegetarianism, with about four million vegetarians in Britain, reduces the demand for meat products, and government action over bovine spongiform encephalopathy (BSE) has depressed the beef market in the UK. The present surge of opposition in Europe to genetically modified crops (Chapter 14) will affect the economics of firms producing such crops.

Selection by breeders may involve measurable characters such as liveweight in meat animals, but care needs to be taken because total body weight is affected by fleece weight in sheep, while in ruminants, 10–25% of liveweight may be gut contents, with large diurnal variations. Selection in animals may directly or indirectly involve economically important **behavioural traits** such as libido in bulls, rams, stallions, boars and cocks, eating persistence, temperament and reactions to handling, and mothering ability in females. In plants especially, the sensitivity of yield to **environmental variables** must be considered when attempting to select good genotypes from good phenotypes.

As well as breeding programmes to increase yield, advances are often through **cost-reductions** and increasing the **harvest index** (the percentage of the harvested organism consisting of the most marketable product, e.g., the grain from wheat, or meat from beef cattle). Thus the very successful semi-dwarf wheats and rice have a higher proportion of valuable grain to less valuable straw, compared to traditional tall varieties. In chickens for egg laying, the number of eggs per hen has usually reached a plateau in most breeds (about 230 eggs to 500 days age, with negative correlations between egg number and egg size), but breeding for better feed utilisation has reduced costs after selection for a smaller body size. Similarly in chickens for meat, breeders have improved the growth rate and the food conversion to meat, with the modern broiler chicken taking only eight weeks to reach a weight of 1.6 kg from only 3.4 kg of feed. In dairy cows, a cow with a poor milk yield, say 1,450 litres per lactation, might use 44% of food for milk and 56% for keeping the cow alive, while a cow with a good yield, say 3,900 litres per lactation, might use 65% of food for milk and only 35% for maintenance. Cost-reductions can improve profitability without causing a glut: see Chapter 19.2.1 for the disastrous effect of gluts on farmers' incomes.

1.3. Revision of basic genetic concepts, definitions and symbols

1.3.1. Alleles, genes, loci, wild-types and mutants

A **locus** is a length of a chromosome occupied by one gene; it is the position of the gene, including all possible alleles, although any one position is occupied by only one allele at a time. Thus the locus controlling red-green colour blindness in humans is in band q28 on the X chromosome, whatever allele it carries, mutant or wild-type. A **gene** is a unit of hereditary function and information, usually being transcribed into RNA or having controlling functions. Alternative forms of a gene are called **alleles**. In peas, there is a locus of major effect controlling plant height, as investigated by Mendel. At that position (locus) on that chromosome, there may be the T allele, giving tall plants about 183 cm high, or the t allele which as $t\ t$ gives short plants about 30 cm high. Confusingly, people use the term **gene** in the sense of locus: "the gene for height in peas", and in the sense of one allele: "the gene for tallness in peas".

Sometimes there is one allele which is much more common than other forms; it is called the **wild-type allele**, and other forms are regarded as **mutant**, especially if they lack the function coded for by the wild-type allele. Thus the allele for not having haemophilia A in humans is regarded as the normal or **wild-type** allele, able to specify the appropriate blood-clotting factor VIII, but alleles giving haemophilia A, an inability to clot blood in wounds, are regarded as **mutant**. What is regarded as the wild-type may even vary between populations. For example, in Negroid Africans the allele for brown eyes is the normal one, with very few blue-eyed individuals, but in Scandinavia, blue-eyed individuals predominate. In Britain, one could not specify which of blue or brown eyes was the wild-type, because both are common.

At the **DNA level**, one can have several or more different wild-type alleles for a locus, all with complete function but with slightly different base sequences and occasionally with slightly different amino acid sequences, where the differences are not crucial. There are very many **potential mutants** for a locus, because for any given wild-type sequence there are three possible different mutant base pairs at any wild-type base pair position, with hundreds or thousands of base pairs in the gene. Many DNA changes from wild-type will have the same phenotype if they give the same loss of function, and may only be distinguishable from each other by DNA analysis or mapping. Thus different small pea plants may all have the phenotype given by $t\ t$, but may have chemically different t alleles. Some DNA changes may give a wild-type phenotype under some conditions and a mutant phenotype under other conditions. For details of when DNA changes give different phenotypes and when they do not, see Chapter 7.1 and Lamb (1975). Some loci have many known alleles, with over a hundred alleles at some higher plant self-incompatibility loci (Chapter 17.1.2). At the human *ABO* blood group locus, there are three main alleles, i^A, i^B and i^o, while **multiple alleles** are very common at the blood group locus *EAB* in cattle, with over 600 alleles, and in the *EAB* locus in sheep, with over 100 alleles.

There are names for **different types of mutation**. In the fruit fly *Drosophila*, wild-type eye colour is red, from dominant allele W; the recessive white-eye mutant, w, has a total loss of eye-colour function, so is an **amorph**, and there are several mutations with partial loss of function, **hypomorphs**, such as white-eosin, w^e, and white-apricot, w^a, with some colour left.

1.3.2. Ploidy

The **ploidy** of a particular stage in the life cycle of an organism is the number of copies of each typical chromosome in the nucleus. Thus the gamete stage of pea plants and maize (pollen, egg cell), cattle and humans (sperm, eggs) is **haploid**, with one copy of each chromosome per nucleus, while the adult stage is **diploid**, with two copies of each typical chromosome. Yeast can grow and multiply vegetatively by mitosis in both the haploid and diploid stages, but meiosis only occurs in the diploid. **Polyploids** have more than two copies of each type of chromosome. Examples are commercial bananas, **triploid**, with three of each chromosome; potatoes, **tetraploid**, four copies; bread wheat, **hexaploid**, with six copies; and strawberries, **octaploid**, with eight copies. The leek, *Allium porrum*, is usually tetraploid but its probable ancestor, the wild sand leek, *A. ampeloprasum*, can be diploid (2n = 16), tetraploid or rarely hexaploid.

A **euploid** has an exact multiple of the basic chromosome number, e.g., three copies of each type of chromosome, but an **aneuploid** has an inexact multiple of that number, e.g., three copies of six types of chromosomes, but one, two or four of a seventh type. In man, Down syndrome (Chapter 13.8.2) is an aneuploid with three copies of chromosome 21 (Plates 35 and 36).

1.3.3. Genotype and phenotype; homozygotes and heterozygotes; hemizygotes

The **genotype** is the genetic make-up of an organism, determined by what alleles it carries in its nucleus and in extra-nuclear DNA, while the **phenotype** is the manifested characteristics of the organism, which depend on both genotype and environment. Thus in the diploid pea plant there are three possible genotypes for the *T/t* locus, and two phenotypes. The *T T* genotype gives a tall phenotype, as does *T t*, and *t t* gives a short phenotype.

If the two alleles in a diploid are alike, the genotype is **homozygous**, e.g., *T T* or *t t* in a **homozygote**, while if they are unlike, the genotype is **heterozygous**, e.g., *T t* in a **heterozygote**; one writes the dominant allele before the recessive. The terms are not applicable in haploids, where there is only one allele for each locus, a condition termed **hemizygous** in a **hemizygote**, which also applies in X Y individuals to those genes on the X sex chromosome which have no counterpart on the Y. Some genes are actually **repeated** many times in one set of chromosomes, or even in one chromosome, such as some genes specifying transfer RNAs, but for simplicity most genetic definitions relate to single-copy genes. The *Alu* sequence in man occurs more than 500,000 times per genome, involving all chromosomes.

1.3.4. Dominance and recessiveness; incomplete dominance and additive action; primes; overdominance; co-dominance; pure breeding

In a heterozygote, the allele which is expressed in the phenotype is the **dominant allele**, and the one not expressed is the **recessive allele**. In pea plants, *T* (tall) is dominant over *t* (short) because the heterozygote *T t* has a tall phenotype. The dominant **allele's symbol** starts with a capital letter, e.g., *T*, *Cn*, *Bw*, but the recessive allele's symbol is all in lower case letters. A particular phenotype may sometimes be dominant and sometimes be recessive, depending on the loci concerned. In blackberry (*Rubus*) for example, thornlessness is dominant in octoploid "Austin Thornless" but recessive in tetraploid "Merton Thornless". In animals, albinism is often recessive but sometimes dominant, and the dark melanic form is dominant in some insects, recessive in others.

Dominance is not always complete. Suppose homozygote *S S* has fruits 20 cm long and homozygote *s s* has fruits 10 cm long. The capital *S* suggests that long fruit is dominant to short, so the heterozygote *S s* may have fruits 20 cm long. If dominance is not complete at this locus, *S s* might have fruits say 18 cm long, so there is **incomplete (partial) dominance** of *S* to *s*. If *S s* had fruits only 10 cm long, short would be dominant to long, but the allele symbols would be wrong. Sometimes there is **no dominance**, with the heterozygote completely intermediate between the two homozygotes, say 15 cm here. This is sometimes called incomplete dominance, but to distinguish it from partial dominance, the term **additive action** is often used. Thus the basic fruit length in this example is 10 cm, with one allele for increased length adding 5 cm to that, and a second allele for length adding another 5 cm to give 20 cm for the long form. To avoid confusion with symbols for dominant and recessive, a scheme of two identical symbols with an initial capital letter is used for additive action, with the allele giving the increase having a **prime** symbol after it, e.g., *S′*. Thus *S′ S′* gives 20 cm (the basic length 10, plus two lots of 5 cm), *S′ S* gives 15 cm (10+5) and *S S* gives 10 cm. Sometimes a system of lower-case superscripts is used for incomplete dominance. Thus in *Antirrhinum*, white flowers and red flowers are incompletely dominant to each other, giving pink in the heterozygote, with allele symbols c^w and c^r for white and red, with *c* standing for flower colour. Incomplete dominance for plant leaf colour is shown in Plates 12 and 13.

An agricultural example of a gene with additive effects is found in Merino sheep, affecting the ovulation rate. There is an average of 1.3 ova per oestrous cycle for the homozygous normal ewe, 2.8 for the heterozygote, and 4.3 for the homozygote carrying the Booroola high fecundity allele, Fec^B (the allele symbol does not follow the pattern given here), with the heterozygote exactly intermediate between the two homozygotes. Because the number of lambs born per litter is less than the number of ova, the three genotypes give respectively averages of 1.2, 2.1 and 2.7 lambs per litter, with proportionately more lambs lost per ovum at higher lambing rates than at lower lambing rates. The litter size therefore shows incomplete dominance of the Booroola allele over the normal one, with the heterozygote nearer to the $Fec^B Fec^B$ homozygote than to the normal homozygote.

Overdominance is where the heterozygote's phenotype lies outside the range of those of the two homozygotes, e.g., if *S′ S* in the above example had fruits 30 cm long, even longer than those of the 20 cm *S′ S′* genotype. It is important in hybrid vigour (Chapter 12.1).

Co-dominance is a condition where two alleles are both fully expressed in the heterozygote, and so co-dominance differs from incomplete dominance where each allele is only partly expressed in the phenotype. One could not have co-dominance for two alleles for height, one giving tallness and one giving shortness, as they cannot both be fully expressed simultaneously. In the human ABO blood groups, alleles i^A and i^B are dominant to i^o and show co-dominance with each other. The heterozygote $i^A i^B$ has full expression of blood antigens A and B, and so shows co-dominance for the two alleles. The three allele symbols have i for immunity group and capital letter superscripts for the co-dominant alleles, and a lower-case superscript for the recessive allele.

An organism is **pure breeding** if it breeds true for a particular character; that is, on self-fertilising or being crossed with another pure breeding individual with the same phenotype, it gives all offspring like each other and like the parent(s) for that character. Heterozygotes will not breed true as they will segregate to give some homozygotes of each genotype and phenotype: $T t$ selfed or $T t$ x $T t$ gives $T T$, t t and $T t$ progeny, but $T T$ selfed or $T T$ x $T T$ gives only $T T$ progeny, so the homozygote is pure breeding.

1.3.5. Additive and multiplicative gene action

In some cases an allele increasing say length might add a fixed increment, e.g., $S S$ is 10 cm long, and with **additive action**, S' might add 5 cm, so that $S' S$ is 15 cm long and $S' S'$ is 20 cm long. In other cases, the existing length might be multiplied by each contributing allele. Suppose we now have **multiplicative action**, with $D D$ being 10 cm long, and each D' allele multiplies the existing length by 2. $D' D$ will then be 20 cm (10 x 2) long and $D' D'$ will be 40 cm long (10 x 2 x 2). Getting three phenotypes for one locus with lengths of 10 and 40 cm for the two homozygotes and 20 cm for the heterozygote could also be called incomplete dominance of short over long.

1.3.6. Mutation

The term **mutation** is used both for a process and the result of the process, when a heritable change occurs in a gene. Thus a change from a wild-type H (non-haemophiliac) allele on a human X chromosome to a mutant h (haemophilia A) allele is a mutation, involving a change in the DNA base sequence and in the amino acid sequence in the polypeptide. Mutations can occur spontaneously (Plate 2) or be induced by man using various mutagens such as X-rays or certain chemicals (Chapter 7, and Plates 11, 12, 13 and 50). In yeast, a wild-type $ade\text{-}1^+$ allele (giving cells not requiring added adenine) might mutate to an $ade\text{-}1^-$ allele, giving cells requiring the base adenine for growth. A "$^+$" superscript symbol indicates a wild-type allele and a "$^-$" superscript indicates a mutant allele. $ade\text{-}1$ stands for adenine locus 1, to distinguish it from other loci controlling other enzymes in the adenine pathway in yeast. Going from the mutant *ade-1*⁻ back to the wild-type is called **reversion** or **back-mutation**. Some books also use the term **mutation** for a change at the level of a chromosome, not just at the gene level, but it helps to keep the two effects separate if one uses the term **chromosome aberration** for changes on a larger scale than that of a single locus, and **mutation** for changes at the level of the gene. See Chapter 7 for mutation and Chapter 9 for chromosome aberrations.

1.3.7. Recombination; linkage; syntenic and non-syntenic loci; coupling and repulsion arrangements

Unlike mutation, where new alleles are produced, **recombination** is the production of **new combinations of existing genes**. For example, a diploid of genotype *A a, B b* formed from the combination of an *A, B* and an *a, b* gamete could produce new recombinant combinations *A, b* and *a, B* in its gametes from meiosis, as well as parental gene combinations *A, B* and *a, b*. The **recombination frequency** is the number of recombinant genotypes x 100, divided by the total number of parental plus recombinant genotypes. Thus if the above *A a, B b* diploid gave the following gamete numbers, 38 *A, B*; 43 *a, b*; 12 *A, b*; 15 *a, B*, the recombination frequency as a percentage (RF %) is $\frac{(12+15) \, x \, 100}{38+43+12+15} = \frac{2,700}{180} = 25\%$. When measuring recombination frequencies, we want the results from meiosis, which in higher organisms means from the gametes. We can measure recombination frequencies directly from meiotic products in fungi such as *Neurospora crassa* and yeast, where meiosis produces haploid ascospores giving rise to haploid colonies, but in higher organisms we usually have to cross the gametes and examine diploid progeny, as most genes cannot be scored in gametes.

Linked genes are preferentially inherited in the parental arrangements rather than in the recombinant arrangements, so genes are linked if they show significantly less than 50% recombination, like loci *A* and *B* in the above example, in which both are say on chromosome 1. Unlinked genes show about 50% recombination from meiosis. Linkage is thus defined in terms of recombination frequencies. Genes close together on a chromosome show low recombination frequencies, increasing to a maximum of 50% for distant genes.

As genes on non-homologous chromosomes also show 50% recombination, it is helpful to use the term **syntenic** for any genes on the same chromosome (or pair of homologous chromosomes) and **non-syntenic** for genes not on the same chromosomes. If locus *C/c* is also on chromosome 1 but far from loci *A/a* and *B/b*, it would show 50% recombination with them, because crossovers and multiple crossovers would be frequent over the long distance. If locus *F/f* is on a different pair of chromosomes, say chromosome 5, its alleles would show 50% recombination with genes *A*, *B* and *C* because they would assort independently at meiosis. Linked loci, with less than 50% RF like *A/a* and *B/b*, must be syntenic, but unlinked loci, with 50% RF, may be syntenic but far apart, like *A/a* and *C/c*, or may be non-syntenic, like *A/a* and *F/f*.

If two loci are segregating in a cross of pure-breeding (homozygous if diploid) parents, one uses the term **coupling arrangement** (or coupling phase) if all recessive alleles at these two loci are together in one parent and all the dominant alleles are together in the other parent, e.g., *A A, B B* x *a a, b b* in diploids or *A, B* x *a, b* in haploids, say in yeast, while in a **repulsion arrangement**, each parent has recessive alleles at one locus and dominant alleles at the other locus, e.g., *A A, b b* x *a a, B B* in diploids or *A, b* x *a, B* in haploids.

One can usefully use these terms for a double heterozygote, as genotype *A a, B b* could have come from *A, B* + *a, b* gametes (coupling combination) or from *A, b* + *a, B* gametes (repulsion combination). If the two loci are non-syntenic, the coupling and repulsion double heterozygotes will be identical, with each of the four alleles on a different chromosome. However, if the loci are syntenic, the coupling double heterozygote has one chromosome carrying *A, B* and the homologous chromosome carrying *a, b*, while the repulsion double heterozygote has one chromosome carrying *A, b* and the other carrying *a, B*. It is essential to register the difference between coupling and repulsion arrangements because they affect the calculation of recombination frequencies for linked loci. Thus gametes *A, B* and *a, b* are parental when coming from a coupling double heterozygote, but are recombinant when coming from the repulsion double heterozygote. As the *A a, B b* double heterozygote mentioned at the beginning of this section gave significantly more coupling gametes (38 *A, B*; 43 *a, b*) than repulsion gametes (12 *A, b*; 15 *a, B*), we can deduce that it is a coupling phase double heterozygote, with *A* and *B* on one chromosome and *a* and *b* on the homologous chromosome, with linked and syntenic loci.

At meiosis, recombination for non-syntenic loci comes from **independent assortment** of pairs of non-homologous chromosomes (Mendel's second law), while recombination of syntenic loci usually occurs by **reciprocal crossing-over** at pachytene. The non-reciprocal process of **gene conversion** can also recombine genes, but is usually only important for recombining alleles within a locus. At very much lower frequencies than at meiosis, one can also have **mitotic recombination** of syntenic and non-syntenic genes (Chapters 8.2 and 18.4). See Chapter 8 for more on recombination and mapping.

1.3.8. Allelism and the cis/trans test

Alleles are alternative forms of a gene. If one isolated 60 *Escherichia coli* bacterial strains all requiring the amino acid histidine, some might be mutant at the same locus, say *hisB*, specifying one enzyme (IGP dehydrase) in the histidine pathway, while others might be mutant at *hisC* (specifying IAP transaminase), or *hisA*, or *hisD*, etc., specifying different enzymes. Different loci controlling the same biochemical pathway are distinguished by their final capital letter, e.g., *hisA*, *hisD* (or sometimes by number symbols as in *ade-1*, *ade-2*), and different mutants at a locus are distinguished by different final superscript numbers, e.g., $hisB^{1-}$, $hisB^{2-}$, with $hisB^+$ representing the wild-type allele. One often needs to find out whether two mutants with similar phenotypes, such as both requiring histidine, are at the same locus (alleles, e.g., $hisB^{1-}$, $hisB^{8-}$) or are at different loci (not alleles, e.g. $hisB^{1-}$, $hisC^{5-}$).

One can map these 60 mutants, where only ones very close together could be alleles, but being close does not prove that they are alleles. In Eukaryotes, two mutations with recombination frequencies more than about 2% are usually too far apart to be alleles. One can also test which precursors of histidine these *E. coli* strains can grow on, since alleles will have the same requirements as each other because they are blocked at the same point in the biochemical pathway.

The best test is the **cis/trans test for functional allelism**, originally devised by Ed Lewis for *Drosophila*, but extensively used in Prokaryotes and lower and higher Eukaryotes. By crossing (or, depending on the organism, using transformation, transduction, or heterokaryons) one makes two diploids or partial diploids, the **cis** arrangement with the two mutants coming from the same parent and the two wild-type alleles coming from the other parent, and the **trans** arrangement, with one mutant and one wild-type from one parent, and the other mutant and the other wild-type from the other parent: see specimen genotypes below, and see Plate 3 for a *cis/trans* test with three adenine-requiring red mutants in yeast. It does not matter whether the two mutants are syntenic or not, but they cannot be alleles if they are non-syntenic or syntenic but far apart.

Let us consider three *E. coli* histidine-requiring single mutations, 1, 2 and 3, where 1 and 2 are allelic, both at the *hisB* locus, while 3 is not allelic with 1 and 2, being at the *hisC* locus. They will have these genotypes and phenotypes:

Single mutants, all requiring histidine:

mutant 1 is $hisB^{1-}$ $hisB^{2+}$, $hisC^{3+}$ genotype, so is hisB⁻, hisC⁺ phenotype, because the *hisB* gene has a defect;

mutant 2 is $hisB^{1+}$ $hisB^{2-}$, $hisC^{3+}$ genotype, so is hisB⁻, hisC⁺ phenotype;

mutant 3 is $hisB^{1+}$ $hisB^{2+}$, $hisC^{3-}$ genotype, so is hisB⁺, hisC⁻ phenotype.

In *E. coli*, one can make partial diploids by transduction with bacteriophage. Provided that the two mutations are both recessive to their wild-types, the *cis* arrangement will give a wild-type phenotype both for alleles and for non-alleles. The phenotype difference comes in the *trans* test, where the *trans* partial diploid will be wild-type if the mutations are not alleles, but will be mutant for two alleles. Non-alleles **complement** (complete each other, make good each other's defect) in *trans* because each brings the wild-type allele which the other lacks. In contrast, alleles of the same locus do not complement in *trans* because both mutants are defective at the same locus.

(i) Partial diploids between mutants 1 and 2, **alleles** at the *hisB* locus.

    (a) In *cis*, one genome has *hisB^{1-} hisB^{2-}, hisC^{3+}*, specifying bad IGPdH and good IAPt; the other genome has *hisB^{1+} hisB^{2+}, hisC^{3+}*, specifying good IGPdH and good IAPt, so the diploid phenotype is wild-type, not histidine-requiring, as both enzymes have some good molecules made.

    (b) In *trans*, one genome has *hisB^{1-} hisB^{2+}, hisC^{3+}*, specifying bad IGPdH and good IAPt; the other genome has *hisB^{1+} hisB^{2-}, hisC^{3+}*, specifying bad IGPdH and good IAPt, so the diploid phenotype is mutant, histidine-requiring, as there is no good IGPdH enzyme.

(ii) Partial diploids between mutants 1 and 3, **not alleles**, as they are mutant at different histidine loci, *hisB* and *hisC* respectively.

    (a) In *cis*, one genome has *hisB^{1-} hisB^{2+}, hisC^{3-}*, specifying bad IGPdH (the gene has an error at site 1) and bad IAPt; the other genome has *hisB^{1+} hisB^{2+}, hisC^{3+}*, specifying good IGPdH and good IAPt, so the diploid phenotype is wild-type, not histidine-requiring, as both enzymes have at least some good molecules made.

    (b) In *trans*, one genome has *hisB^{1-} hisB^{2+}, hisC^{3+}*, specifying bad IGPdH and good IAPt; the other genome has *hisB^{1+} hisB^{2+}, hisC^{3-}*, specifying good IGPdH and bad IAPt, so the diploid phenotype is wild-type, not histidine-requiring, as there is some good IGPdH enzyme and some good IAPt enzyme.

1.3.9. Heritability

Heritability is the proportion of phenotypic variance (*Vp*) attributable to genetic variation in a population, with the term applying to characters with continuous variation. One distinguishes between narrow-sense heritability, *h²*, which is the proportion of phenotype variance attributable to genes with additive effects, and broad-sense heritability, *H²*, which is the proportion of phenotypic variance attributable to all types of genetic variation. The variances due to different types of genes are *Va* for genes with additive effects, *Vd* for genes with dominance, and *Vi* for genes with epistasis or gene interactions (Chapter 2.2). *Ve* represents the amount of variation due to the environment and *Vg/e* represents variation due to genotype/ environment interactions, where different genotypes perform differently in different environments. The basic equations are:

$$\text{narrow sense heritability: } h^2 = \frac{Va}{Va + Vd + Vi + Ve + Vg/e} \; ;$$

$$\text{broad sense heritability: } H^2 = \frac{Va + Vd + Vi}{Va + Vd + Vi + Ve + Vg/e} \; .$$

See Chapter 3.3 for more details and how to determine heritabilities.

It is extremely important to know how much of the observed phenotypic variation is due to genetic variation and how much is due to environmental variation when one is contemplating running **selection programmes**. If nearly all the variation is environmental, or if there is very little phenotypic variation, it is probably not worth running a selection programme to choose individuals with better genes. Those interested in the total causes of variation for a character would use broad sense heritabilities, but practical plant and animal breeders often use narrow sense heritabilities because they give a more reliable estimate of the minimum likely gain from selection. If one considers just one locus, genes with additive action will give three phenotypes, say due to *S′ S′*, *S′ S* and *S S*, and selecting the best phenotype, say that due to *S′ S′*, will select the best genotype. With complete dominance, however, if one wants the dominant allele, selecting the dominant phenotype selects both for the dominant homozygote *T T* (wanted) and for the heterozygote, *T t*, which has an unwanted recessive allele, *t*.

1.3.10. Selection

Selection against a particular phenotype means that that phenotype leaves fewer surviving offspring than corresponding numbers of other phenotypes. Selection acts on phenotypes rather than on genotypes, and may be due to lower viabilities, and/ or to lower reproductive success, and/or to producing fewer viable offspring. It can occur at any stage of the life cycle — in gametes, zygotes, embryos, at birth or germination, youth, middle age or old age, although post-reproductive selection is much less effective than earlier selection. Chapter 5 covers types and uses of selection, with more examples in Chapter 12.

Selection may be **natural** (by the environment), **sexual** (in competition for mates) or **artificial** (by humans, e.g., plant and animal breeders). It is measured by the **selection coefficient**, s, which is the proportional reduction in fitness and/or reproductive success due to selection. If a phenotype has no selection against it compared to others, $s = 0$ and it is **selectively neutral**. If a phenotype is **lethal or sterile**, leaving no surviving offspring, $s = 1.0$. A **deleterious** but non-lethal phenotype will have an intermediate value for s, between 0 and 1.0. Brown versus blue eyes is almost selectively neutral in Britain, but brown eyes have a slight selective advantage in very sunny countries such as Nigeria, giving more protection against strong light, while blue eyes have been favoured in Norway, perhaps allowing better vision in poor light. For males with schizophrenia, $s = 0.51$, so if 100 normal males left 200 surviving offspring, 100 schizoid males would leave 200 $(1 - s)$ offspring = 98. For females with schizophrenia, selection is less, $s = 0.22$, so 100 schizoid females would leave 200 $(1 - s)$ = 156 surviving offspring if 100 normal females leave 200 offspring.

Selection coefficients may vary with the **environment**. Take penicillin as an example. In a drug-free environment, drug-sensitive bacteria (pen^S) usually have a slight selective advantage over drug-resistant bacteria (pen^R), because mechanisms of resistance have metabolic costs, but in the presence of the drug, being drug-sensitive is usually lethal. Wild populations of the bacteria are largely drug-sensitive, with occasional spontaneous mutations to drug-resistance, while in someone being treated with penicillin, nearly all surviving bacteria are drug-resistant. The allele frequency of pen^S therefore drops sharply and that for pen^R increases rapidly when the drug is administered. Once the drug is no longer taken, there is slow selection in favour of drug-sensitive cells, a few of which may have escaped the drug, or they may come from back-mutation from drug-resistance to drug-sensitivity.

1.3.11. Populations and population structure

Populations are of many different kinds, with no single ideal definition; a simple definition is "a community of sexually interbreeding organisms with a common gene pool". "Community" implies existing at the same time and geographically close to each other. Suppose we have a small population of five organisms, with cross-fertilising males (m) and females (f) numbered 1 to 5, but in generation 1, individual 3 does not reproduce. Fig. 1.3.11.1 shows some possible matings over two generations, where the numbers in the second and third generations show the numbers of the individuals in generation 1 from which their genes derive. The bottom left individual comes from a brother-sister mating and the bottom right-hand individual comes from a mating of overlapping generations, with a parent (5,f)-offspring (4,5,m) mating. The diagram shows that there is a common gene pool, with genes "flowing" between individuals and generations.

Consider in contrast a community of five bean plants (*Phaseolus vulgaris*) exclusively with self-fertilisation, when there is no gene flow between individuals, as shown in Fig. 1.3.11.2. Any individual in generation 3 only contains genes from one individual in generation 1, whereas in Fig. 1.3.11.1 an individual in generation 3 could contain genes from two, three, four or five individuals in generation 1. The self-fertilising group of bean plants satisfies the "community" part of the population definition, but not the "common gene pool" part, as there is no common pool of genes being exchanged between individuals.

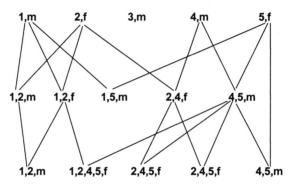

Figure 1.3.11.1. A small population with a common gene pool. See text. Any individual can leave 0, 1, 2 or more offspring and can mate with any member of the opposite sex. The numbers in the first row label individuals, with *f* for female, *m* for male. In the second and third rows, the numbers indicate which members of the first row they have inherited genes from.

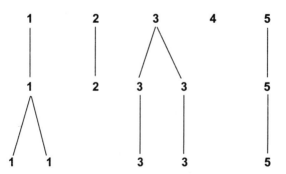

Figure 1.3.11.2. A small community of self-fertilising individuals, with no common gene pool and no flow of genes between individuals. Individuals may leave 0, 1, 2 or more offspring.

Some populations are **discrete**, separate, like a species of fish in a small pond. Other populations **overlap** each other, with breeding within the population and with neighbouring populations, such as rabbits in different but nearby fields. The amount of **gene flow** between different populations depends on proximity, geography and biology, and on the acceptability of mates from different populations. Most urban humans could be described as being part of several non-discrete populations. Thus a Catholic undergraduate from Paris studying at a London college is part of that student population, part of the population of one area of London, of southern England, of France, and of a large religious group. See Chapter 4 for more on population genetics.

1.3.12. Polymorphism

Polymorphism was defined by E. B. Ford as "the occurrence together in the same habitat of two or more discontinuous forms of a species in such proportions that the rarest of them cannot be maintained merely by recurrent mutation." The definition therefore excludes different geographical races and rare, harmful mutations which recur from mutation but do not survive when expressed, such as *albino* in maize, giving white seedlings which cannot photosynthesise and die when seed food reserves have been used up. A locus is sometimes defined as polymorphic if the commonest allele has a frequency of 0.99 or less. Examples of true polymorphisms include human blood groups and eye colours, red, black and intermediate fox fur colour, and purple and cream flowers in the comfrey plant, *Symphytum officinale*. Although the term is usually used for genes of major effect, it is also used for chromosome aberrations. Polymorphism is part of genetic variation, and therefore can be subject to natural, sexual and artificial selection. The topic is dealt with in more detail in Chapters 4 and 15.

1.3.13. Random mating

In **random mating**, mating pairs form independently of genotype and phenotype, and independently of degrees of relatedness between individuals. With **positive assortative mating**, individuals similar with respect to a particular character are more likely to mate with each other than expected by chance, but with **negative assortative mating**, similar individuals are less likely to mate with each other than expected by chance. While humans mate approximately at random with respect to blood groups, there is positive assortative mating for height, with tall males tending to marry tall females and short males tending to marry short females. Although it is often said that opposites attract, it is harder to find examples of negative assortative mating, unless one takes sex as the character.

While positive and negative assortative mating concern particular characteristics, **inbreeding** and **outbreeding** concern degrees of relatedness, not individual characters. **Inbreeding** is a tendency to mating between individuals more closely related (or less closely related for **outbreeding**) than would occur by chance in that population. Suppose that in a small community the average degree of relationship was fifth cousins, then mating averaging out as being between fifth cousins, with some closer and some less close matings, would be random. If mating was on average between second cousins, that would be inbreeding, while if mating was on average between eighth cousins, that would be outbreeding. In a larger population, where the average relationship between individuals was tenth cousins, then mating on average between eighth cousins would be inbreeding, not outbreeding. **First cousins** have one pair of grandparents in common, **second cousins** have one pair of great-grandparents in common, and so on. Mating between relatives is not necessarily inbreeding, and of course all members of a species are ultimately related to each other. The terms inbreeding and outbreeding are sometimes used with different meanings in other contexts. See Chapter 6 for departures from random mating, and Chapter 13.13 for the effects of human inbreeding.

Suggested reading

Agrawal, R. L., *Fundamentals of Plant Breeding and Hybrid Seed Production*. 1998. Science Publishers, Inc., Enfield, New Hampshire.

ASSINSEL. *Feeding the 8 Billion and Preserving the Planet*. (undated, but 1997). ASSINSEL (International Association of Plant Breeders), Nyon, Switzerland.

Cameron, N. D., *Selection Indices and Prediction of Genetic Merit in Animal Breeding*. 1997. CAB International, Wallingford.

Connor, J. M. and M. A. Ferguson-Smith, *Essential Medical Genetics*, 5th ed. 1997. Blackwell Scientific, Oxford.

Dalton, D. C., *An Introduction to Practical Animal Breeding*. 1980. Granada, London.

Falconer, D. S. and T. F. C. Mackay, *Introduction to Quantitative Genetics*, 4th ed. 1996. Longman, Harlow.

Fincham, J. R. S., P. R. Day and A. Radford, *Fungal Genetics*, 4th ed. 1979. Blackwell, Oxford.

Griffiths, A. J. F., W. M. Gelbart, J. H. Miller and R. C. Lewontin, *Modern Genetic Analysis*. 1999. W. H. Freeman and Co., New York.

Hammond, J., J. C. Bowman and T. J. Robinson, *Hammond's Farm Animals*, 5th ed. 1983. Edward Arnold, London.

Hartl, D. L. and E. W. Jones, *Genetics. Principles and Analysis*, 4th ed. 1998. Jones and Bartlett, Boston.

Hayward, M. D., N. O. Bosemark and I. Romagosa, eds., *Plant Breeding. Principles and Prospects*. 1993. Chapman and Hall, London.

Lamb, B. C., Cryptic mutations: their predicted biochemical basis, frequencies and effects on gene conversion. *Molecular and General Genetics* (1975), 137, 305–314.

Mayo, O., *The Theory of Plant Breeding*, 2nd ed. 1987. Oxford University Press, Oxford.

Piper, L. and A. Ruvinsky, eds., *The Genetics of Sheep*. 1997. CAB International, Wallingford.

Poehlman, J. M. and D. A. Sleper, *Breeding Field Crops*, 4th ed. 1995. Iowa State University Press, Iowa.

Simm, G., *Genetic Improvement of Cattle and Sheep*. 1998. Farming Press, Ipswich.

Snustad, D. P. and M. J. Simmonds, *Principles of Genetics*, 2nd ed. 1999. Wiley, New York.

Strachan, T. and A. P. Read, *Human Molecular Genetics*, 2nd ed. 1999. Bios Scientific Publishers, Oxford.

Van Vleck, L. D., E. J. Pollak and E. A. B. Oltenacu, *Genetics for the Animal Sciences*. 1987. W. H. Freeman, New York.

CHAPTER 2

THE INHERITANCE AND ANALYSIS OF QUALITATIVE AND QUANTITATIVE CHARACTERS

2.1. Single-locus qualitative characters: autosomal loci with complete dominance, partial dominance, additive action, overdominance; X-linked and holandric loci

2.1.1. Qualitative characters

Qualitative characters have contrasting phenotypes, with a qualitative difference between them. Thus the two or more alleles at a locus give discontinuous variation, such as white versus green plants in maize, red versus black fur in foxes, the A, B and O blood groups in man, and adenine-requiring or adenine-non-requiring yeast. The term also applies to measurable characters if there are discontinuous differences, as between tall and short pea plants. They are usually controlled by two or more alleles at loci of major effect and show typical Mendelian inheritance in diploid Eukaryotes. We will explore the typical genotype and phenotype ratios produced in different circumstances.

Qualitative characters include biochemical polymorphisms, which may need immunological or electrophoretic methods to determine which alleles are being expressed, and even DNA markers for which there are qualitative differences, such as the presence or absence of a restriction site (Chapters 8 and 14). Such polymorphisms are extensively used in sheep and cattle to check pedigrees — for example, to find out which ram fathered a particular lamb, so that flock pedigree records can be accurate, and so that breeding values based on progeny-testing reflect the correct father. Both biochemical and morphological markers can be used for population studies, using allele frequencies to work out the origin of particular breeds or varieties or human populations, and the genetic relations between them.

2.1.2. Autosomal loci: complete dominance

Autosomal loci are on chromosomes which are not sex chromosomes, so for a locus in diploids there will be two alleles present which may be alike (homozygous) or unlike (heterozygous). A homozygous individual will be pure breeding, e.g., *T T* (tall in pea plants) will give all haploid *T* gametes, so on selfing or on crossing to another *T T* individual, all offspring will be *T T*, tall, while *t t* gives all *t* gametes. Heterozygotes, *T t*, give approximately equal numbers of *T* and *t* gametes, from Mendel's First Law, of Segregation.

Crossing *T t* x *T t* (or selfing *T t*) therefore gives a 1 *T T*: 2 *T t*: 1 *t t* genotype ratio and a phenotype ratio of 3 *T* phenotypes (tall) to 1 *t* phenotype (short) if there is complete dominance.

Crossing *T T* (gametes all *T*) x *T t* (gametes 1 *T*: 1 *t*) gives 1 *T T*: 1 *T t*, all with tall phenotypes.

Crossing *t t* x *T t* gives 1 *T t*: 1 *t t*, with 1:1 genotype and phenotype (tall: short) ratios.

Crossing *T T* x *t t* gives all *T t*, tall.

Thus autosomal loci with complete dominance are characterised by **3:1 phenotype ratios** from crossing two heterozygotes, with three of the dominant phenotype and one of the recessive phenotype. See Plate 4 for a human example, of albinism.

2.1.3. Autosomal loci with partial dominance, additive action or overdominance

The genotype ratios in these cases are exactly the same as for complete dominance, but the phenotype ratios are usually different when there are heterozygotes and homozygotes, because incomplete dominance, additive action and overdominance all give heterozygotes which are phenotypically different from both homozygotes. Let *S' S'* give fruits 20 cm long, *S' S* give fruits 15 cm long, and *S S* give fruit 10 cm long, with additive action.

Crossing *S' S* x *S' S* gives a 1 *S' S'*: 2 *S' S*: 1 *S S* genotype ratio, with a one 20 cm: two 15 cm: one 10 cm fruit phenotype ratio. Similarly in *Antirrhinum* there is incomplete dominance of red flowers over white flowers. The heterozygote has pink flowers, and selfing the heterozygote gives a 1 red: 2 pink: 1 white ratio for plant flower colour. Thus autosomal loci with partial dominance, additive action or overdominance are characterised by **1:2:1 phenotype ratios** from crossing two heterozygotes.

2.1.4. X-linked and holandric loci

Where there are sex chromosomes, the one present twice in one sex and once in the other sex is called the **X chromosome**; the sex chromosome present as one copy in one sex and none in the other is the **Y**. The parts of the X and Y which can pair in meiosis are called the **homologous regions** and can have crossovers between them; genes here are called **incompletely sex linked**. The part of the X with no homologue on the Y is called the **differential region**; genes here are called **sex linked** or **X-linked**. Genes on the Y with no homologues on the X are in the **holandric region**. As this region's loci only occur in the Y chromosome, they are only expressed in the X Y sex, e.g., hairy ears in man. The relative proportions of the three regions vary a lot between organisms but there are usually few active genes in the holandric region. Genes in the holandric region of Y are **pseudodominant**, showing when present, as there are no second copies on another chromosome to dominate them. Thus hairy ears male x normal female will give hairy eared males and normal females, as the females do not inherit the Y, but all sons must inherit their father's Y.

The X X sex is **homogametic** (producing "alike gametes") as all gametes carry one X chromosome. X Y is the **heterogametic** (producing "unalike gametes") sex, producing approximately equal numbers of X-bearing and Y-bearing gametes. In mammals, the X X sex is normally female, producing eggs, and X Y is male, producing sperm. In contrast, the homogametic sex is male in birds (including chickens and turkeys), butterflies, moths, and in some fish, reptiles and amphibia, when symbols Z Z and Z W are usually used for the sex chromosomes. Some plants also have separate sexes, with the male plants producing stamens and female plants producing ovules. The control is usually by an X/Y chromosome system, with the males XY and the females XX, as in *Silene alba* (white campion) and *Humulus* (hop). In *Asparagus*, maleness (XY) is dominant.

Alleles in the differential region of the X are present twice in females but only once in males, which are **hemizygous** for those loci. As discussed in Chapter 13.6, that causes differences between the sexes in the incidence of X-linked recessive diseases. Let us use red-green colour blindness in humans as an example of an **X-linked recessive** condition, when non-colour blindness represents an **X-linked dominant** condition. It is useful to show the chromosomes as well as the alleles, so the recessive allele for colour-blindness is shown as X^{cb} and the wild-type dominant allele as X^{Cb}, and the Y, with no allele, as Y. Males are either hemizygous normal, X^{Cb} Y, or hemizygous colour-blind, X^{cb} Y, while females are normal, $X^{Cb} X^{Cb}$, or carriers but not sufferers, $X^{Cb} X^{cb}$, or sufferers, $X^{cb} X^{cb}$.

Crosses of a normal homozygous female x a normal hemizygous male give all normal, with a 1:1 sex ratio: $X^{Cb} X^{Cb}$ x X^{Cb} Y gives 1 $X^{Cb} X^{Cb}$ (female): 1 X^{Cb} Y (male).

Crosses of a colour-blind female x a colour-blind male give all colour-blind offspring: $X^{cb} X^{cb}$ x X^{cb} Y gives 1 $X^{cb} X^{cb}$: 1 X^{cb} Y.

Crosses of a normal female x a colour-blind male give carrier females (phenotypically normal) and normal sons: $X^{Cb} X^{Cb}$ x X^{cb} Y gives 1 $X^{Cb} X^{cb}$: 1 X^{Cb} Y.

Crosses of a colour-blind female to a normal male give carrier females and colour-blind males: $X^{cb} X^{cb}$ x X^{Cb} Y gives 1 $X^{Cb} X^{cb}$: 1 X^{cb} Y.

Crosses of a carrier female x a normal male give equal numbers of normal females, carrier females, normal males and colour-blind males: $X^{Cb} X^{cb}$ x X^{Cb} Y gives 1 $X^{Cb} X^{Cb}$: 1 $X^{Cb} X^{cb}$: 1 X^{Cb} Y: 1 X^{cb} Y.

Crosses of a carrier female to a colour-blind male give equal numbers of carrier females, colour-blind females, normal males and colour-blind males: $X^{Cb} X^{cb}$ x X^{cb} Y gives 1 $X^{Cb} X^{cb}$: 1 $X^{cb} X^{cb}$: 1 X^{Cb} Y: 1 X^{cb} Y.

With incomplete dominance, additive action or overdominance, the female heterozygote has a different phenotype from both female homozygotes.

An easy way to tell whether a locus is X-linked is to make the two **reciprocal crosses** of pure breeding strains, e.g., normal female x sufferer male, and sufferer female x normal male. The results in the next generation would be identical in both crosses for autosomal loci, with dominant phenotype females and males. For X-linked loci, the first cross gives phenotypically normal carrier females and normal males, but the second cross gives phenotypically normal carrier females and sufferer males (see genotypes above). Differences between reciprocal crosses and between the sexes in one cross are found for X-linked genes.

2.2. Multiple-loci qualitative characters: dihybrid ratios and gene interactions such as epistasis causing modified ratios

2.2.1. Standard dihybrid ratios

Crosses with one, two or three segregating loci are called **monohybrid, dihybrid** and **trihybrid** crosses, respectively. With one segregating locus and complete dominance, selfings give **3:1 phenotype ratios**, while **test crosses** (heterozygote x homozygous recessive) give **1:1 phenotype ratios**, from **Mendel's First Law**, the Law of Segregation. With two segregating loci, both with complete dominance, selfings give **9:3:3:1 ratios** for unlinked loci, while test crosses (e.g., *A a, B b* x *a a, b b*) give **1:1:1:1 ratios** from **Mendel's Second Law**, of Independent Assortment, if there is no interaction between the loci's expressions. Linkage gives more parental gametes than recombinant gametes, modifying these ratios in favour of parental combinations.

In cross diagrams and explanations of what phenotypes correspond to what genotypes, the symbol "-" indicates that either allele could be present at that locus. In any cross diagram when two loci are **segregating** (i.e., two different alleles are present for each of two loci), we must show whether the alleles are in coupling (e.g., *A A, B B* x *a a, b b*) or repulsion phase (e.g., *A A, b b* x *a a, B B*). A cross diagram for a coupling cross with two unlinked and non-interacting segregating loci would look like this for diploid maize, *Zea mays*, where starchy grain (*Su*) is dominant to sugary grain (*su*) and coloured grain (*C*) is dominant to colourless (*c*):

coloured starchy x colourless sugary

C C, Su Su x *c c, su su*	Parents (P)
C, Su *c, su*	Gametes
C c, Su su	First filial generation (F1)

coloured starchy grains

On **selfing**, the unlinked loci will assort independently in the F1 meiosis, giving both in eggs cells and pollen a 1 *C, Su*: 1 *C, su*: 1 *c, Su*: 1 *c, su* ratio in the gametes. We can use a **Punnett chequer board** (a multiplying device named after an early Cambridge geneticist) to see the effects of combining the male and female gametes at random. One just multiplies the items in the top and side gamete panels to get the contents of a square; thus the top left genotype square's contents, 1 *C C, S u Su*, come from combining the male gamete 1 *Cu, Su* at the top with the female gamete 1 *Cu, Su* at the left side. We therefore get an F2 (second filial generation) as shown in Table 2.2.1.1.

Table 2.2.1.1. Punnett chequer board for F2 genotypes and their frequencies, from a dihybrid cross with no linkage.

Male gametes:	1 *C Su*	1 *C su*	1 *c Su*	1 *c su*
Female gametes	F2 genotypes			
1 *C Su*	1 *C C, S u Su*	1 *C C, Su su*	1 *C c, Su Su*	1 *C c, Su su*
1 *C su*	1 *C C, Su su*	1 *C C, su su*	1 *C c, Su su*	1 *C c, su su*
1 *c Su*	1 *C c, Su Su*	1 *C c, Su su*	1 *c c, Su Su*	1 *c c, Su su*
1 *c su*	1 *C c, Su su*	1 *C c, su su*	1 *c c, Su su*	1 *c c, su su*

That gives a ratio of 9 coloured starchy (*C* -, *Su* -, where the "-" indicates that either allele could be present at that locus): 3 coloured sugary (*C* -, *su su*): 3

colourless starchy (*c c, Su -*): 1 colourless sugary (*c c, su su*), which is the standard ratio from a selfing with two unlinked loci with complete dominance at each locus and no locus-interactions, as from Mendel's Second Law, the Law of Independent Assortment. There are 16 F2 genotype squares in Table 2.2.1.1, so a "1" in a square indicates that 1/16 of the F2 will have that genotype.

If the double heterozygote, the F1 in the above cross, had not been selfed, but had been **test-crossed** (i.e., crossed to a doubly homozygous individual, *c c, su su*, all gametes *c, su*), instead of the **9:3:3:1 ratio** we would get a **1:1:1:1** phenotype and genotype ratio. That is because the double heterozygote has a 1:1:1:1 gamete genotype ratio, and each gamete just combines with a fully recessive *c, su* gamete, giving 1 *C c, Su su*: 1 *C c, su su*: 1 *c c, Su su*: 1 *c c, su su*, which is 1 coloured starchy: 1 coloured sugary: 1 colourless starchy: 1 colourless sugary.

If the loci were linked in coupling phase, with 20% recombination, then the gamete frequencies from the double heterozygote *A a, B b* would not be 1:1:1:1, but would be split into 80% parental gametes (as fractions, 0.4 *A, B*; 0.4 *a, b*) and 20% recombinant gametes (0.1 *A, b*; 0.1 *a, B*), and these would be put into the multiplying grid for gamete mating (Punnett square), instead of the 1:1:1:1 values shown for the maize example. A selfing would then give a 0.66 *A -, B -*: 0.09 *A -, b b*: 0.09 *a a, B -*: 0.16 *a a, b b* ratio. A test-cross of the double heterozygote to the *a a, b b* double homozygote would give a 0.4 *A a, B b*: 0.1 *A a, b b*: 0.1 *a a, B b*: 0.4 *a a, b b* ratio, from which the recombination frequency can be calculated as 20% (0.2 as a fraction). There is a way of calculating recombination frequencies from the F2 of a selfed F1: see Stephens (1939 - reference in Strickberger, 1976) or Strickberger (1976) (the table of Z is not given in the 3rd ed., 1985).

Linkage modifies the standard Mendelian dihybrid ratios of 9:3:3:1 and 1:1:1:1, giving more parental combinations and fewer recombinant combinations, with different ratios depending on the particular recombination frequencies. Thus if selfing or test-crossing a double heterozygote does not give the standard dihybrid ratios, linkage is one explanation to explore, where the test-cross is ideal because dominance does not affect the ratios and one can calculate recombination frequencies directly from F2 ratios. Plate 5 shows a maize example with complete linkage (0% recombination frequency), with half the grains of each parental type, coloured full and colourless shrunken, and no recombinants, coloured shrunken or colourless full.

2.2.2. Gene interactions such as epistasis causing modified dihybrid ratios

Let us now examine the effects of **gene-interactions** (more accurately, gene-product interactions) between loci on dihybrid ratios for unlinked loci with complete dominance. By extension, one could also work out expected results with interactions and linkage, or gene interactions and other forms of dominance.

Epistasis is a common form of interaction, where the presence of a particular allele at the **epistatic** ("standing above") locus prevents the expression of any alleles at the **hypostatic** ("standing below") second locus. Epistasis can be caused by the dominant allele or by homozygous recessive alleles at the epistatic locus. For example in maize the *A* allele gives anthocyan colour to the grains, while *a* gives colourless, with *a* showing **recessive epistasis** to a second locus, where the *Pr* allele modifies the red colour from *A* to purple, and the *pr* allele has no effect. If we have a repulsion cross, *A A, pr pr* (red grains) x *a a, Pr Pr* (colourless), the F1 are purple, *A a, Pr pr*. Selfing this double heterozygote, with two unlinked loci, gives the expected genotype ratio of 9 *A -, Pr -* ; 3 *A -, pr pr*; 3 *a a, Pr -*; 1 *a a, pr pr*. The 9 *A -, Pr -* will be purple, with *Pr* modifying the *A* red product to purple. The 3 *A -, pr pr* are red, unmodified, and 3 *a a, Pr -* and 1 *a a, pr pr* are all colourless. There is no red product for the *Pr* allele in *a a, Pr -* to modify, giving a **9:3:4 recessive epistasis ratio** of purple: red: colourless, a modified dihybrid ratio alerting us to recessive epistasis - see Plate 6.

A possible metabolic pathway to explain the results is to have the epistatic locus controlling an earlier step than does the hypostatic locus, as shown in Fig. 2.2.2.1. This shows *A* and *Pr* alleles promoting particular steps, and *a* and *pr* blocking particular steps. It is understood that the recessive allele will only act when homozygous, and only one copy of it is conventionally shown in such diagrams. A horizontal arrow shows the allele promoting the step and a vertical arrow shows the allele blocking that step.

Figure 2.2.2.1. A pathway for recessive epistasis for maize grain colour: see text and Plate 6.

The epistatic allele is dominant in the case of fruit colour in *Cucurbita pepo*, e.g., marrow or pumpkin, where *W*, giving white fruit, is epistatic to the unlinked *Y/y* locus, with *w w, y y* giving green and *w w, Y -* giving yellow as in most commercial pumpkins. A repulsion phase cross with selfing of the F1 would be white fruit (*W W, y y*) x yellow fruit (*w w, Y Y*), giving *W w, Y, y*, white F1, as *W* shows **dominant epistasis** to *Y/y*. Selfing the F1 gives a Mendelian genotype ratio and a modified phenotype ratio, 9 *W* -, *Y* - (white); 3 *W* -, *y y* (white); 3 *w w, Y* - (yellow); 1 *w w, y y* (green), that is, a **12:3:1 dominant epistasis phenotype ratio**, quite different from the 9:3:4 for recessive epistasis. The F2 ratio does not depend on whether the cross was in coupling or repulsion, if the loci are unlinked. A possible pathway, with the epistatic allele controlling the earlier step of the pathway, is shown in Fig. 2.2.2.2.

Figure 2.2.2.2. A pathway for dominant epistasis in *Cucurbita* fruit colour: see text.

One can combine the basic dihybrid F2 phenotype ratio, 9:3:3:1, in other ways such as 15:1 (i.e., the double recessive gives one phenotype, and a dominant allele at either or both loci gives a different phenotype), 13:3 (e.g., a dominant epistatic allele inhibits the dominant allele at the hypostatic locus, but has no phenotype of its own, Plate 7), and 9:7 (Plate 8). All of these have known examples.

The 9:7 ratio comes from **complementary dominant alleles** at two unlinked loci. In sweet pea, *Lathyrus odoratus*, purple flower colour requires the dominant alleles *R* and *C*. A repulsion phase cross would be *C C, r r* (white because it lacks *R*) x *c c, R R* (white because it lacks *C*), giving purple-flowered F1, *C c, R r*, having both *C* and *R*. Selfing the F1 gives the modified dihybrid ratio of 9 purple (*C* -, *R* -): 7 colourless (3 *C* -, *r r*; 3 *c c, R* -; 1 *c c, r r*). This case can be described as having two complementary dominant genes or as mutual recessive epistasis. As in Fig. 2.2.2.3, one can represent it by a pathway, but from the data we could not tell whether locus *C/c* controls the first step and *R/r* controls the second step, or whether the order is reversed. Plate 8 shows this in a maize grain example.

Figure 2.2.2.3. A pathway for complementary dominant genes' action in sweet pea flower colour: see text.

This cross shows several important general points. First, both parents have white flowers yet they have different genotypes and the white phenotype is also given by the double recessive, *c c, r r*, so the same phenotype can be produced by more than one homozygous genotype, as well as by genotype variations concealed by dominance. Second, new phenotypes can appear in the F1 (this cross) and/or the F2 (green fruit in the *Cucurbita* cross). Third, parental phenotypes can differ according to whether the cross is in coupling (*C C, R R*, purple x *c c, r r*, white) or in repulsion (*C C, r r*, white x *c c, R R*, white), but the F1 genotypes and phenotypes are the same from coupling and repulsion crosses, *C c, R r*, purple.

Some characters are determined by a very large number of loci, with many possible gene interactions such as epistasis between them. In maize, more than 20 loci have qualitative effects on grain colour. In sheep, 50 to 100 loci control wool and hair growth, with the genes often clustered into families on chromosomes.

2.3. Quantitative characters; quantitative trait loci and polygenes

Many important characters in many organisms are quantitative ones, with **continuous variation**, such as grain yield, milk yield, meat yield, growth rate of micro-organisms, and human intelligence. Although there are exceptions, quantitative characters typically show greater effects of the **environment** than do qualitative ones. Thus in maize, the quantitative character of grain yield is greatly affected by light, rain, temperature, soil and disease, while the qualitative character of colourless grain versus purple grain is not affected by such environmental conditions. Environmental variation can make working out the genetics of quantitative inheritance very difficult, so making crosses in controlled environments is recommended if practicable.

The simplest way to understand the inheritance of quantitative characters is to start with simple situations and then to build up to more complex ones. Take **grain colour in wheat**, as studied by Nilsson-Ehle in 1909. He crossed a strain with pure breeding white grain (*A A, B B*) with a strain with pure breeding dark red grain (*A' A', B' B'*), obtaining an F1 with intermediate light red grain (*A' A, B' B*). On allowing the F1 to self and to cross amongst themselves, he obtained a discontinuous distribution in the F2, ranging from dark red to white. The F2 phenotypes, proportions, numbers of primes and genotypes, respectively, were:

dark red, 1/16, 4′, *A' A', B' B'*;
red,   4/16, 3′, *A' A', B' B; A' A, B' B'*;
light red, 6/16, 2′, *A' A', B B: A' A, B' B; A A, B' B'*;
pale red, 4/16, 1′, *A' A, B B; A A, B' B*;
white,   1/16, 0′, *A A, B B*.

The proportions, in sixteenths, suggest dihybrid inheritance, and getting five phenotypes suggests that there is additive action. We have two unlinked loci with additive action of the alleles at each locus, and cumulative action of the two loci. Thus one prime (′) adds an equal amount of pigment whichever locus it is at, so the amount of pigment is determined by the total number of primes in the genotype, summed over both loci. This **1:4:6:4:1** is another modified dihybrid ratio. While dominance gives **asymmetrical distributions**, e.g., 3:1, 9:3:3:1, 27:9:9:9:3:3:3:1, for mono-, di- and tri-hybrid crosses respectively, additive action gives **symmetrical distributions** with equal proportions at the two extremes, e.g., 1:2:1 and 1:4:6:4:1. Nilsson-Ehle's results fitted the theory well, with no effects of the environment, but the distribution was discontinuous and qualitative.

Consider a case of height in a diploid, controlled by three unlinked loci, with a basic height of 50 cm, and each primed allele adding 2 cm to that. Initially suppose that the plants are grown in a constant environment, with no environmental effects. The cross $A\ A,\ B\ B,\ C\ C$ (50 cm) x $A'\ A',\ B'\ B',\ C'\ C'$ (62 cm = 50 + plus the effect of 6 primes, 2 cm each) gives an F1 of height 56 cm [$A'\ A,\ B'\ B,\ C'\ C$ has three primes = 50 + (3 x 2) = 56 cm]. One can work out the expected F2 distribution for additive and cumulative gene action from a Punnett chequer board, but the easiest way is to use **Pascal's triangle**, which gives the coefficients of the binomial expansion. As shown in Fig. 2.3.1, it starts with the number one, then two ones on the next line, then it expands each successive line by a one at each end, and fills in the intervening numbers by adding the two numbers diagonally above each place, so to get the threes in line four, one adds the ones and twos diagonally above them.

```
            1                    (no segregating alleles)
          1   1
        1   2   1                (two segregating alleles, one locus)
      1   3   3   1
    1   4   6   4   1            (four segregating alleles, two loci)
  1   5  10  10   5   1
1   6  15  20  15   6   1        (six segregating alleles, three loci)
```

Figure 2.3.1. Pascal's triangle: see text.

One takes alternate rows, starting with the third row, which gives the F2 proportions for one locus (e.g., 1/4 red, 2/4 pink, 1/4 white for flower colour in *Antirrhinum*, where for red, 1 is the figure at the left end of the row, and 4 is the total of the numbers in that row). The fifth row gives the F2 proportions for two loci, 1/16, 4/16, 6/16, 4/16, 1/16, as for wheat grain colour. The case we are currently considering has six segregating alleles, two at each of three loci, so the seventh line gives us the F2 proportions. These are shown in Table 2.3.1.

Table 2.3.1. F2 results for three unlinked loci with additive and cumulative action.

Frequency	Number of primes	Height, cm
1/64	6	62
6/64	5	60
15/64	4	58
20/64	3	56
15/64	2	54
6/64	1	52
1/64	0	50

This gives a symmetric discontinuous distribution, which we can plot as a histogram, Fig. 2.3.2.

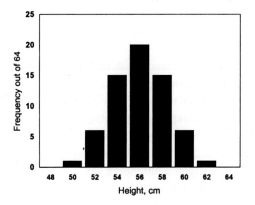

Figure 2.3.2. Discontinuous height distribution in the F2 from selfing an F1 with three segregating loci with additive action at each locus and cumulative action between loci, and no environmental variation.

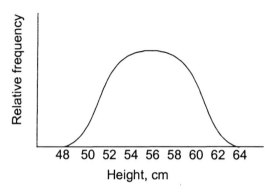

Figure 2.3.3. Continuous height distribution for the conditions of Fig. 2.3.2 but with some environmental variation.

So far we have been considering a case with no environmental effects and getting a discontinuous distribution with seven separate phenotypes, each depending entirely on the number of primed alleles at the three controlling loci. Let us now introduce some environmental variation. A genotype with one prime, giving so far only a height of 52 cm, will now give a spread of possible phenotypes, say 50 to 54 cm, as some environmental conditions will increase height by varying amounts and others will decrease it, giving a potentially continuous set of possible phenotypes within that limited range. All seven genotypes will now give a range of possible phenotypes, so what was a discontinuous set of phenotypes now becomes continuous variation because of the environmental variation. In this case we will get a symmetric continuous distribution approximating to a Normal (Gaussian) Distribution, Fig. 2.3.3.

The basis of quantitative inheritance with continuous variation is therefore one of **discontinuous variation in genotypes** but **continuous phenotype distributions** brought about mainly by two factors. The first is **effects of the environment** and the second is the presence of **segregating polygenes**, each of which gives very small phenotypic effects, hence tending to give continuous variation. A plant in the present case with a height of 51.3 cm might thus have no primes, but a favourable environment for increased height, contributing say 1 cm, and a number of height-increasing polygenes, contributing say 0.3 cm, or it might have one prime but a slightly unfavourable environment or some height-decreasing polygenes.

We considered additive action, which gives symmetrical distributions, and three phenotypes per locus (assuming diploidy and two segregating alleles per locus). If we have dominance, it only gives two phenotypes per locus and can give asymmetrical distributions, with phenotypes at one extreme much more frequent than phenotypes at the other extreme. Consider a situation where four unlinked loci with complete dominance and cumulative action control height, with all loci having the dominant allele increasing height, say *A, B, C* and *D* all adding 10 cm to a basic height of 10 cm for *a a, b b, c c, d d*. Then *A -, b b, c c, d d* is 20 cm and *A A, B B, C C, D D* is 50 cm high. On selfing the quadruple heterozygote, *A a, B b, C c, D d*, one expects a 3:1 phenotype ratio at each locus, over four loci, giving (3/4 dominant phenotype + 1/4 recessive phenotype)^4 = 81 (*A -, B -, C -, D -*, 50 cm): 27: 27: 27: 27 (all 40 cm, each with dominants at three loci, e.g., *A -, B -, C -, d d*; *A -, B -, c c, D -*; *A -, b b, C -, D -*, etc.): 9:9:9:9:9:9 (all with dominants at two loci, 30 cm, e.g., *a a, B -, C -, d d*): 3:3:3:3 (all with a dominant at one locus, 20 cm, e.g., *a a, B -, c c, d d*): 1 (10 cm, *a a, b b, c c, d d*). With 81 at 50 cm and only 1 at 10 cm, the distribution is clearly skewed in the direction of the dominant phenotypes. The overall phenotype ratio is **81: 108: 54: 12: 1**, with five phenotypes from having dominant alleles at 4, 3, 2, 1 or 0 loci respectively.

Where all dominant alleles at different loci give phenotypes all in the same direction for a character, e.g., all increasing or all decreasing height, we have **directional dominance**. If some dominants act in one phenotypic direction and other dominants act in the opposite direction, we have **non-directional dominance**, e.g., *A* and *C* increase height, but *B* and *D* decrease height. Directional dominance tends to give strongly skewed distributions, but with non-directional dominance, skewing is less, depending partly on how many dominants act in each direction. Another simplifying assumption used so far is that each major locus has an **equal effect** on the phenotype, which need not be true. Those loci having significant individual amounts of effect on the phenotype are called **quantitative trait loci**, often abbreviated to **QTL**.

In contrast to QTL we have **polygenes**, which are genes whose individual phenotypic effect is very small in relation to the total phenotypic variation for the character. If a plant has heights ranging from 10 to 50 cm, then quantitative trait loci might have individual effects on height of perhaps several cm, while polygenes might have effects of a few millimetres or less. That is why segregating polygenes can give continuous phenotype variation when the underlying genotype variation for major genes would on its own give a discontinuous set of phenotypes. **Major genes** are ones giving fairly large qualitative phenotypic differences such as black versus red cattle coat colour, or 10 versus 15 cm height, so QTLs are one kind of major gene.

Major genes and polygenes are alike at the DNA level, only differing in their extent of effect. Studies in the fungus *Penicillium chrysogenum* gave the interesting finding that for penicillin yield, many major genes with no direct connection with penicillin production act as polygenes with minor effects on penicillin yield. Thus major gene loci affecting white versus blue-green conidia, or a requirement for methionine or for adenine, acted as polygenes on penicillin yield. Many genes controlling growth rates affect very many other characters, to differing extents.

Polygenes can mutate and recombine, and may show linkage, gene interactions, and dominance or additive action. They can be mapped, but they are difficult to study individually because of their small phenotypic effects which are easily masked by environmental variation. They are generally treated as showing additive action, and considered collectively rather than individually.

If the organisms are grown in constant environments, we can often estimate the number of segregating QTL if each locus has an equal phenotypic effect and additive action. We saw earlier that with one such segregating locus, selfing the heterozygous F1 gave a 1:2:1 genotype and phenotype ratio, two loci gave a 1:4:6:4:1 ratio and three loci gave a 1:6:15:20:15:6:1 ratio. The proportion of the F2 at just one extreme is therefore 1/4 for one locus, 1/16 for two loci and 1/64 for three loci. The proportion of the F2 at just one extreme is therefore $1/(4^n)$, where "n" is the number of segregating loci. We can use this to analyse data as in the three following examples.

Example 2.3.1. In a constant environment, one pure breeding line of diploid rye had a height of 30 cm and was crossed with another pure breeding line of height 60 cm. The F1 had a height of 45 cm and were crossed amongst themselves. In the F2, the 1,000 plants included 16 as short as 30 cm and 20 as tall as 60 cm, with the rest of a variety of intermediate heights. Analyse these data.

We can see that there was no environmental variation because these was no variation within each parent or within the F1. Two lines of evidence suggest additive action: the F1 were intermediate between the two parental types, and the numbers at both extremes of the F2 were similar (dominance, especially directional dominance, gives skewed distributions). With 16 of the F2 at one extreme and 20 at the other extreme, the average number at one extreme was 18. There were 1,000 F2 plants, so the proportion of the F2 at one extreme was $\frac{18}{1,000} = 0.018$. This is close to 1/64, 0.016, so we can deduce that there were three segregating loci. The formula $1/(4^n)$ gives 1/64 for n = 3. We can give the parents' genotypes, 30 cm = *A A, B B, C C*, and 60 cm = *A' A', B' B', C' C'*, using primes for additive action alleles. A six primes difference between the parents gives a height difference of 30 cm, so one prime gives 5 cm additional height, with a basic no primes height of 30 cm. We can therefore deduce that F2 heights were 30, 35, 40, 45, 50, 55 and 60 cm, in ratio 1:6:15:20:15:6:1 from Pascal's triangle. There were seven F2 phenotypes, corresponding to 0, 1, 2, 3, 4, 5, and 6 primes, and 27 genotypes because there are three possible genotypes at each of three loci, giving 3 x 3 x 3 variations. From simple data we can deduce quite a lot, providing that our basic assumptions hold, especially that different loci have equal effects and are unlinked.

Example 2.3.2. Yield in barley (a diploid) was studied in a controlled environment. Two inbred lines had yields of 50 and 70 units. When these lines were crossed with each other, the F1 had a yield of 60 units. The F1 were selfed. The 2,000 F2 plants had 8 as low yielding as 40 units and 11 as high as 80 units, the rest being intermediate. Explain these results.

Inbred lines are pure breeding, so each parent was homozygous. The constancy of the parents and F1 shows that there were no environmental effects. We can deduce additive action from the intermediate F1 and the approximately equal numbers at the two ends of the F2 distribution. In this case, however, the F2 extremes were more extreme than either parent. Getting offspring more extreme than the original parents is called **transgression**, with 40 being lower than the lower yielding parent and 80 being more extreme than the higher-yielding parent, so the F2 extremes must have more extreme genotypes than the parents. With 8 at one extreme and 11 at the other, there was an average of $\frac{9.5}{2,000}$ at one extreme = 0.0048, which is close to 1/256 (= 0.004 = 1/4^4), so we can deduce 4 segregating loci. If the parents are not the most extreme types, then we take the proportion of the F2 **at each extreme**, not just the proportion of the F2 as extreme as the parents. Possible pure-breeding genotypes would be *A A, B B, C C, D′ D′* for the 50 unit parent and *A′ A′, B′ B′, C′ C′, D D* for the 70 unit parent, with each prime adding 5 units to a basic value of 40 (*A A, B B, C C, D D*). The F1 were *A′ A, B′ B, C′ C, D′ D*; in the F2, the highest yielders, 80 units, had eight primes, and the lowest yielders, 40, had no primes. We can therefore deduce four segregating unlinked loci with additive and cumulative action, each prime adding 5 units of yield, with the parents not having the most extreme genotypes or phenotypes.

Example 2.3.3. In maize (a diploid), two pure-breeding lines averaged heights of 122 cm (limits 117-127 cm) and 186 cm (limits 179-193 cm). The F1 plants averaged 154 cm (limits 148-160 cm), and were allowed to cross amongst themselves. In 1,000 F2 plants, 8 were shorter than 128 cm, and 10 were taller than 175 cm. What can you deduce?

From the variation within each parent and within the F1, we can deduce environmental variation affecting height. We deduce additive action as the F1 were intermediate between the parents, and the F2 extremes were about equally frequent.

The number at one F2 extreme averages 18/2 = 9, with $\frac{9}{1,000}$ = 0.009. This lies between the expected value for three loci (1/64 = 0.0156) and the value for four loci (1/256 = 0.0039). If there are four loci, we have too many at the extremes, and if there are three loci, we have too few at the extremes. Which can we explain by environmental effects, too few or too many at the extremes?

Suppose there were four loci, giving a height difference of 64 cm from 8 primes, or 8 cm a prime. A few of the genotypes and phenotypes are shown in Table 2.3.2.

Table 2.3.2. A few of the genotypes and phenotypes in the F2 from Example 2.3.3.

Number of primes	Average height, cm	Expected limits, cm
0	122	117–127
1	130	125–135
2	138	133–143

Looking at the expected limits, one sees that the ranges (limits) of the 0 and 1 prime phenotypes overlap, so that plants of 125-127 cm could have either genotype, so the extreme phenotype of less than 128 cm includes all those with no primes and some, but not all, of those with one prime. We will therefore get more plants at this extreme than the number calculated for no primes alone, and can therefore explain the observed excess over expectation in the data, with a similar explanation at the upper extreme, which will include plants with seven and eight primes. We cannot explain having too few plants at the extremes, so must conclude that there were four, not three, segregating loci. In general, having environmental effects can blur different genotypic classes and cause more individuals in the extreme classes than one expects in the absence of environmental effects.

These three examples have been of genes with additive action, while others might include dominance, as in Example 2.3.4.

Example 2.3.4. When Aberdeen-Angus beef cattle (coloured face, no horns, black coat) are crossed with Hereford cattle (white face, horns, red coat: see Plate 9 for a hornless example), the F1 hybrids show useful hybrid vigour (Chapter 12.1) for beef characters, and have a white face, no horns, and a black coat. When some F1 hybrids were crossed amongst themselves, the following numbers of different types were obtained in the F2: coloured face, horns, black coat, 28; coloured face, horns, red coat, 8; white face, no horns, black coat, 278; white face, no horns, red coat, 85; white face, horns, black coat, 96; white face, horns, red coat, 35; coloured face, no horns, black coat, 89; coloured face, no horns, red coat, 27. Explain these results.

The dominant characters are white face, no horns and a black coat, as shown in the F1, which show one or other parental phenotype, not incomplete dominance. The parents each show a mixture of dominant and recessive traits, so the cross is partly coupling and partly repulsion, e.g., Aberdeen-Angus could be *wf wf, Nh Nh, bc bc* and Hereford could be *Wf Wf, nh nh, Bc Bc*, where the symbols are named for each phenotype, e.g., *Wf* for dominant white face. The F2 shows a 278 (same as the F1, showing all three dominants): 96: 89: 85 (all with two dominants showing): 35: 28: 27 (all with one dominant showing): 8 triple recessives, which is clearly a good fit to the classic Mendelian **27:9:9:9:3:3:3:1 trihybrid ratio** for three unlinked loci with complete dominance at each locus.

In classic **simple quantitative analysis** for traits with additive action, one compares **means, ranges and variances** in the parental, F1 and F2 generations, after crossing pure-breeding parents. The most basic analysis has five assumptions, some of which are tested in the analysis and which may not hold. All held for the wheat grain colour considered earlier.

- Each locus affecting the character has an equal effect.
- Each contributing allele has additive effects: there is no dominance between alleles at a locus.
- There are no interactions between loci such as epistasis, just cumulative action.
- There is no linkage between loci.
- Environmental effects are negligible.

Let us analyse two sets of data, from Emerson and East (1913) and East (1916), given in Srb, Owen and Edgar (1965).

Set 1. Cob (ear) length in maize, in cm.

The pure-breeding short-eared parent averaged 6.6, limits 4-9, while the long-eared parent averaged 16.8, limits 13-22. The F1 from intercrossing them gave an average of 12.1, limits 8-16. On selfing of the F1, the F2 ranged from 6 to 20 cm. As there was variation within each parent and within the F1 (the F1 are all of uniform genotype), we detect environmental variation. At 12.1 cm, the F1 average is between the parental values of 6.6 and 16.8, but is nearer to the long parent value, suggesting either incomplete dominance of long to short, or dominance for long at some loci and additive action at other loci. The F2 have a wider spread of values than the F1 because the F1 has only one genotype, but the F2 show segregation for the genes by which the two parents differ and for which the F1 is therefore heterozygous.

Those F2 in the range 6 to 9 cm overlapped the short-parent range of 4–9 cm, and those in the range 13 to 20 cm overlapped the long-parent range of 13–22 cm. More than 1/64 of the F2 therefore overlapped one parental extreme, so there were three or fewer loci segregating. Possible genotypes which would account for these data are short parent, *A A, B B, c c*, long parent *A′ A′, B′ B′, C C*, F1 *A′ A, B′ B, C c*, and the F2 with a range of different segregants. Postulating one locus with dominance of long and two loci with additive action helps to explain the incomplete dominance in the F1, although other genotypes and explanations are not ruled out.

Set 2. Corolla (flower tube) length in tobacco, *Nicotiana longiflora*, in mm.

The pure-breeding short-flowered parent averaged 40 mm, limits 34–43, while the long-flowered parent averaged 93, limits 86-97. The F1 from intercrossing them gave an average of 61, limits 53–72. On selfing of the F1, the F2 ranged from 51 to 84 mm. As there was variation within each parent and within the F1, we detect environmental effects. At 61 mm, the F1 average was about midway between the parental values of 40 and 93, so we can deduce no dominance. In the F2, none of the plants had flowers as short as the short parent, and none had flowers as long as the long parent. With none of the F2 at either extreme for 444 plants, we cannot estimate the number of segregating loci accurately. An estimate of more than four loci would be reasonable: the numbers are too small to use in the $1/(4^n)$ equation, which gives an answer of infinity.

While this kind of analysis can be done by computer, one needs common sense in interpreting the data. Genetic analysis is easiest if environmental influences can be minimised, and if one can deduce that all segregating loci have complete dominance, or all have additive action. If all segregating loci have dominance, one can look at the F2 from selfing a multiple heterozygote when the phenotype ratio (e.g., 3:1, or 9:3:3:1) should show the number of segregating loci. Note the number of phenotypes as well as the ratio; the rarest phenotype should be that of the multiple recessive. Gene-product interaction and linkage can affect ratios, as we have seen, with linkage and recombination frequencies being readily estimated from test crosses (heterozygote x homozygous recessive).

One can have characters controlled both **qualitatively and quantitatively**, with discontinuous and continuous variation. In cattle coat colour, animals may have no spots (*s s*) or some spots (*S* -), with continuous variation for the amount of spots if present, from a series of polygenic loci. In some plants, there may be white flowers, *c c*, or coloured flowers, *C* -, with the amount and type of colour, when present, determined by other loci. Chapter 3 has further information on quantitative characters. For examples of working with QTL, see Georges et al. (1995) for milk yield in dairy cattle, and DeVincente and Tanksley (1993) for tomatoes. For much more mathematical treatments of quantitative inheritance, see Cameron (1997), Falconer and Mackay (1996), and Kearsey and Pooney (1995).

In forest trees, many commercially important quantitative traits such as height, basal area, stem proportions, and leaf characters such as time of spring bud break, show normal distributions in segregating populations. Their genetics could be polygenic, with many loci having very small effects, or there could be a few controlling QTLs of large effect, with environmental variation masking discrete phenotypes. Bradshaw and Stettler (1995) studied the parents, F1 and F2 from an interspecific cross of two poplars, *Populus trichocarpa* x *P. deltoides*. Instead of polygenic control, most of the traits measured had one to five QTLs of large effect responsible for much of the genetic variance. For example, the timber trait of stem volume (measured after two years growth in this study) had 45% of the genetic variance and 30% of the phenotypic variance controlled by just two QTLs of large effect, even though the F2 showed continuous variation with a roughly normal distribution. There was also hybrid vigour (Chapter 12.1) for this character, with the F1 value (34 dm³) clearly higher than the two parental values, 25 dm³ for *P. trichocarpa* and 3 dm³ for *P. deltoides*, showing transgression (Chapter 3.1), although none of the F2 had values as high as those of the larger parent or the F1, showing regression (Chapter 3.1) from the F1 value. The QTLs were mapped.

Suggested reading

Bradshaw, H. D. and R. F. Stettler, Molecular genetics of growth and development in Populus. IV. Mapping QTLs with large effects on growth, form, and phenology traits in a forest tree. *Genetics* (1995), 139, 963–973.

Cameron, N. D., *Selection Indices and Prediction of Genetic Merit in Animal Breeding*. 1997. CAB International, Wallingford.

Dalton, D. C. *An Introduction to Practical Animal Breeding*. 1980. Granada, London.

DeVincente, M. C. and S. D. Tanksley, QTL analysis of transgressive segregation in an interspecific tomato cross. *Genetics* (1993), 134, 585–596.

Falconer, D. S. and T. F. C. Mackay, *Introduction to Quantitative Genetics*, 4th ed. 1996. Longman, Harlow.

Georges, M. et al., Mapping quantitative trait loci controlling milk production in dairy cattle by exploiting progeny testing. *Genetics* (1995), 139, 907–920.

Kearsey, M. J. Biometrical genetics in breeding, pp 163-183, in: Hayward, M. D., N. O. Bosemark and I. Romagosa, eds., *Plant Breeding. Principles and Prospects.* 1993. Chapman and Hall, London.

Kearsey, M. J. and H. S. Pooney, *The Genetic Analysis of Quantitative Traits.* 1995. Chapman and Hall, London.

Mayo, O., *The Theory of Plant Breeding*, 2nd ed. 1987. Oxford University Press, Oxford.

Srb, A. M., R. D. Owen and R. S. Edgar, *General Genetics*, 2nd ed. 1965. W. H. Freeman and Co., San Francisco.

Strickberger, M. W., *Genetics*, 2nd ed. 1976. Macmillan, New York (the table of Z is not given in the 3rd ed., 1985).

CHAPTER 3

REGRESSION, TRANSGRESSION, ENVIRONMENTAL EFFECTS AND HERITABILITY; CORRELATIONS BETWEEN CHARACTERS; GENOTYPE, PHENOTYPE AND BREEDING VALUES

3.1. Genetic and environmental causes of regression and transgression

Regression (filial regression) is the tendency of the progeny of extreme parents to be less extreme than their parents; that is, they show regression towards the population average value (mean). This applies to quantitative characters and has three causes: dominance, gene interactions such as epistasis, and environmental effects. Dominance causes regression because a heterozygote will segregate both dominant and recessive phenotypes.

Transgression is the production of offspring (in the F1 or F2) more extreme than their parents for a particular quantitative character. It can be caused by genes with dominance or additive effects, by gene interactions such as epistasis, and by environmental effects.

Regression and transgression are very important in applied genetics. The breeders often want transgression to more extreme types than either parent, say offspring with higher yields or more disease resistance, but sometimes they want regression to an average type, say to eliminate extreme types to get uniformity for sowing-to-ripening time for a mechanically harvested seed crop such as wheat.

The **environment** can produce regression. An excellent year giving a high yield for rice in an area might be followed by an average year giving an average yield, and therefore there is regression, high yielding plants having average yielding offspring because of environmental changes. Similarly, an average year with average yields might be followed by a poor year giving low yields, so there has been transgression, average parents giving offspring with a more extreme (which can be in an upwards or downwards direction) yield.

For regression, one needs to distinguish between cases in which there is a general tendency for offspring to be less extreme than their parents, and cases where there is no such general tendency, although some individuals show regression. If the segregating loci all have additive action, then there is no general regression, and the offspring average is expected to equal the parental average. In the wheat grain colour example in Chapter 2.3, if a strain with red grains ($A'A'$, $B'B$, three primes, gametes 1 A', B': 1 A', B) is crossed with one with pale red grains ($A'A$, $B B$, one prime, gametes 1 A', B: 1 A, B), we expect a ratio of 1/4 red (3 primes), 1/2 light red (2 primes), 1/4 pale red (1 prime). The parental average is two primes and the offspring average is also two primes, so there is **no overall regression**, although we have some individuals - the light red type - less extreme than either parental type, showing **individual regression**.

Take a case with complete and non-directional dominance, with A and b increasing yield and a and B decreasing yield so that a a, B - has yield 10 units, a a, b b and A -, B - have yield 20 units and A -, b b has yield 30 units. In some cases the offspring average yield equals the parental average yield, but heterozygosity can cause overall regression. In A a, b b x A a, b b, the parental average yield is 30 units and we get offspring in the ratio 3 A -, b b (30 units each): 1 a a, b b (20 units). The offspring average, $110/4 = 27.5$, is therefore less than the average of the extreme parents, so there is overall regression. Similarly if we take two plants of lowest yield, with heterozygosity, a a, B b x a a, B b, average 10 units, the offspring are in ratio 3 a a, B - (10 units each): 1 a a, b b (20 units), then the offspring average yield is $50/4 = 12.5$, showing regression from the extreme low parental average value. Dominance can therefore give **overall regression** and **individual regression**, while additive action can give individual regression, but no overall regression. Gene interactions like epistasis can be similar to dominance between alleles of different loci and can cause regression and transgression in a similar way to dominance.

Transgression can come from genetic segregation for genes with dominance, additive action or epistasis, or from environmental effects. Thus in the above case with non-directional dominance and no linkage, A a, B b (20 units) x A a, B b (20 units) gives 9 A -, B - (20 units each), 3 A -, b b (30 units), 3 a a, B - (10 units) and 1 a a, b b (20 units). Of 16 offspring, 10 will have the same yield as the intermediate yield parents, 3 show transgression in an upward direction and 3 show transgression in a downward direction.

Suppose instead that we have additive and cumulative action, with four unlinked loci controlling height, that the breeder wants transgression to increased height and has pure breeding lines available of differing heights. None is of maximum height, which would be $A'A'$, $B'B'$, $C'C'$, $D'D'$. If there is no environmental variation, then the phenotypes reflect the genotypes as far as the number of primes, but give no direct information on which loci carry the primes. If one does not know the full genotypes, one does not know the best crosses to make.

Breeders would normally first try a tall line x a different tall line (say both have 6 primes), intercrossing the F1 to get some recombinant homozygotes in the F2, as loci differing between the parents will be heterozygous in the F1. If it turns out that both tall lines have the same genotype, say $A'A'$, $B'B'$, $C'C'$, $D D$, then there will be no recombination, no segregation and no transgression. If the two tall lines have different genotypes, say one is $A'A'$, $B'B'$, $C'C'$, $D D$ and the other is $A'A'$, $B B$, $C'C'$, $D'D'$, then the F1 will be heterozygous at the B and D loci. The F2 will be a mixture of some heterozygotes and both parental genotypes (6 primes) but recombination will also give $A'A'$, $B B$, $C'C'$, $D D$ (4 primes, showing regression) and $A'A'$, $B'B'$, $C'C'$, $D'D'$ (8 primes, the required tallest type from transgression) homozygotes, so the cross will have been a success.

One can easily postulate genotypes from other crosses which will give the desired F2 genotype with eight primes. They include $A'A'$, $B'B'$, $C'C'$, $D D$, (6 primes) x $A A$, $B B$, $C'C'$, $D'D'$, (4 primes), $A'A'$, $B B$, $C'C'$, $D D$ (4 primes) x $A A$, $B'B'$, $C C$, $D'D'$, (4 primes) and even $A A$, $B'B'$, $C'C'$, $D'D'$ (6 primes) x $A'A'$, $B B$, $C C$, $D D$ (2 primes). Therefore not all tall x tall crosses will give increased height, but some will, and so will some tall x medium, some medium x medium and some tall x short. This shows very clearly that getting the desired recombinants with transgression is not just a matter of the height or the number of primes of the strains which might be crossed, but is also very much a matter of **specific genotypes**, of which particular loci carry the primed alleles. If the obvious crosses of tall x tall do not work, this consideration of genotypes shows that trying some tall x medium, medium x medium or even tall x short crosses could give the desired result of transgression. Genetic transgression requires heterozygosity at some stage, either in the parents or the F1, or there can be no segregation or recombination.

The most **extreme transgression** is possible from the most average genotypes, e.g., $A'A'$, $B B$, $C'C'$, $D D$ (4 primes) x $A A$, $B'B'$, $C C$, $D'D'$, (4 primes) giving F2 ranging from 0 primes to all eight primes. Of the F2 from selfing the F1 in this cross, 186 out of 256 are expected to show transgression, with 93 taller and 93 shorter than the parents. The most extreme regression of the offspring average comes from having dominance at all segregating loci and the maximum amount of heterozygosity.

For a particular quantitative character in a population, regression and transgression will be **opposing forces** with transgression giving more extreme types, especially from average types, and regression giving less extreme types, especially from extreme types. This is shown for size in Fig. 3.1.1, with R for regression and T for transgression, and showing the R and T effects for different types of natural cross, assuming random mating. For example, if one has a small x small cross, as with the left-hand group of three arrows, then if these are not the smallest individuals possible, a few offspring will be even smaller, showing transgression from genetic or environmental effects (light arrow, labelled T), while the main tendency (heavy arrow, labelled R) is some regression, but some offspring could show a lot of regression (rightmost light arrow from small x small). Medium x medium will mainly have medium offspring (thick unlabelled arrow) but some will show transgression (light arrows). There will also be large x large, large x small, large x medium and small x medium crosses. In populations where regression and transgression generally balance out, the size distribution in generation 2 will be similar to that of generation 1, as shown in the generation 2 right-hand distribution, unless the environment differs between generations, when the environment might give a population with reduced individual sizes (shown in the left-hand, dashed, distribution) or increased individual sizes.

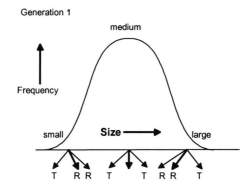

large x small gives largely regression, with small amounts of transgression

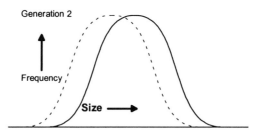

Figure 3.1.1. Size distribution in two successive generations in a random mating population. See text for details. R = regression; T = transgression; thick arrow, average response, e.g., from small x small; thin arrow, extreme response, e.g., from small x small. In generation 2, the unbroken curve is the expected one if the environment is similar to that for generation 1, but the dashed one is for a changed environment, giving poorer growth.

We can now ask: from what kinds of parent in the previous generation do extreme individuals come, say very large individuals? A few will have come from regression from extremely large x extremely large, or have come from very large x very large, while the majority will have come from transgression from large x large, medium x medium, large x small, medium x large, or even a few from small x small. On average, therefore, the parents of very large offspring will be less large, and the parents of very small offspring will be less small, than their offspring.

Example 3.1.1. Compare the average intelligence of university students with that of their parents and that of their own offspring, stating any assumptions which you make.

Assumption 1: university students are on average towards the upper end of the intelligence distribution for their populations. Assumption 2: intelligence is controlled by the environment and by genes with additive action and dominance, and will therefore show regression and transgression. Although some students will have come by regression from even more intelligent parents, the majority will have come from transgression from parents less intelligent on average, as shown in the above discussion about size. As students are near the extreme for intelligence, their own offspring should show regression and on average be less intelligent than their parents. Both assumptions are supported by the available data. The whole answer is framed in terms of averages, and some students will have parents more intelligent than themselves, and some students will have offspring more intelligent than themselves. The answer remains qualitatively the same whether there is positive assortative mating for intelligence, which there is, or whether mating is random.

3.2. Environmental effects on phenotypes

3.2.1. Sex-limited characters

Although one usually considers effects of the external environment on a character, there are effects of the **internal environment**, especially hormone levels. For example, in humans the expression of genes for hereditary early baldness depends on the levels of male hormones such as testosterone in the blood. The character is therefore mainly expressed in males, but can be expressed in females with unusually high levels of male hormones, e.g., through glandular malfunction. Characteristics normally only expressed in one sex are called **sex limited**. They must not be confused with sex-linked characters, because sex-limited characters are often autosomal. Penis characters in males and ovarian characters in females are obviously sex limited.

3.2.2. Phenocopies, conditional mutants and environmental effects

The environment can sometimes change qualitative phenotypes, changing the normal phenotype for one genotype to the phenotype normally associated with a different genotype, causing a **phenocopy**. A woman with genetically dark hair may apply bleach to get fair hair, resembling someone with genetically natural blonde hair. In the fruit fly *Drosophila*, putting silver nitrate in the medium of wild-type larvae gives yellow-bodied adults, resembling the mutant *yellow body*. The genotypes remain unchanged in phenocopies, which will breed as if they had their normal phenotype.

As we have seen, quantitative characters such as grain yield are usually affected by environmental factors such as light, rain, soil, temperature and disease. Whether some qualitative gene mutations are expressed at all may be controlled by the environment. In micro-organisms and higher organisms one can often obtain such **conditional mutants**. They include **temperature-sensitive** mutants which are expressed (giving the mutant phenotype) at one temperature (**restrictive conditions**) but appear wild-type at another temperature (**permissive conditions**). For example in the fungus *Neurospora crassa*, the mutant *rib-1^{ts}* requires riboflavin for growth at 37° but not at 25°. In the rabbit, c^h c^h gives the Himalayan pattern with off-white body fur but black ears, nose, feet and tail, while C C gives black fur all over. In the fungus and the rabbit, the mutation causes a change in amino acid sequence in an enzyme which is denatured at 37° but works normally at lower temperatures. The rabbit's extremities are cooler than the body, so the enzyme works in those extremities, giving black pigment there, but is inactivated by the main body temperature. This can be shown by shaving off some body fur, applying a cooling pack for some weeks, and getting black fur in the regrowth. Siamese and some other cat breeds often have dark points on their ears, from a similar temperature effect. In maize, the allele "sun-red" gives red colour when parts are exposed to direct sun, but green in shaded parts, while the alternative allele, "not sun-red", gives green in sun and shade.

Environmental effects may be temporary, such as climate one season affecting apple yield, or may be more permanent, such as light intensity during leaf development in some plants causing leaves to have one of two different types of structure, "sun leaves" and "shade leaves", each anatomically suited to the high or low light intensities respectively which they received during early development. Another example of plasticity during development is aquatic plants which have morphologically quite different submerged and aerial leaves.

3.2.3. Phenotype plasticity

This is the amount by which different environments can change the expression of a particular phenotypic character in a given genotype. Some quantitative characters are quite plastic, like the amount of melanin pigment in a "white" person's skin, varying with sun exposure, often reversibly. Other characters may be plastic under fairly extreme environments. Thus wild-type birch trees might grow to a height of 25 metres in a wide range of environments, but might be stunted, under two metres, if grown at high altitudes. The number of noses on a person's face, or the number of stamens on an iris flower, is constant, so those are non-plastic characters.

Breeders will be very concerned about **phenotype plasticity**, wanting to optimise the environment for the genotypes they are using for their commercial characters. If they are breeding lettuces, for example, they will be concerned about how different environmental factors affect the yield of leaves. If a genotype reacts unfavourably to a particular local environment, then either the genotype must be changed appropriately, or some other crop should be grown there. Unusually extreme conditions can be devastating. A high-yielding wheat had been bred for use in Britain and after being tested for a number of years it was released for general use. Then one summer was unusually hot and this variety failed to pollinate properly, giving a very low yield, so farmers ceased to grow it.

3.2.4. Variable expressivity

Some alleles are expressed to different extents in different individuals, for reasons which are often unknown - they could be genetic, environmental, or both. In humans with the usually dominant allele for **polydactyly**, some individuals have a small extra knob on the outer sides of both hands, showing low expressivity for this disease, while others might have two complete extra fingers on each hand and perhaps extra toes, showing high expressivity. The reasons for this difference in expressivity are usually unknown. **Variable expressivity** is a quantitative effect.

3.2.5. Incomplete penetrance

In contrast to variable expressivity, **incomplete penetrance** is qualitative, when an allele in a particular genotype is sometimes completely expressed and sometimes is not expressed at all, like a dagger either penetrating the skin or not penetrating! Most characters have 100% penetrance, but in *Drosophila*, the *i i* genotype gives 90% of individuals with fully interrupted wing veins and 10% with the wild-type non-interrupted wing-vein phenotype. The genotypes must be identical in both types of individual, as interrupted x interrupted and non-interrupted x non-interrupted crosses from that *i i* population both give 90% interrupted: 10 % non-interrupted progeny. The character is either fully expressed or not expressed at all: perhaps there is some above-threshold concentration of a chemical during development which gives interrupted wings, and those individuals which fail to reach that threshold have normal wings, like those of wild-type genotype lacking that compound completely.

In man, Huntington's disease (Chapter 13.3) has 95% penetrance, because about 5% of individuals with the dominant allele never develop the symptoms. With dominant inherited colon cancer, individuals may sometimes have the gene but show no symptoms because of incomplete penetrance, but they can pass it on and have affected offspring. Incomplete penetrance is the main reason for a dominant character sometimes "skipping a generation".

3.3. Narrow and broad sense heritabilities; equations, estimation and use; correlations between characters

We saw in Chapter 1.3.9 that **heritability** is the proportion of phenotypic variance (symbol *Vp*) attributable to genetic variation, with the term applying to characters with continuous variation in a population. For an account of variances, covariances, correlations, linear regressions and other mathematical background, see Cameron (1997) if necessary. Narrow sense heritability, h^2, is the proportion of phenotypic variance attributable to genes with additive effects, while broad sense heritability, H^2, is the proportion of phenotypic variance attributable to all types of genetic variation. The variances due to different types of genes are *Va* for genes with additive effects, *Vd* for genes with dominance, and *Vi* for genes with epistasis or gene interactions. *Ve* represents the variance due to the environment and *Vg/e* represents the variance due to genotype/environment interactions, where different genotypes perform differently in different environments.

$$\text{Narrow sense heritability: } h^2 = \frac{Va}{Va + Vd + Vi + Ve + Vg/e} .$$

$$\text{Broad sense heritability: } H^2 = \frac{Va + Vd + Vi}{Va + Vd + Vi + Ve + Vg/e} .$$

The equations for heritability therefore contain one or more expressions for genetic variation in the numerator, and for genetic variation and environmental variation in the denominator. A population with more genetic variation than another will therefore tend to have higher heritabilities if their environmental variations are similar. A population with more environmental variation than another will tend to have lower heritabilities if their genetic variation is similar. Of great importance to the breeder is that fact that during selection programmes, heritabilities will decrease, as genetic variation is reduced by selecting the better alleles and removing the poorer ones. Genetically uniform populations will have heritabilities of zero for all characters, as there is no genetic variation. Heritabilities range from zero, all variation environmental, none genetic, to 1.0, all variation genetic, none environmental.

Although one finds published tables of heritabilities for various characters in commercial organisms (e.g., Dalton, 1980), it is essential to remember that while these may be typical values, they may vary widely, say from herd to herd, depending on how variable that herd is genetically and on how variable its environment is. Heritabilities in a herd of pure line sheep or a field of clonally propagated tea (all of one selected genotype) will be much lower than those in a herd of crossbred sheep or a field of seedling tea (which has a wide range of genotypes).

One way to estimate narrow sense heritabilities is from:

$$h^2 = \frac{\text{observed correlation between relatives}}{\text{theoretical correlation between relatives}}.$$

The **observed correlation** is the statistical correlation coefficient found for a quantitative trait for a given degree of relationship, often **half-sibs** (which have one parent in common, usually the father, and one parent different). So one would take a number of different pairs of half-sibs, measure the character concerned, and work out the correlation coefficient. The **theoretical correlation between relatives** (also called the coefficient of relationship) is the proportion of genes (alleles) expected to be in common by recent descent. Thus one parent and an offspring have half their genes in common, as do **full sibs** (full brothers and/or sisters, with both parents in common), while half-sibs have one quarter of their genes in common, as do uncle/niece or aunt/nephew. So long as the parents are not themselves inbred, the theoretical correlation is twice the inbreeding coefficient, F, (see Chapter 6.3) of their offspring. For example, the offspring of half-sibs have F = 1/8, so the theoretical correlation for half-sibs is 0.25.

Example 3.3.1. The correlation between half-sibs for 160-day weight in a herd of pigs is 0.08. What is the heritability?
As half-sibs have 1/4 of their genes in common, the theoretical correlation is 0.25, so $h^2 = \frac{0.08}{0.25} = \underline{0.32}$.

There is a series of equations relating the **covariances** for a character between particular types of relative to estimated proportions of genetic variance due to genes with additive effects, Va, to variance due to genes with dominance, Vd, and to variance for genes with interactions, Vi. For the mathematical background, see Falconer and Mackay (1996), Chapter 9. Assuming that Vi and Vg/e are relatively small, we get coefficients for the covariances of relatives as shown in Table 3.3.1.

Table 3.3.1. Coefficients of the variance components in the covariances of relatives.

Relatives	Estimated proportion of	
	Va	Vd
Monozygotic twins	1	1
Parent-offspring	½	0
Full sibs	½	¼
Half-sibs	¼	0
Uncle-niece	¼	0

Example 3.3.2. In a population of chickens, phenotype variance for egg weight, Vp, was 16. The covariances between relatives were 1.6 for mother-daughter, 2.8 for sisters, 0.75 for half-sisters. Calculate the heritabilities.

To calculate narrow and broad sense heritabilities, we need to calculate Va and Vd, and will have to assume that Vi and Vg/e are negligible so that we can use the values in Table 3.3.1. We are given the denominator of the heritability equations, the total phenotypic variance, Vp. We can calculate Va from cases where Vd is zero. Parent-offspring (mother-daughter for eggs) covariance = $1/2$ Va as Vd has a coefficient of zero, so $1/2$ $Va = 1.6$, so $Va = 3.2$. We can also calculate Va from the half-sib data, from which $1/4$ $Va = 0.75$, so $Va = 3.0$, in quite good agreement with the value calculated from parent-offspring. We can take an average of the two Va values, 3.1, to use in other calculations. Now knowing Va, we can use the full sib data to calculate Vd, because covariance for full sibs = $1/2$ Va (when we know Va to be 3.1) + $1/4$ Vd, so $2.8 = 1/2 (3.1) + 1/4$ Vd, so $Vd = 5.0$. Putting these values of Va and Vd into the heritability equations, and assuming that Vi and Vg/e are negligible, we get $h^2 = \frac{3.1}{16} = \underline{0.19}$, and $H^2 = \frac{3.1+5.0}{16} = \underline{0.51}$. Providing that Vd and/or Vi are not zero, the broad sense heritability is always larger than the narrow sense heritability.

The most important use of heritabilities is as a guide to whether it is worth running selection programmes to improve a character, bearing in mind the costs of running a **selection programme** and the likely commercial returns from higher yield, better quality or lower costs. For tables of heritabilities, repeatabilities, breeding aims and hybrid vigour in farms animals, see Dalton (1980). Suppose we take some typical narrow sense heritabilities from the 1950's for American dairy cattle, 0.01 for fertility, 0.3 for milk yield and 0.6 for milk butterfat. We should realise that the actual figures would vary widely between herds, depending on how genetically variable each herd was and on how variable their environments were. We could predict that heritabilities would be lower today, as in 40 years the cattle breeders would have selected better genotypes, getting rid of poorer alleles, and so reducing genetic variation.

We see that fertility had a very low heritability, 0.01, so that only 1% of variation was genetic and 99% was environmental. With so little genetic variation, it would probably not be worth running a major selection programme to improve fertility as there would be so little genetic response, though one would **cull** (kill off or not breed from) any individuals of low fertility. It would probably be more useful to try to improve the environment (e.g., the feeding regime and disease control) to increase fertility. As high fertility has been encouraged by natural selection before cattle were domesticated and by man since domestication, it is no surprise that heritability for fertility is low.

In dairy cattle, there must have been natural selection in the wild for moderate milk yields, enough to feed the calves, but man has selected for increased yields. If milk yield heritability is 0.3, 30% of the variation is genetic; because milk yield is of supreme importance in dairy cattle, it would certainly be worth running a selection programme for increased milk yield, with the prospect of a good response, unless in the herd concerned the heritability is much lower than 0.3 now.

There would have been natural selection for a moderate milk butterfat content, with selection for increased values since domestication. In many countries, there are legal minimum butterfat levels for particular types of milk. The breed of cattle and the major uses of their milk - for drinking, cream, butter, cheese, yoghurt, etc. - are relevant to the desired butterfat level for commerce. Reduced-fat milks, such as semi-skimmed, are now important. If the heritability is still high, approaching 0.6, one could expect an excellent response to selection for increased milk butterfat, in grams of fat per litre of milk, as most of the variation is genetic.

We must, however, be conscious that selecting for one character can affect other characters if there are **correlations** between them. If by selection for higher milk butterfat (in grams per litre of milk) one gets more metabolic energy diverted to making butterfat, there would be less energy left for making other milk components. One might therefore predict a negative correlation between milk yield (litres per lactation) and milk butterfat (g/l). One way of getting a response to selection for increased milk butterfat in g/l might be for the cow to produce the same total amount of butterfat per day, but a reduced volume of milk. Breeders would try to select simultaneously for increased butterfat and increased or at least maintained levels of milk yield. If there were to be a negative correlation in spite of dual selection, the breeder would need to work out whether the economic value of the increased butterfat outweighed the reduced value of milk volume.

One cannot always predict **correlations between characters**. Suppose you successfully selected for increased leaf length in lettuce or tobacco, what might happen to leaf width and hence to leaf area? It would all depend on how the genes controlling leaf width acted. If the segregating genes just affected the number of cell divisions in the leaf meristems, in all directions, then selecting for increased length would also select for increased leaf width, giving a positive correlation between length and width, and perhaps a large overall increase in leaf area. If the genes only affect cell divisions in the plane giving length, with no effect on divisions in other planes, then as leaf length goes up, there will be no change in leaf width, so there is no correlation between length and width, and leaf area will increase in proportion to the increase in length. If the genes selected change the number of divisions in different planes, with no effect on the total number of divisions, then as one gets more divisions in the plane giving length, one will get proportionally fewer divisions in the plane giving width, and a negative correlation between the two characters, length and width, perhaps with no overall change in leaf area. There could thus be a positive or a negative correlation, or no correlation, between leaf length and leaf width.

Correlations between characters can result from physiology, e.g., negative correlations between milk yield and butter fat, or between bean weight and the number of beans per plant, because both characters compete for the same physiological resources. Correlations could also be due to linkage between syntenic loci which are in linkage disequilibrium, but such correlations would be broken in time by recombination. Unbreakable correlations could arise from **pleiotropy**, where two characters - even superficially unrelated ones such as deafness and blue eyes in white cats - are caused by the same allele. Some **examples of correlations** are: in yeast, red colony colour with a requirement for adenine; in cereals, small leaves with many leaves per plant, large grains with few grains per ear, rapid autumn growth with poor winter hardiness; in alfalfa, thick leaves with a high photosynthetic rate; in maize, poor seedling vigour with late maturity. In sheep, there are positive correlations between resistance to fly-strike and to fleece rot, and between food intake and wool production, and a negative correlation between hair follicle density and wool fibre diameter. Cows with the largest udders and biggest milk production are the most susceptible to disease.

3.4. Genotype value, phenotype value and breeding value

For a particular character, an individual's **genotype value** is its value as judged by its genotype; its **phenotype value** is its value as judged by its phenotype, and its **breeding value** is its value as judged by the genes it hands on, which is judged by the mean performance of its progeny, allowing for the merits of the individuals it was crossed with. If the character is entirely controlled by genes with additive action and if there are no environmental effects, then an organism's genotype, phenotype and breeding values are identical.

We can take the case of height from Chapter 2.3, determined by three unlinked loci with additive and cumulative action, a base height of 50 cm and each prime adding 2 cm. If there are no environmental effects, we get a simple linear relation between genotype and phenotype, as shown in Fig. 3.4.1. Each genotype has only one phenotype, ranging from 50 cm with no primes to 62 cm with six primes. Selecting the tallest individuals, say those of 61 cm or more, is extremely efficient as only individuals with six primes are chosen, which have the best breeding value for height.

If we now introduce some environmental effects, each genotype can then have a range of phenotypes, giving continuous variation. This is shown in Fig. 3.4.2, where in addition to the phenotypes previously shown (large "X"s), each genotype has other possible phenotypes shown as small "x"s, ranging equally above and below the original points.

If environmental effects are symmetric, giving equal deviations upwards and downwards, then the large X represents the mean phenotype value for each genotype. The phenotype value, P, for any individual is its genotype value, G, plus an environmental deviation, E, which may be positive, zero or negative. In Fig. 3.4.2, points on the diagonal line have $E = 0$; the points above the diagonal line have positive environmental deviations, and those below that line have negative deviations. For a whole population, it is usually assumed that the mean environmental deviation is zero, with positive deviations equalling negative ones. For a whole population, the mean phenotype value then equals the mean genotype value and can be used to estimate the latter.

If we consider the efficiency of selection in the presence of environmental variation, we see from Fig. 3.4.2 that selecting the tallest phenotypes, of 61 cm or more, as before, results in selecting individuals for breeding whose genotypes have

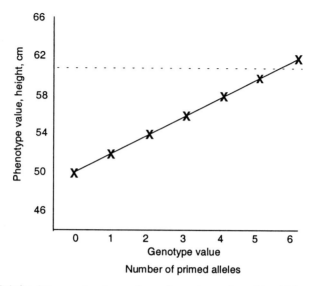

Figure 3.4.1. Relation between phenotype value and genotype value with additive action and no environmental effects. The dashed line indicates the effects of selecting the tallest individuals, of 61 cm or more.

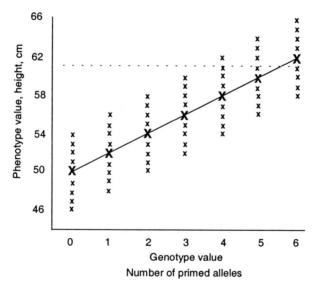

Figure 3.4.2. As for Fig. 3.4.1, but with environmental variation, so that each genotype now has, in addition to its original phenotype (large X), a range of possible phenotypes (small x's).

four, five or six primes, instead of only those with six primes, so the mean breeding value of the selected individuals will be less than six primes. In addition, Fig. 3.4.2 shows that we will fail to select some individuals with the ideal genotypes of six primes, as some have phenotypes in the 58 to 60 cm range. As we have seen before, the greater the amounts of environmental variation, the lower the heritabilities, and the less reliable is the selection based on phenotypes. From Fig. 3.4.2, it can be seen that one could still select only individuals with six primes by taking individuals of heights between 64 and 66 cm, although that is inefficient as it fails to select the majority of individuals with six primes.

Ideally one selects on an individual's breeding value, judged by the mean value of the offspring when it (usually he, as this takes too long to do for females) is crossed to a typical range of individuals of the opposite sex. That is **progeny testing** (Chapter 5.7). The tested individual's breeding value is twice the mean deviation of the progeny from the population mean; the deviation is doubled because the individual provides only half the genes in the progeny. While the breeding value could be given in absolute units, it is usual to express it in terms of deviations from the population mean. Over a whole population, the mean breeding value = the mean phenotype value = the mean genotype value.

In the UK, the Meat and Livestock Commission has a Beefbreeder Service, with a method called BLUP (Best Linear Unbiased Prediction) for more accurate estimates of the genetic merit of pedigree cattle, which allows comparison across herds. BLUP can help in selecting a beef bull to breed replacement heifers with higher maternal performance and higher profitability, possibly worth several thousand pounds over the lifetime of a bull's daughters (based on each bull producing 10 heifers per year by natural service, over four years of breeding).

The home-bred beef replacement heifers should make a direct contribution to the genetic potential of their calves in terms of growth rate, muscling score, etc. They should successfully rear a calf each year, through good fertility, ease of calving, and good milk yield. The cow makes as much genetic contribution to the calf as does the bull, as she passes on half her genes. For example, a purebred cow with an **estimated breeding value** (EBV) of +20 kg for 400 day weight will pass on an average of +10 kg weight at 400 days, compared to calves sired by the same bull out of cows with an EBV of 0. The important direct EBVs are those for birth weight, 200 day growth, 400 day growth, backfat depth, muscling score and beef value. It is more difficult to estimate breeding values in crossbred cows, as an unknown part of their performance will be from hybrid vigour.

The early growth of the suckled calf, and therefore its 200 day weight, depends on its own genetic potential for growth and on the milk yield and milk quality of its mother. In the BLUP scheme, the cow's maternal merit is measured as the cow's ability to rear a calf to 200 days of age; e.g., a cow with a +15 EBV for 200 day milk will wean calves 15 kg heavier, on average, than a cow with 0 EBV for this trait.

This information is taken from Simm, Crump and Lowman (1993). Their final advice was: select a shortlist of bulls with the highest positive EBVs for 200 day milk; delete bulls with EBVs for 200 day growth of below +10 kg, EBVs for birth weight of above + 2 kg, and EBVs for beef value of below £10; visually assess all bulls remaining on the shortlist in terms of physical fitness, rejecting any unsound animals. Selecting for below average birth weight is done to reduce calving difficulties.

BLUP is used with different models to predict breeding values and to estimate environmental effects. It is most often used to predict sire breeding values from progeny testing (the sire model), or to predict breeding values from animals with repeated records (the repeatability model), or to predict breeding values for all animals in a pedigree (the individual animal model): see Cameron (1997) for equations and examples. In BLUP, the B (Best) refers to maximising the correlation between the true breeding value and the predicted breeding value; the L (Linear) refers to predicted breeding values being linear functions of the observations; the U (Unbiased) refers to the estimates of fixed effects being unbiased and the unknown true breeding values being distributed around the predicted breeding values; the P (Prediction) refers to the procedure predicting the true breeding values.

Different types of selection, and their suitabilities for different heritability ranges, are covered in more detail in Chapter 5, including the use of selection indices (Chapter 5.10.3). Further aspects of selection in plants and animals are given in Chapter 12, in humans in Chapter 13, and in fungi in Chapter 18.

Suggested reading

Most of the book references at the end of Chapter 2 also apply here.

Cameron, N. D., *Selection Indices and Prediction of Genetic Merit in Animal Breeding.* 1997. CAB International, Wallingford.

Dalton, D. C., *An Introduction to Practical Animal Breeding.* 1980. Granada, London.

Simm, G., R. E. Crump and B. G. Lowman, *On Choosing a Beef Bull to Breed Replacement Heifers.* SAC Technical Note T336. Undated and with no publisher given, but is 1993, Scottish Agricultural College, Edinburgh.

Strickberger, M. W., *Genetics*, 3rd ed. 1985. Collier Macmillan, London.

CHAPTER 4

POPULATION GENETICS: ALLELE FREQUENCIES, GENETIC EQUILIBRIA, POPULATION MIXING, GENETIC DRIFT AND GENE FLOW

4.1. Introduction

Population genetics is extremely important to the understanding of applied genetics, where plant and animal breeding often consists of the manipulation of populations (e.g., Pirchner, 1983; Brown et al., 1990). In human genetic counselling, a knowledge of allele frequencies is essential in calculating the risks of having affected children for various inherited diseases.

Unless otherwise stated, this chapter refers to diploid Eukaryotes. In Mendelian genetics one usually considers crosses of pure-breeding individuals, with the F1 being selfed or test-crossed, with F2 ratios being examined as in Chapter 2. In population genetics we normally consider populations in which several different crosses may be occurring randomly, e.g., $A A$ x $a a$, $A A$ x $A A$, $A A$ x $A a$, $a a$ x $a a$, $A a$ x $A a$ and $A a$ x $a a$. Even if we start with only two genotypes, they may not be equally frequent and must be considered as crossing with their own genotype as well as with the other genotype, unless sex or incompatibility barriers prevent that.

Allele frequencies may vary from 0 to 1.0 for any allele in a population but the sum of the allele frequencies at one locus is always 1.0. **Genotype frequencies** must also sum to one, as must **phenotype frequencies**, and checking that they do so in problems is a useful safeguard against arithmetical errors. While Mendelian genetics usually is concerned with set-piece crosses of different homozygotes over two or three generations and the ratios produced in the F2 or F3, in population genetics we may be concerned with many generations, with frequencies of alleles, genotypes and phenotypes, with whether mating is random, and with the effects of selection, chance, population size, linkage, mutation, recombination and migration.

4.2. The Hardy-Weinberg equilibrium for one locus and two loci; linkage disequilibrium, and population mixing

4.2.1. One locus, two alleles

The basis of gene frequency analysis is the **Hardy-Weinberg equilibrium**, independently derived in 1908 by Hardy in England and Weinberg in Germany. One can deduce it from Mendelian genetics. If a large population has p as the frequency of one allele at a locus, say A, and q as the frequency of the alternative allele, say a, so that $p + q = 1$, then the genotypes will reach in one generation, and remain at, the frequencies p^2 for $A\,A$, $2pq$ for $A\,a$, and q^2 for $a\,a$. One can obtain this from a simple table, combining the gametes at random, as in Table 4.2.1.1.

Table 4.2.1.1. Derivation of the Hardy-Weinberg equilibrium frequencies from random mating.

Allele frequencies in male gametes:	p of A	q of a
Allele frequencies in female gametes	Zygotes and their frequencies	
p of A	$A\,A,\ p^2$	$A\,a,\ pq$
q of a	$A\,a,\ pq$	$a\,a,\ q^2$

Obtaining this equilibrium depends on there being only two alleles in the population for the locus concerned: cases of multiple alleles and more than one segregating locus are considered later. The population needs to be large or chance fluctuations in allele frequency may affect genotype frequencies as described later under genetic drift. Achieving Hardy-Weinberg equilibrium also requires that mating be random, not assortative, and without inbreeding or outbreeding; that there are no selection, viability or fertility differences, and that migration and mutation frequencies are negligible.

The genotype frequencies in one generation depend on the previous generation's allele frequencies rather than directly on the latter's genotype frequencies. Thus if one has a population with 100 $A\,A$ and 100 $a\,a$ individuals mating randomly ($A\,A$ with $A\,A$ and equally with $a\,a$, and $a\,a$ with $a\,a$ and equally with $A\,A$), it gives the same genotype frequencies in the next generation as a population of 200 $A\,a$ individuals, because both populations have the same allele frequencies, 0.5 for both A and a. The genotype frequencies will be p^2 for $A\,A$ = $(0.5)^2$ = 0.25; $2pq$ for $A\,a$ = 2 x 0.5 x 0.5 = 0.5; and q^2 for $a\,a$ = $(0.5)^2$ = 0.25. The genotype frequencies, like the allele frequencies, total 1.0, because they sum alternatives. Hardy-Weinberg analysis enables one to calculate allele and genotype frequencies from phenotype frequencies, even in the presence of complete dominance.

Example 4.2.1.1. In maize, the character rust-susceptible, *rp* (to the fungus *Puccinia sorghi*) is recessive to resistance to that fungus, *Rp*. If 5 maize seedlings are susceptible to rust in a population of 50,000 seedlings, what are the allele and genotype frequencies?

As susceptibility is recessive, the susceptible genotype must be *rp rp*. Let the *rp* allele have frequency *q*, and the *Rp* allele have frequency *p*, with $p + q = 1.0$. If random mating and the other conditions of the Hardy-Weinberg equilibrium are met, then *rp rp* has frequency q^2. As we are told that the recessive phenotype has a frequency of 5 in 50,000, $q^2 = \frac{5}{50,000} = 0.0001$, so $q = \sqrt{0.0001} = 0.01$. As $p + q = 1.0$, $p = 1.0 - 0.01 = 0.99$. The allele frequencies are therefore 0.99 for the rust-resistance allele, *Rp*, and 0.01 for the rust-susceptibility allele, *rp*. We know that genotype *rp rp* has frequency 0.0001. *Rp Rp* has frequency $p^2 = (0.99)^2 = 0.9801$, and *Rp rp* has frequency $2pq = 2 \times 0.99 \times 0.01 = 0.0198$. The three calculated genotype frequencies add up to 1.0, as expected.

Even though there is complete dominance of resistance to susceptibility, with *Rp Rp* and *Rp rp* having identical phenotypes, just knowing the frequency of the recessive homozygote enables us to calculate the frequencies of these different genotypes. With such a low frequency of the recessive allele, the dominant homozygote is much more frequent (49.5 times) than the heterozygote. As the recessive homozygote is in turn much rarer (198 times) than the heterozygote, we see the very important point that rare recessive alleles are present mainly in the heterozygote, where they are protected by dominance from selection, rather than in the recessive homozygote where they are exposed to selection.

4.2.2. One locus, more than two alleles

We can easily extend the basic Hardy-Weinberg analysis to **multiple alleles** at a locus. Let the alleles be a^1, a^2, a^3 to a^n, with respective allele frequencies *p, q, r* to *n*, which must add up to 1.0, so $p + q + r + ... + n = 1.0$. Many different random matings could occur between homozygotes and heterozygotes. By using a Punnett chequer board as in Chapter 2, Table 2.2.1.1, we could show that random mating gives the frequency of any homozygote as the square of its allele frequency, and of any heterozygote as twice the product of the two allele frequencies. Thus $a^1 a^1$ has equilibrium frequency p^2 and $a^3 a^3$ has frequency r^2; $a^1 a^3$ has frequency $2pr$ and $a^2 a^3$ has frequency $2qr$.

Example 4.2.2.1. A human population has ABO blood group allele frequencies of $i^o = 0.5$, $i^A = 0.4$, $i^B = 0.1$. Alleles i^A and i^B are dominant to i^o, and show co-dominance to each other. What are the expected phenotype frequencies?

As i^o is recessive, all people of blood group O must be $i^o\ i^o$, with a frequency of $(0.5)^2 = \underline{0.25}$.

Blood group A has two genotypes, $i^A\ i^A = (0.4)^2 = 0.16$, and $i^A\ i^o = 2 \times 0.4 \times 0.5 = 0.4$, total $\underline{0.56}$.

Blood group B has two genotypes, $i^B\ i^B = (0.1)^2 = 0.01$, and $i^B\ i^o = 2 \times 0.1 \times 0.5 = 0.1$, total $\underline{0.11}$.

Blood group AB has one genotype, $i^A\ i^B$, frequency $2 \times 0.4 \times 0.1 = \underline{0.08}$. The four phenotype frequencies add up to 1.0, as expected.

4.2.3. Allele segregation at two loci and the importance of recombination frequencies

Although Hardy-Weinberg equilibrium is reached from states of disequilibrium in one generation of random mating for two or more alleles at one locus, there is a gradual approach to equilibrium over many generations for alleles at **two different loci**, because recombination between the loci is involved in decreasing the disequilibrium, and recombination has a maximum frequency of 50%. The maximum reduction in the departure from equilibrium in one generation is therefore 50%, with smaller reductions for linked genes. **Linkage equilibrium** occurs when the alleles of different loci in a population are present in gametes in proportion to the product of the allele frequencies, e.g., the frequency of gamete *A, B* = frequency of allele *A* x frequency of allele B, and **linkage disequilibrium** is the amount of departure from this state.

Let *A* and *a* be alleles at one locus, with frequencies *p* and *q*, with $p + q = 1.0$, and let *B* and *b* be alleles at a second locus, with frequencies *r* and *s*, with $r + s = 1.0$. The equilibrium frequencies for a genotype are calculated by multiplying the separate genotype frequencies for the two loci, e.g., *A A, B B* = p^2 x r^2; *A a, B b* = $2pq$ x $2rs$; *A A, B b* = p^2 x $2rs$. The equilibrium gamete frequencies will be *A, B* = *pr*; *a, b* = *qs*; *A, b* = *ps*, and *a, B* = *qr*. The **recombination fraction** (the recombination frequency as a percentage divided by 100) is *c*, and *d* is the **departure from equilibrium**, measured by the product of the frequencies of the two coupling allele combinations in the gametes (*A, B* and *a, b*) minus the product of the frequencies of the two repulsion allele combinations in the gametes (*A, b* and *a, B*):

$$d = (A,\ B \ \text{x} \ a,\ b) - (A,\ b \ \text{x} \ a,\ B).$$

Departure d is zero at equilibrium, when the products of the two types of gametes, coupling and repulsion, are equal; it can have a positive or a negative sign when there is linkage disequilibrium. Any departure from equilibrium is reduced each generation by a proportion corresponding to the recombination fraction, c. So with d_t as the departure from equilibrium at generation t and d_{t-1} as the departure in the previous generation, we get:

$d_t = d_{t-1} \times (1 - c)$.

For two unlinked loci (50% recombination), $c = 0.5$, so $d_t = d_{t-1} \times (1 - 0.5) = 0.5$ d_t, halving the disequilibrium each generation. For two closely linked loci (5% recombination), $c = 0.05$, so $d_t = d_{t-1} \times (1 - 0.05) = 0.95\ d_t$, with a very small reduction in disequilibrium in each generation.

Example 4.2.3.1. In part of the USA the frequency of the O blood group allele is 0.62 and that of the Rhesus positive factor is 0.7. What is the expected frequency of people who are blood group O, Rhesus positive?

Blood group O are homozygous recessive, so will be $(0.62)^2 = \underline{0.3844}$. Rhesus positive is dominant, so Rhesus positives will include $Rh\ Rh = (0.7)^2 = 0.49$ and Rh $rh = 2 \times 0.7 \times 0.3 = 0.42$, total for Rhesus positive, $\underline{0.91}$. The equilibrium frequency of people who are group O and Rhesus positive is therefore the product of the two phenotype frequencies, $0.3844 \times 0.91 = \underline{0.35}$.

We can multiply phenotype frequencies as well as genotype frequencies for the two loci. The allele frequency of Rhesus negative, rh, was calculated as one minus the frequency of Rh. As well as calculating the numerical answer, one could add that it depended on all the assumptions of the Hardy-Weinberg equilibrium at each locus, and on the population having reached equilibrium for the two loci. The USA has had much immigration and incomplete mixing of different racial groups, so this population might well not have reached equilibrium for the two loci. Because the ABO locus is on chromosome 9 and the Rhesus locus is on chromosome 1, the loci are unlinked, so any disequilibrium will be halved each generation.

4.2.4. What happens when two pure-breeding but different populations mix?

It helps to clarify many genetical issues, including ones connected with the equilibrium for two loci, if we consider what happens when two pure-breeding but different populations mix, say two small herds of cattle.

Example 4.2.4.1. Suppose 15 males and 15 females of cattle herd 1, all genotype *A A, B B*, are mixed with 5 males and 5 females of herd 2, all *a a, b b*, with random mating in the mixed herd, which is counted as generation zero. The herd size then stays constant at 40 animals, with random mating each generation. What happens to the genotype, gamete and allele frequencies, and to the departure from equilibrium, in successive generations?

The initial allele frequencies (*p* for *A*, *q* for *a*, *r* for *B*, *s* for *b*) are therefore 0.75 for *A* and *B*, and 0.25 for *a* and *b*, and the initial gamete frequencies in the mixed herd are 0.75 for *A, B* (all from herd 1) and 0.25 for *a, b* (all from herd 2). With no repulsion gametes produced in generation zero, the initial departure from equilibrium in their gametes, $d_0 = \{(A, B) \times (a, b)\} - \{(A, b) \times (a, B)\} = (0.75 \times 0.25) - (0.0 \times 0.0) = \underline{0.1875}$.

I run a class simulation experiment on this mixing, using different coloured beads to represent the four gamete genotypes, so the role of chance in the population usually causes deviations from the expected results and helps to instruct students in the role of chance. For this book let us use theoretical results, ignoring chance. With generation 0 having gamete frequencies of 0.75 for *A, B* and 0.25 for *a, b*, the expected genotypes frequencies in generation 1 calves and adults for *A A, B B* are $(0.75)^2 = 0.5625$, for *a a, b b* are $(0.25)^2 = 0.0625$, and for *A a, B b* are $2 \times 0.75 \times 0.25 = 0.375$; we can write these particular *A a, B b* individuals as *AB/ab* as all these heterozygotes have come from gamete fusions of coupling gametes, *A, B + a, b*; none of the generation 1 double heterozygotes is from the fusion of repulsion gametes to give *Ab/aB*, as there are no repulsion gametes yet.

When the adults of generation 1 produce gametes, the homozygotes will only produce one type of allele, so the 0.5625 *A A, B B* will contribute 0.5625 *A B* gametes to the gamete pool and the 0.0625 *a a, b b* will contribute 0.0625 *a b* gametes, as we assume no fertility or viability differences. In the first instance, let us assume that the two loci are **unlinked, with 50% recombination**. The double heterozygotes, 0.375 *A a, B b*, will with no linkage produce equal frequencies of the four types of gamete, that is $\frac{0.375}{4} = 0.09375$ each of *A, B; a, b; A, b; a, B*. The total gamete frequencies from homozygotes and the double heterozygotes will then be 0.65625 of *A, B*, 0.15625 of *a, b*, 0.09375 of *A, b*, and 0.09375 of *a, B*, adding up to 1.0. The amount of disequilibrium is now $d_1 = (0.65625 \times 0.15625) - (0.09375 \times 0.09375) = \underline{0.09375}$, exactly half of d_0 because we have 50% recombination, so the disequilibrium has been reduced by proportion $(1 - c) = 0.5$ per generation.

The adults of generation 2 can be worked out by combining those gametes at random, e.g., using a chequer board. We expect $(0.65625)^2$ of *A A, B B* = 0.4307, etc., but in addition to homozygotes and coupling (*AB/ab*) double heterozygotes, we will now get single heterozygotes, e.g., *A a, B B* and *a a, B b*, from combinations of coupling and repulsion gametes, and we will get repulsion (*Ab/aB*) double heterozygotes. The chance of getting a repulsion double heterozygote is twice the frequency of *A, b* gametes times the frequency of *a, B* gametes - the factor of two arises as the double heterozygote could come from sperm *A, b* + egg *a, B* or sperm *a, B* + egg *A, b*. That gives 0.0176 as a fraction, or 0.7 of a cow out of 40. The *a a, B B* class would only be 0.4 individuals out of 40. In a herd of 40 cows, some of the rarer genotypes would probably absent, as one cannot have viable fractions of a cow.

We can now work out the gamete frequencies produced by generation 2 adults. For double homozygotes and the two double heterozygotes, we can proceed as in generation 1. But now we have some single heterozygotes. In these, the recombination frequency is irrelevant; they will segregate at the heterozygous locus in a 1:1 ratio, and as the two alleles are identical at the other locus, it does not matter whether there is recombination between the two loci. For example, *A a, B B* will give 1 *A B*: 1 *a B* gametes whether there is recombination between the two loci or not, and irrespective of the recombination frequency.

Table 4.2.4.1. The consequences for genotypes frequencies of mixing 30 *A A, B B* animals with 10 *a a, b b* animals, from another population, with random mating; (i) with 50% recombination; (ii) with 5% recombination. One would not get fractions of a cow in reality, but these are calculated values.

Generation	Genotypes and their frequencies									
	A A, B B	A a, B B	a a, B B	A A, B b	A A, b b	A B/ a b	A b/ a B	A a, b b	a a, B b	a a, b b
0*	0.75									0.25
1*	0.563					0.375				0.063
(i), with 50% recombination										
2	0.4307	0.1230	0.0088	0.1230	0.0088	0.2051	0.0176	0.0293	0.0293	0.0244
Numbers out of 40	17	5	0.4	5	0.4	8	0.7	1	1	1
(ii), with 5% recombination										
2	0.5485	0.0139	0.0001	0.0139	0.0001	0.3564	0.0002	0.0045	0.0045	0.0579
Numbers out of 40	22	0.6	0.004	0.6	0.004	14	0.007	0.2	0.2	2
...										
Many*	0.3164	0.2109	0.0352	0.2109	0.0352	0.0703	0.0703	0.0234	0.0234	0.0039
Numbers out of 40	13	8	1	8	1	3	3	0.9	0.9	0.2

* The values for generations 0, 1 and "many" apply to both recombination frequencies.

disequilibrium populations will tend to go to $d = 0.0$ over many generations, they may overshoot the equilibrium point by chance, so a value of d of $+0.001$ in generation 12 in a small population might be followed by -0.0003 in generation 13 and $+0.0004$ in generation 14.

- The **equilibrium values** are the same whether one starts with all coupling gametes, all repulsion gametes, or of mixture of coupling and repulsion gametes, because equilibrium gamete and genotype frequencies depend only on allele frequencies.

- Recombination is only effective in **double heterozygotes**, e.g., *AB/ab*, not in single heterozygotes, e.g., *AB/aB*. If there is linkage, the two types of double heterozygote, *AB/ab* and *Ab/aB*, give different gamete frequencies.

- Provided that there is some recombination, the equilibrium genotype and gamete frequencies are unaffected by recombination frequencies, although recombination frequencies determine the rate of approach to equilibrium in disequilibrium populations. At linkage equilibrium, the product of the coupling gametes $\{(A, B)(a, b)\}$ is equal to the product of the repulsion gametes $\{(A, b)(a, B)\}$.

- The presence of **linkage disequilibrium** in a population suggests, but does not prove, fairly recent immigration into that population by organisms with different allele frequencies. It can also be caused by selection.

- In small populations, it may take many generations for **rare genotypes and phenotypes** to show, even when only two loci are segregating. As several hundred loci may be segregating when two populations mix, it may take hundreds of generations for the rarest types to be present in a small population. The breeders should therefore grow as large a population as practical to find beneficial new types, and should **continue looking** over many generations. Closely linked genes will take much longer than unlinked ones to come into equilibrium, and to produce all possible genotypes in a finite population.

4.3. Genetic drift, fixation and effects of population size

Genetic drift is a change in allele frequencies by chance, not by selection. It is non-directional, unlike selection, with genetic drift equally likely to increase or decrease the frequency of a particular allele in a population. It may change allele frequencies temporarily, or it may change them permanently if it leads to **fixation**, with one allele being eliminated and another reaching a frequency of 1.0.

Genetic drift has little effect in very large populations (i.e., ones with high numbers of individuals) and is most severe in very small populations. Suppose that a small soil pocket in a rock face has room only for two individuals of an annual grass, with one plant *A A* and one *A a*, in one generation. By chance, the next generation might have two *A a* plants, in which case there has been a chance change in allele frequency in favour of *a*. By chance, the next generation might have two *a a* plants, with fixation to *a* and elimination of *A*. The population could only become polymorphic again by mating with another population which had *A*, or if a suitable seed from a previous generation germinated to give a mature plant, or if an *a* allele mutated spontaneously to *A*.

We can get some idea of how effective genetic drift might be in populations of different sizes by using a property of the Normal Distribution, that σ, the standard deviation of a proportion, $= \sqrt{\frac{p.q}{N}}$, where *p* and *q* are fractions adding up to one, such as the frequencies for two alleles at one locus, and *N* is the total number, e.g., of alleles, which in diploids is twice the number of individuals (parents). Suppose the frequencies of two alleles are 0.75 and 0.25, in a population of two individuals (e.g., one *A A* and one *A a*). Then $\sigma = \sqrt{\frac{0.75 \times 0.25}{4}}$, as two parents have 4 alleles in total. So $\sigma = 0.217$. A property of the Normal Distribution is that 68% of a population lie within one standard deviation of the mean, and 95% of the population lie within 1.96 standard deviations of the mean. Of all possible daughter populations from this population of two individuals with one *A A* and one *A a*, 68% should have an allele frequency for *A* of $p = 0.75 \pm 1\ \sigma = 0.75 \pm 0.217$, so 68% of possible daughter populations should have *p* within the range 0.533 to 0.967, and the remaining 32% will have *p* values higher than 0.967 or lower than 0.533. Similarly 95% of possible daughter populations should have *p* within 1.96 standard deviations from the mean, which is 0.75 ± 0.425, and the remaining 5% will have even more extreme allele frequencies, where possible. **Fixation**, with $p = 1.0$ or 0.0, is therefore quite likely in such a small population. Fixation is most likely when a population is very small and/or when one allele is already rare.

If another population had the same allele frequencies, 0.75 and 0.25, but was large, with 20,000 individuals, $\sigma = \sqrt{\frac{0.75 \times 0.25}{40,000}} = 0.00216$, so 68% of possible daughter populations will have allele frequencies of $p = 0.75 \pm 0.00216$, or 0.748 to 0.752, and 95% will have $p = 0.75 \pm 0.00423$, or 0.746 to 0.754, so most daughter populations from this large population will have very small chance changes in allele frequency. Fixation within one generation is extremely unlikely in the large population but common in the very small one.

The **rate of fixation** (the proportion of populations fixing in one generation) is proportional to 1/(twice the number of parents) and depends on the allele frequencies, where the alleles with the lowest frequencies are the most likely to be eliminated by drift. The probability of fixation to a neutral allele, *A*, in one generation is p^{2N}, where N is the number of parents and p is the allele frequency. Thus in a population of 10 ($2N = 20$), the chance of fixing to A is $(0.9)^{20} = 0.122$ when p is 0.9, 0.012 when p is 0.8, 0.0008 when p is 0.7, and 1×10^{-20} when p is 0.1.

Because not every member of a population necessarily reproduces, the genetic drift calculations should be based on the number of parents (N_{par}) rather than on the total population size. One needs to be more precise, as not all parents contribute equal numbers of offspring to the next generation, so one uses the **effective population size**, N_e, which takes this into account. If a population of 1,000 diploids contains 300 mating couples all contributing equally to the next generation, $N_e = 600$. Inequalities between the sexes and between individuals reduce the effective population size. Four bulls mated to 400 cows would give N_e of greater than 4 but less than 404, because some individuals are providing many more genes than others to the next generation.

Sewell Wright gave the formula $N_e = \dfrac{4 \times \text{number of females} \times \text{number of males}}{\text{number of males} + \text{number of females}}$,

so 4 males mated equally to 400 females would give an effective population size of 15.8, so drift would be as severe as in a population of about 16 even though 404 individuals are reproducing. Different reproductive success of different individuals would further reduce the effective population size. In populations of adult farm animals such as dairy cows and sheep, there are often far more females than males, so the number of males has much more effect on N_e than does the number of females. When one is trying to conserve genetic variability, say in small populations of a rare breed of farm animal (see Chapter 15), keeping too few males promotes genetic drift and loss of variation.

If one has a large population and a few members migrate to colonise a new area, those founder members may by chance have a different allele frequency from the main population. The allele frequencies of new populations reflect that of their founder members, so a series of new populations founded from a large one may differ between them in their allele frequencies, and may differ from the parent population (the **founder effect**). This can happen for plants, animals, microbes or humans. If the new populations remain small, they will probably be affected by genetic drift; if fixation occurs, it will reduce their genetic variation and their adaptive capacity. Small populations say of farm animals or animals in zoos can easily lose genetic variability and polymorphisms purely through chance genetic drift. Special breeding schemes are now followed in many zoos, to move breeding animals between sites to preserve genetic variation and minimise inbreeding (Chapter 15.3).

In organisms with big **seasonal variations** in number, say with a population size of 1,000,000 in summer but only 200 individuals successfully overwintering, then drift in winter will have a much larger effect than in summer. "Bottlenecks" in population numbers can cause severe drift. When population sizes vary between generations over t generations, $N_e = t/(1/N_1 + 1/N_2 + \ldots + 1/N_t)$, where $N_1, N_2 \ldots N_t$ are populations sizes in generations 1 to t. Two successive populations with sizes 1,000,000 and 200 therefore have $N_e = \frac{2}{\frac{1}{1,000,000} + \frac{1}{200}} = 400$, which is much closer to the small value than to the large one.

Applied geneticists have to consider the numbers, biology and amount of variation in their particular organisms and populations to assess by how much genetic drift will reduce genetic variation. For selection, one usually wants a high amount of genetic and phenotypic variation, but the amount of additive genetic variation decreases by $\frac{1}{2N_e}$ per generation, by drift.

4.4. Gene flow and population structure

Gene flow, the physical movement of genes over a distance within or between populations, depends on an organism's biology as regards movement and mating arrangements, and on population structure. Suppose one has a population of snails, which are hermaphrodite but cross-fertilising, spread throughout a large wood, then all snails are potentially able to mate with all other snails. However, because snails move slowly by crawling, two snails 100 metres apart have a very low chance of ever meeting and mating, but two snails living one metre apart have a high chance of meeting and mating. We will therefore get gene flow restricted by distance: we get **isolation by distance**, which is also called **viscous gene flow** - gene flow can happen but is increasingly less likely as distances become greater. We still have one population, not a series of discrete sub-populations.

Most plants are literally rooted to one spot, but their genes can travel in pollen on the wind or on insects, bats or birds, and there are dispersal mechanisms for fruits or seeds, and sometimes for vegetative propagules, such as runners or stolons. Animals may crawl, walk, swim, burrow or fly, and humans have many forms of transport to use, including long-distance flights. The movement of adults, young or gametes can all contribute to gene flow. Gene flow can be **omni-directional or directional**. For example, birds can fly in all directions, but some species have well-defined annual migration patterns. If we have a wind-pollinated plant in a valley with a strong prevailing wind, then plants in the middle of the valley can be pollinated by those upwind of them and can pollinate plants downwind of themselves. Directional gene flow also arises from river flows, ocean currents (for fish and coconuts, for example), prevailing winds affecting insect flight, fruits rolling down hills, etc.

One can also have gene flow between discontinuous populations, say by pollen flow between plants in different pockets of soil on a mountainside, or by the hopping of frogs between ponds or streams. For a human example, consider people in Great Britain, Ireland, France and the Channel Islands. Most marriages (or matings) are within each of these four geographical entities, but migration (temporary, such as holidays, or permanent) between them occurs, with marriages or matings occurring between individuals from different communities. On a ship, mice in five different grain holds might mate mainly within their own hold, but migrants might walk to other holds and mate there. This kind of population structure, with mating mainly within discrete populations but occasionally between different populations is called an **island population structure**.

A variant of this is the **stepping-stone population structure**, where individuals can move between populations, but in a fixed order so that only members of adjacent populations can mate. Imagine a linear chain of islands, in the order A, B, C, D and E, in which a species can move only between adjacent islands because the distances are too great to reach more distant islands by swimming or by wind pollination, say. The population on island B can receive migrants from A or C, but not directly from D or E. Populations A and E could therefore only indirectly exchange genes if the genes moved between them one island at a time, using each island as a stepping-stone.

In any population structure, gene flow is determined not only by the movement of adults, young or gametes, but also by reproductive considerations. A mouse coming from a different hold might smell different from mice in the same hold, so the **acceptability of immigrants** for settlement and/or mating is crucial to gene flow. This might be a matter of choice in humans, but be determined by instinct or incompatibility genes in some other organisms. Applied geneticists need to understand gene flow between populations in their organisms, but are often able to control it by fencing or other barriers, especially for animals. Specialist books on population genetics have mathematical models of different population structures.

4.5. Effects on allele frequencies of selection, mutation, migration and gene conversion; equilibria between forces in populations

Deviations from the Hardy-Weinberg equilibrium can be caused by selection, mutation, migration, gene conversion, chance variation in small populations (giving genetic drift), and deviations from random mating. We will consider the first four of those here, and the others elsewhere.

4.5.1. Selection

This may be **natural** by environment, **artificial** by plant and animal breeders, or **sexual**, in the choice of mates. The effectiveness of selection depends strongly on **dominance** considerations. A dominant allele is almost always expressed, even in heterozygotes, unless it is in a hypostatic locus with an epistatic allele present, or unless the environment is unsuitable for its expression, e.g., it is inducible in non-inducing conditions. Selection against **dominant alleles** can therefore be completely effective in one generation, e.g., for an expressed lethal dominant allele, all copies of which are eliminated. All the next generation will be homozygous recessives.

Complete selection against **completely recessive alleles** usually takes many generations because they are not expressed in heterozygotes, where they are protected from selection even if the recessive homozygote is lethal. As we saw in Example 4.2.1.1, when recessive alleles are rare, they exist largely in heterozygotes, so even complete selection against the very rare recessive homozygote has little effect on the allele's frequency. Selection against **alleles with incomplete dominance** or additive effects is more effective than against recessives, but not as effective as against dominants. Deleterious recessive alleles can therefore persist for many generations, which can useful in evolution as they may be beneficial in different environments or in combination with other genes. Thus a might be deleterious in combination with B - and C -, but advantageous in combination with b b and c c.

Let us consider **complete selection against a recessive lethal allele**, a, frequency q, while A has frequency p. We want to know the frequency with which a a is produced in each generation, before selection. With complete selection against a a, all a a genotypes must come from A a x A a crosses. At Hardy-Weinberg equilibrium, the frequency of these heterozygotes would be $2pq$ out of the total, or

$$\frac{2pq}{p^2 + 2pq + q^2}$$ but after elimination by selection of the q^2 of a a this becomes

$$\frac{2pq}{p^2 + 2pq} = \frac{p(2q)}{p(p + 2q)} = \frac{2q}{1 - q + 2q} \text{ (because } p = 1 - q) = \frac{2q}{1 + q}.$$

The only way to get a a in the next generation is from the mating of two heterozygotes, and 1/4 of their progeny will be a a, so the frequency of a a before selection will be $1/4 \left(\frac{2q}{1+q}\right)^2 = \left(\frac{q}{1+q}\right)^2$.

If q_0^2 is the frequency of a a in generation zero, before any selection, then after one generation of selection we get from the above equation that $q_1^2 = \left(\frac{q_0}{1+q_0}\right)^2$. The same considerations apply each generation, so after n generations of complete selection against the recessive homozygote we get $q_n^2 = \left(\frac{q_0}{1+nq_0}\right)^2$. One can show that the number of generations, n, to reduce q_0 to a desired value q_n is given by

$$n = \frac{1}{q_n} - \frac{1}{q_0}.$$

Example 4.5.1.1. A strain of maize should have unstriped leaves but the recessive homozygous striped form occurs at a frequency of 1% of the plants. With what frequency would the striped form appear (i), after one generation, (ii), after 10 generations, if all plants with striped leaves were destroyed before flowering?

If we assume random mating and other provisos of the Hardy-Weinberg equilibrium, we can say that $q_0^2 = 1\%$ or 0.01, so taking square roots we get $q_0 = 0.1$. We then substitute this in the above formula, $q_n^2 = (\frac{q_0}{1+nq_0})^2$, so for (i) we get

$q_1^2 = (\frac{q_0}{1+1q_0})^2 = (\frac{0.1}{1+0.1})^2 = 0.00826$, or about <u>1 plant in 121</u>.

For (ii) we get $q_{10}^2 = (\frac{q_0}{1+10q_0})^2 = (\frac{0.1}{1+(10x0.1)})^2 = 0.0025$, or <u>one plant in 400</u>.

These figures show that selection against a recessive allele, even with complete selection in each generation, only slowly reduces the frequency of the recessive allele, because it is present largely in heterozygotes where it is protected from selection. The rarer the recessive, the slower the progress of selection against it. With complete selection against a recessive, it takes only two generations of selection to reduce its frequency from 0.5 to 0.25 (i.e., by 0.25), but 9,000 generations to reduce its frequency from 0.001 to 0.0001 (i.e., by 0.0009).

We have just considered complete selection, $s = 1.0$. The **selection coefficient**, s, is the fraction by which the genotype is reduced in a generation, so if $s = 0.6$ for completely dominant allele A, genotypes $A\,A$ and $A\,a$ would each leave 40 offspring for every 100 left by the same number of $a\,a$ individuals. Selection intensity can vary from 0 (selectively neutral) through increasing levels to 1.0 (complete selection, e.g., for a lethal phenotype). It is common sense that the number of generations required to bring about a particular change in allele frequency is inversely proportional to the selection intensity. Thus to reduce the frequency of a recessive allele from 0.001 to 0.0001 takes 9,000 generations for $s = 1.0$ but takes 90,023 generations if $s = 0.1$, and 9,002,304 generations if $s = 0.001$.

It is not necessary to be able to derive all the selection formulae; derivations are given in books such Strickberger (1985), but the applied geneticist needs to be able to use them. For a **deleterious dominant allele**, A, frequency p, selection coefficient s against $A\,A$ and $A\,a$, Δp, the change in frequency of A in one generation of selection, is:

$$- \frac{sp(1-p)^2}{1-sp(2-p)}.$$

As p will decrease, the expression has a minus sign, and the expression shows that the rate of progress of selection depends on the allele frequency p, as well as on the selection coefficient. We therefore have **frequency-dependent selection**.

For a **deleterious allele a with no dominance** (additive effects), frequency q, selection coefficient s against $A\ a$ and $2s$ against $a\ a$, the change in the frequency of a, Δq, in one generation of selection, is:

$\frac{-sq(1-q)}{1-2sq}$. Again, we have frequency-dependent selection.

For **deleterious recessive allele a** with selection coefficient s against $a\ a$, the change in the frequency of a, Δq, in one generation of selection, is:

$- \frac{sq^2(1-q)}{1-sq^2}$. Again, we have frequency-dependent selection.

This is a very important formula and will be used to derive other formulae. If selection has already made q small, then we can ignore sq^2 in the denominator, $(1-q)$ in the numerator will be approximately one, so the change in allele frequency will be approximately $-sq^2$, which we know must be small, because if q is small, q^2 must be much smaller again. See Chapter 15.3 for selection where there is overdominance, when heterozygote advantage leads to an equilibrium between the two alleles.

Selection obviously affects **Hardy-Weinberg equilibrium frequencies** because it changes phenotype and therefore genotype frequencies and allele frequencies. Note that selection works directly on phenotypes, and only indirectly on genotypes. Because selection can operate at many stages of the life cycle - gamete formation, mating, fertilisation, embryo, germination or birth, and various stages of youth and adulthood - one will get different phenotype frequencies according to whether one scores a population **before, during, or after selection** in that generation. The recessive *albino* mutant of maize gives white seedlings, unable to photosynthesise. Crossing two heterozygotes gives the Mendelian monohybrid ratio of 3 green: 1 albino at germination, but the albino seedlings die within three weeks, leaving a post-selection ratio of all green: no albino.

If selection occurs in **haploid gametes**, or in a **haploid vegetative stage** as in some fungi, one needs special equations: recessives will usually show fully as they are not hidden in heterozygotes. For a deleterious recessive allele a in haploids, frequency q and selection coefficient s against a, the change in the frequency of a, Δq, in one generation of selection is:

$- \frac{sq(1-q)}{1-sq}$.

For **X-linked genes**, the formula also need modifying because the heterogametic sex, X Y, has only one copy of alleles on the X chromosome (it is hemizygous for them). This will be considered later under human genetics, Chapter 13.6.

4.5.2. Migration

Migration can be natural, but the placing of one group of organisms with another group, say by an animal breeder, is formally equivalent to migration. One has **emigration** from a population and **immigration** into a population. If the emigrants or immigrants have the same allele frequencies as the resident population, migration only changes population size, not allele frequencies. If the migrants have different allele frequencies from the resident population, then the changes in allele frequency in the resident population caused by migration are proportional to both the proportion of migrants to non-migrants, and to the difference in allele frequencies between the resident population and the migrants.

The case of immigrants having different allele frequencies from the resident population is much more common than that of emigrants having different allele frequencies from the non-migrants of the resident population, so the former case will be considered. Let r_0 be the frequency of an allele in the resident population before migration, and let that allele have frequency i in the immigrants, with m being the proportion of newly introduced genes into the resident population from immigration each generation. In a generation, the resident population will lose gene frequency mr_0 by dilution and gain mi from the migrants. After n generations of immigration,

$$r_n - i = (1 - m)^n (r_0 - i), \text{ and } (1 - m)^n = \frac{r_n - i}{r_0 - i}.$$

These equations have been used to work out the proportion of genes (m) from "whites" introduced into "blacks" in America, using the present allele frequencies in East African "blacks" (r_0), the present allele frequencies in American "whites" (i), the present allele frequencies in American "blacks" (r_n), and an estimate of 10 generations over the 300 years since African "blacks" were introduced into America and subject to the introduction of genes from "whites". The calculations by Glass and Li, based on the frequencies of Rhesus allele R^0, suggest that about 3.6% of genes in the "black" population were introduced from "whites" each generation, with a total of about 30% of the genes in present American "blacks" having come from "whites" (see Strickberger, 1985).

The plant or animal breeder will usually only be concerned about "migration" over one generation, when the allele frequency will be $r_0(1 - m) + mi$. If one adds 10 animals of genotype $a\,a$ (i for $A = 0.0$) to a resident population of 30 animals of genotype A A (r_0 for $A = 1.0$), m is $10/40 = 0.25$, then the allele frequency for A in the mixed population is $r_0(1 - m) + mi = 1(0.75) + 0.25(0.0) = 0.75$, although one does not need this equation to work that out.

4.5.3. Mutation

Mutation will affect Hardy-Weinberg equilibrium frequencies only to a very minor amount unless mutation is unusually frequent, or if very rare alleles are being considered, because typical mutation frequencies are about 10^{-4} to 10^{-9} per locus per generation. Mutation is however very important as a source of new genetic variation, and can affect allele frequencies over a long time. Let us take an oversimplified case, of allele A, frequency p, mutating with frequency u to allele a, frequency q, which mutates to A with frequency v. Initially we will have

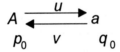

The change in frequency for a in one generation, Δq, will be the gain from mutation from A to a ($= up_0$) and the loss from mutation from a to A ($= vq_0$), so $\Delta q = up_0 - vq_0$. At mutational equilibrium, there will be no change in allele frequencies, so then $up = vq$, so equilibrium values (sometimes represented by showing p and q with a $\wedge$ symbol over them) are $\Delta q = 0.0$ and $p/q = v/u$. If there are no other factors such as selection, drift or migration, the equilibrium frequencies of two alleles are just determined by their relative mutation frequencies, but with typical low mutation frequencies it can take very many generations to achieve equilibrium.

If A is a functional, active allele, it can mutate at many different positions to give a loss of function mutation; if a is such a loss of function mutation, any one a allele needs an extremely specific mutation (a reverse or **back-mutation**), at or near the site of the original mutation, to restore it to the A base sequence. Forward mutations, from A to a, will therefore be much more frequent than reverse mutations, a to A. Suppose that u is 10^{-5} and v is 10^{-7}, then at equilibrium the above equation $p/q = v/u$ suggests that $p/q = 1/100$, with the mutant allele a 100 times more frequent than the wild-type allele A, which would make nonsense of the definition of a wild-type allele. The reason that mutant alleles are not much more frequent than wild-type alleles is that mutant alleles are usually selected against and there is an **equilibrium between mutation and selection**.

Consider selection against a deleterious rare recessive, a, frequency q, selection coefficient s against a a. We saw earlier that the loss by selection of such an allele in one generation was $\frac{sq^2(1-q)}{(1-sq^2)}$, and as q will be small for a rare allele this is approximately $sq^2(1 - q)$. As a is rare, we can ignore losses to it from mutation to A, and the gain from mutation from A is pu per generation, or $(1 - q)u$, as $p + q = 1.0$. At the equilibrium between mutation and selection, the loss from selection $\{sq^2(1 - q)\}$ must equal the gain from mutation $\{(1 - q)u\}$. As $sq^2(1 - q) = (1 - q)u$, we can cancel the $(1 - q)$, giving $sq^2 = u$, so $q^2 = u/s$ and equilibrium $q = \sqrt{\frac{u}{s}}$.

This result for rare deleterious recessives is important. If we know any two values, we can use this formula to calculate the third value. So if $s = 0.1$ and equilibrium $q = 0.001$, we can calculate the mutation frequency u as 0.0000001 or 1×10^{-7}. For a deleterious dominant, the corresponding equilibrium formula is $p(1 - p) = v/s$.

Although these equations are useful, they do not take account of **gene conversion**, whereby one allele can convert another to being of its own kind at meiosis in a heterozygote. Gene conversion can give allele segregation ratios of 3 A: 1 a or 1 A: 3 a in tetrads, or in octads (as in some fungi) can give 8:0, 7:1, 6:2, 5:3, 3:5, 2:6, 1:7 and 0:8 allele ratios for A:a or wild-type:mutant, from meiosis. The direction of gene conversion is often biased, with more conversions to one allele rather than to another, and in the long term that can affect allele frequencies in populations. The **strength of the force of gene conversion** on allele frequencies depends on the **frequency of gene conversion** and on the **amount of disparity in conversion** to different alleles. See Lamb and Helmi (1982) for equations taking gene conversion into account in the equilibria involving selection and mutation rates, and Lamb (1998) for references on amounts of gene conversion disparity.

Since most new mutations are harmful, mutations reduce the immediate fitness of a population by an amount called the **mutational load**, although mutations are useful for long-term adaptation and evolution. The mutational load for recessive alleles is equal to the mutation frequency, and for dominant or partly dominant alleles is equal to twice the mutation frequency (see Crow, 1986). The **Haldane-Muller principle** is that the amount of harm (loss of fitness) caused to a population by mutation is not dependent on how deleterious the mutations are, but depends solely on the mutation frequency. Over all loci, the mutational load is the mutation frequency multiplied by a factor of between one and two, depending on the proportions of recessive and dominant mutations. Although this principle initially seems illogical, it makes sense when one considers that a very harmful mutation will do a lot of harm while it is present but will be quickly eliminated by selection, while a mildly deleterious mutation will do a smaller amount of harm at any time, but over a much longer time as selection against it is much less.

Suggested reading

Brown, A. H. D., M. T. Clegg, A. L. Kahler and B. S. Weir, eds., *Plant Population Genetics, Breeding, and Genetic Resources*. 1990. Sinauer Associates, Sunderland, Mass.

Clarke, B. C. and L. Partridge, eds., *Frequency-dependent Selection*. 1988. The Royal Society, London.

Crow, J. F. *Basic Concepts in Population, Quantitative, and Evolutionary Genetics*. 1986. W. H. Freeman and Co., New York.

Falconer, D. S. and T. F. C. Mackay, *Introduction to Quantitative Genetics*, 4th ed. 1996. Longman, Harlow.

Griffiths, A. J. F., W. M. Gelbart, J. H. Miller and R. C. Lewontin, *Modern Genetic Analysis*. 1999. W. H. Freeman and Co., New York.

Hartl, D. L. and E. W. Jones, *Genetics. Principles and Analysis*, 4th ed. 1998. Jones and Bartlett, Boston.

Lamb, B. C., Gene conversion in yeast: its extent, multiple origins and effects on allele frequencies. *Heredity* (1998), 80, 538–552.

Lamb, B. C. and S. Helmi, The extent to which gene conversion can change allele frequencies in populations. *Genetical Research* (1982), 29, 199–217.

Pirchner, F., *Population Genetics in Animal Breeding*, 2nd ed. 1983. Plenum, New York.

Strickberger, M. W. *Genetics*, 3rd ed. 1985. Collier Macmillan, London.

CHAPTER 5

TYPES AND USES OF SELECTION

5.1. Natural, artificial and sexual selection

Natural selection is selection by the environment. For crops and farm animals this includes the man-influenced environment in which the organisms grow. Natural selective factors include diseases, pests and predators, temperature, rainfall and sunshine, wind and soil. **Artificial selection** is selection by man, e.g., selection for higher yields or for uniform ripening. **Sexual selection** is selection in getting and keeping mates, which may result in different colouring between the sexes, showy crests in some amphibians or showy tails in peacocks. Extreme adaptation in the battle for mates within a species may lead to that species being less competitive against other species, e.g., the long and showy tails of peacocks which attract peahens may reduce the peacocks' fitness relative to competing species.

The artificial selections imposed on populations by plant, animal or microbial breeders may interact with natural or sexual selection. Human geneticists are aware of natural and sexual selection in mankind, but for ethical reasons they cannot apply artificial selection to humans. Medical treatments for infections and infertility may run contrary to natural selection by allowing individuals who would otherwise have died or been sterile to live and reproduce.

Artificial selection may be based on the scorable phenotypic merits of **individuals** if heritabilities are high, but if heritabilities are low so that phenotype is a poor guide to breeding value, it is often better to select on **whole-family averages**, especially for large families of half-sibs. Other selection methods involve an individual and its ancestors (**pedigree selection**) or an individual and its offspring (**progeny testing**). Some other methods, such as **agricultural mass selection**, are covered in Chapter 12.

There can be unwitting human selection against the best types of plant and animal if the best types are chosen for consumption instead of for breeding. For example, in Malaysia and Indonesia local people often pull down the most productive durian trees (*Durio*) in order to collect the fruits; with timber, often the best trees are cut down for wood. In some religions, the best animals are chosen for sacrifice.

In situ **selection** means selection of a new variety well adapted to the local climate, soils and diseases, e.g., with drought-, cold- or salt-tolerance, but those varieties may do poorly when tried in other areas; that is, there are large genotype x environment interactions. It may not be cost-effective to breed many varieties to suit a large number of different niches. *Ex situ* **selection** is done in a small number of good environments, with the plants or animals then tried in various other areas.

Breeding and selection may sometimes be for a very **specialised type**, e.g., fast-growing, high food-efficiency broiler chickens for factory farms. At other times, one might want a generally hardy and **adaptable type**, such as pack animals for tribes who migrate seasonally between different environments. Different breeds of an animal may be selected for different properties and environments. For example, a Merino sheep bred for wool in Australia is good in extreme environments, provides fine-fibre wool suitable for light to medium weight clothing, and about 80% of the farmer's income will be from the wool. A Southdown sheep (Plate 25) in the south of England is suited to fairly rich pasture, provides coarser wool suited to heavy weight clothing, upholstery and interior textiles, and is mainly for meat, with only 10 to 30% of the income generated being from wool. It is used as a sire breed for meat and short wool. The hardy Blackface sheep in the Outer Hebrides islands, off the northwest coast of Scotland, provides wool for the Harris Tweed weaving industry, for heavyweight garments (see Chapter 12.7.4 for the use of Blackface sheep in crossbreeding). In Scotland it provides meat and carpet wool.

Selected characters may remain **stable** over long periods, such as resistance to woolly aphid from the apple variety Winter Majetin, which has remained effective for over 150 years. In contrast, resistance to stripe rust in the wheat cultivar Clement was overcome after only one year by mutation or recombination in the fungus.

The aim of selection is to increase value. Highly selected **pedigree Texel rams** for use in breeding flocks are worth about £5,000 to £32,000 each, but ordinary Texel rams to father lambs for the meat trade are only worth about £550 (figures from G. Crust, Skendleby, Lancashire, *The Daily Telegraph*, 5/6/1999).

5.2. Stabilising selection, towards uniformity

In wild populations, natural selection tends towards producing **optimum phenotypes**, ideally suited to their environment. Departures from optimum will usually be less well adapted and thus will be selected against, so that natural selection often stabilises a population around the average type, which is the ideal for that environment. That is true of height in mice, deer, man, mushrooms, mosses, wheat, etc.: organisms which are much smaller than average have various disadvantages, e.g., small wheat plants being shaded out by taller ones, or small male deer being poor at fighting for mates. Organisms much bigger than average also have disadvantages, e.g., very tall wheat plants being easily knocked down by wind and rain.

Selection in favour of average types and against organisms towards the two extremes is called **stabilising selection**. It often occurs in nature and is used by plant and animal breeders to encourage uniformity and typicalness in a breed. For example, with a mechanically harvested crop such as barley, the farmer wants all plants to ripen their seed simultaneously. When the combine harvester goes through a field, it will harvest all plants, whether ripe, under-ripe or overripe, so **selection for uniformity** of ripening time makes commercial sense, giving the best financial return. Breeders will cull plants or animals not conforming to the desired and recognised form for an animal variety or a plant breed, e.g., maize plants giving brown or purple grains in a strain with yellow grains. Fig. 5.2.1 illustrates stabilising selection for a quantitative character giving a Normal Distribution.

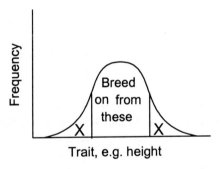

Figure 5.2.1. Stabilising selection, with "x" indicating the organisms to cull (kill, or not to breed from).

There is evidence for stabilising selection for **human births weights**. In data of Karn and Penrose (1951) from a London hospital, the optimum birth weight for survival from birth to one month of age was 3.6 kg, close to the population average of 3.2 kg, with only 1.2% mortality, compared with a mortality of 4.1% averaged over all weights. Very light babies, say below 1.8 kg, had greatly increased mortality, and very heavy babies, say over 4.2 kg, also had a somewhat increased mortality (see Fig. 35-1 in Strickberger, 1985). The optimum weight for minimal mortality in this period just after birth was slightly higher than the average birth weight, 3.2 kg; possibly a lighter weight is optimal earlier, especially at the time of birth when slightly smaller babies may cause fewer complications.

5.3. Directional selection, favouring one extreme

To improve a breed, breeders often use **directional selection**, breeding from organisms at one end of a distribution, such as dairy cows with the highest milk yield, or rye plants with the most disease resistance. In pigs, breeders reduced back-fat thickness from about 6 cm to about 1 cm at market weight in only a few generations. Directional selection occurs naturally for the highest fertility and fitness.

The **direction of selection** may vary in different environments. Light skins have been selected in humans in less sunny climates where they permit more photosynthesis of vitamin D in the skin, and dark skins have been selected in sunnier climates where they provide more protection from damaging and skin-cancer-causing ultraviolet rays, though selection is slow and incomplete.

The breeder may occasionally select in different directions for different purposes. With tomatoes, for example, the small, sweet cherry tomatoes have been selected for smallness, whereas large beef tomatoes for stuffing have been selected for large size; other varieties have been selected for medium size. Directional selection is illustrated in Fig. 5.3.1.

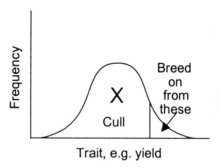

Figure 5.3.1. Directional selection, breeding from one extreme only.

One can have selection for **qualitative traits** as well as for quantitative ones. Selection is much simpler when traits are controlled by major genes with Mendelian inheritance. In Sri Lanka, a mutant variety of rice with altered growth properties was discovered, which could bend upwards at the nodes; if the stem is knocked down by wind or rain, the part of the stem above a node can be raised again. This lodging-resistant strain is shown in Plate 10. Animal examples of selectable characters controlled by genes of major effect include the Booroola gene in sheep for increased ovulation and lambing rate (Chapter 1.3.4), a double muscling gene in cattle, and in pigs, two genes associated with meat quality and one which alters the susceptibility to porcine stress syndrome.

5.4. Cyclic selection, alternatively favouring different extremes

With **cyclic selection**, one has different phenotypic extremes being favoured alternately. This often occurs through seasonal variation for temperature or light intensity. For human skin pigmentation in Britain, there is very mild selection in winter for pale skins which permit more photosynthesis of vitamin D, and in summer there is very mild selection for dark skins which give more protection from ultraviolet light, which can cause malignant skin cancers.

Cyclic selection favours the preservation of genetic variation in a population, by alternately favouring different alleles. Breeders use cyclic selection, usually growing organisms alternately in different habitats or regions, to get environmental tolerance. This has been extremely successful in wheat breeding (Chapter 12.7.1) in getting varieties which do well in a wide range of countries and environments, just needing local adaptation for resistance to local diseases.

Suppose a population of wheat contains some alleles giving specialised success in environment 1 (say sea level), some alleles giving specialised success in environment 2 (say 2,500 metres above sea level) and some alleles give general environmental tolerance. By growing the plants alternately in the two environments for say 10 generations, one selects in every generation for genes giving general environmental tolerance. The genes for specialised success in one environment will be selected for in that environment, but be selected against in the alternate environment, so the genes most selected for are those for general environmental tolerance.

5.5. Disruptive selection, selecting against the average type

With **disruptive selection**, both extremes are favoured, with selection against the average type. This is mainly a laboratory experimental technique. If breeders want small and large tomatoes for different purposes, they will use directional selection, selecting one population for reduced size and another population for increased size.

Disruptive selection does occur in nature, especially by predation or disease. For example, birds might readily recognise the average type of a butterfly species as being good to eat, but butterflies at either extreme away from the normal type might not be recognised as being of that good food species. Diseases might adapt to the commonest form of an organism, and be less well adapted for more unusual types. Sexual selection can lead to increased differentiation between the two sexes, so can be a kind of disruptive selection, selecting one sex in one direction and the other sex in the other direction. Disruptive selection is also likely between different sub-populations for traits giving different survival in different niches or environments.

5.6. Pedigree selection

Pedigree selection is selecting on an individual's performance and on that of its **recent ancestors**, especially its parents and grandparents, as it should on average have half the genes of each parent and one quarter of the genes of each grandparent. Its genetic basis is that individuals of great phenotypic merit are more likely to possess and to transmit good alleles than are individuals of inferior merit. The higher the heritability for a character, the more reliable is pedigree breeding.

One looks up the **performance records** of ancestors in **herd books, stud books** or farm or breeding station **records**, and also assesses the performance of the individual being considered for breeding. In a recognised cattle breed with registration schemes, one might be able to check the milk yield, milk protein and butterfat content, and calving record, of recent ancestors, and to find out what prizes they won at shows. Since 1800, all UK foxhounds have been entered into the Foxhound Kennel Stud Book, so pedigrees can be traced back over very many generations. In sheep, the Border Leicester breed (Chapter 12.7.4) was produced by crossing Leicester Longwools with Cheviots around the beginning of the 19th century, with the new breed being recognised in 1869 for shows, and with the herd book starting in 1898.

Pedigree breeding has been used extensively for all major farm animals, for race horses, dogs and cats. At a national cat show, the author asked a successful cat breeder what genetic techniques he used. He replied: "I don't know no genetics. I just breeds champion tom with champion queen!" This policy has been described as "crossing the best with the best and hoping for the best!"

With a number of organisms including horses and coconuts, the term **prepotent** is used for individuals and for breeds which are excellent and are known to transmit excellent gene combinations (i.e., have a high breeding value, not just environmentally-determined merit). In horses, the Thoroughbred breed is generally prepotent and has been used to improve other breeds such as Cleveland Bays and Quarter Horses. A prepotent Thoroughbred stallion, "Man O'War" (born in 1917), raced for two years and in 21 starts was first 20 times and second once, establishing five world records and winning much prize money. His immediate progeny were also very successful, showing his high breeding value, with transmissible genetic merit. A top stallion in Britain in 1996 had a stud fee of £12,000 a mating because of his proven performance in racing and in breeding excellent offspring; much higher fees have been obtained since then. In 1998, a yearling colt from champion stallion Rainbow Quest x brood mare Silver Lane sold for £2,310,000. The record price for a yearling in Britain was £2,520,000 in the 1980's, with Northern Dancer as a parent, with prices depending on individual and ancestral merit. In Britain there is a National Stud, set up in 1916, and now at Newmarket. The values of stallions range from £400,000 to £5 million, costing £3,000 to well over £12,000 a mating. It includes Shaamit, the Derby winner in 1996, who "covers" about 70 mares a year. A top stallion can receive £80,000 a mating, plus a further £80,000 if the mare becomes pregnant. The pollen of prepotent oil palms is very valuable for increasing oil yields.

Pedigree selection is less useful for characters of low heritability. One of its merits is for selection of opposite-sex characters. Thus in dairy cattle one could select bulls on the milk performance of their female ancestors, such as mothers, grandmothers and great-grandmothers, or select cocks for egg-laying genes on the basis of records of their female ancestors. See also Chapter 12.2.4, 12.7.4 and 12.8.

5.7. Progeny testing

In **progeny testing**, one judges an individual's merit by its own performance and by that of its **offspring**. It is most useful for individuals capable of having many progeny (e.g., bulls, by artificial insemination, AI; Chapter 17.2) - which means males for farm animals - and having a long reproductive life which allows time for producing and assessing the offspring, and then for further breeding from the selected individual.

A typical progeny testing programme for dairy cattle would be to cross a small number of young bulls of good pedigree to a range of unselected different cows in a common environment. The daughters of each bull are then assessed for their average milk yield and quality. The bull with the best average daughter performance is therefore selected to continue breeding, while the others are culled or just raised for meat. The chosen male can have thousands of daughters by AI and still has years of reproductive life.

A disadvantage is that this testing is expensive and lengthy, keeping bulls of low breeding value alive for several years and producing daughters with poor milk yield. It is seldom worth progeny-testing females in farm animals, as they have comparatively few offspring during their reproductive lives.

Like pedigree breeding, progeny testing is excellent for opposite-sex character testing, such as bulls for milk yield and cocks for egg-laying. The method is easily used in plants, using pollen from the plants to be tested on the stigmas of a range of other plants. In annuals such as wheat, one would need to preserve pollen (Chapter 17.1) for use in following seasons once the progeny testing established which plants were prepotent, but there are no such problems with perennials like coconuts and oil palms. The higher the heritabilities, the better progeny testing works, as then phenotypic merit is more likely to reflect genotype merit.

Traits such as body weight or milk yield can be easily measured on **potential parents**, but many **meat carcase characters** can only be assessed after slaughter, such as the flavour, fat content and marbling, and the killing-out percentage (usable proportion of the carcase, by weight): see Chapter 5.13, and Chapter 12.7.5 for beef examples. For characters which can only be measured late in life or after killing, generation intervals are rather long, and the rate of gain from selection *per unit time* (years, as opposed to generations), is higher from pedigree selection than from progeny testing or individual selection on an organism's own merits.

"Genus" is a large cattle firm in the UK. Its Genus Sire Improvement Programme (GSIP) is the largest dairy progeny testing programme in the UK, and was started in 1992 with the intention of using the best genes from anywhere in the world. In 1996, there were more than 2,400 participating farmers, carrying out over 50,000 test inseminations a year. They collect semen from young bulls as early as possible, collecting 1,000 doses between 12 and 14 months. With calving heifers at two years of age, progeny test results are achieved in five years.

5.8. Half-sib and family selection

Selection based on individual merits is appropriate for traits with relatively high **heritabilities**, say 0.25 to 1.0, and where important traits can be measured on potential parents, not after killing. When heritabilities are low, say 0.0 to 0.25, the phenotype of an individual is a relatively poor guide to its genotype and breeding values. It is then more efficient to base selection on whole-family averages, rather than on individuals, as some of the environmental variation should average out over the family, some having better and some having worse environments. This is **family selection**, often in the form of **half-sib selection**.

For farm animals it is easy to obtain large families of half-sibs, with the same father and different mothers. Half-sibs are more useful than full sibs as the families can be larger, can average out differences in mothering ability, and can be contemporary, whereas full sibs are usually born over a number of seasons, except in litter-bearing species. One takes different families of half-sibs and compares their family averages for particular traits, using the best families as parents in breeding programmes. For carcase characters, only some of each family need be slaughtered, leaving others available for breeding. The best family of half-sibs for litter-size in pigs could be used to select males for breeding for this opposite-sex character. One could use information on half-sibs, full sibs and ancestors, but low heritabilities would make records of parental merits inaccurate as regards breeding values, for which a large family of half-sibs can be used.

The family selection method is thus clearly suitable for characters with low heritability, for opposite-sex characters, and for characters scored late in life or after killing. It has been very useful for improving pig carcase quality. Some carcase characters such as mid-back fat depth can now be scored non-destructively, e.g., by use of echo-sounders.

5.9. Selecting for correlated characters

Correlations between characters were considered in Chapter 3.3. Some characters are difficult or expensive to measure, so that it is sometimes better to select for a correlated character rather than the one you are most interested in, once the direction and strength of the correlation have been established. If a character is only scorable late in life or after slaughter, generation intervals can become too long. Selecting for correlated characters which are easier and cheaper to score, or which can be measured earlier in life, is therefore desirable.

For example, in most animals the growth rate is highly correlated (correlation coefficient, r, = 0.6 to 0.8) with **feed conversion efficiency**, which is how efficiently the animal converts food into usable parts of itself, such as meat, wool or leather. Measuring food conversion efficiency on individual animals is expensive, as all feeding has to be monitored individually, while growth rate is easy to measure by weighing the animal at intervals. It is therefore much more convenient to select for an increased rate of weight gain rather than for increased feed conversion efficiency.

5.10. Selection for more than one character: tandem selection, independent culling levels and index selection

In plants, animals and micro-organisms, the breeder usually wants to improve several characters, not just one. In pigs for example, one might want to increase growth rate, food-conversion efficiency, disease-resistance and litter size, and decrease carcase fat. In beef cattle, one mainly selects for birth weight, 200 day growth, 400 day growth, backfat depth, muscling score and beef value. In sheep, one might select for increased yields of meat and/or wool, number of lambs born, meat and wool quality, and reduced susceptibility to fly-strike, fleece-rot, dermatophilosis, foot-rot and facial eczema. In wheat one might want to improve grain yield, disease-resistance, heavy-metal- and cold-tolerance, and to reduce straw length and the need for high nitrogen inputs. There are several methods for selecting several characters.

5.10.1. Tandem selection

With **tandem selection**, one selects for just one character over a number of generations, until the desired improvement has been obtained. Then one selects just for a second character for a number of generations, then for a third character, etc. This method is inefficient and is rarely used. It takes too many generations to finish selecting all the desired characters, and any negative correlations between characters would mean that an initial favourable response for one character might be reversed during later selection for another character.

5.10.2. Independent culling levels

For **independent culling levels**, one sets separate selection thresholds for each character. In pigs one might select for breeding only animals above a certain threshold for growth rate (which one wants to increase), and which are simultaneously below a threshold for mid-back fat (which one wants to reduce). Suppose that one needed to keep 1/16th of the herd for breeding, and that growth rate was four times more valuable commercially than mid-back fat, then one could take just the individuals in the best eighth for growth rate and the best half for fat. That would weight the selection for different characters according to their relative value, and would mean taking 1/16th of the herd, culling the rest.

This is shown in Fig. 5.10.2.1, where the individuals' values for the two characters are shown by a single "x". The vertical line is the threshold for fat, taking only animals to the left of this line, with low backfat, and the horizontal line is the growth-rate threshold, taking only animals above this line. The selected animals for breeding are therefore those in the top left sector, where they are shown in bold type. Independent culling methods are easy to use and are common.

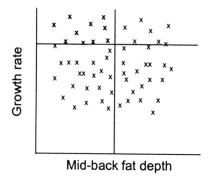

Growth rate

Mid-back fat depth

Figure 5.10.2.1. Illustration of independent culling levels for growth rate and mid-back fat: see text for details. Each "x" represents one pig, and only pigs in the top left quadrant are used for breeding, with the best growth rate and the least fat.

5.10.3. Index selection, including PTA, PIN and £PLI

For just two characters, one might be able to arrive at a single figure combining them, such as the growth rate of the non-fat part of a pig for the above two characters. Usually, however, one constructs a figure called **an index** which is based on adding together the merit scores for several key commercial characters, weighting them in terms of relative economic value, heritability, phenotypic variances and genetic correlations between them. It gives an estimate of an individual's overall merit as a single number. See Cameron (1997) for animal selection indices and Baker (1986) for plant indices. Using **index selection** is more complicated than independent culling levels, but is more efficient. The calculations of index scores are normally computerised.

A **selection index** is a linear function of observable phenotypic values for a number of traits. The observed value for each trait is multiplied by its own **index coefficient**, based on the economic value of the trait, its heritability, its genotypic and phenotypic variances, and covariance with other traits. Suppose it was desired to improve a line of wheat by increasing grain yield, decreasing the days from sowing to maturity, and increasing the protein concentration. A suitable simple index might be: I = (0.3 x yield in kg/ha) + (-1.4 x days to maturity) + (4.7 x protein concentration in g protein per 100 g grain). A particular genotype giving 2,000 kg/ha, maturity at 99 days and a protein content of 11.7% would then have an index score of (0.3 x 2,000) + (-1.4 x 99) + (4.7 x 11.7) = 516.4. This genotype could then be compared with others just using the single merit criterion of the index scores.

In the British National Improvement Scheme, 1987, for **pigs**, the average index score for boars was 200, with most boars in the region of 165 to 235, where higher scores indicate more overall merit. Only boars scoring 250 or over were eligible for use in artificial insemination centres, which was just the top four per cent. A rise of 10 points in a boar's index score worked out at an average improvement worth 23 pence per piglet sired.

In dairy cattle, the **Predicted Transmitting Ability** (Production PTA95) is a genetic index for individual production traits, i.e., milk (kg), fat (kg and %) and protein (kg and %), expressing the production potential an animal will pass to its progeny. PTAs for males and females are comparable and in the same format, and form the basis of other indexes such as PIN, £PLI and ITEM. PTAs are calculated from the individual animal model, using the performance of the animal and its relatives. They allow for as many non-genetic variables as possible by standardising actual production for differences between herds (one may be more high yielding) and within herds (some have larger spreads of values), differences due to time of year of calving, etc. All PTAs are expressed against the same genetic base, set as the average of all cows born in 1990 being zero. As this was set in 1995, they are PTA95 values and the base is changed every five years, so PTA95 will be replaced in year 2000.

For dairy cattle, the **Profit Index (PIN)** is an economic index (expressed in pounds sterling in the UK), giving the additional profit margin over all feed and quota costs per daughter per lactation expected from mating with an individual with a PIN value of £0. It weights the PTA values economically. Currently, PIN = (-0.03 x PTA milk, kg) + (0.60 x PTA fat, kg) + (4.04 x PTA protein, kg). Its calculations include the amount of feed energy to produce each milk component, current and expected milk prices, the cost of a milk quota allocation, and the cost of cooling and transporting milk (see Holstein 1999a and b).

A related index, **ITEM (index of total economic merit)** combines the PIN with a financial value for type, including traits associated with long herd life. Financial values ascribed to key daughter traits for feet and udders are incorporated into the Profit Index to give a single tool for selection of bulls for dairy cattle. For example, Holstein bull Atrius from Italy had a PIN of £80 and an ITEM of £86 in 1995, having given 57 daughters in 49 herds, with milk yield at + 876 kg, fat at + 20.9 kg and protein at +23.3 kg, and strongly transmitting genes for good udder attachment and teat placement (figures taken from the 1996 Genus Dairy Directory).

ITEM was replaced by **£PLI** (Profitable Lifetime Index) in February, 1999, also taking account of the contribution of longevity to profitability. To the PIN value, it adds weightings for specific type traits as indirect measures of longevity, then actual longevity information when available. Weightings include 0.043 x foot angle, 0.123 x fore udder attachment, 0.023 x udder depth and -0.096 x teat length, as the main linear traits associated with survival (see Holstein, 1999a and b). Increased longevity is important to profitability because the age profile increases, with fewer immature non-milking cows, and fewer replacement cows are needed.

$$£PLI = £PIN + (38 \times \text{Lifespan PTA})$$

Genetic gains have been impressive recently, using index selection. For example, the average genetic value of **Holstein Friesian** Society registered females increased by £46 PIN between 1990 and 1997. At £6.44 per £1 PIN, genetic improvements over that period averaged £42 per cow per year, or just under £300 per cow, with the greatest improvement in the top pedigree cows, but even the average cow showed a great improvement: see Table 5.10.3.1.

Table 5.10.3.1. Pedigree cow improvement, Holstein Friesians, 1990-1997. Data from Holstein (1999a).

Group	Average PIN	Average PIN	Annual PIN gain per year	Value of gain, at £6.44 per £1 PIN
	1990	1997	1990–1997	1990–1997
Top 1 %	44.9	99.4	7.79	£351
Top 5 %	31.8	83.8	7.43	£335
Top 10 %	24.8	75.4	7.23	£326
Average	0.0	45.8	6.54	£295

5.11. Selection intensities and rates of response to selection

The **maximum selection intensity** which a breeder can impose on a population depends very much on the organism's biology, especially its reproductive potential. In a micro-organism such as yeast, one can have billions of yeast cells in a culture and select hard for rare improved forms, say more alcohol-tolerant wine yeasts. In cattle, one needs to keep a half to a third of all female calves, just to maintain herd size, so selection on those females is usually weak. One can select much harder on bulls, where one bull can serve 30,000 cows a year by AI, so fewer than about one bull in 10,000 needs be kept. It is easy to monitor weight gain for selecting beef bulls in large herds, but one could not test a large herd of dairy bulls for milk-yield genes by individual progeny testing, as it would require too many cows.

The **generation times** may differ between males and females, depending on the age at which they start and finish useful reproductive life. It is useful to think in terms of genetic gain *per year*, not just *per generation*. The relation between **genetic gain per year**, generation time, selection intensity, phenotype variation and heritablilty is shown in Fig. 5.11.1. It is obvious that there is no genetic gain if there is no phenotype variation or if the heritability is zero.

genetic gain per year =

$$\frac{1}{2}\left(\frac{\text{selection intensity in male}}{\text{generation time in male}} + \frac{\text{selection intensity in female}}{\text{generation time in female}}\right) \times$$

phenotype variation × narrow sense heritability.

Figure 5.11.1. The equation for genetic gain per year.

Simultaneous selection for several characters reduces the maximum possible selection intensity for any one character. Suppose we were selecting for eight different characters in a plant where we can grow and test 20,000 plants a year, and that 80 plants were needed to give sufficient seed to grow 20,000 plants in the next generation. If we selected the best half of the population of 20,000 plants for each of the eight characters, we would be left with $(0.5)^8 = 0.0039$ of the population, or just 78 plants, when we need 80 to give us 20,000 plants for the next generation. If we selected equally for all eight characters, we could not select harder than the best half for each character, if we needed to maintain the population size. In organisms with a lower reproductive potential, we could select even less hard.

Maintaining and testing 20,000 wheat or barley or maize plants is not too difficult, but maintaining a similar number of pigs, horses, goats, sheep or cattle is more than most individual breeders could manage. It is therefore quite common to have national or regional breeding schemes, like the British National Improvement Scheme for pigs, to help share the load between breeders.

5.12. *In vitro* selection

Although selection is usually carried out on whole animals or plants, it can also be tried in plants on germinating pollen (Chapter 17.1.7) or cells or tissue culture. For example, resistance to the weedkiller glyphosate was selected in callus tissue from *Petunia*, and persisted into whole plants, from which it was transferred to several crop species. Sugar cane tissue cultures were used to isolate resistance to Fiji virus disease and to the fungus *Helminthosporium sacchari*. Wheat seedlings grown *in vitro* were selected for resistance to *Fusarium culmorum*. See Wenzel and Forough-Wehr (1993).

Unfortunately, variation selected in tissue culture often proves not to be inherited or to be unstable. Thus one can easily select salt-resistant calli, but either the calli fail to regenerate to give whole plants, or the whole plants lack the resistance.

5.13. Selection for meat characteristics

Because **meat characters** can often only be assessed after slaughter, they require special consideration. We saw in Chapter 5.8 that **family selection**, such as using families of half-sibs, with some slaughtered for assessment at the appropriate age and some still available for breeding, is a good method for carcase characters of low heritability. In sheep, Thompson and Ball (1997) concluded that there was genetic variation for body composition within breeds and between breeds, mainly affecting mature size. Lean-growth selection indexes are used for sheep in New Zealand,

Australia and the UK, usually weighting growth rate: fat depth: eye-muscle depth at a 3:1:1 ratio of importance, respectively.

While midback fat depth can be assessed by echo-sounding on live animals, a much more detailed study of **fat distribution** within live animals can now be made using **computer-aided tomography**. It is better than echo-sounding, but its high capital and running costs rule out its routine use. **Ultrasonics** can be used for muscle depth as well as fat depth. In sheep, the Texel breed has a low-fat carcase, measured as total body fat. One has to consider both the visible ("dissectible") fat and the intramuscular fat (marbling). In beef, there is a slight correlation of increased marbling fat with increased tenderness. According to Wood and Cameron (1994), selection for decreased backfat in pigs has led to a positively correlated decline in intramuscular fat and hence to a decline in the eating quality of meat from modern pigs. Fat often has a lot of pleasant flavour in most farm animals, but health advice is usually to cut off any visible fat. Because **ostrich meat** is very low in fat, it cooks faster than most other meats and shrinks less during cooking, but low fat meats burn more easily during cooking and can dry out if overcooked. Less fat joints often require more basting with fat during cooking than do leaner meats. See Chapter 19.2.4 for health aspects of food, and Table 5.13.1 for a comparison of cholesterol, protein and fat content of different meats.

Table 5.13.1. Meat composition, based on 85 g portions. Undated information provided in 1999 from "Ostrich, the Healthy Alternative", Ashdown Foods, Tenterden, Kent.

	Beef	Chicken	Lamb	Ostrich	Pork	Turkey
Calories	240	138	205	97	235	135
Protein, g	23	27	22	18	23	25
Fat, g	15	3	13	2	19	3
Cholesterol, mg	77	72	78	9	82	59

Important meat characters are the proportions and appearance of muscle, bone and fat in retail cuts, the **yield of usable meat per carcase**, and the **sensory properties** of meat at the time of purchase and when it is eaten. When eaten, which is usually after cooking, meat should have a pleasing appearance as regards colour, surface texture and the amount of fat. It should smell and taste good, and be tender, juicy and convenient to eat. People prefer large chunks of flesh to small amounts which have to prised away from the bone, as with some rabbit joints and some very bony fish. See Chapter 12.7.5 for the critical carcase characters in beef cattle.

Meat colour generally depends on the amount of myoglobin present and on light-scattering effects which depend on the myofibrillar volume. The colour of meat can change on exposed surfaces with time, so that butchering and meat treatments can influence colour at the time of sale. Poultry often has white breast meat and dark leg meat. Tenderness depends on the proportion and composition of connective tissue, the amount of intramuscular fat, and on myofibrillar tenderness: see Thompson and Ball (1997).

In sheep, there are three major genes affecting muscling. The *callipyge* gene gives greater muscling in the leg, loin and shoulder, lower fat thickness and greater feed efficiency, but the meat is tougher. Most meat characters are multifactorial, with low heritabilities and many loci involved. For tables of heritability, repeatability, breeding aims and hybrid vigour in farm animals, see Dalton (1980).

Suggested reading

Abbott, A. J. and R. K. Atkin, eds., *Improving Vegetatively Propagated Crops*. 1987. Academic Press, London.

Baker, R. J., *Selection Indices in Plant Breeding*. 1986. CRC Press, Boca Raton, Florida.

Cameron, N. D. *Selection Indices and Prediction of Genetic Merit in Animal Breeding*. 1997. CAB International, Wallingford.

Chrispeels, M. J. and D. E. Sadava, *Plants, Genes, and Agriculture*. 1994. Jones and Bartlett, Boston.

Dalton, D. C. *An Introduction to Practical Animal Breeding*. 1980. Granada, London.

Holstein 1999a: *Using Genetic Information to Improve Profitability*. Undated but probably 1999. No author named. Holstein UK & Ireland, Rickmansworth.

Holstein 1999b: *Holstein Sire Summary 1999*. No author named. Holstein UK & Ireland, Rickmansworth.

Karn, M. N. and L. S. Penrose, Birth weight and gestation time in relation to maternal age, parity, and infant survival. *Annals Eugenics* (1951), 161, 147–164.

Lupton, F. G. H., ed., *Wheat Breeding: Its Scientific Basis*. 1987. Chapman and Hall, London.

Niks, R. E., P. R. Ellis and J. E. Parlevliet, Resistance to parasites, pp 422–447, in: Haywood, M. D., N. O. Bosemark and I. Romagosa, eds., *Plant Breeding. Principles and Prospects*. 1993. Chapman and Hall, London.

Piper, L. and A. Ruvinsky, eds., *The Genetics of Sheep*. 1997. CAB International, Wallingford.

Poehlman, J. M. and D. A. Sleper, *Breeding Field Crops*, 4th ed. 1995. Iowa State University Press, Iowa.

Strickberger, M. W. *Genetics*, 3rd ed. 1985. Collier Macmillan, London.

Table 6.2.1. The effects on genotype frequencies of selfing, which is extreme inbreeding. In generation 0, with no inbreeding, one gets Hardy-Weinberg equilibrium values.

Generations of selfing	Genotype frequencies in two populations with different allele frequencies						
	Allele frequencies $p = q = 0.5$				Allele frequencies $p = 0.8, q = 0.2$		
	A A	*A a*	*a a*		*A A*	*A a*	*a a*
0	0.2500	0.5000	0.2500		0.6400	0.3200	0.0400
1	0.3750	0.2500	0.3750		0.7200	0.1600	0.1200
2	0.4375	0.1250	0.4375		0.7600	0.0800	0.1600
3	0.4688	0.0625	0.4688		0.7800	0.0400	0.1800
...							
Many	0.5000	0.0000	0.5000		0.8000	0.0000	0.2000

frequency of heterozygotes and to increase the frequency of the recessive homozygote phenotype, but it takes more generations to have the same effect as selfing.

Virtually all effects of inbreeding, good or bad, stem from this reduction in heterozygosity and the increase in homozygosity. If strains or stocks or populations contain deleterious recessives, largely hidden in heterozygotes, then inbreeding will make them homozygous and their deleterious effects will show and can be selected out. Beneficial recessives will also be made homozygous and will be exposed. See Chapter 12.1 for the use of inbreeding in making F1 hybrids, and Chapter 12.2 for breeding methods used with inbreeding organisms. Inbreeding will give a deficiency of heterozygotes in Hardy-Weinberg analysis, and outbreeding will give an excess of heterozygotes, compared with random mating.

In almost all non-inbred diploid populations, there are many deleterious alleles, largely hidden in heterozygotes. If inbreeding is imposed on such populations, there is usually a loss of fitness and vigour, called **inbreeding depression**. In naturally inbreeding populations, however, natural selection will usually have already eliminated many deleterious harmful recessives and selected for beneficial ones. Many cultivated vegetables, cereals, flowers, sheep, pig and cattle breeds are highly inbred whilst still being useful commercially, but except in natural inbreeders this has only been achieved by intensive selection to overcome inbreeding depression.

Inbreeding depression does not occur in naturally inbreeding species and may cease to occur after some generations of inbreeding in formerly random-mating species, as natural selection will tend to adapt them by eliminating deleterious recessives exposed to selection in homozygotes. The classic example is the **pharaohs** of ancient Egypt, where brother-sister marriage was practised for many generations. As we shall see in Chapter 13.12, humans carry on average the equivalent of at least three recessive lethal alleles, but usually in the heterozygous form. Presumably the pharaohs suffered inbreeding depression when the inbreeding started, but must have adapted as they were very successful for thousands of years.

Inbreeding is often associated with the following: annual plants or short-lived life forms, the extremes of a species' distribution, marginal habitats and stressful environments. If the organism is genetically well adapted to its environment, inbreeding provides a high proportion of adapted genotypes, while occasional outcrossing allows the production of altered genotypes which might be better adapted if conditions change or permit colonisation of ecologically different areas. In contrast, long-lived and perennial species are often outbreeding or random-mating: the existing genotypes can survive vegetatively for years and can tolerate high variability from sexual reproduction which can give long-term adaptation. Outbreeding is also frequent in short-lived organisms with restricted recombination, which preserves favourable gene combinations.

6.3. Wright's inbreeding coefficient, F; Wright's equilibrium for genotype frequencies under inbreeding; calculation of F from pedigrees

The amount of inbreeding which has taken place is measured by **Wright's coefficient of inbreeding, F**, named after Sewell Wright. This is the proportion by which heterozygosity has been reduced by inbreeding. With no inbreeding, one gets $F = 0.0$, with Hardy-Weinberg equilibrium genotype frequencies. With complete inbreeding, one gets $F = 1.0$, with no heterozygotes at all; all individuals are homozygous, with the two homozygote types in proportion to the allele frequencies.

The F value is definitely not an absolute measure of heterozygosity or homozygosity, as those are features of allele frequencies and genetic diversity in a population. Having $F = 0.0$ does not mean that all individuals are heterozygous, just that heterozygotes have frequency $2pq$. If a random-bred sheep ($F = 0.0$) in a particular flock is on average heterozygous at 400 loci, then an individual with an inbreeding coefficient of 0.25 will be heterozygous at 300 loci, as 0.25 of the heterozygosity has been lost through inbreeding. In a more genetically diverse flock, a random-bred sheep might be heterozygous at 4,000 loci, and one with $F = 0.25$ would then be heterozygous at 3,000 loci.

Coefficient F can also be described as the probability that two alleles at a locus in an individual are **identical by recent descent**, that is, they are both derived by replication from the same allele in a common ancestor by inbreeding. A **common ancestor** of two individuals is an individual in some previous generation which is an ancestor of both of them, so that they can have some alleles in common by recent descent from it. When the two individuals mate, they can both provide an identical allele to an offspring.

Suppose we have two individuals and number with superscripts the individual alleles 1 to 3 for *A,* and 4 for the one *a* allele. If we cross $A^1 a^4$ x $A^2 A^3$, then some of the full siblings produced, such as $A^1 A^2$ and $A^1 A^3$, and $A^2 a^4$ and $A^3 a^4$, can be mated as full sib-matings, which is inbreeding. Some possible offspring from those matings are $A^1 A^1$ and $A^1 A^3$ from the former pair, and $A^2 A^3$ and $a^4 a^4$ from the latter pair. In each of $a^4 a^4$ and $A^1 A^1$, we have two alleles identical by descent, for example the two A^1 alleles in $A^1 A^1$ being directly derived from the one A^1 allele in the common ancestor (of $A^1 A^2$ and $A^1 A^3$), $A^1 a^4$. In contrast, the two A alleles in $A^1 A^3$ and the two in $A^2 A^3$ are not identical; they are both *A*, but are **similar**, not **identical** by recent common descent.

Consider a population with two alleles at one locus, *A*, frequency *p*, and *a*, frequency *q*. Without inbreeding, F = 0, and we expect Hardy-Weinberg equilibrium genotype frequencies, *A A*, p^2; *A a*, $2pq$, and *a a*, q^2. Alleles in homozygotes will normally be similar, not identical, and the two different alleles in heterozygotes will be neither similar nor identical.

With inbreeding, there will then be a chance that alleles in homozygotes are identical, and a correspondingly reduced chance that they are similar. The proportion of similar homozygotes is reduced by the inbreeding coefficient F to $p^2(1 - F)$ for *A A*, and to $q^2(1 - F)$ for *a a*. We get identical homozygotes in proportion to the allele frequencies, *A A*, pF, and *a a*, qF. The heterozygotes, *A a*, will by definition be reduced by fraction F to $2pq(1 - F)$.

At **Wright's equilibrium**, we therefore get:

$A A = p^2(1 - F)$ for similar alleles + pF for identical alleles,
$a a = q^2(1 - F)$ for similar alleles + qF for identical alleles,
$A a = 2pq(1 - F)$.

These can be simplified, e.g., $A A = p^2(1 - F) + pF = p^2 - p^2F + pF = p^2 + pF(1 - p) = p^2 + pqF$.

After simplifying, the Wright's equilibrium values are:

$A A = p^2 + pqF$,
$a a = q^2 + pqF$,
$A a = 2pq(1 - F)$.

These are important formulae and their derivation brings out the difference between similar and identical alleles. A shorter way to derive them is to say that the heterozygote is reduced by inbreeding by fraction F, from $2pq$ to $2pq(1 - F)$, which leaves $2pqF$ to be equally distributed between the two homozygotes, so the p^2 for A A becomes $p^2 + pqF$ and the q^2 for a a becomes $q^2 + pqF$. If F = 0.0, the Wright formulae become the Hardy-Weinberg formulae.

Values of F vary from zero with no inbreeding (with Hardy-Weinberg levels of heterozygosity), to 1.0, fully inbred (with no heterozygosity). If the parents are not themselves inbred, the maximum F from one generation of inbreeding is F = 0.5, from selfing: we earlier saw that selfing reduced heterozygosity by a half, hence this value of 0.5. Values of F for the offspring, if parents are not already inbred, are 0.25 when parents are full sibs (brother and sister) or parent and offspring (father/daughter or mother/son), 0.125 for half-sibs, 0.0625 for first cousins and 0.03125 for second cousins (which are children of first cousins).

Example 6.3.1. The recessive allele for an inherited recessive disease of sheep has a frequency of 1 in 200. How much is the risk of showing this disease increased if the parents are full sibs, compared to unrelated parents?

The recessive has an allele frequency of $1/200 = 0.005$, so the dominant has a frequency of 0.995. The chance of the disease in a lamb born to unrelated parents is $q^2 = (0.005)^2 = 0.000025$, or 1 in 40,000. With inbreeding from full sibs, F = 0.25, and we have Wright's equilibrium formula, a a = $q^2 + pqF$, so the recessive homozygote has a frequency of $0.000025 + (0.995 \times 0.005 \times 0.25) = 0.00127$, or about 1 in 788 of being born showing this disease. The chance of showing the disease has therefore increased by about 51-fold compared with random mating, because of the full-sib inbreeding.

We can use Wright's equilibrium frequencies to work out the effects of inbreeding on **quantitative characters** in a population. If we have **additive action**, say with a a having height 100 cm, A a 150 cm, and A A 200 cm, then with inbreeding some of the heterozygotes (height 150 cm) go equally to the two types of homozygote, the average height of which is also 150 cm. Therefore inbreeding does not change the average value for a population if there is additive action with the heterozygote exactly intermediate between the two homozygotes.

If we have **complete dominance**, say with a a 100 cm high, A a and A A 200 cm high, then we could imagine a basic height of 150 cm, to which the dominant allele adds 50 cm, and from which the recessive allele subtracts 50 cm if homozygous; let us call this amount of 50 cm value v. For each of the three genotypes, one can work out their frequency, their quantitative value ($+ v$ for A A and A a; $- v$ for a a), and by multiplying their quantitative value by their frequency, their **total quantitative contribution** to the population can be calculated. With no inbreeding, with Hardy-Weinberg genotype frequencies, these total quantitative contributions are $+ p^2v$ for $A\ A$, $+ 2pqv$ for $A\ a$, and $- q^2v$ for $a\ a$, totalling $v(p^2 - q^2) + 2pqv = \underline{v(p - q) + 2pqv}$, because if $p + q = 1$, then $p^2 - q^2 = p - q$. With inbreeding, the genotype frequencies are Wright's equilibrium frequencies and the corresponding total quantitative value becomes $\underline{v(p - q) + 2pqv - 2pqvF}$. Inbreeding has therefore caused the total quantitative value for the population to be reduced by $2pqvF$. This is easy to understand because from inbreeding we lose $2pqF$ amount of heterozygotes, phenotype value $+ v$, with the heterozygotes going equally to $A\ A$, value $+v$, and to $a\ a$, value $-v$, so the $+v$ and $-v$ cancel each other out in the homozygotes produced from the heterozygotes.

Because many quantitative traits such as high yield and high fertility are determined largely by dominant alleles, inbreeding - by moving population values towards the phenotype value of the recessives - usually reduces fitness and fertility, giving inbreeding depression. In alfalfa (lucerne, a forage legume, *Medicago sativa*), few inbred lines survive beyond two or three generations of inbreeding, and in pigs, an increase in F of 0.1 decreases average litter size by about one piglet.

We can also calculate the inbreeding coefficient F for an individual from **pedigree diagrams**, in which we include any individuals contributing to the inbreeding in that individual. Here we use the fact that coefficient F can be described as the probability that two alleles at a locus in an individual are identical by recent descent, both derived by replication from the same allele in a common ancestor by inbreeding. We arrange successive generation in descending order in the diagram, oldest at the top, youngest at the bottom, labelling as individual I the individual whose F we wish to determine.

In Fig. 6.3.1, we have woman A having borne two half-sibs, woman D and man E, by different men, men B and C. The two half-sibs then mate to give child I whose F value we wish to calculate, to find the **F for offspring of half-sibs**. F will be the chance of getting identical alleles in child I. Men B and C can only contribute one allele each, as a maximum, to child I, so they are not common ancestors; only woman A can contribute both alleles to child I, so she is the only common ancestor. The inbreeding coefficient of I is the chance of it receiving identical alleles, that is, of receiving *a1, a1,* or *a2, a2,* or *a1* and *a2* if those different alleles in woman A are themselves already identical from woman A being herself somewhat inbred. The chance of I getting identical alleles is the same as the chance that egg 3 and sperm 4 contain identical alleles. This will only happen if eggs 1 and 2 carry identical alleles, and egg 1 and egg 3 have identical alleles, and egg 2 and sperm 4 have identical alleles.

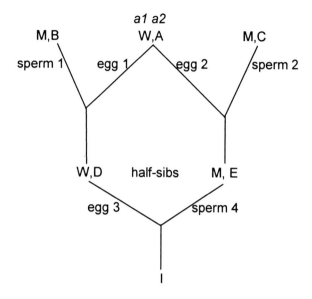

Figure 6.3.1. Inbreeding diagram for the offspring (individual I) of half-sibs. M stands for man, and W for woman.

Let us start by working out the chance that woman A's two eggs, 1 and 2, contain identical alleles: there is a chance of one quarter for them both being *a1*, a quarter for them both being *a2*, that is, a chance of a half that the two alleles will be identical for one or other allele, and a half of being one of each allele, *a1, a2*. If the woman A is inbred, coefficient F_A, then there is a chance F_A of alleles *a1* and *a2* being identical by descent, so the chance that eggs 1 and 2 between them contain one allele *a1* and one allele *a2* which are identical by previous inbreeding is $1/2\ F_A$. The total chance of eggs 1 and 2 having identical alleles is therefore $1/2$ [= chance of *a1 a1* or *a2 a2*] + $1/2\ F_A$, which gives a total of $1/2\ (1 + F_A)$.

The chance of eggs 1 and 3 having identical alleles is $1/2$, because egg 3 has an equal chance of getting its one allele at this locus from egg 1 or from sperm 1. The chance of egg 2 and sperm 4 having identical alleles is $1/2$, as sperm 4 has an equal chance of getting its allele from egg 2 or sperm 2.

The total chance of child I getting identical alleles is therefore $1/2\ (1 + F_A)$ x $1/2$ x $1/2 = (1/2)^3(1 + F_A)$, or in more general terms, $(1/2)^3(1 + F$ of the common ancestor of that pathway). If we count the number of adults in the inbreeding pathway from the individual I to the common ancestor A and back to I, not counting I, we have three individuals, woman D, woman A (the common ancestor) and man E, and our formula has 1/2 to the power of three. One can show that the contribution of any common ancestor (for which the symbol A is usually used) is $(1/2)^n(1 + F_A)$, where n is the number of individuals in the inbreeding pathway, excluding individual I, and excluding non-common ancestors such as men B and C in Fig. 6.3.1. In some pedigrees, there are more than one pathways to individual I, each giving an extra chance of that individual receiving identical alleles, so one has to sum up all the pathways to individual I. We therefore get the general formula, $F_I = \Sigma \{(1/2)^n(1 + F_A)\}$, where n is the number of individuals in a particular pathway and F_A is the inbreeding coefficient of the common ancestor in that particular pathway.

One needs to be able to calculate an individual's F value to see how inbred it is, and so how much heterozygosity it might have lost, and for calculations involving Wright's equilibrium frequencies, as in Example 6.3.1.

Example 6.3.2. Calculate the inbreeding coefficient F for an individual whose parents were first cousins, assuming that the common ancestors are not inbred.

First cousins are descended from full sibs, and thus have one pair of grandparents in common, who will be the two common ancestors of the first cousins' offspring. Although Fig. 6.3.1 showed the sex of the individuals, that is not usually done in pedigree diagrams, and we do not usually show individuals not contributing to the inbreeding, which in Fig. 6.3.1 would be men B and C. Our inbreeding diagram for this example is shown in Fig. 6.3.2, with each grandparent able to contribute both alleles to child I, so both are common ancestors of I.

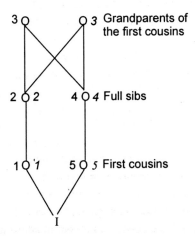

Figure 6.3.2. The inbreeding diagram for the offspring of first cousins. The italic and non-italic numbers represent pathways to different common ancestors.

There are five individuals in the pathway from I to common ancestor 3 and back again, non-italic numbered pathway, and five individuals from I to common ancestor *3* and back, italic numbered pathway; because the common ancestors are not inbred, F_A is zero. From $F_I = \Sigma \{(1/2)^n(1 + F_A)\}$, we get $F_I = \{(1/2)^5(1 + 0)\} + \{(1/2)^5(1 + 0)\} = \underline{1/16}$ or $\underline{0.0625}$ for the F value of the offspring of first cousins.

Example 6.3.3. The pedigree shown in Fig. 6.3.3 is for beef cattle, listing successive generations in order down the page. Calculate F for I if F = 1/8 for B and 1/4 for C.

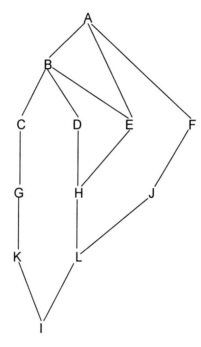

Figure 6.3.3. Pedigree diagram for beef cattle, for calculation of F for individual I.

We see that some individuals, such as G, have only one ancestor shown, because the other parent, such as the individual which mated with C, cannot be a common ancestor or be involved in the inbreeding of I, so is omitted. We see that A is a parent of B but mated with B to produce E, a parent-offspring mating. There are five pathways back to common ancestors, which are not all in the same generation. For example, A could have contributed identical alleles to I, by more than one pathway, and so could B.

The various pathways, with their common ancestor underlined and their contribution to F_I from the equation $F_I = \Sigma \{(1/2)^n(1 + F_A)\}$ are:

KGC<u>B</u>DHL = $(1/2)^7(1 + 1/8) = 9/1024$;

KGC<u>B</u>EHL = $(1/2)^7(1 + 1/8) = 9/1024$;

KGCB<u>A</u>EHL = $(1/2)^8(1 + 0) = 1/256$;

KGCB<u>A</u>FJL = $(1/2)^8(1 + 0) = 1/256$.

The total value for F_I is the sum of these values, <u>26/1024</u>, or <u>0.0254</u>. Answers may be given as fractions or decimals.

There are three points to note in such calculations. First, no individual ever occurs twice in any one pathway. Second, one may be given irrelevant data, such as F of individuals which are not common ancestors, such as C in this example. In any pathway, the inbreeding coefficients of individuals are not used in the calculation unless they are the common ancestor of the pathway being considered. Thus in pathway KGCB<u>A</u>EHL one ignores the known F values of C and B (and the unknown values for K, G, E, H and L), as only A is the common ancestor, although one uses F of B when it is the common ancestor, as in KGC<u>B</u>DHL. Third, pathways go from I to a common ancestor and back to I without zigzagging, so paths such as KGCBDHE<u>A</u>FJL do not count. It is important to identify all common ancestors and all pathways. One can use pedigree diagrams to work out F for the offspring of matings between relatives, such as father-daughter, aunt-nephew, second cousins, etc., or more complex patterns as in Fig. 6.3.3.

Suggested reading

Most of the books listed in Chapter 4 are suitable.

CHAPTER 7

MUTATION AND ITS USES

7.1. Molecular types of mutation and their revertibility; mutation frequencies

Mutations can occur in genes in the nucleus, in organelles such as chloroplasts and mitochondria, and in plasmids. A functioning gene has hundreds of base pairs, and changes in the base sequence can be of a number of types, with different properties and frequencies. Two different mutations in the same gene may give similar or different phenotypes, or even result in no phenotypic change, as mentioned in Chapter 1.3. It helps one to understand phenotypes, functions and alleles if one understands the molecular basis of mutation.

7.1.1. Revertibility

A mutation is **revertible** if it can change back from being mutant to being wild-type, either spontaneously or by induced reversion. For some purposes it is essential to know if a given mutation is revertible. Suppose you were preparing a live but mutation-weakened strain of polio virus for immunising people against polio. If you used a mutation which was revertible, then people immunised with the attenuated virus would occasionally be infected with the disease, following spontaneous reversion from mutant to wild-type. It would be much safer to use more than one mutation, and to ensure that they were non-reverting. For a gene for yield in rice, say, using a revertible mutation would be much less dangerous than for a live vaccine.

Revertibility can also be important in evolution and adaptation. Suppose that allele *A* is best in environment 1, but allele *a* is best in environment 2. An individual from environment 1 could spread to and thrive in environment 2 if there had been an *A* to *a* mutation which had become homozygous, *a a*. If the mutation was a big deletion and was not revertible, then no descendants from the *a a* individuals could produce *A* from reversion, so could not recolonise a type 1 environment. A revertible mutant, say a frame shift, could revert to *A* and hence lead to colonisation of a type 1 environment. Within the type 2 environment, the dominant reversions would be unfavourable and might be considered as imposing a small "reversion load" on

111

fitness, compared to non-reverting mutations, but they would confer adaptive flexibility. These two types of mutation (a frame shift and a big deletion) would probably both give a total loss of gene function and hence the same selection coefficients as each other, but would differ in their reversion ability.

A **forward mutation** from wild-type to mutant can occur at hundreds of different base pairs, in many different ways, and the spontaneous frequency of forward mutation is typically one in about 10^4 to one in about 10^8 or 10^9 per gene per generation, depending on the organism and the circumstances. Because a reversion may require a particular change in a particular base pair, spontaneous reversion frequencies are usually much lower than forward mutation frequencies, by factors of about 1,000 or more.

7.1.2. Germ-line and somatic mutations

It is important to distinguish between **germ-line mutations**, which can be passed on to future generations through the gametes, and **somatic mutations**, in non-germ-line body tissues. In a fungal hypha or a yeast cell or a bacterium, mutations can affect a non-sexual cell and be passed on to future generations by cell division or by sexual processes. In a higher plant, animal or human, mutations in a leaf, leg or head cell will not be passed on, but a mutation in the egg cell, ovum, pollen or sperm, or in tissues in a young organism which later form gametes, can be inherited. A UV-induced mutation in human skin may be important to that individual if it gives rise to malignant melanoma (a form of skin cancer), but it will not be inherited.

7.1.3. Base substitutions

If one base pair is replaced by a different base pair in the same place, that is a **base substitution**. Suppose bases 582 to 584 in the coding strand of DNA of a wild-type functional allele are a codon, CCT, then the mRNA will have GGA, coding for the amino acid glycine. A base substitution at position 582 of AT for CG gives ACT in DNA, UGA in mRNA, which is a stop codon, terminating polypeptide formation, giving a truncated polypeptide. This kind of base substitution is a **nonsense mutation**, giving no amino acid sense. Such nonsense mutations, unless right at the last-translated end of a gene, usually cause complete loss of function as several or many amino acids are missing.

If the wild-type CCT codon had a different base substitution, CG to GC at position 582, we would get GCT in DNA, CGA in mRNA, giving arginine. Such a substitution of one amino acid by another is a **mis-sense mutation**; it gives amino acid sense, but a different amino acid. An amino acid change at a highly specialised part of a enzymatic polypeptide often gives a total loss of function, e.g., for a reactive amino acid at the active site, a fold site, an allosteric regulator-binding point, or a sulphur-containing amino acid at a disulphide bridge point, but a change in unimportant parts could give a partial loss of function or no change of function - the latter would be a **tolerated substitution**. This particular change, from neutral glycine to the basic amino acid arginine, would change the overall charge on the polypeptide, which might well affect its function.

If the wild-type CCT codon had a TA to AT change at position 584, we would get CCA in DNA, GGU in mRNA, giving glycine. This is a **synonymous base substitution** giving no change in amino acid, and so no change in function. The change is completely tolerated and if we are going by function, not DNA sequencing, we will not even know that a base change has occurred. Functionally, we would have another wild-type allele although its base sequence, CCA, differed from that of our originally considered wild-type allele, which had CCT there.

Base substitutions therefore have a wide range of phenotypic effects, from no effect (a **cryptic mutation**, see Lamb, 1975), to a mild loss of function, to a severe or total loss of function. They are always revertible by the opposite base substitution to the one which caused them, e.g., a CG to AT substitution is revertible by an AT to CG substitution at exactly the same place in the molecule. These mutations and reversions can occur spontaneously, or be induced (e.g., by alkylating agents; see Chapter 7.2.1).

7.1.4. Frame shifts

Frame shifts are caused by the addition to or deletion from DNA of single base pairs, or of some other number of pairs which is not a multiple of three. If we had three codons in a wild-type gene's DNA, ACA, GGT, TGA with UGU, CCA, ACU in mRNA, we would get cysteine, proline, threonine in a polypeptide. If we had a deletion of the first A in the DNA, the mRNA would then have GUC, CAA, CU, giving valine, proline, and then a series of wrong amino acids to the end of the molecule, or until the wrong sequence of bases gave a stop codon. Unless right at the end of a polypeptide, frame shifts usually give a complete loss of function because many amino acids are changed and stop codons may be created which truncate the polypeptide. Additions and deletions both **change the reading frame** of the triplets of bases in mRNA, hence their name. A deletion or addition of a multiple of three base pairs is not a frame shift: the reading frame remains the same, but there will be missing or additional amino acids, which may well affect the function of the product.

Frame shifts are revertible, especially if the number of base pairs added or deleted is low. A base pair addition, say of AT, can be reverted by deletion of that AT base pair, with complete restoration of wild-type function. An addition, say of AT, can sometimes also be reverted, wholly or partly, by the deletion of one base pair near to, but not exactly at the original point of addition. This is because the reading frame is correctly restored, so most amino acids in the polypeptide are correct, and the few amino acids between the point of addition and the point of deletion may not be essential for function. Reversion can be spontaneous, or induced (e.g., by intercalating agents; see Chapter 7.2.1).

7.1.5. Large deletions

Large deletions usually cause a complete loss of function as so many amino acids are lost. The loss of 60 base pairs from a coding region will lead to a loss of 20 amino acids. Large deletions which are not multiples of three bases will also be frame shifts, causing further loss of function. Large deletions including "stop" or "start" reading signals will also affect function from that. The deletion of the first 80 bases of a coding sequence can therefore cause loss of function from loss of amino acids, from being a frame shift, and from deleting a "start reading" signal.

Large deletions will not revert, spontaneously or by induction, because the chance of getting the right number of bases inserted in the right place in the right order is vanishingly small. Large deletions are therefore extremely useful if one wants a non-revertible total loss of function, as when preparing a live weakened viral vaccine from a wild-type strain.

7.1.6. Unstable length mutations

Unstable length mutations can be intermediate in size between gene mutations and chromosome aberrations. In humans, adult-onset **myotonic dystrophy**, giving progressive muscle weakness, is due to an unstable length mutation where small length mutations cause no or few symptoms, but amplification in successive generations (possibly from unequal crossing-over) gives increasing severity of disease. It is an autosomal dominant where normal individuals have 5 to 37 repeats of a CTG sequence, with stable numbers, while 50-99 repeats give mild symptoms and greater numbers, up to 2,000 repeats, give severe disease. Larger numbers of repeats are unstable. The condition has a frequency of 1 in 7,500. Affected males have half the offspring affected, half unaffected, but affected females have 50% of offspring unaffected, 29% with late-onset, 12% neonatal deaths, and 9% with severe symptoms at birth, including mental handicap (Connor and Ferguson-Smith, 1997).

The human **fragile-X syndrome** has a frequency of 5 in 10,000 males and is caused by an unstable length mutation, as is Huntington disease (Chapter 13.3), both with a trinucleotide sequence involved. Fragile-X is the commonest X-linked cause of mental handicap. Under thymidine starvation in culture, X chromosomes with fragile-X often show chromosome or chromatid breaks near the tip of the long arm, at Xq27.3. Normal individuals have 6 to 52 copies of a CGG repeat, but small length mutations (premutations) with 60-200 repeats are phenotypically normal. Affected males with the full mutation have 230-1,000 repeats, giving mental handicap, and enlarged testes after puberty. Heterozygous females for the full mutation may be normal, but 20-30% have mild mental handicap, and 1% have moderate mental handicap. The premutation exists in 1 in 1,000 males and about 1 in 400 females. The mutation is unstable in inheritance. All daughters of males with the premutation receive a mutation on their paternal X chromosome, either the unchanged premutation or an expanded form with the full mutation. Further expansion can happen at mitosis, giving mosaics, or in female meiosis. Mothers heterozygous for the full mutation have half their sons mentally handicapped, and the daughters are carriers or have some mental handicap. Prenatal diagnosis from DNA (Chapter 13.12.6) is possible in the first three months of pregnancy.

7.2. Spontaneous and induced mutation; mutagenic agents

Spontaneous mutations (Plate 2) typically have frequencies ranging from about 1 in 10^4 to about 1 in 10^9 per gene per generation for forward mutations (to loss of function), and hundreds or thousands of times less than that for backmutations (restoring the lost function for frame shifts or base substitutions). The frequency of spontaneous mutations can increase with age in seeds and men (see Chapter 17.1.1 and 17.3), and in microbial cultures. Induced mutations can have much higher frequencies than spontaneous ones, but very high doses of mutagen can seriously weaken an organism even when the desired mutation is induced, as there are usually many unwanted deleterious mutants.

7.2.1. Spontaneous mutations

Spontaneous mutations are those occurring spontaneously, without the help of man, even though they have a range of known causes, sometimes involving natural mutagens. **Mutagens** are chemical or physical agents which cause mutation. **Induced mutations** are ones induced by man by various means. Spontaneous mutations can occur through **tautomeric shifts** in bases in DNA, where rare tautomers often have different base-pairing properties from the normal tautomer. For example, for cytosine the normal amino form pairs with guanine, but the rare spontaneous cytosine tautomer in the imino form pairs with adenine. If at replication a cytosine is in the imino form, it will get an adenine opposite it in the new strand. At the next replication, that adenine will get a thymine pairing with it, so a GC base pair can give rise to an AT base pair, a base substitution.

Spontaneous mutations can arise from non-ionising UV light and from various ionising radiations, which in order of decreasing wavelength and increasing penetrating power are **X-rays**, **gamma rays** and **cosmic rays**. There is not space here to deal with their mechanisms of mutation, some of which are not fully understood.

UV light has low penetrating power, being absorbed by water, skin and glass, so while it can cause inherited mutations in microbial cells, it causes superficial somatic mutations rather than germ-line mutations in higher organisms. UV can cause a whole range of mutation types, especially base substitutions and deletions, but including frame shifts and many types of chromosome aberration (Chapter 9). 254 nm is the most mutagenic UV wavelength, being the one most strongly absorbed by DNA.

One gets natural **ionising radiations**, X-rays, neutrons, alpha and beta particles, gamma rays and cosmic rays, from space, rocks, radon gas, and radioactive elements in the environment and within our bodies. These rays have high penetrating power and can therefore cause germ-line and somatic mutations. They cause base substitutions, frame shifts, deletions and chromosome aberrations.

There are also many **chemical mutagens**, some occurring in the environment or in food and drink. **Base analogues** are chemical analogues of DNA bases and get incorporated into DNA at replication. They often have a much higher rate of tautomeric shifts than the four normal DNA bases (the **pyrimidines** thymine and cytosine, and the **purines** adenine and guanine), so give rise to base substitutions at replication. Natural base analogues include caffeine (present in tea, coffee and some colas), a purine analogue.

Some chemicals react directly with DNA, as does nitrous acid, HNO_2. This compound replaces an NH_2 group with OH, changing cytosine (which pairs with G) to uracil (which pairs with A), and adenine (which pairs with T) to hypoxanthine (which pairs with C). The altered bases paired cause base substitutions at replication. Nitrous acid occurs at low concentrations in animals as a metabolic byproduct. It is a more effective mutagen in Prokaryotes than in Eukaryotes.

Alkylating agents add methyl or ethyl groups to bases in DNA, giving altered pairing properties or leading to a loss of purines, especially of guanine, and cause base substitutions and deletions. They can cross-link DNA strands. Like many mutagens, they can be **carcinogenic** as well as mutagenic. They include nitrogen mustards, various nitrosoguanidines such as NMG (N-methyl-N'-nitro-N-nitrosoguanidine), and EMS (ethyl-methane-sulphonate). Some of these compounds occur at low concentrations in the body as metabolic byproducts, e.g., from nitrates or nitrites in the diet, sometimes from chemicals used to treat meat.

Intercalating agents insert themselves between adjacent purines in DNA, causing base pair additions if inserted into an old strand, or base pair deletions if inserted into new strands; a round of replication is needed before the final mutation is formed. They therefore cause frame shift mutations, not base substitutions. Intercalating agents include acridine dyes such as proflavin and acridine orange, which were once used as antibiotics for wounds. There are some natural intercalating agents, but most are man-made.

There are many other kinds of natural mutagen, including extremes of temperature and altered pH. There are also mutagenic genetic elements, so-called "jumping genes", which can replicate themselves from one place in the genome into other places. These **insertion sequences** and **transposons** can therefore transfer themselves to within the coding sequences of other genes, interrupting the polypeptides produced, usually causing complete loss of the gene's function. These transposing elements occur in microbes, plants and animals. It has been estimated that more than half the spontaneous mutations in the fruit fly, *Drosophila melanogaster*, come from such transpositions. The causes of mutation in man are explored further in Chapter 13.7.

7.2.2. Induced mutations

Induced mutations can be obtained with a much higher frequency than spontaneous mutations. As with spontaneous mutations, most induced ones are harmful, but a few are useful. Higher doses give more harmful mutations as well as more useful ones, and may seriously weaken, kill, or sterilise the organism treated. For any organism, one needs to establish a safe but effective dose for a particular mutagen. One normally wants to induce germ-line mutations, not somatic ones. Dominant mutations are easier to spot than recessive ones, for which selfing or other close inbreeding is needed to make them homozygous to get expression in diploids. As mentioned in Chapter 12.4, recurrent backcrossing of mutant progeny to a non-mutant stock can be used to remove unwanted harmful mutations.

For something as unprotected as yeast cells or bacteria, one can irradiate them in a thin film of water, being careful to remove the petri dish lid before irradiation with UV, or briefly expose cells to chemical mutagens. See Chapter 13.7.2 and Plate 11 for the chemical induction of histidine-non-requiring backmutations in the

Ames test for mutagenicity. In animals, one can use mutagens directly on sperm for AI, or inject chemical mutagens into gonads, or one can used penetrating radiation such as X-rays or gamma rays on the gonad or whole-body areas. With plants, one can treat pollen, seeds or meristems. In all cases, too high a dose can cause sterility. Plate 12 shows the increasingly damaging effects of 45,000 to 60,000 rads of gamma-ray irradiation on tomato seeds. See Chapter 7.5 for more details of the use of induced mutations in agriculture and horticulture.

Some of these treatments will cause a mixture of somatic (Plate 13) and germ-line mutations. When treating seeds, for example, a germ-line mutation will not be visible in the somatic tissue of the plant grown from the seed: one has to cross or self the plants grown up, and look for mutants in the next generation. If one gets a desirable mutation expressed in somatic tissue in a plant, one can often propagate it vegetatively, e.g., from buds or callus, to get whole plants and germ-line mutants.

In some organisms it is more efficient to screen existing populations for spontaneous mutations, but in others it is better to use induced mutations. Some mutants which are useful in agriculture are disadvantageous in the wild. For example, the wild ancestors of wheat and barley had a brittle ear stem (rachis), aiding grain dispersal by animal contact. Early man must have selected a spontaneous mutant with a non-brittle rachis, so the ear held together during the harvest instead of shattering in the field, and it could be threshed back at home. Such a mutation would be disadvantageous in the wild, because of poor grain dispersal.

Although induced mutations have been used very extensively in laboratory studies of animal genetics, physiology and development, they have not been used much with farm animals. They have been used extensively with micro-organisms and quite a lot for higher plants, for which gamma rays from a cobalt-60 source are usually used. **Commercially used induced mutations** in barley include earlier ripening, shorter stiffer straw, greater diastatic activity (diastase is important in starch breakdown in malting barley for beer, gin and whisky), larger grains, and various kinds of disease resistance. Induced mutations for polygenes would be difficult to detect, and for quantitative characters it is usual to use existing genetic variation. Much of animal breeding is quantitative, with little reference to individual loci.

If generation times are long in plants, as with most trees, one can mutate buds, then vegetatively propagate them on mature or young plants. An advantage of vegetatively propagated mutants is that the favourable gene combinations are not lost through gene reassortment at meiosis. Chapter 17.1.5 covers various propagation methods and **somaclonal variation**, which involves altered forms from vegetative reproduction, and which can produce good new cultivars.

7.3. Mutation control and repair systems

Organisms can have **partial control** over their mutation frequencies, by various means. They can evolve to influence their mutation rates either through **access of mutagens** to the DNA or through the **efficiency of repair mechanisms**, or in specialised cases, through mutations which suppress the expression of other mutations.

7.3.1. Mutagen access

Many fungi have dark asexual or sexual spores, or fruit bodies, giving some protection against UV light (Lamb et al., 1992). Dark melanin skin pigments in humans give some protection against skin mutations which could cause cancer. The effectiveness of the enzyme nitrite reductase in the liver and kidneys affects levels of endogenous nitrous acid, which we saw in Chapter 7.2.1 acted directly on DNA as a mutagen. There can be genetic differences in the amount of uptake of environmental mutagens from soil by plant roots, or from food by the gut, or in excretion of mutagens by the liver and kidneys.

Mutation rates are very temperature-sensitive, so in seasonally breeding mammals (Chapter 17.2.3) such as deer, the time of descent of the testicles from the body cavity to the scrotum affects gonad temperature during spermatogenesis, and hence mutation rates in sperm. Ionising radiations are hard to control because of their penetrating power, although behaviour affecting what kind of rocks an animal lives amongst or burrows under might have minor effects. People living in Cornwall, in the west of England, get exposed to higher levels of radioactive radon gas than those in other regions, so migration to other areas could have small effects on natural radiation dosage (Chapter 13.7).

7.3.2. Repair systems

An organism can partly control its mutation rates by the efficiency of its **repair systems**, with mutations which decrease their efficiency increasing mutation rates, and mutations increasing their efficiency decreasing mutation rates.

For UV damage, only a **pre-mutational lesion** is initially produced, mainly thymine-thymine dimers, or cytosine-thymine. If not corrected, such within-chain dimers can cause mutations at replication by getting wrong bases inserted opposite them. Many organisms (micro-organisms, plants, animals and humans) have repair systems which can prevent such pre-mutational lesions from becoming full mutations. One system is **photo-repair**, where a photoreactivating enzyme uses visible light (especially blue-green) to break these dimers back to two monomers. In **dark-**

repair, an endonuclease recognises the dimer, cuts the affected strand, and an exonuclease enzyme erodes the exposed ends, including the dimer. DNA polymerase resynthesises the missing bases, using the other strand as a template, and DNA ligase joins up the ends. Mutations affecting sensitivity to ionising radiations and to UV are known in many organisms. People with the inherited autosomal recessive disease **xeroderma pigmentosum**, which has at least nine subtypes, have defective DNA excision repair of UV damage and suffer from multiple skin cancers. The combined frequency of the different types is 1 in 70,000, and sufferers are advised to keep out of bright light, to use strong sun-block creams, and to wear hats and opaque clothing.

Calculations of the expected mutation frequency on the basis of known frequencies of tautomeric shifts in the DNA bases show that there must be a **proof-reading enzyme system** at replication which preferentially reduces the incorporation of rare tautomers into the new strands. The efficiency of proof-reading and correction systems will obviously affect mutation rates.

7.3.4. Suppressor mutations

There are mutations which on their own give mutant phenotypes, but which can restore, partly or wholly, a wild-type phenotype to other mutations. The deletion of a base pair gives a frame shift mutation, but the addition of a base pair (which on its own gives a mutation) near to that deletion mutation restores the reading frame and may give whole or partial restoration of the wild-type phenotype.

One can have mutations which suppress a whole class of mutations. Amber stop codons with UAG in mRNA, whether normal or in a nonsense mutation, can be suppressed by amber-suppressor mutations. In these, there has been a mutation in a gene for a transfer RNA (those genes are duplicated, so wild-type genes still remain for that tRNA), changing its anticodon from one binding to an amino acid codon in mRNA to one binding with the amber stop codon, inserting an amino acid, and permitting reading through of the stop codon, producing unusually long polypeptides. Because they suppress normal stop codons as well as mutant ones, such amber-suppressed strains tend to grow slowly. While of considerable use in the laboratory, **suppressor mutations** are not normally used in commercial organisms to control mutant expression.

7.3.5. Optimum mutation rates

There is some evidence of **selection for optimum mutation rates** for an organism according to its habitat. Strains of the fungus *Sordaria fimicola* from a stressful exposed environment had higher inherited rates of spontaneous mutation and of induced mutation than strains from a less stressful environment, as if the production of more genetic variation was useful in adapting to the stressful environment (Lamb et al., 1998). Because most mutations are harmful, an ideally adapted organism in a stable environment would do best with a low mutation frequency, but higher mutation frequencies would be appropriate for organisms in unstable environments, or for colonising different environments.

7.4. How different types of mutation can complicate population genetics calculations

In equations in population genetics, we have generally used a very simple system of two alleles, with A mutating with frequency u to a, which backmutates to A with frequency v. Suppose that A is an essential, functional, wild-type allele of a given base sequence, with no selection against it. It could mutate in many different ways. Some DNA changes such as synonymous mutations can change the DNA sequence without changing the polypeptide or the phenotype. We would probably not detect that the mutation had occurred, so would just regard it as a normal A allele, with $s = 0.0$, but it would mean that there were in fact a number of different A alleles, which might differ in mutation frequencies, and which might be able to recombine at meiosis to produce further variants.

If we had a large deletion, it would almost certainly give a complete loss of function, $s = 1.0$, and it would not revert, so $v = 0.0$. If we had a frame shift or a nonsense mutation, it would probably give a complete loss of function, $s = 1.0$, but either type could revert spontaneously, so v is not zero. If we had mis-sense mutations, some might give no loss of function, $s = 0.0$, some might give a partial loss of function, say $s = 0.1$, while others, say at an active site, might give a complete loss of function, $s = 1.0$. All those mis-sense mutations could revert, but probably at different frequencies. If all alleles giving a loss of function were called a, we would have a whole range of different a alleles, differing very much in selection coefficient and in reversion frequency.

It will now be clear that in a very large population one seldom has just two different alleles at a locus, at the molecular level, even if one can describe the phenotypes as falling into two classes, such as tall and short pea plants. Mutation will eventually produce a whole **range of alleles**, although a new small population may stay genetically uniform for a number of generations if started with only one type of homozygote for a particular locus.

The simple equations involving only two alleles at a locus are very useful for showing the principles, say of an equilibrium between mutation and selection. If one takes into account that there are many possible alleles with different mutation and reversion frequencies, and different selection coefficients, the equations would become extremely complicated. It is important to realise that we have been using over-simplified equations.

7.5. Using induced mutations

Because of its low penetration, UV has restricted use, mainly to bacteria, fungal spores and hyphae, pollen or thin tissue cultures. X-rays and gamma rays are widely used in plants and animals as they have good tissue penetration and can be given in precise doses. Chemical mutagens can give high mutation frequencies; the dosage received can be uncertain if tissue penetration to the target tissue is needed, leading to poor reproducibility, especially with variable persistence of the mutagen after treatment. It is difficult to find an **ideal dosage** with a mutagen, since low doses give too few mutations and high doses give too many unwanted mutations, which can seriously weaken an organism's viability and/or fertility even if it does carry a favourable mutation. All mutagens need strict safety procedures in storage and use.

The following examples of the **plant part treated**, the mutagen and the dosages given are taken from Micke and Donini (1993). Dry seeds, in seed-propagated plants: oats, rice and maize, gamma rays, 14-28 krad; tomato, EMS, 0.8%, 24 h at 24°; *Triticum durum* (wheat), gamma rays, 10-25 krad, or fast neutrons, 600-800 krad, or EMS, 3.8%; *Triticum durum* pollen, gamma rays, 0.75-3 krad. Vegetatively propagated plants: potato, tubers, EMS, 100-500 ppm, 4 h, 25°, or shoot tips, gamma rays, 2.5-3.5 krad; *Dianthus caryophyllus* (carnation), nodal stems, X-rays, 1.5-2 krad; *Malus pumila* (apple), dormant graftwood, gamma rays, 6-7 krad, dormant buds, gamma rays, 2.5-5 krad; *Musa* (banana), shoot tips *in vitro*, gamma rays, 1-2.5 krad; *Citrus sinensis* (orange), ovular callus *in vitro*, gamma rays, 8-16 krad.

As seeds are multicellular, mutation of one cell in them gives a **chimeric M1 generation**, with some parts mutant and most parts wild-type (Plate 13). The mutation will only be passed on to progeny if it gets into the germ-line tissue. Gross chromosome aberrations may be selected against during the growth and differentiation of meristematic tissues. Selfing the M1 generation is done to get recessive mutations homozygous for expression and selection. Recurrent backcrossing (Chapter 12.4) is often used to get rid of unwanted deleterious mutations, while selecting for the wanted mutation. The use of mutagens on gametes such as pollen is one way of avoiding chimeras. In chimeras, adventitious buds may develop from single cells, and so give genetically uniform (homohistont) plants. Vegetatively propagated plants are often highly heterozygous, with hybrid vigour, so mutating vegetative parts avoids sexual reassortment of the genes but leads to chimeras.

In 1990, a FAO/IAEA symposium recorded 1363 cultivars produced by mutagenesis, with more than 90% from X-rays or gamma rays. The mutations involved characters such as crop yields, flowering time, flower shape and colour, fruit size and colour, resistance to pests and pathogens, the oil content of sunflowers and soya beans, the fatty acid content of linseed, soya and rapeseed, the protein content, amino acid composition and starch quality of rice, barley, wheat and maize, and fertility restorers and male sterility in maize. In Italy there was a successful mutation in pasta wheat to lodging resistance, and in Pakistan there were mutationally improved varieties of cotton, mungbean and chickpea. In vegetatively propagated crops, there were improvements by mutation to apples, cherries, olives and apricots for characters such as compact tree shape, earlier flowering, self-incompatibility, or seedlessness. Ornamental plants had mutations affecting leaves, flowers, temperature requirements, flowering periods, etc.

Induced mutations have hardly been used at all in farm animals, although they have been widely employed in lab organisms such as *Drosophila* and mouse, and can be used in human tissue cultures.

Suggested reading

Connor, J. M. and M. A. Ferguson-Smith, *Essential Medical Genetics*, 5th ed. 1997. Blackwell Scientific, Oxford.

Friedberg, E. C., G. C. Walker and W. Siede, *DNA Repair and Mutagenesis*. 1995. Blackwell Science, Oxford.

Lamb, B. C., Cryptic mutations: their predicted biochemical basis, frequencies and effects on gene conversion. *Molecular and General Genetics* (1975), 137, 305–314.

Lamb, B. C. et al., Interactions of UV-sensitivity and photo-reactivation with the type and distribution of ascospore pigmentation in wild-type and mutant strains of *Ascobolus immersus*, *Sordaria brevicollis* and *Sordaria fimicola*. *Genetics* (*Life Sci. Adv.*) (1992), 11, 153–160.

Lamb B. C. et al., Inherited and environmentally induced differences in mutation frequencies between wild strains of *Sordaria fimicola* from 'Evolution Canyon'. *Genetics* (1998), 149, 87–99.

Micke, A. and B. Donini, Induced mutations, pp 52-62, in: Haywood, M. D., N. O. Bosemark and I. Romagosa, eds., *Plant Breeding. Principles and Prospects*. 1993. Chapman and Hall, London.

Venitt, S. and J. M. Parry, eds., *Mutagenicity Testing: a Practical Approach*. 1984. IRC Press, Oxford.

CHAPTER 8

RECOMBINATION AND MAPPING

8.1. Recombination, genetic distances and the numbers of progeny needed to get particular recombinants

As we saw in Chapter 1.3, **recombination** is the production of new combinations of existing genes. Recombination frequencies in Eukaryotes are measured from crosses in per cent recombination, which is the same as **centiMorgans (cM**, named after the American *Drosophila* geneticist, Thomas Hunt Morgan), where one cM is one per cent recombination. Plate 5 shows a maize example with no detected recombination between two very close loci. The dihybrid ratios in Plates 6, 7 and 8 are modified from 9:3:3:1 from selfing the F1 (50% RF) by gene interactions between the unlinked loci.

The breeders of microbes, plants and animals will often want to produce new combinations of existing genes. If the relevant loci have been mapped in terms of recombination frequencies, the breeder can then work out the approximate **number of progeny to raise** to be reasonably sure of getting a particular desired recombinant. As shown below, if there is dominance, it may take three generations to get and identify the desired genotypes.

We saw in Chapter 4.2 that recombination for two loci can only occur in double heterozygotes, so if one wants the pure-breeding recombinant *a a, B B* from *A A, B B* and *a a, b b*, one must cross them to get the coupling double heterozygote, *AB/ab*, which can then be selfed. If the two loci are unlinked, with 50% recombination, then one quarter of the gametes from *AB/ab* will be *aB* and one sixteenth of the progeny from the selfing will be *a a, B B*. If there is complete dominance at the *B/b* locus, one would need further genetic testing to distinguish *a a, B B* from double that number of *a a, B b* genotypes, which will share the *a B-* phenotype. If one only reared 16 offspring, then one might or might not find the desired genotype whose expected frequency was 1/16, so it would be safer to rear perhaps 60 offspring, or 120 if one wanted at least one *a a, B B* of each sex. If we were using yeast or another fungus with haploid progeny, 1/4 would be of the desired type, *a, B*, and there would be no complications from dominance in identifying the desired genotype just from the phenotype, so fewer progeny need to be raised.

If the two loci were syntenic and closely linked, with 5% recombination instead of 50%, only 2.5% of the gametes would be of the desired type, *a, B*, and therefore only 0.0625% of the progeny would be *a a, B B*, or 1 in 1,600. One would have to raise several thousand offspring to be reasonably sure of getting one of the desired recombinants, and in diploids with dominance for *B/b* one would have to test individuals of the desired phenotype to separate the *a a, B B* types from *a a, B b*.

If one does not know how far apart the loci are, or whether they are syntenic, then one does not know how many progeny to raise to get a desired recombinant. It is therefore very useful to be able to do **genetic mapping**, finding out the recombination frequencies between various pairs of loci, finding which are linked and which are unlinked, and constructing **genetic maps** showing the loci in physical order in different linkage groups. All loci within a linkage group are syntenic, but the more distant ones may be unlinked to each other, with 50% recombination. In a well-mapped organism, linkage groups should be correlated with visually identified chromosomes, with as many linkage groups as there are different types of chromosome (one of each type of chromosome in a haploid, two in a diploid, three in a triploid, etc., counting X and Y, if present, as one type only).

One can obtain **extremely rare recombinants** quite easily in many micro-organisms, and estimate even very small map distances, if one can use **selection** to identify recombinants. For example, if one wanted to obtain a wild-type recombinant between two auxotrophic mutants within a locus, say *lysA⁺* from *lysA⁵* and *lysA⁹* in a fungus such as yeast or *Neurospora*, one just plates out millions of ascospores (from a repulsion-phase cross) on a series of petri dishes of minimal medium, containing no lysine. The two parental types of spore could not germinate, nor could the double mutant recombinant, because they all require lysine for growth, and only the prototrophic recombinant wild-type ascospores could germinate and grow into colonies. By estimating on minimal medium plus lysine the number of viable spores plated out, and using the number of wild-type colonies growing up on minimal medium, one can estimate the recombination frequency between the two alleles. Allowing for the fact that the wild-type *lysA⁺* and double mutant *lysA⁵ lysA⁹* recombinants should be equally frequent, the recombination frequency would be twice the frequency of wild-type germlings. It would be very difficult to identify the double mutant recombinants amongst so many parental single mutants, because they have the same *lysA⁻* phenotype.

Recombination and chiasma frequencies are under **genetic and environmental control** in plants, animals and micro-organisms. If a population is segregating for genes affecting recombination frequencies for syntenic loci, selecting for desired recombinants could accidentally also select for genotypes with alleles for more recombination.

8.2. Types of recombination and their effects; meiotic and mitotic crossovers; interference and map functions

We covered the basic aspects in Chapter 1.3. Non-syntenic loci recombine in meiosis by **independent assortment,** because the members of different pairs of homologous chromosomes line up independently of each other at metaphase, giving 50% recombination. Syntenic loci can recombine reciprocally at pachytene of meiosis if a **crossover** occurs between them. The chance of a crossover increases as the distance between loci increases, so close loci show low recombination frequencies and distant loci may show 50% recombination. With reciprocal recombination in a double heterozygote, if one type of recombinant is produced, say *a, B* from an *AB/ab* meiosis, then the reciprocal product is also produced in the same meiosis, *A, b*in this case.

With **gene conversion,** recombination is generally non-reciprocal, and non-Mendelian ratios can be produced in meiotic tetrads or octads. For a heterozygous base substitution mutant, *A/a,* let the *A* chromatid have base pair AT and let the *a* chromatid have a CG base pair at the corresponding position. If one has an *AB/ab* meiosis, and one *A* chromatid forms hybrid DNA by invading an *a*chromatid, then in the formerly *a* chromatid one might get a mispair such as CT in that chromatid, while the other three chromatids have AT, AT and CG. See Lamb (1996) for an account of gene conversion mechanisms. The CT mispair has three possible fates: it may stay uncorrected, being resolved at the next replication into one AT (allele *A*) chromatid and one CG (allele *a*) chromatid; it may be corrected by enzymes to AT; it may be corrected to CG. Figure 8.2.1 shows the consequences for the products of meiosis, and thus how gene conversion can give non-reciprocal recombination.

One can also have recombination at mitosis, but with a much lower frequency than at meiosis. At mitosis, there is no regular pairing of homologous chromosomes, and no synaptinemal complex. At mitosis one can have **mitotic crossing-over** and **mitotic gene conversion,** but at a frequency reduced by a factor of at least a thousand compared to meiosis. The distribution of crossovers (affecting the relative genetic distances in different ‚parts of a chromosome) may also differ between meiosis and mitosis. See Chapter 18.4 for details of mitotic recombination in fungi, where it can be used by breeders in species where sexual reproduction is absent.

Alleles	Base pairs at *A/a*, at the point of the base substitution.	Base pairs at *A/a* after hybrid-DNA formation in one *a* chromatid by invasion by the T-carrying strand of one *A* chromatid.
B A	<u>A</u> T	<u>A</u> T
B A	<u>A</u> T	<u>A</u> T
b a	<u>C</u> G	mispair <u>C</u> - can go to <u>C</u> or <u>A</u> T G T from correction
b a	<u>C</u> G	<u>C</u> G

B/b is a segregating locus syntenic to *A/a*, with *B* on the *A* chromatid and *b* on the *a* chromatid. If mispair C/T at the *A/a* locus is corrected to C/G, it gives an *a* chromatid, with a normal 2*A*:2*a* allele ratio from that meiosis, or 4*A*:4*a* in octads, and no recombination with *B/b*. If it is corrected to A/T, it gives an *A* chromatid, with a 3*A*:1*a* gene conversion ratio, or 6*A*:2*a* in octads. Correction of the mispair C/T to A/T, an *A* allele, would give a non-reciprocal recombinant, *A, b*, with no corresponding *a, B* recombinant. If it is uncorrected, we will get' postmeiotic segregation, with a 5*A*:3*a* gene conversion ratio in octads; in tetrads, the meiotic product carrying the mispair would give one *A* and one *a* product after the first mitosis. This can cause a mosaic of two different genotypes; for example, if the post-meiotic segregation occurred in an egg which was fertilised by an *a* sperm, the first zygotic mitotic division would give one cell *A a*, and one cell *a a*.

Figure 8.2.1. The consequences of gene conversion for allele ratios and recombination.

Mitotic recombination occurs in all Eukaryotes which have two or more copies of each chromosome; in haploids, there is no homologue to cross over with. It was demonstrated in the fruit fly *Drosophila melanogaster* by Stern in 1936, long before it was known in fungi. He used mutants *y*, yellow body, and *sn*, singed bristle, which are linked on the X chromosome. Fig. 8.2.2 shows how a single crossover between the centromere and the two loci can give two genetically different nuclei after mitosis, if the centromere segregation is favourable. If this mitotic recombination occurs early in development, further cell divisions by the daughter cells can give observable twin-spots on the adult fly, where a spot with yellow body and non-singed bristles is next to one with non-yellow and singed bristles, when the general body appearance is non-yellow and non-singed bristle.

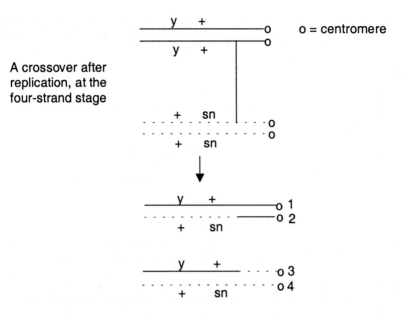

A crossover after
replication, at the
four-strand stage

o = centromere

If centromeres 1 and 3 go to the same pole, and 2 and 4 go to
the other pole, then one nucleus will get y + /y +, giving a yellow,
non-singed, phenotype, and the other nucleus will get + sn/+ sn,
giving a non-yellow, singed phenotype, which after enough further
cell divisions would give visible twin spots. If centromeres 1 and 4
go to the same pole, and 2 and 3 go to the other pole, both nuclei
will give wild-type phenotypes.

Figure 8.2.2. The origin of twin-spots in *Drosophila* from a mitotic crossover between the X
chromosomes after replication.

In meiosis or mitosis, there can be **interference between crossovers** affecting
either the occurrence of double or multiple crossovers, or the distribution of crossovers
between strands in double crossovers. See Foss et al., (1993), Fincham, Day and
Radford (1979) and Chapter 18.3. With **chromosome interference** (chiasma position
interference) the occurrence of one crossover in an interval may reduce (positive
interference), increase (negative interference) or have no effect on the chance of a
second crossover in that interval. In **chromatid interference** (strand interference),
the occurrence of a crossover between two particular chromatids may affect which
chromatids which are involved in a second crossover in that interval. If the same
strands are involved again less often than one would expect by chance, there is

positive chromatid interference, giving a deficiency of two-strand doubles and an excess of four-strand doubles; if the same strands are involved more often in the second crossover than expected by chance, there is negative chromatid interference, with an excess of two-strand double and a deficit of four-strand doubles. With no chromatid interference one expects a ratio of 1 two-strand double crossover: 2 three-strand doubles: 1 four-strand double. Usually there is no chromatid interference, but it requires tetrad analysis to detect it (see Lamb, 1996).

Because **undetected crossovers** in double or multiple crossovers reduce the observed genetic map distance in relation to the physical map distance, especially over long distances, it is useful to be able to correct map distances for undetected crossovers. This can be done mathematically by using a **map function** such as **Haldane's mapping function**. That particular function assumes no chromosome or chromatid interference. If d is the true map distance as a fraction (true map units/100), and ρ is the observed recombination distance as fraction (recombination %/100), then $d = - 1/2 \ln(1-2\rho)$. From the other form of the equation, $\rho = 1/2\ (1-e^{-2d})$, one can see that even very large map distances can only give a maximum recombination frequency of 50% for syntenic loci, because e^{-2d} becomes very small at large d values.

Chromosome aberrations can also give new combinations of existing alleles (Chapter 9).

8.3. Numbers of types of gamete, and of offspring genotypes and phenotypes, for different numbers of segregating loci

If one has a multiple heterozygote, with two alleles segregating at each of n loci, there are **general formulae for its gametes and offspring**, as follows.

The number of different types of gamete $= 2^n$.
The number of different gamete combinations $= 4^n$.
The number of different offspring genotypes $= 3^n$.
The total number of offspring phenotypes are:

(i), with complete dominance, 2^n;

(ii), with anything giving three phenotypes per locus, e.g., additive action, incomplete dominance or overdominance, 3^n;

(iii), with p loci with complete dominance and $(n - p)$ loci where the heterozygote is distinguishable, $2^p + 3^{(n-p)}$.

The offspring phenotype ratio with complete dominance at all segregating loci and no linkage is: (3/4 dominant + 1/4 recessive)^n; e.g., locus one has starchy grains dominant, sugary grains recessive, and locus two has coloured grains dominant and colourless grains recessive. We expect a Mendelian dihybrid 9:3:3:1 phenotype ratio of (3/4 starchy + 1/4 sugary)(3/4 coloured + 1/4 colourless) = 9/16 starchy coloured + 3/16 starchy colourless + 3/16 sugary coloured + 1/16 sugary colourless.

If the heterozygote is distinctive at all segregating loci, as with additive action, we get a phenotype ratio of (1/4 homozygote 1 + 1/2 heterozygote + 1/4 homozygote 2)^n. If some loci have complete dominance and some have distinctive heterozygotes, the offspring phenotype ratio is (3/4 dominant + 1/4 recessive)^p (1/4 homozygote 1 + 1/2 heterozygote + 1/4 homozygote 2)$^{(n-p)}$. If one has linkage between any two or more loci, these phenotype frequencies will have more of the parental combinations and fewer of the recombinant ones.

8.4. Calculation of the frequencies of particular genotypes and phenotypes

To estimate how many offspring should be reared to get particular genotypes or phenotypes, one needs to calculate their expected frequencies. The probability of a series of independent events all occurring together is the product of the probabilities of each separate event. If all the loci are unlinked to each other, one just multiplies together the expectations at the different segregating loci.

The **expected frequencies** of different genotypes and phenotypes at any one locus are a simple matter of Mendelian genetics. Thus if one has say *A a* x *A a*, the genotype expectations at that locus in the next generation are 1/4 *A A*, 1/2 *A a*, 1/4 *a a*, and the phenotype expectations with complete dominance are 3/4 *A* and 1/4 *a*. For *A a* x *a a*, the genotype expectations are 1/2 *A a*, 1/2 *a a*, with phenotypes 1/2 *A*, 1/2 *a*.

Suppose we had seven unlinked segregating loci in a polyhybrid *A a, B b, C c, D d, E e, F f, G g*, which was selfed, and we wanted the expected frequency of *A A, b b, C -, D -, e e, F f, G g*. Working out the expectations at each locus and multiplying them together, we get 1/4 x 1/4 x 3/4 x 3/4 x 1/4 x 1/2 x 1/2 = 9/4096 = 0.0022, or about 1 in 455.

Suppose we had *A A, B b, c c, D d, E E, f f* x *A a, B b, C c, D d, E E, F F*, we could again work out the expected frequency for any genotype or phenotype by working out expectations at each locus and multiplying them together. *A a, b b, C c, D D, E E, F f* would have frequency 1/2 x 1/4 x 1/2 x 1/4 x 1 x 1 = 0.015625 = 1 in 64. Linkage between any loci will give an excess of parental types over recombinants. If in this cross *B/b* and *D/d* were linked, with 10% recombination and with *B* and *d* on one chromosome, and *b* and *D* on the homologue, the chance of getting a *b D* gamete from *Bd/bD* is no longer 1/4, but is 0.45 (because there will be 90% parental gametes), so the chance of getting *b b, D D* is no longer 1/4 x

1/4 but is 0.45 x 0.45, so the frequency of the desired genotype is 0.050625, or about 1 in 20 instead of 1 in 64.

8.5. Mapping, including physical mapping

Genetic maps are traditionally based on recombination frequencies from crosses, and can include markers giving phenotypic effects and molecular markers. Much progress is being made with many organisms, microbial, plant, animal and human, with major sequencing programmes under way for whole genomes. **Physical maps** are based on examination of DNA bands on stained gels after electrophoresis. In many organisms, a combination of genetic and physical methods is adding greatly to the detail of maps. For example in 1990 there were only 30 loci on genetic maps of sheep, but by 1996 this had increased to about 600 loci and molecular markers. The total number of sheep genes has been estimated at about 100,000. A catalogue of sheep genes and inherited traits can be found on the Internet at http:// www.angis.su.oz.au. There is also a catalogue of inherited traits and disorders in a wide range of animal species, Mendelian Inheritance in Animals, at http:// morgan.angis.su.oz.au (Nicholas, 1997). For genome mapping in plants, see Paterson (1996), and for a forest tree example, see Bradshaw and Stettler (1995).

A physical map is a diagram of a DNA molecule showing the positions of physical landmarks; these commonly include restriction sites and other particular DNA sequences. A **restriction site** is the short base sequence, usually of four to six nucleotides, at which a particular restriction endonuclease enzyme makes a cut. For example, the restriction enzyme *Eco*RI from the bacterium *Escherichia coli* recognises the base sequence 5′ -GAATTC- 3′ and cuts that strand between the G and the A. On the other strand, there must be the complementary sequence 3′ -CTTAAG- 5′, which is also cut between the G and the A, giving **staggered breaks** in the two strands, with **complementary single-strand ends**, which we will see in Chapter 14 are very useful for genetic engineering. Some restriction enzymes cut both strands of DNA at the same point, giving **flush ends**, also called **blunt ends**.

There are several hundred different restriction enzymes, all with different restriction sites (recognition sequences). They occur in all bacteria and Eukaryotes, with the best known and most used ones being from bacteria. The restriction sites involve **palindromic** sequences, running in opposite directions in the two DNA strands, as in the above example of GAATTC in one strand and CTTAAG in the other.

Much of physical mapping is based on the lengths (judged by migration speed, with smaller fragments migrating faster than long ones) of fragments of DNA after digestion with different restriction enzymes. Pieces of DNA with no, one, two or three *Eco*RI restriction sites would respectively give one, two, three or four fragments of DNA which could be detected after staining on gels. A piece with one such site near the middle would give two fragments of approximately equal length, while a piece with the site near one end would give two very unequal fragments. Having found out where the *Eco*RI sites are, one could then try a series of different restriction enzymes, say *Hind*III, *Bam*I, and others.

If a mutation changes the base sequence within a restriction site so that the restriction enzyme no longer recognises that sequence, then using that enzyme on the wild-type DNA gives two fragments, but with the mutant DNA only one fragment is found. Less frequently, mutation can cause restriction sites. Differences in DNA fragment length and number caused by the presence or absence of restriction sites are called **restriction fragment length polymorphisms (RFLPs)**.

If one has two cleavage sites for a particular restriction enzyme in a given region of DNA, one might expect that DNA treated with that enzyme should produce the same fragment lengths from different individuals. They will produce different numbers of fragments if the individuals (or the two homologous chromosomes in diploids) differ in restriction sites, as mentioned above. Another kind of difference arises when different individuals (or haploid genomes) give different lengths of fragment from the two cleavage sites. A frequent cause is the existence of **tandemly repeated DNA sequences**, repeated say 10 to 300 times, with different numbers of repeats in different individuals. One individual might have the sequence repeated 15 times, while another has it repeated 60 times, giving a different fragment length from the DNA between the two restriction sites which are cut with that enzyme. The general term is a **variable number of tandem repeats, VNTR**. See Chapter 9.3 for a general account of duplications.

If the repeat length is just of two bases, one has **simple tandem repeat polymorphisms**, STRPs, where pairs of bases make repeating sequences in tandem, such as 5′ -TGTGTGTGTGTG- 3′. Different "alleles" (the quotation marks are because these are not really different forms of a functional gene locus) have different numbers of repeats of the pairs of bases, which vary very frequently, so that each STRP site is on average heterozygous in 70% of humans tested. STRPs have been very important in mapping the human genome. Because there is such a high degree of polymorphism between human individuals for STRPs and other VNTRs, they have been extensively used in forensic science, for confirming or eliminating suspects on the basis of DNA samples, e.g., comparing sperm DNA in a rape case with DNA from different suspects' hair roots.

8.6.2. Parasexual methods

Other methods for chromosome/linkage group correlations included **parasexual methods**, using hybrid cell cultures rather than crosses. In the 1970's, one could get tissue-culture-adapted mice cells, for example, and fuse these with human cells such as skin fibroblasts or peripheral blood leucocytes, using chemical agents or Sendai virus for the fusion. Say the human cell has 2nH representing its diploid number of 46 human chromosomes, and the mouse cell has 2nM chromosomes, then the fusion of cell membranes and cytoplasm gives a **heterokaryon** (which has unlike nuclei in common cytoplasm) with 2nH + 2nM chromosomes. There can then be rare chance fusion between the two kinds of nucleus to give a **synkaryon**. As the synkaryon proliferates in cell culture to give a hybrid cell population, various chromosomes are lost by **non-disjunction**, a failure to segregate regularly and equally to both daughter cells from a cell division. Different chromosomes, especially human ones, are lost in different cell lines. Typically after 30 generations in the synkaryon proliferation, about 7 out of 46 human chromosomes remain, with a range of about one to 20. By correlating the loss of certain genes which are expressible in tissue culture with the loss of visible chromosomes, one can correlate visible chromosomes with linkage maps. It is essential to use tissue-culture-expressed genes or DNA markers, not something like eye colour genes.

When a whole chromosome is lost, all genes on it will be lost, so one can find out which loci are syntenic. If only part of a chromosome is lost, say from a spontaneous big terminal deletion, one can correlate genes to a particular end or region of a chromosome. In the 1980's, physical mapping largely took over from other methods in humans, with massive sequencing of all chromosomes in the human gene project from the late 1990's.

One can fuse sheep lymphocytes or fibroblasts to mutant cell lines of mouse or hamster, using polyethylene glycol or Sendai virus, using rodent cell lines with multiple enzyme deficiencies. During hybrid cell growth, sheep chromosomes are preferentially eliminated. This can be used for mapping by testing hybrid cells for growth in a range of nutrient media, checking which sheep chromosomes can complement particular mouse enzyme deficiencies. For sheep physical maps, see Broad, Hayes and Long (1997); for sheep linkage maps, see Montgomery and Crawford (1997).

One can also use cell fusions between individuals of the same species, or between an individual and a reference cell line, to do *cis-trans* tests for functional allelism (Chapter 1.3.8), to see at which locus an individual carries a mutation.

8.6.3. Hybridisation probes

In molecular genetics, the term **probe** is used for a piece of single-stranded DNA or RNA used in DNA-DNA or DNA-RNA hybridisation assays. It usually has a radioactive label, with hybridisation products being identified through autoradiography, where the radioactivity causes dark spots on sensitive film. Fluorescent or luminescent labels may be used instead, where the probe can be detected under the microscope with the right lighting conditions, rather than by autoradiography. The basis of probing is that the single-stranded DNA or RNA probe will, under the right conditions, anneal by base pairing to a complementary sequence in DNA which has been made single-stranded.

If one has cloned a particular gene and wants to find its chromosomal location, it can be grown up in radioactive medium to get it labelled, then it is denatured into single strands by heating or chemical treatment. A preparation of chromosomes at mitosis is then made where individual chromosomes can be seen. The chromosomal DNA is denatured to single strands, then the DNA probe is added. Conditions are made to favour annealing of the probe to complementary sequences in the chromosomes, then the excess probe is washed off, so that only base-paired probe remains. For radioactive probes, autoradiography follows, to see where the silver grains from radioactive decay occur on the chromosome. In a diploid with a single-copy gene, one would expect one point to be labelled on two homologous chromosomes, showing where the DNA sequence of the probe is on the chromosomes. If the gene has many copies, as for tRNA genes, many points will be labelled, which helps to identify such multi-copy genes and to find whether they are limited to particular chromosomes or regions.

8.7. A practical use of molecular markers in agriculture

Molecular markers can be used for the **indirect selection** of phenotypic agricultural traits if they are closely linked to that character, and if both markers are segregating in a population. The European Apple Genome Mapping Project (EAGMAP) is led by the Horticultural Research International at East Malling, Kent, and Wellesbourne, Warwickshire, developing a reference linkage map for *Malus*. The molecular markers used are isoenzymes and several kinds of DNA markers, including randomly amplified polymorphic DNA (RAPD), restriction fragment length polymorphisms (RFLPs) and microsatellites. The microsatellites are often simple tandem repeat polymorphisms (see Chapter 8.5), which are ubiquitous, highly polymorphic and co-dominant, needing very small amounts of DNA for analysis (King et al., 1996).

The EAGMAP reference population is a cross between Prima, a cultivar susceptible to rosy leaf curling aphid (*Dysaphis devecta*; see Chapter 16.3), and Fiesta (formerly Red Pippin), a resistant cultivar which carries the dominant resistance allele Sd_1 from Cox's Orange Pippin, which is heterozygous for it. Segregation in the reference population gave 75 resistant plants to 62 susceptible, a good 1:1 allele ratio from $sd_1 sd_1$ x $Sd_1 sd_1$. Roche et al. (1996) also scored segregating RFLP and RAPD markers. Three RFLP markers from Fiesta were very closely linked to Sd_1, within 2 cM. Any one of them could therefore be used to screen plants from a segregating population for the aphid resistance soon after germination (Plate 20), before growing plants to maturity. Roche et al. (1996) comment, however, that the use of these markers is too laborious for routine screening, and that a PCR-based assay might be more convenient for **marker-assisted selection**. A visible morphological marker which is closely linked to aphid resistance and which could be scored in seedlings would be even simpler to use. Susceptibility to the aphid can be scored at the seedling stage anyway, but having large populations of pests for testing can be inconvenient.

Suggested reading

Bradshaw, H. D. and R. F. Stettler, Molecular genetics of growth and development in Populus. IV. Mapping QTLs with large effects on growth, form, and phenology traits in a forest tree. *Genetics* (1995), 139, 963–973.

Broad, T. E., H. Hayes and S. E. Long, Cytogenetics: physical chromosome maps, pp 241–295, in: Piper, L. and A. Ruvinsky, eds. *The Genetics of Sheep*. 1997. CAB International, Wallingford.

Fincham, J. R. S., P. R. Day and A. Radford, *Fungal Genetics*, 4th ed. 1979. Blackwell, Oxford.

Foss, E. et al., Chiasma interference as a function of gene distance. *Genetics*(1993), 133, 681–691.

Griffiths, A. J. F., W. M. Gelbart, J. H. Miller and R. C. Lewontin, *Modern Genetic Analysis*. 1999. W. H. Freeman and Co., New York.

Hartl, D. L. and E. W. Jones, *Genetics. Principles and Analysis*, 4th ed. 1998. Jones and Bartlett, Boston.

Khush, G. S. and C. M. Rick, Cytogenetic analysis of the tomato genome by means of induced deficiencies. *Chromosoma* (1968), 23, 452–484.

King, G., N. Periam and C. Ryder, Development of microsatellite markers in the Rosaceae. *Annual Report* 1995–1996, p 64. Horticulture Research International, Wellesbourne, Warwickshire.

Korol, A. B. and I. A. Preygel, *Recombination Variability and Evolution*. 1994. Chapman and Hall, London.

Lamb, B. C., Ascomycete genetics: the part played by ascus segregation phenomena in our understanding of the mechanisms of recombination. *Mycological Research* (1996), 100, 1025–1059.

Montgomery, G. W. and A. M. Crawford, The sheep linkage map, pp 297–351, in: Piper, L. and A. Ruvinsky, eds. *The Genetics of Sheep*. 1997. CAB International, Wallingford.

Nicholas, F. W., Genetics of morphological traits and inherited disorders, pp 87–132, in: Piper, L. and A. Ruvinsky, eds. *The Genetics of Sheep*. 1997. CAB International, Wallingford.

Paterson, A. H., ed., *Genome Mapping in Plants*. 1996. Academic Press, London.

Roche, P. et al., Molecular markers linked to aphid resistance. *Annual Report* 1995–1996, pp 64–65. Horticulture Research International, Wellesbourne, Warwickshire.

CHAPTER 9

STRUCTURAL CHROMOSOME ABERRATIONS: THEIR ORIGINS, PROPERTIES AND USES

9.1. Introduction

In Chapter 1.3, we saw that **chromosome aberrations** were changes in the chromosomes on a scale larger than a single locus, with changes within a locus being called mutations. They may involve changes to just one chromosome, with the loss of a region of the chromosome (**deletion**), reversal of the order of loci in part of the chromosome (**inversion**), or duplication of parts (**duplication**). They may also involve two non-homologous chromosomes, as in **translocations**, which may be reciprocal or non-reciprocal. Some chromosome aberrations are usable by plant and animal breeders, while others cause infertility or other problems in micro-organisms, plants, animals and humans. Male gametes tend to be more sensitive than female ones to the adverse effects of chromosome aberrations which are therefore often transmitted more through females than through males. This chapter is concerned with changes in chromosome structure; changes in chromosome number are covered in Chapter 10.

All the kinds of aberration described here occur in fungi, plants, animals and humans. For example, in man Down syndrome (an unbalanced aberration) can be caused by a translocated chromosome 21 on chromosome 13 or 14, in addition to two normal 21s, or by a 21 on 21 translocation plus one normal 21 (Chapter 13.8.2). Aberrations may be **balanced**, with no change in the total amount of genetic material, as in a translocation of part of chromosome 1 onto chromosome 12 (Plates 30 and 31), or **unbalanced**, as in the addition of part of chromosome 5 to chromosome 17, in the presence of two normal chromosome 5's (Plate 32). Unbalanced aberrations usually have harmful phenotypic effects from gene-dosage imbalance, while balanced ones may be neutral or harmful.

The production of chromosome aberrations involves spontaneous or induced **double-strand breaks** of DNA (see Chapter 7.2 for which types of mutagen cause these). Aberrations may also involve the rejoining of broken ends of chromosomes; this occurs frequently because the broken ends act as if they were "sticky". In the following diagrams the wild-type chromosome is usually considered to have a sequence of regions (not loci, but longer stretches) *ABCDoEFGH*, where *o* represents the centromere. The regions involved in the aberration will be shown in **bold typeface**. Any **acentric** fragment (without a centromere) cannot segregate properly at mitosis and will eventually be lost, with the **centric** fragment being retained and usually segregating normally.

Some aberrations arise from **illegitimate crossing-over**, where supposedly non-homologous regions cross-over, say a crossover starting between *B* and *C* in one homologue of *ABCDoEFGH* and ending between *F* and *G* in the other homologue (see Fig. 9.4.1). Some small genetic elements, such as transposons and insertion sequences, are present many times in some genomes and can thus provide points of homology in otherwise non-homologous regions within and between chromosomes, as can repeated gene sequences; they may promote **illegitimate crossovers** (i.e., between generally non-homologous regions), as in Fig. 9.4.1.

Chromosome aberrations change map distances, or even whether two loci are syntenic or non-syntenic. Some balanced aberrations may have phenotypic effects just because they move genes into different parts of the chromosome (**position effects**). Thus moving a gene from the **euchromatic** middle of a chromosome arm to near the centromere, into a **heterochromatic region** (one with a different time of DNA condensation in the cell cycle from normal euchromatic DNA), may affect function and therefore affect the phenotype even if there is no change in what genes are present or how many times they are present. Aberrations changing the **gene dosage** (the number of times a gene is present) can change the phenotype in ways which are difficult to predict. Aberrations are normally only microscopically visible if extending over at least 4 Mb, although probes can detect much smaller aberrations.

9.2. Deletions

A **deletion** involves the loss of a region of the chromosome and is also called a deficiency. A single break (spontaneous or induced) causes a **terminal deletion**, with the acentric fragment being lost. For example, wild-type chromosome *ABCDoEFGH* would give centric deletion *CDoEFGH* if there were a break between *B* and *C*, and would give deletion *ABCDo*, with an almost terminal centromere, if the break occurred between the centromere and *E*.

To get an **interstitial** (non-terminal) **deletion** requires two breaks and the rejoining of the outer ends. Thus if *ABCDoEFGH* has a break between *E* and *F*, and another between *G* and *H*, one could get the interstitial deletion form, *ABCDoEH* if the *E* end joins up with the *H* end. As interstitial deletions require two independent breaks and a rejoining, whereas terminal deletions only require one break, interstitial deletions will occur more rarely than terminal deletions. The map will change, as regions *E* and *H*, previously separated by *F* and *G*, are now adjacent, so will have fewer crossovers between them. Heritable deletions cannot involve the centromere, because acentric fragments are lost.

Because deletions lack lengths of chromosome, they usually involve the loss of some essential genes, making them lethal when homozygous. Even in heterozygotes (one chromosome normal, one with a deletion), the deletion may be lethal in the gamete or other haploid stage if expression of some of the genes lost is needed in the haploid stage. Because a heterozygote for a deletion is hemizygous for those loci on the wild-type chromosome which are missing from the deleted chromosome, any recessive lethals or deleterious or beneficial recessives in that region of the wild-type chromosome will be **pseudodominant** (Chapter 8.6), and will show. Except for inbreeding organisms, most diploids carry some deleterious recessives hidden in heterozygotes, so if there are any in the hemizygous part of the wild-type chromosome, there will be a loss of fitness and possible lethality.

In Chapter 8.6, we saw how induced terminal deletions could be used in mapping loci to visible chromosomes. Deletions are not much used by breeders as they are generally deleterious. Unless they are deleterious in the gametes, they should not directly affect fertility, although they may affect viability. Unless very small, they can often be detected cytologically at meiosis in heterozygotes because the wild-type undeleted chromosome will have no homologue to pair with in the region of the deletion, so the wild-type chromosome forms an **unpaired loop**, as shown in Fig. 9.2.1. Such loops can also be seen in *Drosophila* salivary gland chromosomes, as they show somatic pairing of homologues even though they are not in meiosis.

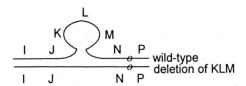

Figure 9.2.1. The unpaired loop formed at zygotene of meiosis in a heterozygote for an interstitial deletion, of *KLM*. The loop is in the wild-type chromosome, *IJKLMNoP*, with *o* representing the centromere. Each strand represents two chromatids if at meiosis, but only one chromosome if this is somatic paring as in *Drosophila* salivary glands. Recessive alleles in regions *KLM* can show pseudodominance and be expressed as they are hemizygous.

An example of a terminal deletion in humans is the **Cri-du-Chat syndrome**, caused by heterozygosity for a deletion of the end of the short arm of chromosome

5. Affected individuals give a persistent cat-like cry, are severely abnormal physically and mentally, and usually die young. It is very rare, 1 in 100,000, normally arising anew in the germline of a parent of each sufferer, and would be lethal if homozygous. Plates 30 and 31 show a terminal deletion of chromosome 1, with the missing part translocated non-reciprocally on to chromosome 12, giving a balanced genotype.

9.3. Inversions, paracentric and pericentric; their effects on fertility

An **inversion** involves the turning around of a section of chromosome. It requires one break if terminal or two breaks if interstitial, and the joining of ends. Thus *ABCDoEFGH* with a break between *C* and *D*, and joining of the unbroken end *A* with broken end *D* after the fragment has turned round, gives *CBADoEFGH*. This is a **paracentric inversion** (para = beyond) as the centromere, *o*, is not within the inversion. If instead *ABCDoEFGH* had two breaks, one between *B* and *C* and one between *F* and *G*, with turning round of the broken central section and joining of the broken ends, one gets *ABFEoDCGH*. This is a **pericentric inversion**, (peri = around) because it includes the centromere. The map distances between various loci, such as *B* to *F*, or *B* to *C*, change considerably.

Inversions can affect fertility and fitness, and are common in many wild organisms and in evolution. Species often differ from each other by inversions, which can act as partial isolating mechanisms through reducing fertility in heterozygotes. Within a species, inversions may be **fixed** in some populations, with all individuals homozygous for the inversion, and **floating** in others, where the population is polymorphic for an inversion. Wild tulips, flies and grasshoppers have frequent inversions, and they occur in all kinds of organism.

Their effects on **fitness** (not fertility) are variable, sometimes being deleterious, sometimes beneficial, sometimes neutral, and fitness may differ between homozygotes and heterozygotes for the inversion. For example in *Drosophila pseudoobscura* an inversion called Chiricahua, in the third chromosome, had a relative fitness of only 0.4 when homozygous, but 1.0 when heterozygous with wild-type (called Standard), compared with 0.9 for wild-type when homozygous, so there was heterozygote advantage (see Chapters 12.1 and 15.3) for fitness with this inversion, even though it had a reduced fitness when homozygous. There must be cyclic selection during the year, since at Pinon Flats, California, the relative frequencies of Standard, Chiricahua and another third chromosome rearrangement, Arrowhead, changed during the year. Although Standard was the most common type in most months, Chiricahua was the most frequent in June. See references in Ayala and Kiger (1980).

Inversions do not usually affect fertility in homozygotes because in inversion/ inversion diploids, both homologues can pair normally with each other, and crossovers have no adverse effects. **Inversion heterozygotes**, however, usually have reduced fertility unless they have adapted by suppressing crossovers in the inverted section. We saw that with heterozygous deletions at meiosis, one chromosome looped out into an unpaired region. In heterozygous inversions, however, both homologous chromosomes are almost fully paired because **both chromosomes loop**, enabling homologous pairing within the inverted region. Segregation is normal if there is no crossover within the inverted region. A crossover within the inverted region, however, produces 50% of inviable gametes, as shown in Fig. 9.3.1. For simplicity, this shows the two chromosomes as *ABoCD/acobd* (a **pericentric inversion**) and as *oABCD/oacbd* (a **paracentric inversion**), with heterozygosity for inversion of region *bc*, and different centromere positions relative to region *ABCD* in the two kinds of inversion.

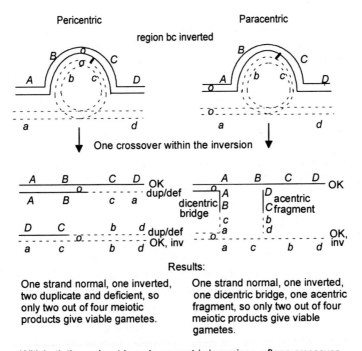

Results:

One strand normal, one inverted, two duplicate and deficient, so only two out of four meiotic products give viable gametes.

One strand normal, one inverted, one dicentric bridge, one acentric fragment, so only two out of four meiotic products give viable gametes.

With both the pericentric and paracentric inversions, after a crossover within the heterozygous inverted region, only the non-recombinant products give viable gametes.

Figure 9.3.1. The consequences of crossing-over at meiosis within heterozygous inversions, pericentric and paracentric. Each line represents a chromatid, except for the heavy short line which indicates a reciprocal crossover at pachytene.

Duplications in some **fungi** such as *Neurospora* and *Ascobolus* get severely attacked immediately before meiosis, with premeiotic deletions or repeated induced point mutations inactivating the duplicated region. In any organisms, duplications could cause further chromosome aberrations from unequal crossing-over, whether they are homozygous or heterozygous, as there are more than two regions of homology which could pair and cross-over. Such illegitimate cross-overs could occur at meiosis, or with much lower frequencies, at mitosis. They can be between or within chromosomes.

9.5. Translocations, single and multiple

A **translocation** occurs when part of one chromosome is inserted in or attached to a non-homologous chromosome. One or two breaks could cause a terminal or interstitial part of one chromosome to break off, say *XYZ* from *PQoRSTUVWXYZ*, and a break in a non-homologous chromosome, *ABCDoEFGH*, could allow it to be inserted with rejoining of the ends, giving say *ABXYZCDoEFGH*, or it could attach to one end of an intact non-homologous chromosome, as in *ABCDoEFGHXYZ*. Those would be **non-reciprocal translocations**. Plates 14, 30, 31 and 32 show human non-reciprocal translocations. Many translocations are **reciprocal**, caused by illegitimate crossing-over between non-homologous chromosomes. For example, a crossover could occur between *F* and *G* on *ABCDoEFGH*, and *W* and *X* on *PQoRSTUVWXYZ*, giving two reciprocally translocated products, *ABCDoEFXYZ* and *PQoRSTUVWGH*. Reciprocal translocations are also called **interchanges**, and do not change the number of copies of each locus present.

Translocations will obviously change genetic maps, as previously non-syntenic loci are now syntenic and may even be closely linked, as for *W* and *G* in the above example. Even balanced reciprocal translocations which involve no change in which genes are present may affect phenotypes and fitness through **position effects**: see Chapter 9.1. Wild populations polymorphic for translocations are common in snails, grasshoppers, and the plants *Datura* (thorn apple) and *Oenothera* (evening primrose).

Whether or not translocations affect phenotypes, they have strong effects on **fertility** when heterozygous. In homozygotes for translocations, pairing is usually normal, but recombination frequencies between pairs of loci may be altered. In heterozygotes for reciprocal translocations, complete homologous pairing at zygotene of meiosis requires that all four chromosomes come together to form a **quadrivalent** as in Fig. 9.5.1. In the quadrivalent shown, all unbroken lines are paired with homologous unbroken lines, and all broken lines have broken line partners.

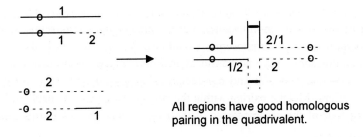

All regions have good homologous pairing in the quadrivalent.

At anaphase, the two centromeres of type 1 must go to opposite poles, and the two centromeres of type 2 must go to opposite poles. If the centromeres of 1 and 2 go to one pole, the centromeres of 1/2 and 2/1 must go to the other pole; both daughter cells have balanced genomes, one untranslocated, 1 and 2, one translocated, 1/2 and 2/1. If the centromeres 1 and 2/1 go to one pole and centromeres 2 and 1/2 go to the other pole, both daughter cells will be duplicate/deficient, giving inviable gametes. For example, a cell with 1 and 2/1 is missing the acentric end of 2 and has two copies in the haploid gamete of the acentric end of 1.

Figure 9.5.1. Pairing in meiosis in a heterozygous reciprocal translocation between chromosomes of pair 1 and pair 2, so that there is one normal chromosome 1, one normal chromosome 2, a centric part of 1 plus an acentric part of 2, and a centric part of 2 plus an acentric part of 1.

Figure 9.5.1 shows why half the meiotic products of a heterozygous reciprocal translocation are inviable duplicate/deficient combinations. One gets partial sterility, with distorted pollen grains or sperm. Two crossovers are shown in the figure, in positions having no effect on chromatid segregation. In a diploid with one heterozygous reciprocal translocation, one would get just one quadrivalent at meiosis, plus normal bivalents for the other pairs of chromosomes, so partial sterility plus one quadrivalent at meiosis can be used to identify this condition. More than one reciprocal translocation, if heterozygous, will give more than one quadrivalent, but homozygous translocations just give bivalents (two chromosomes paired).

Extraordinary cases of **multiple translocations** have been identified in a number of plants, especially in the evening primrose genus, *Oenothera*. For example, *Oenothera erythrosepala* cells at mitosis have 14 separate chromosomes, so 2n = 14. In meiosis one would expect seven bivalents. Instead one gets one bivalent for chromosomes 1 and 2, and a **ring** of twelve chromosomes because there have been six successive interchanges. Even more surprising is that the species is **permanently heterozygous** for the interchanges, with a system of **balanced lethal recessives**.

The 12 chromosomes making up the ring form two separate units, the *velans* and *gaudens* sets of chromosomes. Let us number the chromosome arms, so the untranslocated set of chromosomes would be 1/2, 3/4, 5/6, 7/8, 9/10, 11/12, 13/14. Some chromosomes have no translocation, but there have been reciprocal (5/6 with 7/8 to give 5/8 and 6/7) and non-reciprocal (e.g., 3/4 with 11/12 to give 4/12) translocations. Meiotic pairing within the ring at zygotene and anaphase I is shown for just four of the chromosomes of the ring in Fig. 9.5.2. All arms can be fully paired.

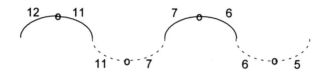

Only four of the 12 chromosomes making up the ring are shown.
Unbroken lines: part of the velans set of chromosomes.
Broken lines: part of the gaudens set of chromosomes.
The numbers represent chromosome arms, so that 5/6 and 11/12 are untranslocated chromsomes and 7/11 and 6/7 have translocations.
o represents a centromere. The four chromosomes are shown above at early anaphase I of meiosis.
At zygotene, they would have been paired as follows:

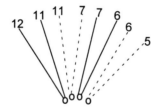

Figure 9.5.2. Meiotic pairing within part of the ring chromosome complex in *Oenothera erythrosepala*.

The *velans* complex is made up of chromosomes 3/4, 12/11, 7/6, 5/8, 14/13 and 10/9, and is *l1⁻*, *L2⁺*, with a recessive lethal allele, *l1⁻*, and a dominant non-lethal allele at a different locus, *L2⁺*. *Gaudens* is made up of 4/12, 11/7, 6/5, 8/14, 13/10 and 9/3, and is *L1⁺*, *l2⁻*, carrying the opposite recessive lethal, *l2⁻*. These lethals only act in the diploid, not the haploid, and the only viable gamete combination for the zygote is *velans* with *gaudens*, genotype *L1⁺ l1⁻*, *L2⁺*, *l2⁻*, when each recessive lethal is hidden by the corresponding non-lethal dominant allele. Homozygous *velans* plants die because they are homozygous for lethal *l1*; homozygous *gaudens* plants die because they are homozygous for lethal *l2*. One would therefore expect only 50% of seeds to be viable, whereas only about 40% are viable. The extra deaths are caused by unbalanced gametes, e.g., from non-terminal crossovers.

For these ring-chromosome systems to survive such high lethality, they must have some major compensatory advantages. The most obvious is that one can have **permanent hybrid vigour** from heterozygote advantage, say with the *velans* set carrying one allele and the *gaudens* set carrying a different allele, for many different loci. The ring system is excellent for hybrid vigour (see Chapter 12.1 and 15.3) and maintaining already well-adapted lines. It usually prevents favourable gene combinations from being split up by recombination.

The ring can be broken up by certain crossovers, giving new variants. In fact, about 2% of the seedlings from *Oenothera erythrosepala* are so different from their parents that they could be classified as new species or subspecies. Different ways of **breaking the ring** give different forms. Rings can break down into bivalents, and bivalents can re-evolve rings. In *Rhoeo discolor*, there are no bivalents at meiosis, just a ring of 12 chromosomes. Some other plants, such as *Campanula percicifolia* and *Paeonia*, can have rings but without a balanced lethal system, so homozygous individuals can occur for each ring complex.

Special multiple translocation stocks are used in some fungi for mapping, especially for assigning loci to chromosomes, e.g., in *Neurospora crassa*. See for example *Catalogue* (1998, especially page 159).

Suggested reading

Ayala, F. J. and J. A. Kiger, *Modern Genetics*, 2nd ed. 1980. Benjamin Cummings, Menlo Park, California.

Bahl, P. N., ed., *Genetics, Cytogenetics and Breeding of Crop Plants. Vol. 2. Cereals and Commercial Crops*. 1997. Science Publishers, Inc., Enfield, New Hampshire.

Catalogue of Strains, 7th ed. *Supplement to Fungal Genetics Newsletter*, no. 45, 1998. Fungal Genetics Stock Center, Kansas.

Connor, M. C. and M. Ferguson-Smith, *Essential Medical Genetics*, 5th ed. 1997. Blackwell Science, Ltd., Oxford.

Griffiths, A. J. F., W. M. Gelbart, J. H. Miller and R. C. Lewontin, *Modern Genetic Analysis*. 1999. W. H. Freeman and Co., New York.

Hartl, D. L. and E. W. Jones, *Genetics. Principles and Analysis*, 4th ed. 1998. Jones and Bartlett, Boston.

Jahier, J. et al., *Techniques of Plant Cytogenetics*. 1996. Science Publishers, Inc., Enfield, New Hampshire.

CHAPTER 10

CHANGES IN CHROMOSOME NUMBER: THEIR EFFECTS AND USES

10.1. Background

As we saw in Chapter 1.3, the **ploidy** is the number of copies of each typical chromosome in the nucleus. **Haploids** have one copy of each chromosome, **diploids** have two, **triploids** three, **tetraploids** four, **pentaploids** five, **hexaploids** six, etc., with those having three or more copies being **polyploids**. The number of different types of chromosome (counting the sex chromosomes as one type) in the nucleus is the **basic number of chromosomes**, and a set of one of each type of chromosome makes up a haploid **genome**. The **basic number** has the symbol **X**, unfortunately the same as for a sex chromosome. Symbol **n** represents the gametic number of chromosomes and **2n** is the adult number unless the adult is haploid. So 2n can = 2X, 3X, 4X, etc. In the rose genus, *Rosa*, different species have 2n chromosome numbers of 14, 21, 28, 35, 42 and 56. Clearly, X = 7, so these chromosome numbers represent 2X, 3X, 4X, 5X, 6X and 8X, with all but the diploid being polyploid. Some people use a different system, with 3n for the triploid, 4n for the tetraploid, etc.

Euploids have an exact multiple of the basic number of chromosomes, but **aneuploids** do not have an exact multiple. For example in hyacinths, varieties with 16 chromosomes are euploid diploids, 2n = 2X, but the variety Rosalie has 17, so is an aneuploid of the diploid, with 2n = 2X + 1. Another aneuploid has 23 chromosomes, so 2n = 3X - 1, an aneuploid of the triploid.

Polyploids make up about 60% of Monocots and 40% of Dicots, and more than one third of all domesticated plants. In animals, there are some polyploid snails and worms, fish, amphibia and reptiles, but polyploidy is very rare in higher animals. In humans, triploids and tetraploids occur at high frequencies in spontaneous abortions (Chapter 13.9).

There is an important distinction between two types of tetraploid. **Autotetraploids** have four of each type of chromosome, from a self-doubling in a diploid, so if each type of chromosome is represented by a letter, they are AAAA, BBBB, CCCC, etc. They tend to have reduced fertility as all four of each type of chromosome try to

153

pair at meiosis, often forming quadrivalents and having irregular segregations. **Allotetraploids**, however, come from crosses of different species or very different strains in which previously homologous chromosomes have diverged in evolution. If two diploid species with different genomes, AA in one and BB in the other, cross by chance, they usually form a sterile diploid, A,B; this is sterile because the two sets of chromosomes are no longer homologous and therefore cannot pair at meiosis, giving **univalents** (unpaired chromosomes), not bivalents. Sometimes in this sterile diploid hybrid there is a failure of cell division after DNA replication and hence a doubling of the chromosome number. The resulting allotetraploid, with genomes AA, BB, is usually fertile because each chromosome of the A genome now has a homologue with which to pair, and so does each member of the B genome; therefore there are **bivalents** (pairs of chromosomes) at meiosis, and correct segregation of chromosomes. See Chapter 12.5 for the use of allopolyploids in interspecific hybrids.

Polyploids may have different ecological niches and distributions from diploids. If annual diploids form polyploids, these are often perennial. Polyploids are often more vigorous than diploids, and allopolyploids may have a kind of permanent, pure-breeding hybrid vigour, with say genome AA homozygous for one allele of a locus, say *P P*, and genome BB homozygous for an alternative allele, say *p p*. Polyploids tend to be hardier than diploids in extreme conditions, and are often short-day flowering (Chapter 17.1.1). Woody plants tend not to be polyploid. Polyploids are often evolutionarily a short-term success, but then tend to be stagnant, with a reduced chance of new recessive mutations showing in the phenotype.

In plant breeding, it is often best to do the basic work with diploids or even monoploids (Chapter 10.3), especially if you seek new beneficial mutations. In *Dahlia*, most garden plants are octaploid, 8X, double allotetraploids, but behave like double autotetraploids. According to Lawrence (1968), not once in 170 years of cultivation has an abrupt new mutation been found. Only dominants with **xenia** (one dominant shows fully in the presence of any number of recessive alleles) would show, and the chance of *A A A A A A A a* ever giving *a a a a a a a a* is remote. Similarly, a mutation from *A A A A A A A A* to *A A A A A A A A'*, with additive action, might go unnoticed. There is a lot of genetic variation in garden dahlias but it is released primarily through recombination and segregation, not mutation. Their crossovers tend to be localised to the ends of chromosome arms.

10.2. Changes in ploidy

Increases in ploidy usually result from a failure of cell division (mitosis), so that the chromosomes replicate but the nucleus fails to divide, so the chromosome number doubles. Thus a haploid yeast cell could fail to divide after DNA replication, giving a diploid vegetative cell, or a diploid rye cell could give a tetraploid cell. These events could affect the next generation if occurring in germline cells. This mechanism produces **even-number multiples** of the basic chromosomes number. The plant breeder can treat seeds or growing points (Plate 15) with certain chemicals, such as colchicine from the autumn crocus, or synthetic colcemid, to double the chromosome number, identifying affected shoots by altered morphology, and then checking their chromosome numbers by microscopy.

Odd-number multiples can be produced by fertilisation of a diploid gamete by a haploid gamete, or by fertilisation of a haploid gamete by two haploid gametes instead of by one. A diploid gamete could come from a tetraploid, or by division failure in a gamete from a diploid. In higher plants, two of the four products of female meiosis usually fuse to form a diploid polar body in the ovule. This diploid nucleus is then fertilised by one of the two gametic nuclei from the pollen, to form a triploid nucleus. In maize and other plants, this divides rapidly to form the **triploid endosperm** which surrounds the **diploid embryo** and forms a major food reserve for the embryo at germination.

Unreduced chromosome numbers in gametes have spontaneously given rise to various cultivars. An unreduced gamete with 14 chromosomes from the diploid raspberry (*Rubus idaeus*, 2n = 2X = 14) fertilised a normal gamete with 28 chromosomes from the octoploid *R. vitifolius* (2n = 8X = 56), giving the hexaploid loganberry, 2n = 6X = 42. The blackberry "John Innes" is tetraploid, 2n = 4X = 28, coming from an unreduced gamete with 14 chromosomes from the diploid *R. rusticanus inermis* (2n = 2X = 14) when crossed to the tetraploid, *R. thyrsiger* (2n = 4X = 28); see Lawrence (1968).

In some *Hymenoptera* (bees, wasps, ants), fertilised eggs give diploid females but unfertilised eggs give **haploid (monoploid) males**. In honey bees, 2n = 32 for females, but male bees (drones) only have 16 chromosomes. In the haploid male, meiosis is modified to be more like mitosis, with gametes still having one complete genome, 16 chromosomes, not 8. The sex ratio is therefore not the usual 1 female: 1 male, but is strongly biased in favour of females, and is determined by the proportion of eggs which the queen bee allows to be fertilised by sperm which she stores after mating. The difference between the sterile female diploid worker bees and the fertile diploid female queen is a matter of nutrition, with the feeding of royal jelly to larvae to produce queens.

In plants, one occasionally gets unfertilised female gametes developing (**parthenogenesis**), giving haploid progeny, but male gametes do not have enough cytoplasm to form an embryo. This kind of effect has been exploited by potato breeders, with a kind of phantom pollination. **Potatoes** are derived from wild Andigena autotetraploids of *Solanum tuberosum*, but autotetraploids are inconvenient for basic breeding work, especially with recessive alleles. It is possible to go from the tetraploid to the diploid by pollinating the autotetraploids with pollen from selected clones of the diploid *Solanum phureja*. Both the pollen gametic nuclei fuse with the polar bodies, starting endosperm development which stimulates the unfertilised egg to divide and develop parthenogenically, giving the required diploids of *S. tuberosum*. Those diploids can then be crossed with other diploid species to introduce desirable genes, e.g., additional disease resistance. Once the diploid has been suitably improved, it can be returned to the autotetraploid state by treatment with colchicine. Crosses of diploid *S. tuberosum* to compatible strains of *S. phureja*, and making autotetraploids with colchicine from the sterile diploid, could be used to exploit hybrid vigour (see Chapter 12.1 and 12.6). Parthenogenesis in tetraploid *Medicago* (alfalfa) and in potatoes has sometimes given diploid embryos without pollination.

10.3. Monoploids

Monoploids, where the adult is haploid, are sometimes useful to plant breeders because deleterious and beneficial recessives can be detected since dominance does not hide them. Monoploids were first reported in 1924 from natural populations of the thorn apple, *Datura stramonium*, arising from unfertilised eggs. Like most monoploids, they were weak, small and had very low fertility, and soon died out. As there is only one copy of each homologue, meiosis gives only univalents, with irregular segregation of chromosomes, so most gametes have a number of essential chromosomes missing and are inviable.

Monoploids have been used in breeding maize and tomatoes, particularly for selecting against deleterious recessives. In maize, $2n = 2X = 20$, so the monoploid has a complete genome of 10 different chromosomes. Gametes by chance get from 0 to 10 chromosomes, with intermediate numbers being most common. Getting a fertile seed depends on a very rare fertile ($n = 10$) male gamete fertilising a very rare fertile ($n = 10$) female gamete. Crossing monoploids is much less successful that restoring the diploid state by doubling up the chromosome number by using colchicine or other agents, as previously discussed. That gives fully homozygous diploids without the necessity of having to make inbred lines over several generations. They can be used to make F1 hybrids (Chapter 12.1).

Monoploids can now be made for most plants by **anther culture** (androgenesis). Anthers at a particular stage are put on nutrient agar with certain growth hormones, and often are given a mild heat-shock, so that some microspore (not pollen) cells which have undergone meiosis divide to produce small plantlets. By changing the nutrient agar and hormones, the plantlets can be made to grow roots, and eventually whole rooted plants or explants can be cultured. After selection of the best haploid phenotypes, colchicine is applied to the growing points, and diploid side-shoots (see Plate 15) are selected by eye, then tested cytologically. For a commercial *Brassica* example, see Ockendon (1984). He obtained yields of up to 357 embryos per 100 anthers at the 4 to 5 mm anther length stage, using a thermal shock of 16 hours at 35°. Complete plants were regenerated mainly from hypocotyl explants, and haploids (smaller flowers, < 20% pollen stainability) and diploids (>70% pollen stainability) were easily distinguished at flowering (Plate 16). Although a few plants were triploid or tetraploid, about half were haploid and about half were diploid. Anther culture of diploid potatoes (see above) can produce haploids. In many species, the success rates from anther culture vary widely between different varieties, and many unwanted albinos are produced.

Haploids can sometimes be produced by **chromosome elimination**, as in haploidisation in the fungal parasexual cycle (Chapter 18.4). When two diploid barleys were crossed, *Hordeum vulgare* x *H. bulbosum*, the *H. bulbosum* chromosomes were preferentially eliminated from the diploid hybrid, giving haploid *H. vulgare* plants. Embryo culture is needed because the endosperm fails to develop. This method was used to make the barley variety "Mingo", after doubling up the chromosomes of the selected haploid.

Even in plant monoploids or haploid fungi, not all genes are necessarily expressed. This is true of hypostatic genes in the presence of an epistatic allele at another locus (Chapter 2.2) and applies to inducible genes in the absence of inducers, and to conditional mutants under permissive conditions (Chapter 3.2.2).

10.4. Diploids

Diploids are very frequently used for the basic breeding work in plants and in all commercial higher animals. Unless there are chromosome aberrations or sterility mutations, they are usually fertile, with only bivalents at meiosis. If diploids are not inbred, problems in identifying the best genotypes can arise because of dominance.

10.5. Triploids

Triploids can be produced by crossing tetraploids with diploids, or by chance double fertilisation of gametes, or fusion of a haploid gamete with an unreduced gamete from a diploid. Not all tetraploids have fully fertile gametes, but even some fertile gametes will be enough in crosses to diploids to get triploids. Triploids are sexually isolated from their diploid or tetraploid parents, because triploids are usually largely sterile. With three copies of each chromosome trying to pair at zygotene, there are **trivalents**, sometimes plus bivalents and univalents, with most gametes receiving unbalanced sets of chromosomes and being sterile.

This sexual sterility can be useful if viability, growth and fruiting are satisfactory. In diploid bananas there are hard black seeds to spit out, but the triploids set fruit very well, with aborted seeds forming small dots in the flesh. The triploid bananas are easily propagated vegetatively (see Chapter 17.1.5). Similarly, diploid water melons have lots of seeds, and consumers generally prefer triploid (or sometimes aneuploid) varieties, and almost seedless watermelons are usually cultivated now. Some apples are also triploid.

10.6. Tetraploids

The different origins of **autotetraploids** and **allotetraploids** were discussed in Chapter 10.1. Man-made allotetraploids, from doubling up the chromosome number of a diploid hybrid, are sometimes also known as **amphidiploids**. Tetraploids are the commonest natural polyploids, especially autotetraploids, and allotetraploids have been important in evolution. Tetraploids are often hardier and more vigorous than their parental diploids, with larger cells. For example in the saxifrage, *Saxifraga pensylvanica*, the lower leaf epidermis cell area is 1,600 μm^2 in diploids and 2,740 in tetraploids.

Tetraploids sometimes have larger and showier flowers and fruits than diploids, so are useful in horticulture, e.g., tetraploid Easter lilies and *Antirrhinums*. There may also be useful physiological changes, such as more ascorbic acid (vitamin C) in tetraploid cabbages and tomatoes, while potatoes are normally tetraploid.

If one has a mixture of ploidies and wishes to identify particular types, then initial karyotyping from root tips to find say diploids, triploids and tetraploids is essential. Once they have been characterised, then **other features** which are quicker to determine than chromosome numbers can be used. In sugar beet, the tetraploid beet has a shorter, thicker and darker leaf than the diploid or triploid. If one stains a strip of the lower leaf epidermis with dilute iodine in dilute potassium iodide solution, the tetraploid has more plastids (shown by starch-grain containing bodies) per pair of guard cells than the triploid, which has slightly more than the diploid. In rye-grass (*Lolium perenne*) and turnip (*Brassica campestris*), one can distinguish diploids from tetraploids from the length of the leaf stomata. In members of the

Solanaceae such as potatoes (*Solanum*) and tobacco (*Nicotiana*), one can treat pollen grains with sulphuric and acetic acids to get swelling from the germ pores; the diploid plants usually have pollen grains with three germ pores while tetraploids mainly have four germ pores (some grains have 3, 5 or 6 pores).

The crops best suited for autotetraploidy are those with a low chromosome number, which are harvested for their vegetative parts, which are cross-pollinated, perennial, and are reproducible vegetatively, e.g., rye-grass (*Lolium*), where the tetraploid usually has vegetative reproduction in the grazed sward but has enough sexual seed for planting out and for breeding experiments.

Although autotetraploids tend to have reduced fertility, breeders can sometimes improve fertility by selection. For example, the forage grass *Dactylis glomerata* is in the wild a rather variable, partly-fertile autotetraploid, but breeders at Aberystwyth, Wales, have selected for fertility, producing commercial strains with nearly 100% sexual fertility. In autotetraploids at meiosis, an **adjacent** (four in a ring) arrangement of chromosomes usually gives 2:2 segregation of the four chromosomes of a quadrivalent, giving fertile gametes. In contrast, a **linear** arrangement of chromosomes in pairing may give unbalanced gametes from 1:3 segregations to the poles of the cell at anaphase.

Tetraploids of all kinds are **genetically isolated** from their diploid relatives because even if crosses with them are fertile, they generally produce sexually sterile triploids. Diploid pollen, from tetraploids, cannot usually compete in the style with the faster-growing haploid pollen. Male gametes are much more sensitive to chromosomal unbalance than are female gametes, so unbalanced chromosome combinations are more likely to be transmitted by females rather than males.

If a plant has no known relatives for chromosome number comparisons, how does one tell if it is an autotetraploid or an allotetraploid? Having a chromosome number which is a multiple of four is expected in tetraploids but is not diagnostic, as many diploids have that. At mitosis, an autotetraploid should have four copies of each visible type of chromosome, such as the nucleolar-organiser chromosome which is often long and has a centromere and a non-staining nucleolus-organising region. Unless the chromosomes of its two parental lines have visibly diverged, allotetraploids should also have four of each type of chromosome. Looking at pollen grains is helpful because autotetraploids usually have many aborted grains, but so do triploids, aneuploids, and plants with pollen-lethal alleles. Looking at meiosis is best as diploids and allotetraploids have bivalents, with the chromosomes in pairs, while triploids have trivalents, with the chromosomes in threes, and autotetraploids have quadrivalents, with sets of four chromosomes. Another method is to look at **segregation ratios** in various crosses.

Autotetraploids have four copies of each chromosome and of each locus, giving five possible genotypes per locus, instead of the three for diploids, and will show **tetrasomic inheritance**. If one allele is dominant to any number of the alternative allele, it shows **xenia**; e.g., *A A A A*, *A A A a*, *A A a a* and *A a a a* all have the same phenotype, where *A* shows xenia. The names of the five tetraploid genotypes and the ratios of tetrasomic inheritance are shown in Table 10.6.1.

Table 10.6.1. Tetrasomic inheritance in autotetraploids.

Genotype	Name	Gametes if balanced (random combinations)	Test-cross phenotype ratio (when crossed to *a a a a*)	Selfing phenotype ratio
A A A A	quadriplex	all *A A*	all *A*	all *A*
A A A a	triplex	1 *A A*: 1 *A a*	all *A*	all *A*
A A a a	duplex	1 *A A*: 4 *A a*: 1 *a a*	5 *A*: 1 *a*	35 *A*: 1 *a*
A a a a	simplex	1 *A a*: 1 *a a*	1 *A*: 1 *a*	3 *A*: 1 *a*
a a a a	nulliplex	all *a a*	all *a*	all *a*

The ratios in this table are correct when no crossing-over occurs in the interval between the locus and the centromere, but there are additional complications if all possible crossovers are considered. Monohybrid ratios such as **5:1 from test-crosses** and **35:1 from selfings** are clearly different from standard diploid ratios.

Allotetraploids are normally fertile and include many cultivated plants such as tobacco, plum, macaroni wheats and sugar cane. They are functionally diploid for most characters, unless the two different genome types interact. By incorporating genetic material from two different species, they are wonderful new material for selection and evolution. Even the inbreeding allohexaploids like bread wheat can be highly heterozygous and breed true for that heterozygosity, because different alleles were often present (as homozygotes) in the three species which formed the allohexaploid. See Chapter 12.5.1 for more on wheat, interspecific hybridisation and allopolyploids.

One can get **segmental allotetraploids**, where some segments of the two component genomes are in common, and others differ. At the Royal Botanic Gardens, Kew, *Primula verticillata* (genomes VV, 2n = 18) from the Arabian peninsula spontaneously crossed with *Primula floribunda* (genomes FF, 2n = 18) from the Western Himalayas. The seed gave a hybrid, genomes VF, 2n = 18, which was sexually sterile because the V chromosomes could generally not pair with the F chromosomes. The chromosome number spontaneously doubled to give *Primula kewensis*, 2n = 36, VV,FF (Plate 17). This is only partially fertile because V chromosomes occasionally pair with F chromosomes; there are occasional VVFF quadrivalents giving irregular chromosome segregation. There is clearly some homology left between the V and F genomes, so the new species is a segmental allopolyploid, part-way between an autotetraploid and an allotetraploid.

Lotus corniculatus, bird's foot trefoil, forms regular bivalents, with no quadrivalents, so is strictly allopolyploid at the cytological level. A few loci, however, show tetrasomic inheritance, with a locus for cyanogenesis giving a 35:1 ratio from a duplex selfing, so there must be a few chromosome segments in common to the two genomes of this segmental allopolyploid, but not long enough to cause homologous pairing. Thus two homologous chromosomes might carry segments *MNOAP* while two different homologous chromosomes, which do not pair with them at meiosis, carry *CDEAF*, so that there are four copies of the *A* segment or locus between the two genomes, giving tetrasomic inheritance for *A* but not for *MNOP* or *CDEF*.

10.7. Higher polyploids

In some plant groups, polyploidy does not go higher than tetraploidy in nature, but in *Rumex* (docks), it goes up to 20X. Many Pteridophytes are highly polyploid.

Bread wheat, *Triticum aestivum,* is a double allohexaploid, 2n = 6X = 42, X = 7, genomes AA, BB, DD; see Chapter 12.5 for its origin as an interspecific hybrid. Normally there is only homologous pairing at meiosis, say A1 with A1, A2 with A2, B6 with B6. There is however a locus, *Ph*, on chromosome B5 which determines whether there is also **homeologous pairing** (pairing of related chromosomes from different genomes) at meiosis I, in which A6 can pair with B6 or D6, A1 with B1 or D1, chromosomes for which the residual homology is normally too little to cause pairing. With homeologous pairing, one can get hexavalents (six chromosomes attempting to pair) in hexaploid meiosis. This usually causes sterility, but genes such as the one on wheat's B5 could control exchanges between genomes in polyploids. In Chapter 11.2 we shall see how homeologous paring in *Lolium* can cause fertility in a diploid hybrid and sterility in an allopolyploid. The *Ph* gene is sometimes called a diploidising gene because the normal allele makes the polyploid have chromosome pairing in bivalents, like a diploid.

10.8. Loss or gain of single chromosomes: aneuploids, monosomics and trisomics

Aneuploids have an inexact multiple of the basic chromosome number, with **monosomics** of diploids having 2X - 1, **trisomics** having 2X + 1, **double trisomics** having 2X + 2, etc. Aneuploids are often used in horticulture, especially in bulbs such as hyacinths which can be propagated vegetatively. They are also used as intermediates in some chromosome-substitution breeding techniques (Chapter 10.9). Human examples include Turner syndrome, 2n = 45, with one sex chromosome missing, and Down syndrome, 2n = 47, with three copies (trisomy) of chromosome 21 (Chapter 13.8 and Plates 35 and 36). Turner syndrome sufferers are normally sterile, as are Down syndrome males, but Down females can be fertile. There cannot be complete chromosome pairing at meiosis in individuals with an odd number of chromosomes.

Monosomics of diploids tend to be weak, with reduced fertility, but trisomics of diploids sometimes, but not always, have increased vigour. In *Datura stramonium*, the thorn apple, there are twelve pairs of chromosomes, 2n = 2X = 24, so there are twelve different trisomics, each with a different chromosome in excess. All twelve trisomics can be distinguished from each other and from the diploid from their fruit morphology. In *Datura*, the extra chromosome is rarely transmitted through the pollen because the n pollen (2.6 mm/h) grows faster than the n + 1 pollen (1.9 mm/h from the "Cockleburr" trisomic) down the style.

Compared with diploids, polyploids in plants are usually better balanced against the effects of the loss or gain of a chromosome, so 4X - 1 is often as vigorous as 4X, although its fertility is usually reduced. In animals, polyploidy and aneuploidy are usually very harmful, and are not normally used in breeding techniques.

10.9. Chromosome manipulations and substitutions

Although artificial allopolyploids have been of limited use in developing agricultural cultivars, they make excellent intermediates in transferring genes from alien species into cultivars. One usually only wants the **desired allele** transferred, but if there is little recombination between the cultivar and the alien chromosome, then the whole chromosome may be retained, or one can try transferring just part of the alien chromosome to a cultivar chromosome by translocation, perhaps following irradiation. Several backcrosses of a cultivar x alien species hybrid to the cultivar may be necessary to get rid of the unwanted alien genes, while selecting for the desired alien gene.

The main techniques involve **chromosome additions** or **chromosome substitutions**, or a combination of both methods, and may involve manipulation of recombination control of homeologous pairing (see Chapter 10.7 above, and Thomas, 1993). An example is the production of yellow-rust-resistant bread wheat cultivar "Compair" by Riley and others. Hexaploid "Chinese Spring" wheat, 2n = 6X = 42, is susceptible to yellow rust, *Puccinia striiformis*, but the wild diploid *Aegilops comosa*, 2n = 2X = 14, has a gene for resistance to this rust. The two species were crossed, giving a fertile hybrid which was backcrossed for several generations to "Chinese Spring". The backcrossing, with selection each generation for rust resistance, got rid of all *Aegilops* chromosomes except for the one carrying the resistance. There were therefore all 42 "Chinese Spring" chromosomes and the one alien one, M2, which corresponded to homologous group 2 in "Chinese Spring".

The *Ph* locus on wheat chromosome 5B suppressed homeologous pairing, so there was no crossing-over between M2 and wheat chromosomes which could introduce the resistance locus into a wheat chromosome while reducing the presence of unwanted alien genes. The M2 addition line was then crossed to *Aegilops speltoides*, a diploid, 2n = 2X = 14, in which the *Ph* locus is inhibited, giving a 29 chromosome hybrid which was repeatedly backcrossed to "Chinese Spring". A rust-resistant plant with 42 chromosomes and 21 bivalents at meiosis was isolated, heterozygous for the *Yr8* yellow-rust-resistance gene from *Aegilops comosa*, and not expressing undesirable other properties from *A. comosa*. The plant was selfed to obtain plants homozygous for the resistance, thus producing the useful new cultivar, "Compair". It did not carry *A. speltoides*'s suppression of *Ph*, which would have made "Compair" have homeologous pairing and thus be mainly sterile.

Chromosome addition lines were produced from crossing *Agropyron intermedia* (perennial intermediate wheat grass, a hexaploid with 2n = 42, but with genomes not closely related to those of *Triticum*) with hexaploid bread wheat. The hexaploid hybrid was weak, with low fertility. It was backcrossed twice to bread wheat, with selection for fertility and for rust resistance. A stable, fertile, rust-resistant octaploid line was obtained with 42 wheat chromosomes and 14 *Agropyron* chromosomes but it was low yielding. Further backcrosses to bread wheat gave a series of single chromosome addition lines of *Agropyron* chromosomes to wheat, including one with the rust gene. Reducing the alien chromosome contribution improved yield. Addition lines tend to be unstable, often tending to lose the alien chromosomes, especially from monosomic lines such as 2n = 43 = 6X wheat + 1 alien.

Although addition lines are not usually stable enough to use as cultivars, they can be used in making substitution lines, where an alien chromosome is substituted for one from the crop species. Wheat cultivars grown in Germany in the 1930's were shown to be wheat/rye substitution lines, with mainly wheat chromosomes. A *Triticale* (wheat x rye) line was used in a crossing programme, with selection for rye's rust resistance, giving spontaneous substitution lines. Alien chromosomes will only substitute for ones in the same homeologous group.

The following scheme shows how a stable substitution line was obtained with 2n = 42, made up of 40 wheat chromosomes and one pair of rye chromosomes in the same homeologous group as the missing wheat chromosome pair (i.e., 6X wheat - 2 wheat + 2 rye). A monosomic line, 2n = 41, gave 20 bivalents and one univalent at meiosis, and gametes with 20 or 21 chromosomes. A disomic addition line had 2n = 44, with 21 wheat bivalents and one rye bivalent, giving gametes with 22 chromosomes, made up of 21 wheat and one rye chromosome. When such a gamete fertilised a gamete with 20 chromosomes from the monosomic line, the plants had 2n = 42, with 40 pairing wheat chromosomes, one unpaired wheat chromosome and one unpaired rye chromosome. This produced gametes with 20 wheat, 21 wheat, 20 wheat + 1 rye, and 21 wheat + 1 rye chromosomes. It was backcrossed to the disomic addition line, and when the latter's 21 wheat + 1 rye chromosome gamete fused with a 20 wheat + 1 rye chromosome gamete, a hybrid was produced with 2n = 43, from 40 paired wheat chromosomes, one unpaired wheat chromosome and pair of rye chromosomes. When this was selfed, some plants had 40 paired wheat chromosomes, no unpaired wheat chromosomes and one homologous a pair of rye chromosomes, giving a **stable substitution line** with good agricultural properties.

By crossing monosomic lines of say cultivar A (e.g., 2n = 41 = 6X - 1) to a euploid cultivar B (e.g., 2n = 42 = 6X), and backcrossing about seven times to the monosomic line, one can **transfer single chromosomes** from B to A, getting rid of the other B chromosomes to see if any substitution lines have desirable properties.

Often one wants to transfer a single gene between species. Addition lines were used to transfer a whole chromosome from *Aegilops umbellulata* to wheat, but only a gene for leaf-rust resistance was wanted. Dry seeds were irradiated to induce a translocation of part of the *Aegilops* chromosome to a wheat chromosome. A reciprocal translocation achieved this, but gave a duplicate/deficient chromosome. This was tolerated in the hexaploid but would not be tolerated in the diploid. It is better to get recombination between the alien chromosome and a wheat chromosome, by suppressing gene *Ph* on chromosome 5B, rather than having a translocation, if possible.

Autopolyploids can also act as **bridge species** in interspecific gene transfers. Tetraploid rye-grass (*Lolium*) x hexaploid fescues (*Festuca*) can give pentaploid hybrids, from fusion of a diploid gamete with a triploid gamete. These pentaploids can be crossed as males to diploid rye-grass, with backcrossing to diploid rye-grass.

Whole fescue chromosomes are eliminated during the backcrossing, but some fescue genes are retained on rye-grass chromosomes after crossovers.

Suggested reading

Bahl, P. N., ed., *Genetics, Cytogenetics and Breeding of Crop Plants. Vol. 2. Cereals and Commercial Crops*. 1997. Science Publishers, Inc., Enfield, New Hampshire.

Ignacimuthu, S., *Plant Biotechnology*, 1997. Science Publishers, Inc., Enfield, New Hampshire.

Jellie, G. J. and D. E. Richardson, eds., *The Production of New Potato Varieties: Technological Advances*. 1987. Cambridge University Press, Cambridge.

Lawrence, W. J. C., *Plant Breeding*. 1968. Edward Arnold, London.

Lupton, F. G. H. ed., *Wheat Breeding: its Scientific Basis*. 1980. Chapman and Hall, London.

Mayo, O., *The Theory of Plant Breeding*, 2nd ed. 1987. Oxford University Press, Oxford.

Ockendon, D. J., Anther culture in Brussels sprouts (*Brassica oleracea var. gemmifera*). I. Embryo yields and plant regeneration. *Annals of Applied Biology* (1984), 105, 285–291.

Poelman, J. M. and D. A. Sleper, *Breeding Field Crops*, 4th ed. 1995. Iowa State University Press, Iowa.

Thomas, H., Chromosome manipulation and polyploidy, pp 79–92 in: Hayward, M. D., N. O. Bosemark and I. Romagosa, eds., *Plant Breeding. Principles and Prospects*. 1993. Chapman and Hall, London.

CHAPTER 11

SUPERNUMERARY ("B") CHROMOSOMES

11.1. Definition, origins, numbers and size

Supernumerary ("B") chromosomes are additional to the normal "A" chromosomes and differ from them in a number of ways, especially in being inessential and variable in number. Their effects on phenotypes and fertility are so variable that it is difficult to generalise about them.

11.1.1. Origin

B's arise by accidents to normal A chromosomes, especially by breakage and loss of genetic material, when centric fragments may become B chromosomes.

11.1.2. Numbers

B's may be present in some members of a population and absent in others, and may differ between populations too. When they are present, they may be present in different numbers in different individuals. Even in plants and insects regularly having B's, they are not essential. Selection tends to favour a given number of B's in a particular environment, but irregular segregation means that an optimum number cannot be uniformly maintained.

They occur widely in insects such as beetles and grasshoppers, in hundreds of higher plant species, especially cereal grasses, and have been found in some fungi. Wild populations polymorphic for low numbers of B's are found in organisms as diverse as platyhelminths, shrews, insects and many grasses. The number of B's per cell can even vary within a plant; e.g., in *Sorghum* and *Poa alpina* B's tend to get lost in root mitosis but are stable in the stem and reach the germ-line tissues. Karyotyping for B chromosomes from root-tip squashes could therefore be misleading.

Even numbers of B's are more common than odd numbers, and low numbers are more common than high numbers. Diploid rye, 2n = 14, can have up to 10 B's, while the autotetraploid rye, 2n = 28, can have up to 12 B's, although increasing numbers of B's usually give increasing loss of fertility. Maize, 2n = 20, is unusual in that it can have more than 20 B's per cell. B's can be a significant part of the total genetic material: in rye, 2n = 14, with 8 B's, the B's increased the nuclear DNA by 40%.

B's do not usually pair with A's, not even those they were derived from. They may or may not pair with other B's, and will not do so unless they have regions of homology. Often they do not segregate regularly even when present in even numbers, and they may be transmitted differently through male and female gametes, in plants and animals. A parent with one B may give 0, 1 or 2 B's in different offspring, because of irregular segregation at meiosis.

11.1.3. Size

Arising by loss of genetic material, B's are on average shorter than A's. They are often, but not always, **heterochromatic**; i.e., their DNA condenses at different times of the cell cycle compared with normal DNA.

11.2. Effects on phenotype and fertility

B's were originally thought to be inactive, but C. D. Darlington and colleagues published a paper on "The activity of inactive chromosomes", showing that B's did affect phenotypes. In general, cell volume increases with increasing numbers of B's, but many quantitative characters of economic or ecological interest show strange zigzag patterns when phenotype values are plotted against the number of B's. Odd numbers of B's seem to depress phenotype values, and even numbers tend to increase them, as shown in Fig. 11.2.1.

Breeders who find unexpected amounts of phenotypic variation between individuals, especially in cereal grasses, should check cytologically for differences in the numbers of B chromosomes between individuals. Ecological effects of B chromosomes have been reported. For example, white spruce (*Picea glauca*) has different typical numbers of B's in different geographical locations.

B's usually affect **fertility**, often with high numbers of B's giving sterility, although maize can tolerate 20 B's without losing fertility. In many plants, increasing numbers of B's increasingly delay flowering, so those with high numbers of B's could be partly isolated reproductively from the main population, but with an increased chance of crossing with each other. In *Plantago coronopus*, a plantain, having one B gives male sterility but not female sterility. Pollen with the most B's is sometimes the most successful in fertilisation, as if B's can be advantageous in

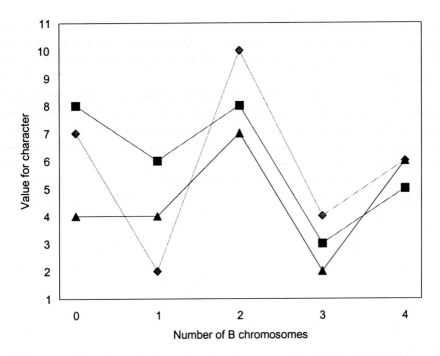

Figure 11.2.1. The effect of different numbers of B chromosomes on plant quantitative characters in rye (*Secale*). The lines represent different quantitative characters such as straw weight, plant weight and tiller number: the aim is to show typical effects, so it does not matter which line relates to which character. All measures are in arbitrary units.

the gametophyte even though they are often disadvantageous in the sporophyte.

One important evolutionary effect of B chromosomes, and one which can be of use to plant breeders, is the effect of B's on **chiasma** and crossing-over frequencies. B's sometimes increase chiasma frequencies, as in maize and rye, and sometimes decrease them, as in *Lolium*. If one wanted recombinants in rye, one could obtain them by using plants with B chromosomes, which raise chiasma frequencies, then selecting for recombinants lacking B chromosomes for a more stable product.

In *Lolium*, a widely used meadow rye-grass genus, B's suppress homeologous pairing (pairing between related chromosomes of different genomes), with important effects on fertility. *Lolium temelentum* (genomes TT) and *Lolium perenne* (genomes PP) are both diploids with 2n = 14. When crossed, they give a hybrid, TP, 2n = 14. With no B's, there is good homeologous pairing, e.g., T5 with P5, giving a fertile hybrid with seven bivalents at meiosis, with T and P genomes able to exchange genes through crossing-over. If the diploid hybrid has B chromosomes, homeologous pairing is suppressed, giving little pairing of T and P genomes, mainly with univalents at meiosis, so the hybrid is sterile. If one doubles up the hybrid's chromosomes, say with colchicine, the allotetraploid has TT, PP, 2n = 28. If there are no B's, the allotetraploid behaves like an autotetraploid, with quadrivalents from homeologous pairing of TT with PP chromosomes, and so is largely sterile. If the allotetraploid has B's, then these suppress homeologous pairing, just leaving bivalents from homologous pairing of T with T, P with P, so the allotetraploid is then fertile, behaving like an allotetraploid.

B's could in nature or cultivation convert largely sterile autotetraploid-behaving hybrids to fertile allotetraploid-behaving hybrids. In planned breeding programmes, one has to take account of the irregular distribution of B chromosomes to the offspring.

Suggested reading

Jones, R. N. and H. Rees, *B Chromosomes*. 1982. Academic Press, New York.

CHAPTER 12

BREEDING METHODS AND EXAMPLES

This chapter is about various methods used in plant and animal breeding. For reproductive physiology and crossing methods, see Chapter 17. For types and uses of selection, see Chapter 5. For genotype, phenotype and breeding values, see Chapter 3.4. For blood group in farm animals, see Chapter 13.10.3.

12.1. Using hybrid vigour

12.1.1. Definition of hybrid vigour (heterosis)

Hybrid vigour (heterosis) occurs when the hybrid offspring are better than either parental line for a particular trait or traits. The parents considered are usually inbred lines, but could be different species. The classic case is the mule (2n = 63) which is hardier than either parent, a mare (female horse, 2n = 64) and a he-ass (male donkey, 2n = 62), although it is sexually sterile as not all chromosomes can pair at meiosis. Nearly all maize grown in the USA is hybrid maize, from crossing selected inbred lines, because the F1 hybrids have faster growth and higher yields than pure lines.

F1 hybrids are used extensively in maize, kale, sprouts, sheep, cattle, pigs, tomatoes, and many garden flowers and vegetables. The inbreeding of plants which normally have self-incompatibility, such as many *Brassica* species, may require special techniques such as hand-pollination at the bud stage, before the self-incompatibility develops. Hand-pollination is expensive to do.

12.1.2. Explanations of hybrid vigour

There are two explanations for hybrid vigour, both of which are correct some of the time; they are not mutually exclusive. On the **dominance hypothesis**, the masking of harmful recessive alleles by dominant alleles improves yield. For example, inbred line 1 might have genotype *A A, b b, C C, D D* and inbred line 2 might have *A A, B B, C C, d d*, where *b* and *d* are harmful recessives. The F1 hybrid would then be *A A, B b, C C, D d*, in which harmful recessives *b* and *d* are both hidden by the

171

corresponding dominant alleles, giving the F1 hybrid a better yield than either inbred parent. On selfing the F1 hybrid, the average yield would be reduced because the harmful recessives would, in some of the offspring, be homozygous and expressed.

On the **overdominance (heterozygote advantage) hypothesis**, the hybrid is fitter than either parent because the heterozygote is phenotypically different from and superior to both homozygotes; i.e., *B b* is fitter than either *B B* or *b b*. This could happen if the two alleles specified different products, with one dose of two different products being better through complementation than are two doses of a single product. It could also happen if the recessive allele is inactive, and one dose of the dominant allele product is better than two doses. On selfing the F1 hybrid, the average yield would again fall, as on the dominance hypothesis, but because heterozygosity was being lost.

While the practice of using F1 hybrids works whichever hypothesis is most often applicable, the two hypotheses have different long-term implications for breeding aims and breeding economics. On the dominance hypothesis, one could breed an ideal high-yielding pure-breeding type by breeding out harmful recessives, to get say *A A, B B, C C, D D*. On the overdominance hypothesis, that would be impossible, as the ideal type would be highly heterozygous and could not be pure breeding, e.g., *A A, B b, C C, D d*. The hybrid would have to be made afresh each generation, unless it could be propagated vegetatively.

If plant or animal breeding firms produced a pure-breeding ideal type, they could sell it initially to farmers, but the farmers would never again have to buy that line from the breeders, as they could propagate their own stock for future generations. If the ideal type was highly heterozygous, the farmers would have to buy new stock each generation from the breeding firms which grew the parental inbred lines. One Dutch spinach breeder was describing the advantages of his F1 hybrid spinach seed, and his first-listed "benefit" was that "The hybrid excludes the growing of further generations independent of the original breeder." The breeding firms therefore prefer the overdominance hypothesis to be right, while the farmers would prefer the dominance hypothesis! The experimental evidence generally suggests that the dominance hypothesis is more often right than the overdominance one. For plant breeders' rights, see Chapters 12.9 and 19.2.6.

12.1.3. Typical F1 hybrid breeding programmes

The breeders obtain a number of lines of the chosen crop, from a range of diverse habitats and geographical areas. They propagate and inbreed these lines, selecting for desirable characters, qualitative and quantitative, and against undesirable ones.

The inbreeding exposes desirable and undesirable recessive alleles, which might have been hidden in heterozygotes, to selection. Over a number of generations this produces a series of more or less homozygous pure-breeding lines with desirable characters, but they are usually fairly low yielding because of inbreeding depression (Chapter 6.2).

Different pairs of inbred lines are then crossed to produce a range of F1 hybrids. These will be genetically uniform within one cross and will be highly heterozygous if the parental inbred lines were quite different genetically. Many will show hybrid vigour, with increased yields compared to the inbred lines. Crossing inbred line 1 with inbred line 2 produces F1 hybrid 1,2; if this has a high yield, it can be tested for commercial use. Such a simple **two-way cross** has the disadvantage that the F1 hybrid seed is itself produced on a low-yielding inbred line, so the seed or stock will be expensive, in addition to all the developmental costs. One solution is to use **three-** or **four-way** crosses, so that the final F1 hybrid seed (it is still called F1 hybrid seed even if it is really F2) is borne on high-yielding hybrids. For example, lines 1 and 2 can be crossed to produce hybrid 1,2, which is used as female in crosses to line 3, producing three-way cross hybrid 1,2,3. Alternatively, 1 can be crossed with 2, and 3 can be crossed with 4, then 1,2 is crossed with 3,4 to give a four-way cross F1 hybrid 1,2,3,4. In both cases, the final seeds for sale, 1,2,3 or 1,2,3,4, are borne on high yielding F1 hybrids, reducing seed costs, but there will be some loss of uniformity compared with a two-way cross, as segregation can occur in the hybrid parent (1,2 and/or 3,4) of the three-way or four-way cross hybrid seed.

Even six-way crosses have been used in some crops such as kale, because of complications due to self-incompatibility. An F1 hybrid **coconut tree**, intermediate in height between its normal and dwarf parents, is shown in Plate 18. Unlike typical F1 hybrids, it is not produced by crossing inbred lines, but its advantages include easier harvesting of coconuts from its intermediate height, and bearing nuts about five years earlier than do tall palms, as well as some hybrid vigour. The yield is similar to that of tall palms. In **farm animals**, one cannot self-fertilise an individual, so mating of close relatives, such as sib-mating, is practised.

12.1.4. Hybrid maize production

Zea mays is a natural outbreeder, producing separate terminal male inflorescences near the top of the plant, and female inflorescences in the lower leaf axils, with the stamens ripening before the stigmas. **Maize** plants of diverse genetic backgrounds are obtained from different areas and are selfed for number of generations, selecting for desired characters such as disease resistance and grain colour and composition. The breeder tries some selection for yield, as inbreeding this natural outbreeder gives inbreeding depression. Deleterious and beneficial recessives are made homozygous and are then expressed, permitting selection. This initial selection within and between inbred lines gives definite improvements, but the big increase in yield mainly occurs at the hybrid stage.

Not all crosses of inbred lines give increased yields, so the breeders test their best selected inbred lines for **general combining ability**, usually by a "top cross" to a common tester pollen stock, itself an inbred line. The inbreds giving the best general combining ability are then crossed to each other in different combinations to find which pairings give the best **specific combining ability**. Two-, three- and four-way crosses are tried. The best combinations are then multiplied up and tested over a number of years in a range of agricultural situations and areas, to find the best commercial F1 hybrids to sell for particular areas.

The results of a typical maize F1 programme are shown in Table 12.1.4.1. It is obvious that the big increase in yield comes at the F1 hybrid stage, with 2-, 3- and 4-way crosses all increasing yield by a factor of about 2.7 compared with the fully inbred parental lines. If the seeds from random pollination of the F1 hybrids are sown, the yield declines, as shown in Table 12.1.4.1, because heterozygosity is lost (overdominance hypothesis) and/or because deleterious recessives can become homozygous (dominance hypothesis). The decline in yield is different for 2-, 3- and 4-way crosses, with the percentage decline roughly matching the inbreeding coefficient after the random pollination.

Table 12.1.4.1. Results of a typical maize F1 hybrid programme.

Cross type	Lines crossed	Yield in bushels per acre (1 bushel per acre is about 90 litres per hectare)				
		Average yield of parents, $F = 1$	Yield of F1 hybrid, $F = 0$	Offspring yield after random pollination of F1 hybrid	Percentage decline in yield after random pollination	F after random pollination
2-way	1 x 2	24	63	44	48	0.500
3-way	1,2 x 3	24	64	49	37	0.375
4-way	1,2 x 3,4	25	64	54	26	0.250

This **decline in yield** after using the original F1 seeds is why fresh hybrid seeds are needed each generation. One drawback of using F1 hybrids in some less developed parts of the world is that farmers often will not buy fresh seeds each year, just using seeds from their F1 hybrids, with declining yields. This is especially true if farmers are initially supplied with free or cheap F1 hybrid seeds under an aid programme, but are later expected to pay commercial rates.

Hybrid maize doubled yields in the USA between 1925 and 1955 and accounts for nearly all maize planted there now, but there were no commercial **hybrid wheats** until 1975, and most wheats are still pure lines, not F1 hybrids. Wheat and maize are both wind pollinated, but pollination is less efficient in wheat. Hand-pollination is uneconomic for cereals, so inbred lines to be hybridised must be planted in close proximity, with some system for avoiding self-fertilisation. In wheat, one has to plant alternate rows of the inbreds acting as male and female, so only half the rows bear seeds. In maize, one only needs one row of males to two rows of females. The system used in wheat to make males and females only gives 75% pollination, and it is difficult to ensure that males and females are fertile simultaneously. For such reasons, F1 hybrid seed in wheat is about four times as expensive as pure line seed, which is usually too much compared with the increased yield of the F1 hybrids.

The system devised to get separate males and females in maize is very ingenious and even with one row of males to two rows of females, it gives almost 100% pollination. It involves cytoplasmicly-determined male sterility, where cytoplasmic factors S (male sterile) and F (male fertile, normal) show **maternal inheritance**, being transmitted through the ovules but not the pollen. There are two nuclear alleles, R, a dominant restorer of male fertility even with S cytoplasm, and r, non-restorer. In a 4-way cross, using 1,2 x 3,4, one could have inbred line 1 with S, r r, line 2 with F, r r, line 3 with S, r r, and line 4 with F, R R.

In an area away from other maize crops, line 1 would be interplanted for crossing with line 2. Because of S with r r, line 1 would be male sterile and with r r, F, line 2 would be male fertile. Seeds collected on line 1 cobs must therefore be hybrid, 1,2, with genotype S, r r, and seeds collected on line 2 must have been selfed. Similarly, line 3 would be interplanted with line 4 for crossing. Because of S, r r, line 3 would be male sterile, while line 4, F, R R, would be male fertile. Seeds collected on line 3 must be hybrid, 3,4, while seeds from line 4 would be selfed. The hybrids 1,2 (S, r r, male sterile) and 3,4 (S, R r, male fertile because of the restorer gene) are interplanted for crossing, again away from any other maize plants. Seeds collected from male-sterile 1,2 are the commercial F1 hybrid seed, 1,2,3,4, and seeds from 3,4 are selfed. Because 3,4 is heterozygous, R r, the commercial seed consists of about equal numbers of male-fertile *S, R r* and male-sterile *S, r r* types, but both are female fertile and having half the plants male fertile is enough to pollinate both genotypes.

Similar cytoplasmic male-sterile plus dominant nuclear restorer systems have been used in **other organisms**, such as onions, sugar beet, sorghum, rye and sunflowers. In biennial crops harvested for their vegetative parts, as in onions and sugar beet, there is no need for the F1 seeds to give sexually fertile plants. In farm animals, self-fertilisation does not occur, so no precautions are needed against selfing: one just has one sex of one line mated to the opposite sex of the other inbred line. As one would expect, male sterility is usually selected against, and it is very difficult to find good combinations of nuclear restorers and cytoplasmic male sterility genes in the wild or in crop organisms. Once a suitable system has been found in a crop, it tends to be used widely, which may have unfortunate consequences from pests or pathogens.

In the early 1970's, most maize stocks had the same cytoplasm, from a Texas strain, and the **Southern Corn Blight** fungus (*Helminthosporium*) developed the ability to attack plants with that cytoplasm, costing about 20% of maize production in 1970 in the USA. As a result, many growers abandoned the cytoplasmic male sterility system for ensuring crossing between inbred lines. Instead, the lines to be used as male-sterile parents had their tops cut off mechanically before the anthers matured, as all the male flowers are in the terminal tassels; hybrid seeds were collected from the detasselled lines. In sugar beet F1 hybrid production, Owen cytoplasm is used world-wide, but fortunately there has been no disease problem like the one associated with maize Texas cytoplasm.

12.1.5. Hybrid sprout production

F1 hybrid Brussels sprouts are widely grown, but so are pure lines. To do selfing requires hand pollination at the bud stage to overcome self-incompatibility, which makes the inbred line seed very expensive.

Although these data are old, from 1974 at Asmer Seeds, Ormskirk, Lancashire, they show the relative costs of different seed very clearly. Non-F1 commercial seeds cost £11 a kilo, needing 1.1 kg per ha. The seeds gave very variable plants, and only the best were transplanted into the fields from the seed beds, so seed costs were about £12 per ha, plus transplanting costs. F1 hybrid seeds cost £176 a kg, but plants were uniform and good, so only 0.14 kg were needed per ha, giving seed costs of about £25 a ha, with no transplanting costs as they could be planted directly in the fields. Unlike with onions, seed costs are not a major part of the sprout production costs and the increased yield of the F1 hybrid compared with non-F1 lines makes it often worthwhile to buy F1 hybrid seed. Because of the expense of hand bud-pollination, the parental inbred lines' seeds for making the F1 hybrids cost about £550 a kilo. To avoid bastardisation of hybrid seed production from wild brassicas, two boys were trained to go out with binoculars to spot and destroy any such plants within a mile (1.6 km) of the seed fields. See Chapter 10.3 and Plates 15 and 16 for the production of homozygous diploid sprout lines from monoploids from anther culture.

12.1.6. Hybrid animals

In **fowls**, first crosses are made between different breeds or strains. The F1 are mated to a third breed or strain (three-way cross) or to another first cross (four-way cross) to produce vigorous second-cross birds for egg or meat production. Alternatively, inbred lines are made within breeds for egg production, and various F1 hybrids are tried between inbred lines. Only the most productive crosses (those which "nick" best) are used commercially.

In **cattle**, the most obvious effects of hybrid vigour are an increase in female fertility and milk yield, and in the growth rate of calves. The use of F1 cattle for meat is partly due to the fact that the improved milk supply gives faster growth and earlier maturity of the offspring. In Britain many beef herds consist of cows from inter-breed crosses mated back to bulls of one parental breed or to bulls of a third breed. In British beef breeds, the F1 hybrids are better than the two parental breeds' average by about 5% for growth and viability, and about 8% for cow fertility. Hybrid vigour is two to three times greater than that when British cattle are crossed with Zebus *(Bos indicus*, Brahman cattle; see Chapter 12.5.1). In American results with Hereford cows x Shorthorn bulls, with the F1 cows mated to Aberdeen Angus bulls, the hybrid calves weighed 212 kg at weaning compared with 176 kg for the pure breeds, with a final weight of 469 kg versus 414 kg for the pure breeds. See Chapter 12.7.3 for hybrid sheep examples.

12.2. Selection methods for inbreeders

12.2.1. Single line selection

Inbreeding, by self-fertilisation or crossing of close relatives, eventually gives pure-breeding lines homozygous for all loci. Not all individuals in habitual inbreeders will be homozygous for all loci, because there might have been some outcrossing or mutation. A diploid individual heterozygous at 10 loci could give $2^{10} = 1024$ different homozygous inbred lines.

If one has a population of inbreeders with some genetic variation, one can take a large number of superior selected individuals from that population, say of plants, and raise progenies from selfing them, each plant's progeny constituting a line. Further selection by the breeder between many lines is practised each generation, for qualitative and quantitative characters, killing off unsuitable individuals and only planting seeds from the best lines. Once a line has become homozygous at all loci, nothing can improve it further except new mutations or outcrossing to other lines.

The best lines after a number of generations of single line selection are used in replicated field trials for possible commercial use. This kind of single line selection has been used to improve the garden bean, *Phaseolus vulgaris*, and other inbreeders.

12.2.2. Using a mixture of lines, with agricultural mass selection

Another method is to plant a mixture of different lines of an inbreeder in the same field, and allow them to self-fertilise each generation. The breeder destroys any inferior looking plants before flowering, and harvests the bulk seed, not recording individual plant or line yields, unlike in the previous method. The seeds are sown for the next generation, with some destruction of visibly poor plants, and seeds are harvested. This continues for several generations and involves **agricultural mass selection**. This works because the highest yielders will be proportionately better represented in the next generation than lower yielders, because those having the most seeds, and seeds with the highest viability, will give rise to the most plants in the next generation.

The selection is therefore a combination of the breeder's deliberate actions and the agricultural environment selecting for plants best adapted to that particular set of field conditions and climate. Unlike the previous method, there is no labour-intensive recording of line yields. The final selected plants may often be a mixture of the best lines, and variation is beneficial from a disease-resistance aspect. Such a line mixture may have a disadvantage in that the breeder may not be able to register breeder's rights (a patent, see Chapters 12.9 and 19.2.6) on a mixture, depending on local registration rules. It would have a mixture of homozygous genotypes from selection between different inbreeding lines.

12.2.3. Bulk population breeding

This is very similar to the previous method except that one starts by hand-cross-pollinating **two different compatible inbreeding lines**, to be sure of having genetic variation on which to select. The F1 are allowed to self for several generations, with the practice of agricultural mass selection on the segregating offspring, plus the breeder "roguing out" undesirable types, such as ones which are small, weedy, deformed or disease-susceptible.

This method is very good for adapting proven varieties to new localities and environments, and has been widely used in wheat, other cereals, and beans. The initial cross between different lines could occur naturally in inbreeders which have occasional outcrossing, such as wheat.

12.2.4. Pedigree breeding

In **pedigree breeding** for inbreeders, one takes account of line averages as well as individual yields. One crosses two varieties which complement each other in useful ways, selfs the F1, then in the F2 one selects single plants with desirable traits, including good yield. One continues with selfing through several more generations, but recording individual yields. In the F3, 4, 5 and 6, one selects the best plants from the parents in the previous generation which gave the best average yields. For example, in the F4, one line might have an average yield of 20 units, with individual yields ranging from 14 to 27. Another line might average 25 units, ranging from 16 to 40, and a third line might average 30, with individuals ranging from 18 to 40. The best plants to select on the pedigree evidence would be those yielding 40 units from the line averaging 30 units as they are superior individuals from a superior family.

By the F6, most lines will be homozygous at nearly all loci, and significant further progress is unlikely. There will be little further response to selection between lines, so the best lines are bulked up for commercial testing. The best lines can also be crossed to some other good line, and another cycle of pedigree selection can be started. See Chapter 5.6 for general aspects of pedigree breeding.

12.3. Selection methods in outbreeders or random-maters

The techniques described here can be used with organisms having some degree of outcrossing, even if they are mainly selfing, such as wheat, as well as with outbreeders or random-maters. Rather confusingly, the term **"outbreeders"** is sometimes used just to mean non-inbreeders, such as random-maters.

In outbreeders, there will often be deleterious and beneficial recessives hidden from selection in heterozygotes. To get these to show for selection in plants or animals, one often uses some enforced inbreeding, by selfing or mating of close relatives, followed by selection as recessives become exposed in the phenotype. During this inbreeding, there will often be inbreeding depression, so one selects for vigour as far as possible to counter it. Having selected for particular desirable characters, one then restores some degree of heterozygosity at other loci by crossing different selection lines. This is what is done in F1 hybrid production, but it can also be used to produce superior lines which are not remade as F1 hybrids each generation.

An important long-term method, which is especially suitable for annual grain crops with at least some outcrossing, is the **composite cross with agricultural mass selection**. To make a composite cross, one plants together several to many different strains of the crop and allows random pollination, even if some individuals self-pollinate. One harvests the seeds and sows them next season, then continues harvesting and sowing for several generations. The random pollination makes an enormous number of different crosses between types and between their descendants. There is no labour of hand-pollination, nor of recording yields.

The agricultural mass selection works because individuals with the highest yields and most vigorous seedlings and plants are proportionately better represented in the seed and in the next generation's plants than are less good types. This easy method can be used to adapt existing strains to new environments or to new or prevailing agricultural practices. For example, if one wants earlier ripening stocks, one harvests the seed early each season, when the earliest maturing types will be the only ones with viable grain.

The composite cross plus agricultural mass selection has even been successful with wheat and barley, which usually self but have a little outcrossing. The more the outcrossing, the faster the method will work, so a regular outcrosser like maize responds faster than wheat. Maize yields went up 22% in four generations in one trial, and 33% in three generations in another. With barley, 25 years of mass selection in California gave 19 important new commercial varieties, and the whole trials paid for themselves from the sale of surplus seed each year. A composite cross initially produces very variable individuals, but mass selection by agricultural practices imposes stabilising selection for critical characters.

Breeding new varieties often consists of crossing two existing varieties and looking for good progeny in the F1, F2 or later generations. When crossing two genetically different **pure-breeding lines**, all first generation plants or animals will be identical to each other except for chance mutations or environmental effects. Segregation and recombination at meiosis in the F1 will then cause a wide range of types to select from in the F2 and later generations. If the parental lines are **not pure-breeding**, the F1 will be genetically heterogeneous and selection can start in the F1. **Apple trees** are not pure-breeding and cannot be selfed because of self-incompatibility mechanisms (Chapter 17.1.2). Crosses between any two compatible apple varieties can therefore give a whole range of new genotypes in the F1 from pips which will be highly heterozygous, and successful new varieties can be propagated vegetatively by bud-grafting onto suitable rootstocks (Chapter 17.1.5). Thus variety "Lord Lamborne" (Plate 19) was selected as a high quality dessert apple from an F1 seedling raised in 1907 by Laxton Brothers from a "James Grieve" x "Worcester Pearmain" cross.

Modern selection methods often involve testing for **molecular markers** as well as for conventional characters such as yield and flavour. Plate 20 shows a tray of apple seedlings awaiting early selection from molecular markers before growing on. Where there are known molecular markers (e.g., DNA, enzyme isozymes or antigenic characters) associated with known commercial characters, say a DNA RFLP marker associated with fruit quality, then selection for that marker **at the seedling stage** reduces the number of plants to be grown for further selection at the fruiting stage. For **marker-assisted selection**, one needs heterozygosity for the agricultural character and the DNA marker, and close linkage between them.

12.4. Recurrent backcrossing for gene transfer

Recurrent backcrossing is repeatedly crossing offspring back to one parental type, so that all genes except the desired ones from one parent are replaced by those of the recurrent parent. It can be used in plants, animals and microbes. It is used to transfer single qualitatively-acting alleles, especially ones for disease resistance. In an inbreeder, it may require hand cross-pollination. In sheep, it was used to transfer a gene with additive action, Fec^B, for a higher lambing rate, from the Booroola Merino into less fertile breeds, backcrossing to those less fertile breeds to maintain their other breed characteristics. See Chapter 1.3.4.

Suppose one has a good line for yield, but which is susceptible to a particular disease ($r\ r$), and that there is a line available which is poor for yield but genetically resistant to this disease, with **dominant resistance** ($R\ R$). One crosses the two lines, and all the F1 should be $R\ r$, resistant to the disease, and one tests this. The F1 are then backcrossed to the $r\ r$ recurrent parent, the one with good yield, so the next generation should be 1 $R\ r$: 1 $r\ r$. The individuals are tested for disease resistance, and resistant ones are backcrossed to the recurrent parent. This is continued for about six generations, or as long it takes to get the progeny yield up to that of the recurrent parent.

During this process, the genes of the poor line are replaced by ones from the recurrent parent, with about half the remaining poor-parent genes being replaced each generation, especially ones unlinked to *R*. The final product is only *R r*, so is selfed or crossed with others of that type to get some *R R* genotypes, which can be distinguished from *R r* by test-crossing to *r r*. This has been used to transfer salt-resistance from a seashore wild tomato into commercial lines for growing in partly saline soils, selecting each generation for salt-tolerance, fruit colour and flavour.

If the character to be transferred is **recessive**, say *a*, one needs a more complicated crossing scheme, with selfing to allow identification of the recessive homozygote. For example, to transfer beneficial recessive allele *a* into *A A* plants, one crosses donor *a a* to recurrent parent *A A*, then backcrosses the resulting *A a* to *A A*. The next generation will consist of 1/2 *A A* and 1/2 *A a*, with the same phenotypes; the plants are selfed. One eighth of the progeny will be identifiable phenotypically as *a a*. Those individuals are backcrossed to the recurrent parent, *A A*, and the cycle of crosses is repeated. In plants, this cycle can be shortened; the selfing does not contribute to the getting rid of unwanted donor parent genes. In the generation with 1/2 *A A* and 1/2 *A a*, each plant can have some flowers selfed and some backcrossed to the recurrent parent. Seeds from the backcross are only used from plants whose selfing showed them to be *A a*, not *A A*.

Recurrent backcrossing is also used in **mutation breeding**, to reduce the number of unwanted harmful mutations, using a wild-type as the recurrent parent. One mutates the wild-type to get the M1 generation (Chapter 7). This can be selfed or intercrossed to get the M2 generation, in which recessive and dominant mutations should be recognisable. Favourable mutations are selected for, and the wanted M2 types are then subjected to recurrent backcrossing to the wild-type to get rid of the many unwanted mutations which will also have been induced by the mutagen. For recessive mutations, generations of selfing need to be alternated with the recurrent backcrossing to identify homozygous recessives.

12.5. Interspecific and intergeneric hybrids

12.5.1. Interspecific hybrids

Interspecific hybridisation has occurred naturally many times (Plate 17, *Primula kewensis*; see Chapter 10.6) and can be practised by the breeder. Many cultivars, especially of garden plants, are man-made interspecific hybrids, including many rhododendrons, roses and dahlias. See Chapter 10.2 for examples with raspberries, blackberries and loganberries. In animals, such hybrids may occur in the wild or zoos, e.g., sheep (2n = 54) x goat (2n = 60) gives a shoat, lion x tiger gives a liger, or horse x zebra gives a zebroid. Although the offspring are usually healthy, the chromosomes are not usually sufficiently homologous to pair well at meiosis, so the offspring are infertile. As we saw in Chapter 12.1, a female horse (2n = 64) x a male ass (donkey, 2n = 62) gives the hardy and useful mule (2n = 63), but it is sterile and has to be remade each generation by interspecific crosses. Even intergeneric hybrids (Chapter 12.5.2) have been used in some conifers, orchids and cereal grasses, e.g., *Triticale* from wheat (*Triticum*) x rye (*Secale*).

There is at least one animal example of a successful interspecific cross, **Bonsmara cattle** in Southern Africa, started in 1937 by Bonsmara. He crossed *Bos taurus* (Hereford/Shorthorn) with *Bos indicus* (Indian Brahman and native cattle), with Bonsmara cattle being genetically about five-eighths *B. taurus* and three-eighths *B. indicus*. They combine the good yield of the European cattle with the hardiness of the African ones.

Interspecific hybridisation occasionally occurs in the wild for plants, where it can lead to the production of successful new types, whether diploid or allopolyploid, but the breeder has access to all kinds of species, from different regions of the world, which would not normally meet or cross. Breeders have many exciting and exotic possibilities as they can cross related species, even from different hemispheres.

With plants, interspecific hybrids may be **sexually sterile**, not setting seed even when they produce wonderful flowers; except in seed crops, sterility may not matter if the plants can be propagated vegetatively, e.g., by cuttings, offsets, grafts, etc.; see Chapter 17.1.5. Grape vines (Plates 46 and 47), apples and roses, for example, are usually propagated by grafts onto genetically different selected rootstocks, not grown from seed. The sterility of the initial hybrid can sometimes be overcome by making allotetraploids, e.g., by treatment with colchicine, so that each chromosome has a homologue with which to pair at meiosis. Although colchicine-induced allotetraploids are excellent intermediates in transferring genes between species (see Chapter 10.9 and the tomato examples below), not many are in wide use. Those which are include tetraploid rye-grass (*Lolium multiflorum* x *L. perenne*) and *Triticale* (wheat x rye; see below).

The best example of natural successful interspecific hybridisation utilised by man is **wheat**. Its history was worked out by Riley and others from cytogenetic studies of possible ancestors and relatives. It is illustrated in Fig. 12.5.1.1. Two different sterile hybrids, giving unpaired chromosomes in meiosis, had a spontaneous doubling of the chromosome number, giving fertile allopolyploids. From a diploid, an allotetraploid and a double-allohexaploid, man selected - with further breeding and agricultural selection - the modern einkorn, durum and bread wheats, respectively.

By knowing the ancestry of the wheats and their genomes, wheat breeders can easily plan further breeding, e.g., by crossing to wild relatives with useful disease resistance.

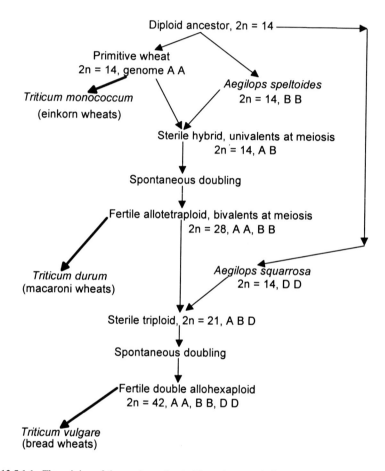

Figure 12.5.1.1. The origins of the modern wheats. Normal arrows indicate natural events and heavy arrows indicate selection by man.

The wild diploid teosinte (*Fuchlaena mexicana*, synonym *Zea mays* ssp. *mexicana*, and there is also ssp. *parviglumis*), 2n = 20, is the presumed ancestor of **maize**. It is fertile when crossed to the modern diploid maize, *Zea mays*, 2n = 20, so useful genes can be bred into maize when required. In crosses between the two species,

Doebley and Stec (1993) used molecular marker loci to map quantitative trait loci. They found that a relatively small number of loci with large effects were involved in the evolution of the key differences between teosinte and maize.

An example of man-made successful interspecific hybridisation is the **strawberry**. An old octoploid, $2n = 56 = 8X$, *Fragaria virginiana*, from Eastern USA, was of poor quality. Another octoploid, *F. chiloensis*, $2n = 56 = 8X$, from Chile, had large fruits of poor quality, and a dioecious habit (separate male and female plants) reduced yield. In the early 19th century, English breeders crossed the two octoploids, selecting among the progeny for large, good quality fruit and hermaphrodite flowers. They succeeded. See Lawrence (1968).

As mentioned in Chapter 10.9, making allopolyploids is an excellent intermediate step in **transferring genes** from wild species into cultivated ones, before repeated backcrossing to get rid of unwanted other genes from the wild relatives. This has been done extensively to transfer genes into the cultivated **tomato**, *Lycopersicon esculentum*. They include genes for fungal resistance from *L. hirsutum*, *L. pimpinellifolium* and *L. peruvianum*, genes for virus resistance from *L. chilense* and *L. peruvianum*, insect resistance from *L. hirsutum*, quality characters from *L. chmielewskii*, and adaptation to adverse environments from *L. cheesmanii*. In the case of nematode resistance, it took 12 years to separate that character from poor fruit characters which were introduced in the same cross to the wild relative. See Esquinas-Alcazar (1981).

One problem with interspecific and intergeneric hybrids is **sterility**; if this arises from a lack of pairing in meiosis, one can try making allopolyploids, as twice occurred naturally in wheat: see Fig. 12.5.1.1. Other causes of sterility can often be treated. One can try reciprocal crosses: with species A and species B, one crosses A as the female (ovule) parent x B as the male (pollen) parent, and B as female x A as male, because sometimes only one of the two reciprocal crosses is fertile, with unilateral incompatibility. One can try hormone treatments. For example, pear tree (*Pyrus communis*) as female x apple tree (*Malus pumila*) as male is usually sterile but can be made fertile by applying 40 parts per million of the hormone naphthoxyacetic acid to the ovary and style. This increases growth of the apple pollen and reduces early abscission of the flowers and fruit. If failure of endosperm development, or of late embryo development, occurs, one can dissect out the early embryo and grow it on suitable nutrient agar, as in *Lilium lankongense* hybrids. In *Lathyrus* (sweet pea) and *Pyrus* (pear), stylar incompatibility within these genera has sometimes been overcome by amputating the style and pollinating the style base directly.

In conifers, many species have the same chromosome number, 24, so that chromosome number differences are often not a problem in natural or man-made hybrids between species or genera. The main forestry **larch** now planted in Britain is the Dunkeld larch, *Larix x eurolepis*, 2n = 24, which excels its parents in vigour. It arose naturally in about 1900 at Dunkeld House, Perthshire, Scotland, from cross-pollination between the European larch, *Larix decidua*, 2n = 24, and the widely planted Japanese larch, *Larix kaempferi*, 2n = 24, which was introduced into Britain in 1861.

12.5.2. Intergeneric hybrids

Natural and man-made **intergeneric crosses** have been used successfully in conifers, orchids and cereal grasses. The **Leyland Cypress**, *x Cupressocyparis leylandii* (Plate 21), arose by natural cross-pollination between its two parents, *Chamaecyparis nootkatensis* and *Cupressus macrocarpa*. In 1888 at Leighton Hall, Welshpool, Wales, six seedlings from *C. nootkatensis* were visibly different from others and were deduced to have *C. macrocarpa* as their pollen parent. From these seedlings came the commonest cultivar, cv Haggeston Grey. In 1911 the cross happened naturally again, but with seedlings from *C. macrocarpa* with *C. nootkatensis* as pollen parent. This reciprocal cross gave the cultivars Leighton Green and Naylor's Blue. Later crosses gave plants with yellow foliage such as cv Castwellan and Robinson's Gold. "Leylandii" is one of the fastest growing conifers, often growing a metre a year and reaching 30 m; it is an excellent shelter-belt tree and hedge, but grows too large for most small gardens, hence its notoriety in disputes between urban neighbours.

Triticale is a man-made allopolyploid from crossing bread wheat (*Triticum*, 2n = 6X = 42) with rye (*Secale*, 2n = 2X = 14), and doubling up the chromosome number from 28 to 56. It combines the high yield of wheat with the hardiness of rye, and is often planted in areas not good enough for wheat.

Many **orchids** grown for showy flowers are hybrids from crossing two or even several different genera. Such intergeneric hybrids include *Vuylstekeara* and *Odontocidium* from South America, which are available in a wide range of colours. Intergeneric hybrid *Oncidium* orchids are widely grown in Asia for the cut-flower market.

12.6. Making polyploids

It is useful to be able to **make polyploids**, to see whether they are better than diploids, or to overcome sterility problems in interspecific hybrids. Whether one is making autotetraploids from a diploid, or allotetraploids from interspecific hybrids, any of the following can be tried.

One can treat seeds, germlings or buds (anything with a growing point, especially one giving rise to both somatic and germ-line tissue) with **chemicals** which allow chromosome replication but sometimes inhibit nuclear division, thus doubling the chromosome number. Colchicine, from the autumn crocus, or synthetic colcemid are commonly used. Colchicine was used to make commercial polyploids in beets and various grasses. Other chemicals which have been used include chloral hydrate, sulphanilamide and ethyl mercury chloride. See also Chapter 10.2.

Heat shock has been used to make tetraploid maize and cold shock was used to get tetraploid fruit flies (*Drosophila*). Even mechanical methods can work, e.g., decapitating young tomato plants, where some regrowths may be polyploid if the damage interfered with a cell's cleavage.

One can identify polyploid shoots from chromosome analysis of dividing cells, but often a polyploid shoot can be identified from its looking different in growth habit, thickness or colour, or by analysis under a microscope for cell area, number of plastids per guard cell, stomatal pore length, etc., where values have been established for diploids (Chapter 10.6).

12.7. Examples of plant and animal breeding programmes

Although people sometimes get the idea that breeding programmes must involve high technology, most of today's commercial plants and animals come from straightforward methods based on the principles considered in much of this book.

12.7.1. Semi-dwarf wheats

In Mexico in 1945, wheat yields averaged 750 kg/ha. This had risen to 2790 kg/ha by 1967, mainly through the use of **semi-dwarf wheats**. After the Second World War, a short-stemmed wheat, **Norin 10**, was introduced from Japan to the USA. It was only one third of the height of normal wheat, but was high yielding because high levels of nitrogenous fertiliser could be used without the plants lodging (being knocked down by wind and rain). It was taken to Mexico and subjected to **cyclic selection** (see Chapter 5.4) to improve environmental tolerance. Seeds were planted just above sea level in October, with new seed being harvested in March. This was taken to a second site at an altitude of about 2,440 metres above sea level, planted in April, and harvested in September. Those seeds were returned to the first site for planting in October, and this cyclic selection with two generations a year was continued for five years, which is 10 generations. This two-centre agricultural mass selection gave daylight-independent flowering, and the cyclic selection gave environmental tolerance, giving good yields when the seeds were tried in diverse places, though the plants often needed breeding to adapt them against locally important diseases.

From 1966, dwarf or semi-dwarf wheats were rapidly introduced in many areas of the world, including Africa and Asia, for example doubling yields in Pakistan. Apart from the higher yield in each generation, the extra benefit of reduced sowing-to-harvest time allowed farmers in warmer areas to grow two cereal crops a year, e.g., wheat and sorghum or wheat and maize. This "Green Revolution" was a big agricultural advance, but did require extra inputs of nitrogenous fertilisers and of pesticides.

12.7.2. Broad beans

In the early 1970's, the procedure for **broad beans** for canning and freezing was to mow the plants with ripe pods, then to leave them for two to three days for evaporation to reduce the bulk of the plants before the vining machines came to separate the beans from the waste to be dumped. The breeding aim was to produce smaller plants so that only a single combined mowing/vining operation was required.

A commercial variety, "Triple white" (white flowers, stipules, etc.) was crossed to a dwarf bean (15 cm high). The breeder selected for intermediate height and the production of several pods per node, as the smaller plants had fewer nodes and "Triple white" only had one pod per node. Selection was also made for more beans per pod. The new variety only needed a single mowing/vining operation.

12.7.3. Semi-leafless combining pea

Cultivar Countess is a semi-leafless white marrowfat protein **pea**, specially bred for combine harvesting to meet the expanding demand for protein for compounding in animal feedstuffs, although it could be used for canning for human consumption. The genes for being semi-leafless convert leaflets into tendrils, which twine together and help support the plants. There is less leaf trash to separate out at harvesting, and the leaf area prone to disease is reduced by more than 50%, allowing light to penetrate the open canopy, and better air circulation. It is fast drying and easy to harvest because it stands up well, and the interlocking tendrils help to prevent lodging in bad weather. Countess is well adapted to a wide range of soil types and is particularly drought-resistant on light soils.

It is high yielding, with yields 109% of control variety Protegra in the 1987 NIAB (National Institute of Agricultural Botany) Recommended List for General Use, and in 27 independent trials from 1982 to 1986 had a yield 116% relative to cultivars Protegra, Maro and Birte. It is resistant to Race 1 of pea wilt (*Fusarium oxysporum*) and to the most common pathotype 4 of downy mildew (*Peronospora viciae*). Countess was bred by Booker Seeds, Essex, England.

12.7.4. Sheep, including cross-breds and Border Leicesters

Sheep are often **cross-bred** for hybrid vigour and to make the best use of particular breeds or hybrids in particular environments. For example in Britain, the **Blackface** breed (about 110 lambs born per 100 ewes mated) is very hardy and does well in poor hilly country, such as much of Scotland's uplands. Surplus ewes are taken from the hills to slightly better land, for mating with rams of breeds such as the Border Leicester which are larger and more fertile, with genes for a higher milk yield and therefore rearing capacity. The **Scottish Greyface** males produced are reared for meat and the ewes are taken to good pasture land in the north of England. The ewes (more than 150 lambs per 100 ewes mated) are crossed to rams of specialised meat breeds such as the Suffolk or Hampshire Down, to produce fat lambs for slaughter.

In 1991, a **Border Leicester** Group Breeding Scheme was set up, with an "Elite Flock" in Caernarfon, Wales, with 48 ewes selected from flocks all over Britain. They were in-lamb to top sires, and had two crops of lambs taken from them to continue the elite flock. Selection is made for prolificacy (number of lambs), mothering and milking qualities, plus body conformation and growth rate. Border Leicester rams give high survival rates of crossbred lambs, which inherit the thick coat of wool. They give rapid growth rate, uniformity of carcase, easy lambing even when used on small breeds such as Welsh Mountain ewes, high prolificacy, abundant milk, good mothering, and a large yield of prime meat. Over one million breeding ewes in the UK are sired by Border Leicester rams. Border Leicesters developed from Leicester Longwool and Cheviot sheep in the England/Scotland borders in the early 19th century, and were recognised as a breed in 1869, with a herd-book from 1898.

The **Southdown** sheep (Plate 25) can be used as a pure bred, with very fine fleece, excellent meat flavour and fast growth. It can also be used as a sire on large ewes of other breeds to give a smaller birth-weight lamb with fast growth and high quality. For an account of common sheep breeds, see Maijala (1997).

Pedigree breeding is very important in pure-bred sheep, including **Suffolks**, which are excellent for weight gain, and are used for meat and short wool. An example of a prepotent "super-ram" is Pankymoor Prelude, a Suffolk ram bought as a six-month-old lamb in 1993 for a then record price of £23,100. By 1999 he had fathered 189 top-flight rams and 350 pedigree daughters, worth more than £1 million, and still had years of reproductive life left. In 1995, one of his sons sold for a record £71,400, while ordinary Suffolk rams only fetch £250-£500.

In Mediterranean areas such as Greece, which has 10 million ewes, sheep are often kept for **milk** for cheese-making, with the owners' incomes coming about 2/3 to 3/4 from milk and 1/3 to 1/4 from meat. Sheep milk fat and milk protein have fairly high heritabilities, 0.5 to 0.6, with a high correlation between them, +0.7, but with a slight negative genetic correlation. As one might expect from competing physiological demands, milk yield is negatively correlated with % milk protein (-0.4) and % milk fat (-0.3): see Barillet (1997).

In **wool breeds**, the critical characters are fleece weight per year, the fineness of the wool fibres, and the number of lambs weaned per year. The phenotypic merit of a breeding ewe could thus be expressed in dollars per year from an index such as (clean fleece weight per year, kg, x $7.2) + (- wool fibre diameter, μm, x $1.2) + (number of lambs weaned per year x $7.2), where cleaned fleece fetches $7.2 a kg and coarser wool is penalised at $1.2 an extra μm diameter (figures from Kinghorn, 1997). Colourless fleeces are usually preferred to coloured ones. For wool fibre characters and for general tables of heritabilities, breeding aims and hybrid vigour in farm animals, see Dalton (1980). For sheep and cattle breeding, see Simm (1998). For transgenic animals, see Bishop (1999).

12.7.5. Cattle, including Ayrshires

See Chapter 3.4 for breeding values, Chapter 5.13 for meat selection and Chapter 17.2 for reproductive physiology in cattle. In Britain in about 1950, farm labour was abundant. Most dairy cattle herds were small, with fewer than 20 cows, and 70 was considered a very large herd. The proportion of herds milked mechanically rose from 5% in 1936 to 10% in 1939, to 50% by 1949, and to 70% by 1956. Today there are far fewer herds, but larger, with scarce labour and much more mechanisation. In one high-intensity system seen by the author near Toronto, Canada, cows stay indoors in stalls and are fed individually by a computerised system which adjusts the amount and composition of each feed in accordance with that cow's recent milking and reproductive record.

Breeding has become more scientific, but bureaucratic interference has become more acute. Under EU rules, Britain cannot produce all the milk it needs, with strict quotas on production, with knock-on effects on dairy products such as cheese. In Britain in the 1950's, most milk was sold to a national body, the Milk Marketing Board, which then paid a pool price for all milk, regardless of its nutritional quality. That gave breeders and farmers no incentive to improve milk protein or fat content, except for farm-bottled milk, where the rich milk of **Guernsey cattle** commanded a price premium. AI (Artificial Insemination, see Chapter 17.2.4) became more widely used, with Guernsey inseminations from the Whiligh Cattle Breeding Centre rising from 1,000 in 1946 to 78,000 in 1964. In the South East, a Dairy Progeny Testing Scheme was set up in 1964.

Ayrshire cattle started in South West Scotland several hundred years ago, and now number more than two million around the world, with breed societies in Australia, Brazil, France, Finland, Kenya, UK, USA, former USSR countries, and several other countries, and with a World Federation of Ayrshire Breed Societies based in Ayr, Scotland. The World Federation organises conferences, promotes the exchange among countries of semen, livestock and embryos, and promotes the breed internationally to bodies such as governments. The exchanges widen the gene pool available for selection.

About 450 young Ayrshire bulls are progeny-tested annually throughout member countries, and the INTERBULL Centre in Sweden takes into account the different environments in which they are tested. Its INTERBULL rankings allow comparison of bulls in different countries without a need for conversion formulae. The Finnish Ayrshire breeders have an Embryo Transfer Breeding Programme, "ASMO", with a target of 900 embryos a year to market in Finland and 300 for export. Only the top 1% of cows are used as bull dams. 70 heifers a year are evaluated for production and functional traits, with the best 20 selected as embryo donors. With the OPU method, collecting oocytes directly from the cow's ovaries, about 60 embryos can be obtained per donor per year. See also Chapter 17.2.5.

On the marketing side, "Ayrshires"[®©] milk is available in the UK as a breed-branded milk, with a six-fold sales increase since the launch in 1996. In South Africa, milk from 2,000 Ayrshires (one third of the country's total) supplies a major retailer, Woolworths. Ayrshires are dairy cattle and the breed's national milk records over a lactation of 305 days include: America, 22,067 kg; Canada, 15,082 kg at 4.1% butterfat and 3.2% protein; UK, 12,441 kg at 4.5% butterfat and 3.2% protein. In Britain, the South East Ayrshire Breeders Group introduced progeny testing of young bulls in 1970. One of their best bulls, Dunlops Shanghai, was sold in 1979 for £6,500 to the Milk Marketing Board's AI service, so his semen was very widely distributed. One of his daughters, Ingram's Gladys Jane, produced well over 100 tonnes of milk over 11 lactations. It was only in 1989 that the South East Ayrshire Club changed the rules of its Herd Competition to allow the protein content of milk to be included as a show criterion.

According to an undated leaflet, *Your Concise Guide to Successful Showing*, published by The Ayrshire Cattle Society of Great Britain and Northern Ireland, Ayr, available in 1999, the following are some **aims of Ayrshire breeding and showing**.

"A head in proportion to the rest of the cow, displaying femininity with a bright, alert appearance and broad strong muzzle. The udder, depending on the age of the cow, of medium size but always carried above the hock. The rear attachment must be high and wide, displaying a well-defined central ligament with the fore attachment firm and strong, blending smoothly into the body; teats should be about 6 cm long, placed in the centre of each quarter. The texture of the udder should be soft and silky with prominent veining and a tortuous milk vein carried well forward from the udder. The shape of the cow should provide generous heart room for strength and durability, ample body capacity through length and depth for a high forage intake and weighing 550-600 kg, with a level topline giving strength through the loin. She should be clean and sharp, with an indication of her production potential; a minimum height at the shoulder of 54" (137 cm) is desired. When viewed from the side, the legs should exhibit a small degree of set but when viewed from the rear they should be straight. The bone should be flat, the hock clean, with short strong pasterns. The cow can be any shade of red or brown including mahogany and white, although either colour may predominate. Each colour should be distinctly defined."

In contrast to Ayrshires, Limousin (Plate 24), Charolais (Plate 23), Saler and Simmental cattle are **beef breeds**. **Simmental** bulls can be finished on cereals (weight gain about 1.5 kg a day) for slaughter at 10 to 12 months, or on a concentrate/high quality silage ration to finish at 12 to 15 months. Males can be left intact (not castrated) and slaughtered at about 600 kg, with very good feed conversion rates. Simmental suckler cows (i.e., ones for feeding calves, not for milk for humans) typically calve in spring, with weaning in autumn. The bulls have good hindquarters and are ideal for the restaurant trade which wants lean meat.

Limousin cattle are early maturing, with medium-weight carcases ideal for butchers and supermarkets, while light-boned frames mean more meat from the same live weight. They can achieve weight gains of 1.5-2.0 kg a day, with high food conversion efficiency. Limousin females are excellent suckler cows, with large pelvic openings and light-framed calves, giving easy calving, and they are fertile and milky. They are hardy, so can be used in upland or lowland conditions. Limousin bulls pass on their conformation when crossed to other breeds, their light-boned frames allowing easy calving, so they make good terminal beef sires. For Limousins, the critical carcase characters are: killing-out percentage about 55%, lean-to-bone ratio about 4.2, saleable meat about 73% of the carcase, and higher priced cuts about 45% of the carcase (figures from the Meat and Livestock Commission).

Saler cattle are one of the oldest and most genetically pure breeds of cattle in the world, from the southern half of the Massif Central in the Auverne, France, and their mountain origins make them suitable for poor conditions and rough ground. The carcase kill-out is greater than 60%, with high-quality marbled meat. They have the largest pelvic area of the major beef breeds and a short gestation period, with easy calving and less need for human intervention. They have milky, fertile cows showing early maturity (with the first calf at two years of age) and capable of calving progeny from more muscled terminal sire breeds. They have a long working life, are hardy, disease-resistant, and can use rough forage.

In 1999, a bull calf, Mr Frosty, was produced using the last sample of 37-year-old frozen semen collected in 1961 from a champion **British Friesian** bull, Horwood Janrol. The latter was born in 1950, before widespread antibiotic use on cattle, and long before BSE. The breeder, Mr Booker of Avon Breeder Services, Bath, is reviving the British Friesian breed to produce cattle for export which can thrive on a variety of forage and are BSE-free. British Friesians went out of fashion as imported black and white Holsteins were used to breed Holstein-Friesians, which produce most of Britain's milk. Mr Booker said that Holsteins were high-performance animals but needed a lot of expensive bought-in rations and had high vets' bills, unlike British Friesians.

12.8. Breeding for shows

The value of an animal is usually greatly enhanced if it has done well in competitive shows with recognised expert judges. See Plates 22, 23, 24 and 25. There are many shows of varying degrees of importance. In England, a win at the Royal Agricultural Show would increase an animal's value considerably, as would a win at the South of England Show for a bull up to two years of age in the Sainsbury's Super Beef Bull competition or for a cow in the Unigate Super Dairy Cow competition.

The **ideal type** is defined in the **show rules** or the **breed handbook**. For example, cat types are specified in Britain by the Governing Council of the Cat Fancy, and dog types by Cruft's. For pets, the criteria are aesthetic, but for farm animals such as Hereford cattle, the required characters are a mixture of commercially relevant ones such as size and body proportions, plus ones of largely aesthetic value such as tail length and body colouration. Each breed society lays down its ideal type against which individuals can be judged on "**breed points**" as well as on performance where that is included. For example, a cow may be judged on her appearance and on her milk record (litres per lactation, percentage protein and butter fat), and reproductive record. See Chapter 12.7.5 for Ayrshire cattle show points.

One problem in cattle is that the cash value of a bull may reflect his performance in winning awards at shows, whereas his capacity as a sire in breeding may not be closely related to that ability. His breeding value may be much less than his phenotype value. There may be conflict between achieving a good level for some historic aesthetic character and for some modern commercial character. Thus for show purposes a bull might need to be very large and deep chested, whereas a lighter body size might be optimum for modern commercial types. Show rules are not always updated to follow commercial developments.

For a dog with no working function, purely a pet, all characters are "fancy", as with a Chihuahua, but some dogs are or have been working animals. The following is a typical show description, from a Cruft's dog show catalogue, of the **Saint Bernard dog**. This used to be a working dog for rescuing people from deep snow in the Alps. Some of the description relates to functional characters for that work, such as size and muscles, but others such as facial expression and thigh feathering are largely irrelevant.

"The expression should be benevolent and kindly. The head is very massive, the circumference of the skull being more than double the length. From the stop to the tip of the nose it is moderately short, and the muzzle is square. The lips are deep; the eyes rather small and deep set, of darker colour and not too close together. Neck long, muscular and slightly arched, with much dewlap. Chest wide and deep; loins slightly arched, wide and very muscular. Forelegs perfectly straight, with huge bone, and hind-legs should not be cow-hocked. Feet large and compact. The minimum height of a dog should be 31 in. [79 cm], but the taller ones are preferred. In the rough variety, the coat should be dense and flat, rather fuller round the neck, and some feathering on the thighs. The colours may be red, orange, various shades of brindle, or white with patches on the body. The muzzle, the blaze on the face, collar round neck, chest, forelegs, feet, and end of tail - white."

12.9. Breeding programmes from crosses to selection, to local and national trials, possible commercial release, and approved lists

The information in this section comes largely from the Home-Grown Cereals Authority and the British Society of Plant Breeders Ltd. Certification procedures differ between countries, especially outside the EU. From making initial crosses to the release of a proven new variety typically takes 10 to 15 years even for an annual plant such as wheat, and up to 18 years for potatoes. **Seed potatoes** are often grown in Scotland where the lower incidence of aphids reduces aphid-transmitted potato viruses.

Breeding will be illustrated with a **winter wheat** example. The initial crosses are planned to get: high yields for cost-effectiveness; quality characteristics for specific markets, e.g., bread-making, biscuit-making, food markets (e.g., domestic flour) or exports; disease-resistance to minimise fungicide inputs; predictable uniform maturity; straw strength for lodging resistance and ease of harvest. In early trials, the wheat breeder looks for improved pest and disease resistance, greater standing power and straw stiffness, early maturity date, drought and frost resistance. Genetic fingerprinting for molecular markers may be used to predict some aspects of agronomic performance (**marker-assisted selection**). In the early years, breeders will sow millions of lines and select them down to a few hundred. The best lines have to be shown to be stable, then after the breeder's own trials, are entered in national trials over two years. Specific milling and baking tests are carried out in Britain by the Campden & Chorleywood Food and Research Association. The National Association of British and Irish Millers, and British Cereal Exports, help to devise suitable testing methods. Trials by the main testing bodies and a range of regional collaborators provide **national and regional recommendations** for particular varieties of a crop. If successful, the new varieties join the national recommended lists of varieties and can be sold commercially.

To make the F1 generation, two or more parental lines with good characteristics are crossed. Hundreds of different matings may be made, and even if only two parents are used, if one or both are not totally homozygous, segregation and recombination will produce many different genotypes in the F1. In the F1, perhaps 1,100 crosses might be made, with perhaps 2,000,000 individual F2 plants grown. In the F2 the best plants are selected, say 50,000. The seed from each selected plant is grown in a small row, giving 50,000 F3 "ear rows". The best plants are again selected, giving say 3,000 F4 families. The selection process is repeated, with pedigree selection as well as individual plant selection. In later stages, yields of lines are accurately measured. Samples of each line are grown in special plots where they are exposed to particular diseases; they may be sprayed frequently with water when testing for resistance to *Septoria* fungus.

Up to the F5, trials are usually on one site, but by the F6 and F7, trials are often on three sites, perhaps with different soils or microclimates. By F8, there may be trials of the most successful lines in eight locations. During the generations F6 to

F10, stabilising selection is used to "**purify**" each new line so that all plants are alike within a line. Each wheat line is **multiplied** during testing, with each kilogram of seed yielding about 30 to 60 kg seed. The breeder only multiplies up the best lines. During National Trials, the breeder will probably grow about 0.15 ha of a line, building up to 3 ha and then 80 ha for commercial sale of the seed if the line is approved.

By about the F9, lines are ready to be entered into official trials undertaken in Britain for the Ministry of Agriculture, Fisheries and Food by the National Institute of Agricultural Botany, the Department of Agriculture for Northern Ireland, and the Scottish Agricultural Colleges. Lines are assessed over several years for merit, as "Value for Cultivation and Use", which for wheat includes yield, straw stiffness (lodging can halve crop yields) and resistance to the main cereal diseases. Seeds and plants are also checked to see whether the line is "Distinct, Uniform and Stable", i.e., that it is a distinct new variety giving a uniform crop.

The **National List trials** were set up under the Plant Varieties and Seeds Act 1964 and the European Communities Act 1974. About 30 wheat varieties tested in the first year might be selected down to 15 in year two, to six in year three and only three might make the final recommended list. Successful performance in the National Trials means that a new variety is added to the UK National List and to the EU's Common Catalogue, allowing them to be marketed anywhere in the EU. These lists help the farmer compare the new varieties with control varieties (see Chapter 12.7.3 for a pea example). It takes two years of National List trials before a variety can be marketed, and at least one more year before a variety can be added to the NIAB/ HGCA Recommended Lists. Most of the costs of the trials are borne by the breeders, so with producing a new variety taking over 10 years and incurring very large costs, breeding is a financially risky business. The breeders return comes from a small royalty charged on seed. For cereals, it is about £2.50 an acre, or just over £6 a hectare (Chapter 19.2.6).

UK wheat **yields**, in tonnes/ha, have risen from 4.05 in 1963/67 to 7.20 in 1990/ 94, with much of the increase being due to plant breeders, and over the same period barley yields have gone from 3.67 to 5.40, and oat yields from 3.10 to 5.16. There has been much better disease resistance, better resistance to adverse weather, better lodging resistance, less premature grain shedding before harvest, and earlier ripening allowing harvesting before autumn rains. Varieties are aimed at much more specific markets, e.g., malting barley for brewing is quite different from animal feed barley. These advances have made the UK the world's sixth largest wheat exporter, instead of being a net importer as previously.

Big advances are being made with other crops. **Oilseed rape** was once grown for limited industrial markets, with 40-50% erucic acid, but since it has been bred for the food oils market, with much less erucic acid, the acreage has increased enormously (from about 48,000 ha in 1977 to over 400,000 ha in 1996 in the UK), making vast patches of yellow across the landscape in spring. Rape oil is now the third most important oil crop in the world. Rape is unusual in including several species, *Brassica napus, B. rapa, B. carinata, B. juncea* and *B. nigra*. It is now being bred for biofuels and for specialised industrial lubricants, by changing its oil composition. As rape is partly (30 - 50%) cross-pollinated and partly selfed, it needs bagging for protection from stray pollen during breeding. Cross-pollination threatens varietal purity, making the use of farm-saved seed risky.

Linseed was grown mainly for quick-drying oils for the paint industry but is now being bred for use in animal and human foodstuffs. More cold-tolerant sunflowers and lupins have been bred for growing in Britain. When farmers are planning what to grow on their land, the availability of **subsidies** (Chapter 19.2.1) is important, with newer crops often being encouraged to avoid overproduction - and hence lower prices - of traditional cereals and dairy products. In the **EU Set-Aside scheme** (see Ansell and Tranter, 1992), farmers can even be paid to grow nothing, to avoid surpluses. Some of the absurdities of the EU system were shown in 1999 (*The Daily Telegraph*, 25/5/99, p 5 and 11/11/99, p 18). A farmer in Wales grew 29 ha of linseed fibre flax, receiving a 100% EU subsidy of £15,000 to produce 100 tons of fibre flax. In previous years he had harvested the crop but was ordered to destroy it as the 350-mile round trip to the processing plant in East Anglia made it uneconomic. In 1999 he was fined £1,500 for ploughing the crop back in, to save £1,300 harvesting costs. In Britain, the crop of flax and hemp increased from 2,200 ha in 1993 to 20,200 ha in 1996, because of the EU subsidy of £550 a ha, which the EU admitted "was extremely high in relation to production costs and the value of the product." Much of the flax in not processed as it is only grown for the subsidy, and fit only for paper pulp.

Suggested reading

Abbott, A. J. and R. K. Atkin, eds., *Improving Vegetatively Propagated Crops*. 1987. Academic Press, London.

Ansell, D. J. and R. B. Tranter, *Set-aside: in Theory and in Practice*. 1992. University of Reading, Reading.

Bahl, P. N., ed., *Genetics, Cytogenetics and Breeding of Crop Plants. Vol. 2. Cereals and Commercial Crops*. 1997. Science Publishers, Inc., Enfield, New Hampshire.

Barillet, F., Genetics of milk production, pp 539–563, in: Piper, L. and A. Ruvinsky, eds., *The Genetics of Sheep*. 1997. CAB International, Wallingford.

Bishop, J., *Transgenic mammals*. 1999. Longman, Harlow.

BSPB. *Cereals. Breeding a Brighter Future*. Undated but about 1998. The British Society of Plant Breeders Ltd., Cambridge.

Cameron, N. D., *Selection Indices and Prediction of Genetic Merit in Animal Breeding*. 1997. CAB International, Wallingford.

Dalton, D. C., *An Introduction to Practical Animal Breeding*. 1980. Granada, London.

Doebley, J. and A. Stec, Inheritance of the morphological differences between maize and teosinte: comparison of results for two F2 populations. *Genetics* (1993), 134, 559–570.

Esquinas-Alcazar, J. T., *Genetic Resources of Tomatoes and Wild Relatives*. 1981. IBPGR-FAO, Rome.

Hammond, J., J. C. Bowman and T. J. Robinson, *Hammond's Farm Animals*, 5th ed. 1983. Arnold, London.

Hartman, H. T. et al., *Plant Propagation: Principles and Practice*, 6th ed. 1997. Prentice Hall International (UK) Ltd, London.

HGCA. *Varieties. Breeding and Testing for Improvement*. Undated but about 1997. Home-Grown Cereals Authority, London.

Ignacimuthu, S., *Plant Biotechnology*, 1997. Science Publishers, Inc., Enfield, New Hampshire.

Kinghorn, B. P., Genetic improvement of sheep, pp 565–591, in: Piper, L. and A. Ruvinsky, eds., *The Genetics of Sheep*. 1997. CAB International, Wallingford.

Lawrence, W. J. C., *Plant Breeding*. 1968. Edward Arnold, London.

Lupton, F. G. H. ed., *Wheat Breeding: its Scientific Basis*. 1980. Chapman and Hall, London.

Maijala, K., Genetic aspects of domestication, common breeds and their origin, pp 13–49, in: Piper, L. and A. Ruvinsky, eds., *The Genetics of Sheep*. 1997. CAB International, Wallingford.

Mason, I. L., *A World Dictionary of Livestock Breeds, Types and Varieties*, 4th ed. 1996. CAB International, Wallingford.

Mayo, O., *The Theory of Plant Breeding*, 2nd ed. 1987. Oxford University Press, Oxford.

Moreno-Gonzàlez, J. and J. I. Cubero, Selection strategies and choice of breeding methods, pp 281–313, in: Hayward, M. D., N. O. Bosemark and I. Romagosa, eds., *Plant Breeding. Principles and Prospects*. 1993. Chapman and Hall, London.

Piper, L. and A. Ruvinsky, eds., *The Genetics of Sheep*. 1997. CAB International, Wallingford.

Poelman, J. M. and D. A. Sleper, *Breeding Field Crops*, 4th ed. 1995. Iowa State University Press, Iowa.

Simm, G., *Genetic Improvement of Cattle and Sheep*. 1998. Farming Press, Ipswich.

Van Vleck, L. D., E. J. Pollak and E. A. B. Oltenacu, *Genetics for the Animal Sciences*. 1987. W. H. Freeman and Co., New York.

CHAPTER 13

HUMAN AND MEDICAL GENETICS

13.1. Introduction

We are diploid Eukaryotes, sharing the principal genetic features of such organisms, e.g., molecular mechanisms, chromosome behaviour and some population genetics. For genetic studies, our life cycle is inconveniently long, family sizes are too small, and desired matings are difficult or illegal to procure. One cannot legally make F1 backcrosses and really close inbreeding is forbidden. Certain types of environmental treatments are forbidden for legal, moral or social reasons.

To balance these disadvantages, there are a few advantages. In some cases, visible features of families over many generations have been recorded through painted portraits of aristocratic families, e.g., King George III, Queen Charlotte and their 13 children. Humans have also been studied in great detail biochemically, physiologically, anatomically and socially, so many techniques are available and much background information already exists.

Another important advantage is that extensive statistical records exist, often made for other purposes, but of interest to human geneticists, e.g., data on human height and weight from army medical records; data from death certificates on causes of death; church or civil records of births, baptisms, marriages and deaths. These records may be inaccurate; e.g., the woman's husband may be recorded as the father of a child, when its biological father was different.

The number of identified human genes is constantly rising, and will rise dramatically as a result of the human genome project. By 1997, more than 5,000 human phenotypes inherited in a Mendelian fashion had been identified, of which about 4,000 were disease-associated. Fewer than 10% are X-linked; more than half are autosomal dominants, and, about 36% are autosomal recessives, often loss-of-function mutations. The main methods used to work out how human characters are inherited are listed below. For an on-line World Wide Web catalogue of human genes and genetic disorders with over 10,000 entries, contact Online Mendelian Inheritance in Man at http:/www3.ncbi.nlm.nih.gov/Omim; it also has mapping details. For human reproductive physiology, see Chapter 17.3 and Sherwood (1995).

13.2.   Finding out how or whether characters are inherited: pedigree studies; twin studies; Hardy-Weinberg analysis; familial incidence

13.2.1. Pedigree studies

Pedigree studies are used to find out how particular traits are inherited, e.g., whether they are controlled by one locus or more than one, by alleles which are dominant or recessive, and X-linked or autosomal. In **pedigree studies**, one finds someone, the **proband** (sometimes called the propositus), showing the character one wishes to study. Then one traces as many relatives of the proband as possible, seeing which members of the kindred have the character and which do not, over as many generations as possible. Because people's memories of relatives may be faulty, one tries to get reliable records of many related individuals from written data or from observations by competent individuals. A set of related individuals is called a **kindred**, which is a larger unit than one family. For the purposes of pedigrees, the term "marriage" can include less formal relationships.

In the **pedigree diagrams**, the earliest generations are nearest the top, with generations having Roman numerals. A horizontal line directly connects two individuals of opposite sex who are "married", with their children, if any, shown coming from a lower horizontal line in age order, the first-born to the left, and they may be numbered for ease of reference, from left to right. Individuals marrying into the kindred have no vertical line above them, and are numbered along with those born into the kindred in that generation. Individuals with the trait have a filled-in symbol and unaffected ones have empty symbols, with a circle for females, a square for males. The proband is indicated by a bold arrow and the letter P. Deceased individuals may be indicated by a diagonal crossing through of the symbol. Identical (monozygotic) twins have divergent downward lines joined by a horizontal line, while non-identical (dizygotic) twins just have the downward divergent lines connecting them. A diamond with a number inside indicates that number of unaffected offspring, of unspecified sex. Some but not all of these symbols are shown in Fig. 13.2.1.1, for a kindred with **severe foot blistering**. For example, in generation II, there were five unaffected children and one affected one, who married an unaffected man, and they were the parents of generation III, in which the last-born male had two marriages, to different females. The proband is shown by the arrow in generation IV.

When trying to work out the genetics of severe foot blistering, we see that once that trait has appeared in the kindred, all affected individuals have at least one affected parent, suggesting dominance. In families where one parent is affected (and probably heterozygous, not homozygous) and one is unaffected, affected and

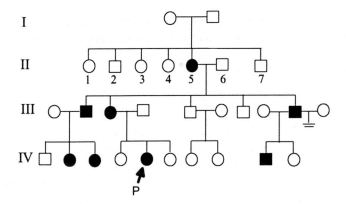

Figure 13.2.1.1. A pedigree for severe foot blistering. The proband is indicated by an arrow. The symbol downwards between the last two individuals in generation III indicates a childless marriage or relationship. From Haldane and Poole, J. Heredity, 1942, 33: 17, but with some individuals omitted.

unaffected children are produced in approximately equal numbers (seven affected, six unaffected), which again fits with dominance, with affected individuals being heterozygous as they have only one affected parent.

The trait can be passed from affected fathers to sons and daughters, and approximately equal numbers of males and females are affected, suggesting that the trait is an autosomal dominant, not X-linked. A dominant X-linked trait could not be passed to a son from his father, as the son inherits the father's Y chromosome, not his X. Additionally, the trait cannot be an X-linked dominant as a father always passes on his one X to his daughters, but an affected father had an unaffected daughter (IV-10).

For a trait so rare in the population, it could not be an autosomal recessive, as that would require nearly all the unaffected spouses to be heterozygous carriers, which is highly unlikely. For an autosomal recessive character, both parents of an affected individual must have at least one copy of the recessive allele.

In generation I, both recorded parents were unaffected, yet they had an affected offspring showing this autosomal dominant trait. One explanation is that the gamete from one parent carried a spontaneous mutation from the recessive wild-type allele to the dominant allele. The other most likely explanation is that in generation I, the recorded father is not the child's biological father, with the mother having mated with an unrecorded affected man, producing the affected daughter. It is a fact of human genetics that **adultery** is much more common than spontaneous mutation, so the second explanation is more likely than the first, though mutation cannot be ruled out. Other less likely explanations include incomplete penetrance of the dominant allele in one parent, or one parent being a mosaic of normal and mutant tissue.

When looking at pedigrees, one looks to see if the traits are dominant or recessive, and then whether they are autosomal or X-linked. Recessive X-linked characters tend to affect males (hemizygous) much more than females, especially if the recessive alleles are very rare (Chapter 13.6).

Many human kindreds are less extensive than this one, but can still be conclusive as to the mode of inheritance. Working out pedigrees is exciting detective work, but can be frustrating if key individuals are not available through distance, or absence of an address, or through death, or if they are unwilling to be investigated. One can make pedigrees using molecular characters as well as morphological ones, but some individuals may be unwilling to submit samples for analysis.

13.2.2. Twin studies

Twin studies have been used in man and cattle in deciding the relative contributions of heredity and environment (i.e., heritability) for particular traits. See also Chapter 3.3 for determining heritabilities from covariances, including using pairs of monozygotic twins. Twins are either **monozygotic** (identical), from one egg, one sperm, one zygote, where the embryo divides into two genetically identical embryos, or **dizygotic** (non-identical, fraternal), from two separate fertilised eggs which are genetically different. Monozygotic twins are almost always of the same sex, but dizygotic twins can be the same or different sexes, just like siblings. Dizygotic twins are no more alike, except in age, than a pair of full siblings born at different times. Because monozygotic twins are of the same sex, one should use dizygotic twins of the same sex when comparing the two types of twins, as sex differences can be present in addition to other genetic differences.

One can compare traits in monozygotic (same genotype) and dizygotic (different genotypes) twins reared together (same environment) and reared apart (different environments). With humans, cases of twins reared apart are fairly rare, by chance, but with farm animals one can do controlled experiments. Because of the difficulty in finding enough pairs of twins, especially twins reared apart, some authorities keep special **"twin registers"** of pairs of twins willing to cooperate in genetic studies, such as that kept at the Twin Research Unit at St. Thomas' Hospital, London.

One method used to find relative contributions of heredity and environment for quantitative characters is **average pair differences**; one measures a number of pairs of twins to find out the average difference within pairs. Typical data are shown in Table 13.2.2.1. With head width, the environment induces small differences within pairs of monozygotic twins, but genetic effects are more important because the dizygotic twins column, showing pairs with individuals differing in genotype, has much bigger average pair differences than do the monozygotic twins columns. Monozygotic twins reared apart are not much more different than are monozygotic twins reared together, again showing only small amounts of environmental effects. The analysis of head width is very similar to that for height, showing small environmental effects but much larger genetic effects. For weight, however, monozygotic twins reared apart are much more different than are monozygotic twins reared together, suggesting a big environmental component of weight. There is a genetic component too, because monozygotic twins reared together are more alike than dizygotic twins or alike-sex siblings reared together.

Table 13.2.2.1. Average pair differences for three quantitative traits, height, head width, and weight.

Trait	Monozygotic twins	Monozygotic twins	Alike-sex dizygotic twins	Sibs of alike sex scored at the same age
Reared:	Together	Apart	Together	Together
Height (cm)	1.7	1.8	4.4	4.5
Weight (kg)	1.9	4.5	4.5	4.7
Head width (mm)	2.8	2.9	4.2	No data

For qualitative characters, one uses **concordance scores**. A pair of twins is concordant if they both have the trait, and are discordant if one has it and the other does not. Table 13.2.2.2 shows concordance scores from monozygotic and alike-sex dizygotic twins.

If pairs of monozygotic twins are not always concordant for a trait, that suggests that there is an environmental or chance element to developing that trait. A high concordance score for monozygotic twins does not necessarily prove a high heritability, because monozygotic twins are usually brought up in the same environment, as are pairs of dizygotic twins. What demonstrates a strong genetic determination for a trait is if monozygotic twins have a high concordance and dizygotic twins have a much lower concordance, as for non-insulin-dependent diabetes mellitus and schizophrenia. Where monozygotic twins are discordant for having a criminal record, the one lacking such a record might not have been criminal,

Table 13.2.2.2. Concordance scores for monozygotic and dizygotic twins for various characters.

Trait	Monozygotic twins Concordance, %*	Dizygotic twins Concordance, %*
Age at first walk	68	31
Stomach cancer	27	4
Scarlet fever	64	47
Schizophrenia	46-80	13
Down syndrome	89	7
Criminal record	68	28
Coffee drinking	94	79
Clubfoot	32	3
Male homosexuality	50-100	25
Handedness (left or right)	79	77
Hair colour	89	22
Eye colour	99.6	28
Diabetes mellitus, non-insulin-dependent	100	10
Diabetes mellitus, insulin-dependent	30-40	6
Rheumatoid arthritis	30	5
Epilepsy	37	10
Tuberculosis	87	26

* Twins are concordant when they both show the same trait, e.g., both develop stomach cancer, or are both phenotypically similar for a measured character within a small range, e.g., for age at first walk. Data from various sources including Connor and Ferguson-Smith (1997) and Strickberger (1985).

or might just not have been caught (a chance or environmental effect). With stomach cancer, a heritable element was shown by monozygotic twins having a much higher concordance than dizygotic twins (27% versus 4%), but the fairly low concordance for monozygotic twins shows a large chance (environmental) element.

Club foot is multifactorial, and the higher concordance of monozygotic twins than of dizygotic twins shows a clear heritable element, but with only 32% concordance for monozygotic twins, there must also be a large element of chance (or environmental factors, some of which are unknown) in whether twins with the same polygenes actually develop club foot. The trait has a frequency of 0.1% and a narrow sense heritability of 68%.

The most unexpected figure in Table 13.2.2.2 is the 89% concordance for monozygotic twins for **Down syndrome**, when one would expect 100% for a condition caused by having three copies of chromosome 21. Perhaps when a zygotic nucleus with three copies of chromosome 21 divides to give monozygotic twins, one extra copy of 21 sometimes gets lost, giving one normal and one Down's embryo, with one chance non-disjunction of chromosomes correcting an earlier non-disjunction.

The data on **schizophrenia** show a strong heritable element, but with some environmental or chance factors also operating, since only 46-80% concordance is shown for monozygotic twins. Coffee drinking is probably largely environmental (influenced by family habits rather than family genes), with a slight heritable element shown by monozygotic twins having a somewhat higher concordance than dizygotic twins. Where monozygotic twins are discordant for a largely inherited character such as eye colour or natural hair colour, one can often attribute the difference to incomplete penetrance. Even where hereditary determinants are shown by concordance scores to operate, the concordance scores do not tell us how the character is inherited, e.g., monogenic or polygenic, dominant or recessive, X-linked or autosomal. Data on concordance may differ sharply in different studies. The concordance for monozygotic twins for schizophrenia was 80% in one study but 46% in another.

13.2.3. Hardy-Weinberg analysis of populations

We considered **Hardy-Weinberg** analysis in Chapter 4.2. It can be used to test how human traits are inherited. For example, a population of 420 subjects was studied for the M and N antigens on red blood cells. 137 had the M antigen only, 87 had the N antigen only, and 196 had both antigens. A simple hypothesis for the inheritance of the M and N blood groups is that they are determined by two alleles at one locus, with codominance, 137 being $M M$, 87 being $N N$, and 196 being $M N$. The allele frequency of M is therefore $\frac{(137 \times 2)+(196 \times 1)}{420 \times 2}$ = 0.56, so the allele frequency of N is 0.44. One therefore expects Hardy-Weinberg equilibrium frequencies of the M blood group to be 0.314, the N group to be 0.194 and the MN group to be 0.493. The expected numbers out of 420 would be M group, 132, N group, 81, MN group 207, giving good agreement between observed and expected numbers (a χ^2 test gives $P = 0.28$). The hypothesis of two alleles with codominance therefore fits the data, and most other hypotheses do not.

13.2.4. Familial incidence

A character with a large heritable element will tend to "run in families", with an affected person often having one or more affected relatives. A character being heritable is not quite the same as being **familial,** tending to run in families, because members of a family tend to have more similar environments than unrelated individuals, so familiality could be caused by environmental effects or heritable effects, or both. Speaking a particular language, for example, is familial, but not heritable. For traits with little effect of family environment, one can use familiality as evidence that a character is inherited. For example, people with **vitiligo** (giving unpigmented skin patches; see Plate 26 and Lesage, 1997) have one or more affected relatives more often than one would expect from the frequency of the trait in the population, suggesting a heritable element to vitiligo. It affects between 1% and 4% of individuals in populations throughout the world, but with different degrees of severity from a few small white patches to complete whiteness all over the face, body and limbs. Vitiligo does not usually follow a simple monogenic pattern of inheritance and various environmental factors are probably involved, so it is multifactorial (see Chapter 13.4); it also has different ages of onset in different people, and different triggering factors. There are probably several different heritable forms of vitiligo.

To use familiality, one studies the frequencies say of diseases or malformations among genetically related individuals and in the general population. If the trait is heritable, there is a higher frequency of the trait amongst the relatives of affected individuals than in the general population, with the increase in frequency being proportional to the degree of relatedness. This is shown in Table 13.2.4.1 for the incidence of cleft lip (with or without cleft palate) in Europe and North America, where the trait has a population incidence of 1/1,000 and a narrow sense heritability of 76%. The frequency of cleft lip among relatives of an affected person is much higher than in the general population, especially for first-degree relatives, but the decrease in second- and third-degree relatives does not fit expectations of an

Table 13.2.4.1. Familial incidence of cleft lip. Data from Carter (1969).

Relatives	Percentage of affected relatives	Incidence relative to the general population
First degree:		
Full sibs	4.1	x 41
Parents/children	3.5	x 35
Second degree:		
Aunts and uncles	0.7	x 7
Nephews and nieces	0.8	x 8
Third degree:		
First cousins	0.3	x 3

autosomal dominant (where a decrease of a half is expected for each degree of relation less) or an autosomal recessive. Concordance scores for cleft lip are 40% for monozygotic twins and 4% for dizygotic twins, suggesting an environmental or chance component in addition to a strong hereditary element. For further information on familiality, especially for multifactorial traits, see Gelherter, Collins and Ginsburg (1998).

13.3. Single gene characters and disorders, and their treatments

A range of different conditions is described, to show different modes of inheritance, different effects, causes, diagnoses and treatments. It is not exhaustive and concentrates on inherited disorders. Some such as polydactyly (extra digits) are expressed early in the embryo, so they show at birth, while others may only be expressed much later in life. For example, **Huntington's chorea** (Huntington disease, an unstable length mutation; see Chapter 7.1.6) has an average age of onset of 37 years, affecting people at any age from the teens to the eighties. It involves progressive nervous and mental deterioration, with chorea (involuntary leg and arm movements). It is eventually fatal. It is an autosomal dominant, affecting about 1 in 20,000 people. The frequency of alleles for inherited diseases often differs sharply between different racial groups. Sometimes the cause for these differences is understood, such as exposure to malaria increasing the frequency of alleles for thalassaemia and sickle-cell anaemia, but the reason for cystic fibrosis being so much commoner in Europeans than in Negroes or Orientals is not clear. **Leber's optical atrophy** (or neuropathy) is unusual in being a mitochondrial mutation and showing maternally inherited blindness.

13.3.1. Autosomal genes

Chapter 2.1 covered the general inheritance of such characters, so this section deals with a range of dominant and recessive traits determined by single autosomal loci, including diagnosis, symptoms and treatment where relevant.

The common form of **polydactyly** (isolated polydactyly) is an example of an autosomal dominant allele in otherwise normal individuals (although there are other aetiologies, e.g., trisomy 13, Meckel syndrome, and some autosomal recessive conditions). One in 2,000 Caucasians has extra fingers and/or toes. It has variable expressivity, so some sufferers have strong expression, with more than one extra complete toe and/or finger, while others just have a lump on one side of the hand. Sufferers may be homozygous, when both parents should have the condition, or heterozygous, which is much more frequent for a rare trait, as expected from the Hardy-Weinberg equilibrium (Chapter 4.2). Homozygotes will have all offspring affected, but heterozygotes, if married to an unaffected person, will have about half their offspring affected. It is slightly disfiguring but not usually serious, and is treated by surgery to remove the extra digits, or by tying them off in the first week of life to cut off their blood supply if they are little more than a small polyp: see Plate 27.

Another rare autosomal dominant is **achondroplasia**, giving **dwarfism**, caused by allele *D*. Such dwarves have short stature (mean adult height 132 cm in males, 123 cm in females), relatively large heads, normal IQ and lifespan and short limbs, especially in the proximal parts: see Plate 28. It is unusual for a dominant allele in that about 84% of cases have two normal parents and arise from new mutations in a parent's germ line. Even more surprising is that almost all mutations are in exactly the same nucleotide, in a gene encoding the type 3 fibroblast growth-factor receptor, which mediates the effect of that factor on cartilage growth. Allele *D* acts as a recessive lethal, although dominant for dwarfism, so that *D d* gives a viable dwarf but *D D* is lethal, giving a very short-limbed individual with a very small chest who invariably dies *in utero*. The frequency of the mutation, and therefore of the trait, is about 6×10^{-5} per gamete. Dwarves are fertile, producing a 1:1 ratio of dwarves to normals if mated with non-dwarves, or 2:1 if mated to dwarves, but females have small pelvises and usually need a Caesarean section when giving birth.

Sickle-cell anaemia is caused by an autosomal recessive allele, *a*; when homozygous this causes severe and usually lethal anaemia, with the red blood cells going sickle-shaped at low oxygen concentrations and being removed in the spleen. Most sufferers die young, but some live 20 years or more, with many medical problems. It can be extremely painful when small blood vessels become blocked by distorted red cells. The heterozygote, with sickle-cell trait, is clinically more or less normal in spite of expressing the mutant β^S globin gene as well as the wild-type one. Only at very low oxygen concentrations does the blood of heterozygotes show sickle-shaped red blood cells, and people with sickle cell trait sometimes need help from oxygen masks on long-distance flights, as planes have reduced air (and therefore oxygen) pressure. During the second world war, aircrew with African ancestors often suffered blackouts in high-flying unpressurised planes, from the low oxygen, if they were heterozygotes for sickle-cell anaemia.

In sickle-cell anaemia, there is a mis-sense mutation (see Chapter 7.1.3) in codon 6 of the β globin gene on chromosome 11, changing GAG in the normal allele to GTG, and so coding for glutamic acid instead of valine. Most people with the *a* gene have exactly the same mutation, and its frequency is much too high in some populations to be due to chance mutation. The Amba people of Uganda have about 40% with sickle-cell trait, and its incidence in black Americans is about 9%. It was A. C. Allison who showed that the heterozygotes with sickle-cell trait, *A a*, were much less susceptible than normal *A A* individuals to malaria caused by *Plasmodium falciparum*. In malarial regions, the *a* allele is maintained by heterozygote advantage, even though the recessive homozygote suffers severe or lethal anaemia. For the population genetics, see Chapter 15.3. The heterozygotes share their falciparum malaria resistance with heterozygotes for beta thalassaemia (see below) and for glucose-6-phosphate-dehydrogenase deficiency (Chapter 13.3.2), which is X-linked, unlike the other two traits. Sickle cell trait occurs mainly in people of African malarial region origin, but also in those from other malarial areas, such as southern India, Greece and Italy.

There are a number of inherited autosomal recessive diseases associated with defective phenylalanine or tyrosine metabolism. In normal individuals, allele *P* specifies the liver enzyme phenylalanine hydroxylase, converting dietary phenylalanine into tyrosine, which can be converted by tyrosinase to the melanin eye, skin and hair pigments, or by other enzymes to thyroxine growth hormone, or to homogentisic acid. People suffering from **phenylketonuria** (**PKU**) are *p p* and cannot convert phenylalanine to tyrosine and thence to melanin, so they have lighter pigmentation but are not albinos as the diet provides some tyrosine. The main effect is that phenylalanine builds up in the blood, with up to a gram a day being excreted in the urine. Most of the amino acid is converted to phenylpyruvic acid and other compounds which damage the developing brain in babies, especially during the first six months, causing severe mental retardation. Two thirds of untreated PKU babies grow up as idiots with IQs of less than 20, plus a range of physical symptoms, including a small head. The frequency of the disease is about 1 in 25,000 in Europe.

Fortunately, PKU is easily detected and treated. In developed countries, babies around three days old have a blood sample taken from a heel-prick. The blood is used in a bioassay with phenylalanine-requiring strains of *Bacillus subtilis*, which can grow when the blood has abnormally high concentrations of phenylalanine. PKU babies are immediately taken off milk and are given a low phenylalanine diet for about 15 years, while the brain is still developing. The treatment is very effective. The heterozygotes, *P p*, can be detected by a test in which phenylalanine is injected into the blood and its rate of removal is monitored, as they have less enzyme than *P P* individuals. Detection of heterozygotes is useful for genetic counselling (Chapter 13.13).

People who lack the tyrosinase enzyme cannot make the melanin pigments from tyrosine and are **albinos**, *a a*: see Plate 4. Albinos have white skin and hair. In strong light, their eyes appear pink from the blood vessels in the retina; the lack of eye pigments usually results in visual problems. There is no treatment but sufferers usually wear tinted glasses and some now wear contact lenses with coloured irises so that they look more normal. Their exposed skin needs sun-block protection from the UV in sunlight. The frequency of albinism is about 1 in 20,000 in Europe, 1 in 3,000 in Nigeria, and 1 in 132 in San Blas Indians in Panama. Like vitiligo (Chapter 13.2.4 and Plate 26), the disease shows up most clearly amongst races with dark skins.

People who lack the enzymes to convert tyrosine to the thyroxin hormone suffer **goitrous cretinism**, with stunted growth and mental defects. Tyrosine is also converted to homogentisic acid, which is converted by the enzyme homogentisic acid oxidase to maleylacetoacetic acid and thence to water and carbon dioxide. People lacking this enzyme have **alkaptonuria**. They excrete homogentisic acid in the urine, which goes black in the light or if alkaline. Deposits can cause degenerative arthritis of the spine and large joints, and black or ochre pigments in cartilage and collagenous tissues, including the palate and the sclerae of the eyes.

The nature of some of these inherited biochemical diseases was first recognised by the English physician Archibald Garrod, who studied alkaptonuria, pentosuria, cystinuria and albinism. His pioneering book, *Inborn Errors of Metabolism*, was published in 1909 (Henry Frowde, Hodder and Stoughton, London).

Cystic fibrosis is one of the commonest severe single gene disorders, affecting 1 in 622 Afrikaners in South Africa, 1 in 2,000 white births in Europe, but only 1 per 250,000 Negroes and Orientals. It is called cystic fibrosis of the pancreas because the part of the pancreas which normally produces digestive enzymes is often replaced by fibrous scar tissue with fluid-filled cysts. An upset glycoprotein metabolism results in abnormal secretions. The **sweat** is very high in sodium chloride, which can result in salt depletion. This symptom must have been known for hundreds of years, as old Northern Europe folklore has a saying: "Woe to that child which when kissed on the forehead tastes salty. He is bewitched and soon must die". The **lungs** produce thick, viscid mucus which does not drain readily, so that sufferers are prone to lung infections such as pneumonia. The **pancreas** fails to secrete adequate amounts of three important digestive enzymes: amylase, which converts polysaccharides to disaccharides; trypsin, which helps to digest proteins to amino acids, and lipase, which converts triglycerides to fatty acids and glycerol. With impairment of carbohydrate, protein and fat digestion, malnutrition results.

Cystic fibrosis is caused by allele *c* on chromosome 7, and the gene was cloned in 1989. There is no cure, but the different **symptoms can be treated**, such as giving salt tablets for salt depletion. Antibiotics and vigorous physiotherapy to help drain the lungs are used for the respiratory problems. Diets can be adjusted to ameliorate the digestive problems, and sufferers can be given encapsulated enzymes (usually from dried pig pancreas extract) with buffers, to take with every meal. Untreated children usually die before the age of 10, with an average life expectancy of five years, but with treatment at least 70% now survive to adolescence, with an average survival of 30 years or more. Those who avoid serious lung infections live to 50 or 60, while those infected by the bacterium *Pseudomonas aeruginosa* have a life expectancy of about 30 years, going down to 16 years for those attacked by the related opportunistic human parasite (but mainly a pathogen of onions), *Burkholderia cepacia*. The females are fertile but not the males, who usually lack a functional vas deferens.

The high frequency of this deleterious allele, with about 1 in 25 white Europeans being heterozygous carriers, has been attributed to **heterozygote advantage**, but without much experimental evidence. One theory is that the heterozygote is better able to resist the lethal effects of diarrhoea-causing diseases such as cholera, as heterozygotes have fewer chloride-ion channels in the gut than dominant homozygotes, and so excrete less water in diarrhoea, resulting in fewer deaths from dehydration. It has also been suggested that the lower incidence of cystic fibrosis in Negroid Africans and in Asians comes from the somewhat saltier sweat of the heterozygote than the normal homozygote, with selection in hotter climates against salt-depletion being more severe than in colder climates.

The **thalassaemias** (meaning "sea blood", from the reduced haemoglobin content of the blood, giving anaemia) are diseases caused by a deficiency, but not an absence, of α-globin, giving **α-thalassaemia**, or of β-globin, giving **β-thalassaemia**. α-thalassaemia is very common in Southeast Asia, while β-thalassaemia is very common in people from Cyprus, Greece, Turkey and other malarial (or formerly malarial) regions of the Mediterranean. In Cyprus, carriers of β-thalassaemia recently made up about 16% of the population, with 0.6% of births being of recessive homozygotes, usually dying within the first year of life. This very high frequency of a harmful recessive is accounted for by heterozygote resistance to malaria. Malaria was eliminated from Cyprus after World War II, which should remove the heterozygote advantage and reduce the frequency of the recessive alleles. Genetic counselling and screening of those coming up to marriageable age is helping to reduce the frequency of recessive homozygotes born.

Normal adult human haemoglobin (haemoglobin A) is a tetramer of two α-haemoglobin and two β-haemoglobin molecules. In α-thalassaemia, there is a large excess of β chains, which form homotetramers, giving inclusion bodies of haemoglobin H in the red blood cells, which have much less oxygen-carrying capacity than normal and are small (50-80 μm^3 instead of 90) and reduced in number, giving severe anaemia. In β-thalassaemia, the excess α chains form homotetramers, but these are insoluble and precipitate, leading to destruction of the red blood cells in bone marrow and spleen. The red blood cells are again small and reduced in number.

Each chromosome 16 carries two α-globin genes in tandem, so that a normal individual has four such functional genes, written αα/αα. If one of the four genes is mutant, αα/α- (where "-" here means nonfunctional, not any allele), one gets the "**silent carrier**" state, which is virtually normal. There are two ways of having two genes inactivated, giving α-thalassaemia 1, also called **α-thalassaemia trait**, where symptoms are mild, although red blood cell volume is somewhat reduced. In Southeast Asia it is common to find αα/- -, with both genes on one chromosome inactivated, but "blacks" usually have one globin gene on each chromosome inactivated, α-/α-. If three genes are inactivated, α-/- -, one gets **haemoglobin H disease**, as there are four times more β globin chains than α chains. Sufferers have moderate to severe, but not lethal, anaemia when born. The case with all four genes inactivated, - -/- -, gives **lethal Hydrops fetalis**, with stillbirth or neonatal death.

There is only one β-globin gene on each chromosome 11, where mutations can affect the amount of functioning β-globin produced, e.g., from having different levels of transcription, from none to almost wild-type levels. Three levels of severity are recognised. In **β-thalassaemia minor**, a heterozygous condition, there are no symptoms and the genotype has one normal allele and one with no or a reduced function. In **β-thalassaemia intermedia**, individuals are anaemic, with symptoms, but do not need blood transfusions. Both β-globin genes are mutant, but one mutation is mild, so that a moderate but reduced amount of β-globin is produced. In **β-thalassaemia major**, both alleles are mutant, with no or severely reduced function. If both are nonfunctional, with no β-globin and therefore no haemoglobin A, one gets $β^0$-thalassaemia, but if one allele has some function, one gets $β^+$-thalassaemia.

People with β-thalassaemia major require frequent blood transfusions, or the condition is eventually lethal. It is not usually marked at birth because foetal haemoglobin is still being produced. As foetal haemoglobin production tails off during the first year of life, symptoms of severe anaemia slowly develop, together with weakened bones and enlargement of the liver and spleen. If transfusions are not given, death usually occurs within 10 years, from anaemia, weakness and infection. Progressive transfusions, however, lead to a build up of iron, which is eventually deposited in the heart, liver and other organs, and can lead to their failure, with death frequently occurring in the teens and twenties. Continuous infusion with iron-chelating agents, especially desferrioxamine, requires the wearing of a pump for subcutaneous injection, which is inconvenient but prolongs life.

13.3.2. X-linked genes

X-linked recessive deleterious alleles typically affect males, because they are hemizygous, much more often than they affect females, who need two copies of the recessive alleles to show the condition (see Chapter 13.6 for equations). The rarer the recessive allele, the bigger the imbalance between the sexes. The severity of such conditions varies from mild for red-green colour blindness (8% of males), to eventually lethal for Duchenne muscular dystrophy (3 per 10,000 males). See Chapter 7.1.6 for **fragile-X syndrome**, due to an unstable length mutation.

Glucose-6-phosphate-dehydrogenase (G-6-P-D) is an enzyme necessary for the stability of glutathione, which plays a role in cellular respiration. A greatly reduced activity for this enzyme is rare in "whites", but occurs in about 9% of American "blacks" and in some people of Mediterranean origin; heterozygosity for the deficiency confers some malaria resistance as mentioned in Chapter 13.3.1. Hemizygous males and homozygous recessive females for this X-linked trait are phenotypically normal most of the time, with the condition only showing in enzyme tests. They become ill, with the sudden destruction of many red blood cells giving severe haemolytic anaemia, when they inhale pollen of broad beans (*Vicia faba*) or eat the raw beans (favism), or on taking certain drugs such as sulphanilamide, the antimalarial primaquine, or naphthalene (used in moth balls). They recover when the causative agents are removed. The heterozygous females usually have intermediate enzyme levels, but their level may vary from low to normal. There are many different alleles at this locus, unlike with sickle-cell trait.

Another X-linked recessive condition is **Duchenne muscular dystrophy**, giving progressive muscular wasting from about age 5. Affected children are usually confined to wheelchairs by the age of 10, with death before 20. Although the functions of the large dystrophin protein are understood and the gene has been cloned, there is no satisfactory treatment at present. The frequency is about 1 in 3,000 male births. See also Chapter 14.1.9.

Haemophilia A is an X-linked recessive condition resulting in spontaneous and wound-induced bleeding in soft tissues and joints, with very poor clotting, so it can be fatal. The gene codes for clotting factor VIII and is large, about 200 kb long. Factor VIII levels in sufferers vary from 30% of normal (mild symptoms) to less than 1% (severe). One particular inversion (see Lakich et al., 1993) accounts for half of all severe cases and for 20% of all cases. Different mis-sense, nonsense, frame shift and deletion mutations of varying severity account for the other haemophilia A cases. Carrier detection and prenatal diagnosis are possible by blood or DNA analyses. It affects 1 to 2 males per 10,000 and accounts for 85% of haemophiliac patients. Treatment with intravenous factor VIII is effective.

Heterozygous females often have less factor VIII than normal but are usually symptomless. **Haemophilia B** has similar symptoms to haemophilia A and is also an X-linked recessive, but comes from mutations in a different gene, giving a deficiency of clotting factor IX. It affects 1 in 30,000 males and is treated by intravenous factor IX injections.

Testicular feminisation (Androgen insensitivity syndrome) results in individuals who are genetically XY developing as females, with excellent breast development and female external genitalia with a short vagina, but they are sterile, with small internal testes. They often marry as females, unaware of their condition. The single X carries a recessive allele resulting in insensitivity to male hormones. They are usually only detected if they report their lack of menstruation and are then karyotyped as XY, 2n = 46. The testes are often surgically removed because gonadal neoplasms are frequent (5–20%). It affects 1 in 62,000 males.

13.4. Polygenic and multifactorial disorders

Polygenic disorders or multifactorial disorders arise from the interactions of genes at many loci, some of which may have major effects but many of which have individually small effects. There are often effects of the environment in the aetiology too, so the term "multifactorial" is used. The terms "polygenic" and "multifactorial" are sometimes used interchangeably, although "polygenic" is sometimes used for traits controlled only by polygenes, and "multifactorial" for ones with environmental effects too. There may also be incomplete penetrance and/or variable expressivity (Chapter 3.2.4 and 3.2.5), making their genetics difficult to be sure about. An example of a polygenic trait presumed to have little environmental effect is the total number of fingerprint ridges counted over all ten fingers, which is determined early in development and which gives a roughly normal distribution of values.

Multifactorial disorders include club foot, cleft lip and cleft palate, coronary artery disease, most congenital heart diseases, hypertension, vitiligo (see Chapter 13.2.4), diabetes mellitus and schizophrenia. They account for about one third of paediatric hospital admissions and about a third of childhood mortality. Cleft lip, with or without a cleft palate, occurs in 1 in 1000 births.

Many multifactorial traits give qualitative phenotype variation, where an individual has the trait or does not have it, though there may be different extents of expression amongst those with the trait. For example, babies either have a cleft lip or do not, but those who have it have it to different extents, and some have a cleft palate as well. People at a given time either have or do not have schizophrenia or insulin-dependent diabetes mellitus (type 1 diabetes). For such qualitative traits with multifactorial control, it is thought that there is a more-or-less continuously distributed genetically determined predisposition to that trait, with additive effects of polygenes and certain environmental factors, with only individuals exceeding some threshold value developing the symptoms.

As an affected individual usually has a genotype with a high predisposition to the condition, his or her genetic relatives are likely to have a greater predisposition to the condition than are unrelated individuals, with the risk of being predisposed being greatest for the most closely related relatives. Suppose that for a given level of environmentally predisposing factors it requires say 10 cleft-determining polygenes to cause a **cleft lip**, and 20 cleft-determining polygenes to give a cleft palate as well, then the risk of siblings (brothers and sisters) of someone with a cleft lip and a cleft palate having just a cleft lip will be higher than the risk of those siblings having a cleft lip and a cleft palate. The actual risk of a cleft lip in siblings of children with just a cleft lip is 2.5%, but is 6% in siblings of children with a cleft lip and a cleft palate. In contrast, with an autosomal recessive single-gene condition such as sickle-cell anaemia or cystic fibrosis, the risk of a sibling of an affected child being affected is one quarter (if both parents are unaffected heterozygotes), and for a dominant autosomal gene like isolated polydactyly, the risk is one half (if one parent is heterozygous and the other is homozygous recessive). For a number of multifactorial diseases, including congenital heart disease and insulin-dependent diabetes mellitus, the risk of a sibling of an affected child being affected (called the **recurrence rate**) is roughly 5% or less. See Gelehrter, Collins and Ginsburg (1998), Chapter 4, for further details.

Multifactorial diseases are much more common than single-gene diseases. In the UK, with about 58 million people, multifactorial disease sufferers number about 400,000 for Alzheimer's, 1 million for diabetes, 1.25 million for ischemic heart disease, 1.5 million for cancer, and 5 million for hypertension, compared with 2,500 for Huntington's, 3,000 for Duchenne muscular dystrophy, and 7,000 for cystic fibrosis single-gene diseases.

Diabetes is caused by a deficiency of insulin or a loss of response to insulin, so blood glucose concentrations rise, leading to glucose excretion in the urine, excessive urine production, persistent thirst, poor carbohydrate utilisation and weight loss. Diabetes is familial and can lead to hypoglycaemia and coma, kidney failure, blindness, and other long-term complications. The three main types of diabetes have a combined incidence of about 1 in 50, varying with age. **Insulin-dependent diabetes mellitus** (IDDM, type I diabetes) affects 1 in about 400 children with a peak age of onset of 12 years, and also affects adults. It is an auto-immune disease, with T-lymphocytes attacking the pancreas β-cells which make insulin. It is polygenic, with at least 10 contributing loci, especially an HLA gene on chromosome 6 and the IDDM2 insulin gene on chromosome 11. Treatment consists of several insulin injections each day, plus dietary restrictions. **Non-insulin-dependent diabetes mellitus** (NIDDM, type II diabetes, adult-onset diabetes) is caused by either insufficient insulin, or more commonly, a reduced responsiveness to insulin in target cells, from changed receptors. It usually occurs after about age 40, becoming more frequent with increasing age. More than 90% of diabetics are type II, and most can control it by diet and exercise alone. Both type I and type II are multifactorial: see Table 13.2.2.2 for their different concordance scores. A third type of diabetes, **maturity-onset diabetes of youth**, is usually inherited as an autosomal dominant.

13.5. X-inactivation and Barr bodies

In diploid mammalian females, one of the two X chromosomes in a cell is inactivated, forming a small, flat structure within the nucleus, a **Barr body**, stainable with dyes for DNA. The inactivation occurs early in the embryo by DNA methylation, and the inactivation persists through subsequent cell divisions. It is a matter of chance which of the two X chromosomes is inactivated in any embryonic cell, so the resulting female is a mosaic of patches of tissue, with one X inactivated in some patches and the other inactivated in other patches. This is very obvious in tortoiseshell (Plate 29) and calico cats, which are females heterozygous for black (*o*) and orange (*O*) alleles at a locus for fur colour; some patches of fur are orange (black-carrying X inactivated) and others are black (orange-carrying X inactivated). Human females heterozygous for X-linked anhidrotic ectodermal dysplasia (lack of sweat glands) have been shown to be external mosaics, with affected patches sometimes covering the whole face and neck, or one shoulder, or part of one thigh or most of one lower leg, with oval or irregular patches varying from a few centimetres to over 70 cm long.

This mechanism helps to solve the problem of unequal gene doses for X-linked loci, where the female has two copies of each locus (but only one active copy) while the hemizygous male only has one copy. Thus for red blood cell glucose-6-phosphate-dehydrogenase activity (see Chapter 13.3.2), the average amounts of this X-coded enzyme are the same in hemizygous males as in females with two copies of the gene. There is only one active X per cell, so an individual with one X has no Barr bodies (normal males and Turner syndrome females - see Chapter 13.8), one with two X has one Barr body (normal females and XXY males), an individual with three X has two Barr bodies, etc. The presence or absence of Barr bodies means that normal females and normal males can be sexed even in non-dividing cells just from the number of Barr bodies. This has been used a sex-test for athletes, being less intrusive and embarrassing than a genital inspection. It can also be used for immediate sexing of foetuses in prenatal diagnosis (Chapter 13.11). Not quite all genes are inactivated on an inactivated X, and inactivation is not always random if one of the chromosomes carries an aberration.

13.6. Sex differences in disease susceptibility

13.6.1. X-linked diseases

With X-linked recessive diseases, one expects **more male sufferers** than female sufferers, because males, being hemizygous, only need the one recessive-carrying X, while females need two recessive-carrying X chromosomes to be affected. One needs modified Hardy-Weinberg equilibrium equations to take account of hemizygous males. With a recessive allele a, frequency q, one gets equilibrium genotype frequencies for females of $A\ A$, p^2; $A\ a$, $2\ pq$; $a\ a$, q^2, but for males one gets A (Y) $= p$, and a (Y) $= q$. The ratio of female to male sufferers is therefore q^2: q, or q female sufferers to one male sufferer. Thus the rarer the recessive allele, the bigger the proportional difference between male and female sufferers. For red-green colour blindness, about 8% of males are red-green colour blind, so $q = 0.08$ and $q^2 = 0.0064$, with only 0.64% of female sufferers, with an imbalance of 1 male sufferer to only 0.08 female sufferers, a 12.5-fold difference. With a much rarer condition, haemophilia A, an inability to clot blood in wounds, the frequency of male sufferers is about 1 in 10,000, so $q = 0.0001$, and the frequency of female sufferers expected is 0.00000001, or 0.000001%, or 1 in 100 million. One thus expects about one quarter of a suffering female in the whole of Britain, but about 2,800 suffering males, a 10,000-fold sex difference. Female haemophiliacs are thus expected to be extremely rare, but in reality have a frequency about 4% of that of male sufferers. This is because true homozygotes arise from consanguineous relationships (so the Hardy-Weinberg frequencies are not obtained), and because of the vagaries of X-inactivation, so that at the extreme the cell line with one particular active X predominates. Female sufferers may also come from having chromosome rearrangements involving the X chromosome, such an X-autosome translocation. Female haemophiliacs do not die from excessive bleeding at menstruation, because menstruation ends by blood vessel contraction, not clotting; they can have children (all sons will be sufferers) but close medical attention at birth is essential to control bleeding.

One can obtain the frequency of the recessive X-linked allele directly from the frequency of male sufferers, which is a. If calculating overall allele frequencies over both sexes for X-linked loci,

$p = \dfrac{(p \text{ males} + 2\ p \text{ females})}{3}$ if males and females are equally frequent.

Thus if we had a non-equilibrium population, 0.2 *A* (Y), 0.8 *a* (Y) in males and 0.2 *A A*, 0.6 *A a*, 0.2 *a a* in an equal number of females, then *p* for males is 0.2 and *p* for females is 0.5, so the overall $p = \dfrac{0.2+(2 \times 0.5)}{3}$ = 0.4, when at equilibrium we would expect 0.4 *A* (Y) and 0.6 *a* (Y) males, and 0.16 *A A*, 0.48 *A a*, and 0.36 *a a* females. Allowances need to be made if there is an unequal number of the two sexes.

13.6.2. Non-X-linked diseases

There are a large number of diseases or conditions which are not X-linked but which show different incidences in the two sexes. Many are multifactorial traits, or have low heritabilities, with large environmental influences. Diseases **more common in males** than in females, with the preponderances as shown, include: alcoholism, 6-1 (i.e., six times more male sufferers than female ones); amoebic dysentery, 15-1; angina pectoris, 5-1; Asperger syndrome, 9-1; autism, 4-1; cancer of the skin, 3-1; cirrhosis of the liver, 3-1; coronary sclerosis, 25-1; duodenal ulcer, 7-1; gout, 49-1; hernia, 4-1; leukaemia, 2-1; pleurisy, 3-1; sciatica, large; suicide, 13-9. Diseases with **female preponderances** include: anaemia, very high; cancer of the genitalia, 3-1; cancer of the gall bladder, 10-1; gall stones, 4-1; hyperthyroidism, 10-1; chillblains, 6-1; influenza, 2-1; rheumatoid arthritis, 3-1; varicose veins, fairly large; whooping cough, 2-1 (data mainly from Montagu, 1963).

The actual values for these preponderances often vary over time and between populations, social classes and different occupations. Social factors are involved as well as biological and purely genetic factors. Exposure during jobs or pastimes to environmental mutagens or infectious agents is involved in some of these differences, such as some largely male manual occupations involving exposure of the skin to oils containing carcinogens. Mothers may often be more exposed to childhood diseases than fathers, through having longer and closer contact with children, which might explain the influenza and whooping cough sex differences. The amounts of stress and exercise, and diet, will affect some of these conditions, but the large excess of females with gall-bladder cancer is difficult to explain. Role-playing by some males as the hard-drinking, go-getting, aggressive stereotype might partly account for the alcoholism difference.

13.7. Causes of human mutation; cancer

Humans suffer mutations from all the causes of spontaneous mutation listed in

Chapter 7, plus additional ones such as medical X-rays. The **mutation rate**, μ, is the frequency of mutations per locus per gamete per generation. It is most easily calculated for rare autosomal dominant mutations, for which $\mu = n/2N$, where n is the number of affected patients and N is the total number of births: one uses 2N because mutation at either allele at an autosomal locus could cause the mutant phenotype. For most human genes, typical mutation rates from wild-type to mutant are between 10^{-6} and 10^{-5} per locus per gamete per generation, although values go up to 10^{-4}. The latter value applies to X-linked Duchenne muscular dystrophy, where the high mutation rate may reflect the large size of the gene as a target for mutation, being more than 2,400 kb long, the longest human gene known (Gelehrter, Collins and Ginsburg, 1998). Mutation rates may rise with increasing parental age. Most mutagens are also carcinogens, so anything causing mutation in humans will probably also cause cancer.

According to Morgan (1989), the approximate proportions of risk of cancer attributable to different classes of environmental agent were: tobacco, 30%; alcohol, 3%; diet, 35%; food additives, < 1%; sexual behaviour, 7%; occupation, 4%; pollution, 2%; industrial products, < 1%; medicines and medical procedures, 1%; geophysical factors, 3%; infection, possibly 10%; unknown, possibly 3%. Morgan (1989) gave the following percentages of average exposures to radiation of people in the UK: radon, 32%; gamma rays, 19%; internal, 17%; cosmic rays, 14%; medical, 12%; thoron, 5%; atomic fallout, 0.5%; occupational, 0.4%; nuclear discharge, 0.1%, with 87% being natural and 13% being artificial.

Figures for the United States from the National Radiation Council, Committee on the Biological Effects of Exposure to Low Levels of Ionizing Radiation, 1990, were, in millisieverts, where 1 mSv = 0.1 rem: radon gas, 2.06; cosmic rays, 0.27, natural radioisotopes in the body, 0.39; natural radioisotopes in soil, 0.28, making a total for natural radiation of 3.00 mSv; medical diagnostic X-rays, 0.39; radiopharmaceuticals, 0.14; consumer products including TV, clocks, and building materials, 0.10; fallout from weapons tests and nuclear power plants were both less than 0.01, making a total from non-natural sources of 0.63 mSv, and a grand total from all sources of 3.63 mSv. See also Chapter 13.7.3.

While the exact figures differ between different surveys and in different countries, it is clear that some causes of cancer and mutation are unavoidable, such as those from cosmic rays; others can be controlled to some extent by choice of human behaviour, or by using appropriate safety procedures. Some, such as exposure to radon gas, are determined by local geology, with Cornwall in Britain having unusually high levels of radon gas given off by the rocks.

13.7.1. UV light

UV light is not a source of human germ-line mutations, because it cannot penetrate to the sperm or eggs. It does however cause many somatic mutations in exposed skin, often leading to skin cancer. There are large racial differences in UV-induced skin cancer, controlled largely by skin colour, where the melanins in yellow, brown and black skins give considerable protection, and where sun-tanning of white skins is a phenotypically plastic (and usually reversible) response. For example, in America the annual incidence of malignant melanoma in the late 1970's was 4.2 per 100,000 for "whites" and only 0.6 per 100,000 for "blacks", with the greatest incidence in both groups in the sunnier states.

Oettlé, when head of the National Cancer Association of South Africa, made extensive studies on the different **racial groups** in that country. Skin cancer was 35 times more frequent in "whites" than in "blacks", although the groups were not matched for equal sun exposure. Cancers of the colon and breast were respectively 10 and 5 times more frequent in "whites" than in "blacks", with the former perhaps being diet-related and the latter perhaps being UV related. Womb cancer was 6 times more frequent in "blacks" than in "whites", possibly related to a greater average number of sexual partners and oncogenic sexually transmitted viruses. Liver cancer was 20 times more frequent in "blacks" than in "whites", possibly from "blacks" eating more food with fungal infections, where fungal aflatoxins are liver carcinogens. There may also be different genetic susceptibilities to various cancers in different races. Chapter 7.3.2 described xeroderma pigmentosum, an autosomal recessive disease causing multiple skin cancers because sufferers have impaired excision repair of DNA damage by UV.

13.7.2. Chemicals

Diet influences cancer and germ-line mutations. Tea, coffee and most cola drinks contain caffeine, which is a natural base analogue (see Chapter 7.2.1): as well as being a mutagen (sometimes used in labs for mutating micro-organisms), it can inhibit the enzymes which repair DNA damage. Fortunately for us, it is less mutagenic in mammals than in micro-organisms. Caffeine induces mutations in human tissue culture cells, and there is evidence for germ-line mutations. In one survey, Americans drinking eight or more cups of coffee a day had significantly fewer sons, relative to daughters, compared with the rest of the population. Lethal induced autosomal recessives would not show in the next generation because they would be heterozygous, but lethal X-linked recessives will kill hemizygous sons but not heterozygous daughters, affecting the sex ratio.

There are many other dietary potential mutagens, and many possible **chemical mutagens in the environment** which may be picked up through touching, breathing, eating and drinking. Some have been identified, such as benzypyrenes in charcoal-grilled steaks, diesel exhaust particles and tobacco smoke. There was high incidence of lip cancer in smokers who used clay pipes, because of a carcinogen in the clay, and various oils used in industry contained carcinogens. Smoking is responsible for about 15-20% of all deaths in Britain, causing deaths mainly through cancer and through effects on the circulatory and respiratory systems.

Environmental mutagenesis is a very active research field, with many man-made products being tested each year for mutagenicity and carcinogenicity (carcinogens are usually also mutagens, and *vice versa*). One testing system uses the induction of resistance to 8-azaguanine in human tissue culture. In the **Ames test**, however, quantitative studies are made of the induced reversion to wild-type of various histidine-requiring mutations of the bacterium *Salmonella typhimurium*: see Plate 11. Both frame shift mutations (revertible by frame shift-causing mutagens) and base substitution mutations (revertible by base substitution-causing mutagens) are used, together with other mutations which inactivate excision-repair or make the cells more permeable. Because some chemicals only become mutagenic after being metabolised in the liver, the medium incorporates a rat-liver extract. Many substances have been tested using the Ames test, including industrial chemicals, cosmetics, hair dyes, food additives and pesticides, resulting in bans on certain chemicals and the reformulation of certain cosmetics and hair dyes to reduce their mutagenicity. Tests for direct carcinogenicity, by assessing tumour formation in laboratory animals, are done but are much slower and more expensive than the mutagenicity tests.

13.7.3. Ionising radiations

These cause mutations, cancer and leukaemia. When ionizing radiation strikes water or living cells, highly reactive **free radical ions** are produced and react with DNA and other molecules. Single-strand breaks in DNA are usually repaired efficiently, but double-strand breaks can cause chromosome aberrations. Free radicals can damage nucleotide bases and cause point mutations in various ways. In human cells in tissue culture, a dose of 0.2 sieverts gives about one visible chromosome break per cell.

X-rays and **gamma rays** are measured in Roentgen (R), where a 1 R dose gives 2.08×10^9 ion pairs per cm^3 of dry air at 0°C and 1 atmosphere pressure. Taking a generation to be 30 years, the natural ionising radiation received per person per generation is about 3 to 5 R, with about 0.8 R from **cosmic rays**, 0.7 R from radioactive elements within our bodies (e.g., potassium isotopes), and 1.5 R from soil and rocks. Radon gas gives about 2,500 cancers a year in Britain. Medical X-rays average about 3 R per generation, with typical doses to the male gonads per irradiation being 0.4 R for a pelvic X-ray if the scrotum is unprotected, but only 0.03 R if protected, 0.001 R for a chest exposure and 0.0001 R for the head, but these figures are decreasing with better application and protection methods. Fallout from atomic weapons testing accounts for only 0.1 R. In contrast, a cancer radiotherapy course might involve about 5,000 R, with an extra 6,000 R for a booster course.

Radiation is also measured in microsieverts (1 Gray = 1 Sievert, and 100 rads = 1 Gray; rads and Roentgens are similar, with 100 R of hard X-rays corresponding to about 93 rads, and 1 rad = 100 ergs of energy per gram of material irradiated), where natural exposure is about 5.5 microsieverts per day. X-ray doses of various treatments, in microsieverts, are about 20 for dental X-rays, 60 for the chest, 100 for joints and limbs, 2,400 for the spine, 1,000 for the lumbar area, 300 for the hip, 5,000 for the stomach and duodenum with a barium meal, 1,000 for the skull, 9,000 for the large bowel, compared with about 100 for a return flight from London to New York in a standard jet, and more for Concorde as it flies at a higher altitude.

There is no "safe level" for ionising radiations, as damage is approximately proportional to dose over a wide range of doses, with no threshold level. When it was found that X-rays cause temporary male and female sterility, they were actually used on human male gonads as a contraceptive, but this was quickly abandoned when their harmful effects were found. Similarly, in about 1950 in Britain, X-ray machines were used in shoe shops to see how well feet fitted inside shoes, with the X-rays directed up through the feet to a green fluorescent screen. They were rapidly abandoned when they were found to harm shop assistants and customers.

13.7.4. Temperature

Humans have constant-temperature devices such as sweating, but **fevers** can cause hyperthermia, and prolonged exposure to low temperatures can cause hypothermia. In *Drosophila melanogaster*, a rise of 3.3° increases the spontaneous mutation rate by 85%, equivalent to a radiation dose of 40 R. **Wearing trousers** raises the scrotal temperature in man by about 3.3°. If this raises the mutation rate by as much as it does in the fruit fly, wearing trousers could have genetic hazards 10 to 1,000 times those of all sources of radiation, and possibly be a factor in the low active sperm count in man. Central heating and hot baths or showers may also raise scrotal temperatures. Men suffering from low fertility have been advised not to wear close-fitting underwear, and have been recommended to spray their testicles with cold water.

13.7.5. Infection

Some **viral diseases** such as measles can cause extensive chromosome breakage. That can be mutagenic and cause chromosome aberrations. There are also **oncogenic** (cancer-causing) **viruses**, and various **transposable elements** can interrupt coding sequences, causing mutations.

13.8. Chromosome number abnormalities and chromosome aberrations

13.8.1. Human chromosome methods

For studying human chromosomes, one needs only a few drops of blood, say from a fingertip or heel or ear prick. The blood is cultured at 37° with a bean extract containing phytohaemagglutinin, plus nutrients and antibiotics such as penicillin and streptomycin. During three to four days of culture, the nucleated white blood cells (leukocytes, T-lymphocytes) divide by mitosis, then are arrested in metaphase by the addition of colchicine or colcemid. The cells are allowed to swell in hypotonic solution to get good separation of the chromosomes. The swollen cells are then fixed and spread to air-dry on slides, after which they are stained for DNA, e.g., with Giemsa stain.

That procedure gives stained but unbanded chromosomes which can be photographed. The photos can be cut up to make **karyotypes** in which the chromosomes are arranged in order of size and centromere position (see Plate 30 for a banded example of a human karyotype). Metacentric chromosomes have the centromere near the middle; submetacentrics have the centromere off-centre, and acrocentrics have the centromere near one end. Telocentrics, with the centromere virtually at the end, are not found in normal human karyotypes. With unbanded chromosomes, one could identify about 10 groups of chromosomes based on size and centromere position, e.g., pair one, the largest metacentrics, pair two, the largest submetacentrics, etc. Metaphase lengths vary from about 2 μm (40 Mb of DNA) for chromosome 21 to about 10 μm (200 Mb) for chromosome 1.

Then various methods of chromosome treatment and staining such as **G-banding** were developed so that each chromosome showed a number of distinctive stained bands: see Plates 30 and 31 for a G-banded karyotype involving a 1:12 non-reciprocal balanced translocation. Giemsa banding gives 300-400 alternating dark and light bands over the whole genome, caused by different condensation, with dark bands being AT rich and containing about 20% of the active genes. All 22 pairs of autosomes plus sex chromosomes X and Y could each then be identified separately from length, centromere position and banding pattern at metaphase of mitosis.

There are various automatic scanning and image analysis systems which can give computerised karyotypes, even if some chromosomes overlap, but checking of the metaphase by an experienced operator is always advisable, for at least three cells.

The next advance involved **fluorescent in-situ hybridisation** (FISH). DNA probes with fluorescent labels were made which could hybridise, under appropriate temperature regimes, to denatured DNA in particular chromosomes or particular regions, of the metaphase spreads; see Plates 14 (whole chromosome paint for 5) and 32 (sub-telomeric probe for 5). Fluorescent light then revealed to which chromosome each probe hybridised, so that a probe for the whole of chromosome 5, with DNA sequences complementary to repeated specific DNA sequences in chromosome 5, would cause all copies of chromosome 5 to fluoresce, in interphase nuclei as well as in metaphase spreads. Using repeated DNA sequences within a chromosome allowed the whole chromosome to be "**painted**", which was very useful when looking for translocations or other chromosome aberrations. For example, if chromosomes 2 and 3 were "painted" green and pink respectively along their lengths, then a reciprocal translocation between the ends of those two chromosomes would show up as one pink and one green chromosome (untranslocated), one with most of the pink chromosome plus a green tip, and one with most of the green chromosome plus a pink tip, the 2/3 and 3/2 translocated chromosomes.

Chromosome prints, as opposed to paints, were developed by Advanced Biotechnologies, based on single copy sequences in particular chromosomes, usually near the centromere, "lighting up" just one region of a specific chromosome, in interphase and mitosis, taking about 1 hour instead of about 13 hours preparation for chromosome painting. Plate 33 shows a normal XY cell in uncultured white blood cells from a baby with ambiguous genitalia; with these particular dual-colour centromeric FISH probes, the X centromere shows red and the Y centromere shows green.

Some kits are specific for detecting particular abnormalities. Thus the TriGen® Assay is for Down syndrome and X or Y numerical abnormalities, with chromosome 21 fluorescing orange, the X is green and the Y is blue. It can be used directly on cells from amniocentesis without culturing, and so gives results in 9 to 20 hours, as opposed to 7 to 21 days if amniotic fluid cells need culturing and conventional preparation.

Applied Spectral Imaging have developed Spectral Karyotyping (SKY), using a "probe cocktail" containing painting probes for all 24 human chromosomes, so that each of chromosomes 1 to 22, X and Y shows up a different colour after using special spectral image equipment and software. See Plate 34 for a related system, R_xFISH. The system can give automatic identification of chromosome aberrations, and even diagnosis of complex chromosomal rearrangements in solid tumours.

13.8.2. Autosomal abnormalities

Down syndrome (Mongolism), see Plates 35 and 36, is due to an extra chromosome 21, 2n = 47. It occurs in about 1 in 600 live births, in both sexes. Intelligence is reduced, with IQs usually in the range 25-75, often 40-50. Affected individuals are of small height, with an extra eye-fold of skin, broad hands with odd palm prints, and nearly half suffer from heart defects in the valves or septa. The mouth is often open with the large tongue protruding. They have a reduced life span, with about one sixth dying in their first year, and an average span of about 20 years. They have a 15-fold increased risk of leukaemia and are more susceptible to infections. They are universally susceptible to Alzheimer's disease, which begins around the age of 35 years. Some females are fertile but males are usually sterile. In 95% of cases, the extra chromosome is of maternal origin, partly because male gametes are more sensitive to chromosome imbalance but mainly because of a maternal age effect. Even in young mothers the eggs are more than a decade old because oogenesis is largely completed before birth, with the primary oocyte suspended in diplotene until sexual maturity, while sperm only form from puberty.

The condition is easily detected by the symptoms, and confirmed by karyotyping. For genetic counselling, it is important to distinguish between ordinary Down syndrome arising by chance non-disjunction in one parent's meiosis, in which there is no special risk of further siblings being affected, and the 4% of cases caused by **translocation** of a chromosome 21 onto another autosome, usually 14 but sometimes 13 or 21. A translocation Down's has two normal 21's and one translocated 21, and often one parent (usually the mother) has 2n = 45, with one normal 21 and one translocated 21 e.g., 14+21, with no external symptoms as all genes are present in the right dose. Such a mother produces four types of gamete in about equal proportions, with the translocated chromosome behaving as if controlled by centromere 14:

normal 14 and normal 21, giving a normal child, 2n = 46;

14+21 and normal 21, giving a translocation Down's child, 2n = 46;

14+21, giving a translocation carrier, unaffected, 2n = 45;

normal 14, with no 21, which is lethal as it gives monosomy for 21, 2n = 45.

A couple who have had a translocation Down's child, and where one parent has been found to be a carrier, would be advised that there is a high risk of having a further affected child, but that an affected foetus could be detected by prenatal diagnosis (Chapter 13.11). The risk depends on the chromosome involved and the parent involved. 13/21 and 14/21 female carriers have about a 10% risk of a Down's baby; for male carriers the risk is about 2%. For 21/21 translocations, the risk is 100% for both male and female carriers.

For normal non-disjunction Down syndrome, there is a clear increase in the frequency of affected children with **increasing maternal age**. The risk is 1 in 1,923 for mothers aged 20, 1 in 885 for age 30, 1 in 139 for age 40, and 1 in 32 for age 45. Once one allows for the fact that older husbands tend to have older wives, there is no effect of paternal age on the incidence of Down syndrome. There is no age effect with translocation Down's. The non-disjunction Down's arises mainly (80%) in the first division of meiosis, which may be related to the fact that older women have fewer chiasmata to hold the bivalents together until normal separation at anaphase I. The frequencies of XXX and XXY abnormalities (Chapter 13.8.3) also rise with maternal age, as do those for most autosomal trisomies, with smaller chromosomes more affected than large ones.

If there is non-disjunction at mitosis in the very early embryo, one can get **mosaics**, individuals with some cells abnormal and some cells normal. If chromosome 21 fails to segregate evenly at the second mitosis in the egg, for example, two cells will be $2n = 46$, normal, one will be $2n = 47$, with three of 21, and one will be monosomic for 21, $2n = 45$, and will die. Depending on what cells are affected and how many, a mosaic's phenotype may vary from normal to somewhat affected.

Other autosomal abnormalities occur, but are rarer than Down syndrome. **Patau syndrome** has three of chromosome 13, with a frequency of 1 in 5,000 live births, increasing with maternal age. Sufferers have many abnormalities, often including brain development failure and cardiovascular defects, and half die within one month. **Edwards syndrome** has three of chromosome 18, with a frequency of 1 in 3,000 live births, increasing with maternal age, but the incidence at conception is much higher, with perhaps 95% aborting spontaneously, especially males. They have multiple malformations and 90% die within one year.

About 1 in 2,000 babies have microscopically visible deletions or duplications, usually giving multiple mental and physical symptoms. Sometimes the mothers carry balanced translocations, but more usually the abnormalities have arisen *de novo*. Heterozygosity for a deletion of about half the short arm of chromosome 5 gives the **Cri-du-Chat syndrome** (Chapter 9.2). Affected individuals are severely abnormal physically and mentally, and give a typical plaintive, continual, cat-like cry.

Robertsonian translocations, also known as centric fusions, are whole-arm exchanges between the acrocentric chromosomes 13, 14, 15, 21 and 22, and are the commonest human chromosome rearrangements. They arise by breaks or crossovers, usually just above the centromeres, with fusion of the products with centromeres and the loss of the now acentric short arms. They have a very high rate of production, about 4×10^{-4} per gamete per generation, an order of magnitude higher than the highest mutation rates for human autosomal dominant genetic diseases, and so they contribute significantly to foetal wastage, birth defects and mental retardation. Rob(13q14q) (i.e., exchange and loss of the short arms, designated "p" as opposed to long arms, "q", of 13 and 14, in a Robertsonian translocation between one 13 and one 14, to give a single translocated centric product) and rob(14q21q) form about 85% of all Robertsonian translocations. Those two have breakpoints localised to very specific regions of the proximal acrocentric short arms, but other such translocations have more diverse breakpoints. Nearly all Robertsonian translocations arise during female meiosis, especially in older mothers: see Page and Shaffer (1997).

13.8.3. Sex chromosome abnormalities

The most common **sex chromosome abnormality** is **Klinefelter syndrome**, about 1 in 1,000 male live births, increasing with maternal age. It results from an XXY, 2n = 47, karyotype, written 47,XXY. The penis is normal but the testes are small, with few sperm. Sufferers tend to be tall, with long limbs. Most individuals have normal intelligence but the average intelligence is reduced because Klinefelter patients are ten times more likely to be mentally retarded than normal XY males. Many sufferers have difficulty with oral communication, even if of normal intelligence. About 40% of sufferers have some breast development. Testosterone replacement therapy can be given from early adolescence to improve sexual characteristics, but XXYs remain sterile unless they are mosaics, with about 15% being 46,XY/47,XXY mosaics.

Turner syndrome, 45,X, has a very high frequency of spontaneous abortion, about 99.7%, with 1 per 5,000 live female births. The intelligence and life span of these females are normal but the height is short, averaging 145 cm. The hands and feet are often swollen. The ovaries start to degenerate after 15 weeks gestation, giving failure of sexual development. About 10% menstruate and a very few are fertile, but these are normally mosaics. Sufferers usually have a broad chest, underdeveloped breasts, a webbed neck, an increased risk of heart disease and other problems. Growth hormone treatment from childhood increases adult height by about 4 cm, and sex hormone therapy allows for improved sexual development, but not fertility. Although 45,X, from maternal or paternal meiotic non-disjunction, is the commonest type (50%), Turner syndrome can also arise from: having one normal X and one **isochromosome** (a chromosome with two identical arms, containing homologous genes) of the X long arm (and thus having only one copy of the X short arm in their karyotype), (17%); being XX/XO mosaics (20%); being

XY/XO mosaics (4%); having one normal X and one ring X (7%); having a deletion in the short arm of one X (2%). The somatic effects are thought to be caused mainly by the lack of a second X-chromosome short arm, while the sexual effects are probably related to the X long arm, as they are less severe in patients with short-arm deletions or with long-arm isochromosomes. Two active X chromosomes are required for proper foetal oogonial development, with inactive X's being reactivated in oogonia when meiosis begins in the foetus.

47,XYY individuals are males. The incidence is 1 in 1,000 live male births, with no effect of maternal or paternal age, but is 3 per 1,000 in mentally handicapped adults and 20 per 1,000 in male criminals. The condition arises most often from the production of YY sperm from the father's second meiotic division. Intelligence is 10 to 15 IQ points less than for normal siblings. Sufferers tend to be tall and are often aggressive. They are fertile, with normal sexual development, and sperm usually only carry one sex chromosome.

47,XXX individuals are females. The incidence is 1 in 1,000 live female births, with a maternal age effect. Many are normal physically and mentally; some are sexually underdeveloped and some have reduced intelligence. About 75% are fertile. The lack of clear symptoms means that many go undiagnosed unless they are karyotyped.

46,XX males occur at an incidence of 1 in 20,000 males, usually with translocation of the sex-determining region (SRY) of the Y to an X chromosome. They are sterile with small testes but are mentally normal. A minority have normal X chromosomes and ovaries, but are externally male because of exogenous androgens or defects in adrenal steroid biosynthesis.

46,XY females are very rare, with a Y chromosome lacking the SRY region, or with it mutated.

13.9. Selection before and after birth

The true frequency of chromosome abnormalities at conception cannot be measured as there is **strong selection** at embryonic and foetal stages against many abnormalities. Probably about half of all conceptions abort, often without the woman realising that she had conceived, but giving a detected miscarriage if occurring later. Of recognised pregnancies, 15-20% end in spontaneous abortion, with about 40 to 60% of these abortuses having chromosome abnormalities if they are lost in the first trimester (three months), and 5% if lost in the second trimester, compared with only 0.6% of the newborn having chromosome abnormalities.

The following figures are from a number of sources, including Jacobs and Hassold (1995), Connor and Ferguson-Smith (1997) and Hartl and Jones (1998), with some differences in the figures between the different sets of data. In studies of thousands of **spontaneous abortuses in the first trimester**, 39% had normal karyotypes and 61% were abnormal. Of the **abnormal types**, about 50% were **trisomics**, with one extra chromosome (2n = 47), especially number 16 (16%, although trisomy 16 never reaches full term). Other common trisomies were for 22 (6%), 21 (5%), 15 and 14 (4% each), and 2 (2%). Trisomy for 1 was never found, so it must be lethal very early in development. There was a small proportion of double trisomics (2n = 48). Monosomics (2n = 45) must also generally be lethal, except for Turner syndrome, 45,X, (18%). The high frequency of Turner syndrome abnormalities in these abortuses, compared to only 1 in 5,000 live female births, shows the very strong prenatal natural selection against Turner syndrome. According to Jacobs and Hassold (1995), only 0.3% of Turner's conceptuses survive to birth, and 3% of trisomy 13, 5% of trisomy 18, and 22% of trisomy 21. **Polyploidy** was common, with 20% triploids (2n = 69, with XXX, XXY and XYY) and 6% tetraploids, showing a very large contribution to abnormal zygotes from unreduced gametes or double fertilisations. There were also about 0.06% each of XYY and XXY, and 0.3% of XXX in these first trimester abortuses. **Translocations** were 3%, mostly unbalanced, and mosaics were 1 to 2%. Structural abnormalities such as translocations and deletions are much less frequent than chromosome-number abnormalities.

In figures for **newborn children**, only about 0.6% had abnormal karyotypes, showing strong selection before birth against abnormalities. There were 0.17% sex chromosome abnormalities (47,XYY, 0.1% of males; 47,XXY, 0.1% of males; 47,XXX, 0.1% of females; XO; 0.02% of females). The autosomal abnormalities were 0.14% trisomy for 21 (Down syndrome), 0.03% trisomy 18 (Edwards syndrome) and 0.02% trisomy 13 (Patau syndrome). There were 0.2% balanced translocations, 0.05% unbalanced translocations, and a surprisingly high 1% (excluded from the 0.6% quoted above) heterozygous for small pericentric inversions of chromosome 9, with no phenotypic abnormalities. Heterozygotes for large inversions have a risk of 8% for a carrier mother and 4% for a carrier father of producing offspring with unbalanced chromosomes.

Abnormalities of chromosome number are not usually passed on to future generations as they often cause sterility. Structural abnormalities such as translocations can be passed on, especially through the mother, but they often reduce fertility and cause abortions.

Selection also occurs, usually after birth (depending on when the deleterious allele is expressed) for X-linked and autosomal genes, such as Duchenne muscular dystrophy and cystic fibrosis, respectively. Some disorders cause premature death and some cause sterility, as in testicular feminisation. Others, such as achrondroplastic dwarfism (Chapter 13.3.1), allow viability and fertility when heterozygous but are lethal when homozygous, so there will be overall selection against them.

One of the worries about genetic counselling is that it might increase the frequency of harmful alleles in human populations by reducing selection. For example, if one reduces the frequency of recessive homozygous sufferers for a deleterious autosomal recessive allele, by counselling heterozygous carriers not to marry other heterozygotes but to marry dominant homozygotes, then that reduces selection against the bad allele, while new bad alleles continue to arise by mutation.

13.10. Blood groups, especially ABO and Rhesus

About 400 blood group antigen systems have been described and these antigens on the surface of red blood cells determine a person's **blood group**. The most important ones are the ABO (chromosome 9) and Rhesus (chromosome 1) groups, which are critical for successful blood transfusions because humans can make antibodies to their antigens. For some other groups, such as MN, the antigens are only detected by using antibodies from other organisms, so are not usually a problem in transfusions, but the groups can be used forensically. Groups are named after the antigens. Some other blood groups should be matched for recipients of **repeated transfusions**, who can eventually develop antibodies to the Duffy, Lewis, MN and S groups. Plates 37 and 38 show card and tube tests for the ABO and Rhesus groups, using monoclonal antibodies.

13.10.1. The ABO blood group

People of **blood group O**, 46% of the UK population, are $i^o\,i^o$ genotype, have no specific ABO antigen on the red blood cells, and have anti-A and anti-B antibodies in their serum. **Group A**, 42%, are $i^A\,i^A$ or $i^A\,i^o$, with the A antigen on red blood cells and the anti-B serum antibody. **Group B**, 9%, are $i^B\,i^B$ or $i^B\,i^o$, with the B antigen and the anti-A antibody. **Group AB**, 3%, are $i^A\,i^B$, with A and B antigens, but neither antibody. The A and B traits are codominant, and O is recessive to A and B. Different countries have different blood group frequencies, with most native South Americans being group O, most Asians being group B, and A being commonest in Norway.

The three alleles determine the activity of a glycosyl transferase enzyme which in A and B modifies the H surface antigen, adding N-acetylgalactosamine for A, D-galactose for B, and nothing for O, which has a single base deletion (of G at nucleotide position 258) giving a frame shift and an inactive antigen protein. Blood transfusions between members of the same ABO blood group are safe, but if say A blood is put into a B recipient, the anti-A antibody in the serum will react with the A antigen on the red blood cells, agglutinating the cells, perhaps with fatal consequences. In emergencies, certain other transfusions are usually safe, because antibodies in the incoming plasma are partly adsorbed by the tissues of the recipient, as well as being diluted by the recipient's plasma. Group O are sometimes called **"universal donors"** because they lack an active antigen, so their donated red cells are not agglutinated in the recipient, and group AB are called **"universal recipients"**, because they lack the two antibodies which could agglutinate donated red blood cells. Thus if O blood was transfused into a group B patient, there would be no A antigen for B's anti-A to combine with, hence no agglutination of O's red blood cells, and the donated blood's anti-A and anti-B antibodies would be diluted and largely adsorbed in the recipient, with little effect on the recipient's red blood cells.

Alcohol-soluble ABO antigens occur in many tissues besides blood, and people carrying the dominant **secretor allele**, *Se*, also have water-soluble antigens in their secretions such as saliva, urine and spermatic fluid, which has been useful forensically, especially in rape cases before DNA fingerprinting was developed. The secretion only applies to ABO and Lewis-group antigens, and is absent in *se se* individuals. ABO antibodies are effectively constitutive, not needing red blood cell antigens to stimulate their production; common bacteria may trigger their formation.

13.10.2. The Rhesus blood group

This is called **Rhesus** because the antibodies were first found in Rhesus monkeys. The antibody is not constitutive in humans and is produced only after a challenge by Rhesus positive blood. Rhesus positive people have the Rhesus D antigen on their red blood cells and Rhesus negatives do not. Rhesus negatives only have the anti-D Rh antibody if previously exposed to Rhesus positive blood. Rhesus positives have the *D* gene, as *D D* or *D d*, while Rhesus negatives are *d d*, but eight Rhesus alleles can be distinguished using three antisera (anti-D, anti-C, anti-E).

The Rhesus group is very important because children born to Rhesus negative women (*d d*) and fathered by Rhesus positive men may develop **haemolytic disease of the newborn**, giving anaemia and jaundice, with much agglutination and breakdown of the red blood cells. This used to cause the death of most affected children, though survivors usually recovered fully. In the UK each year, more than 80,000 Rhesus negative women have Rhesus positive babies.

In "white" populations such as in the UK, about 16% of individuals are Rhesus negative, $d\ d$, so from Hardy-Weinberg q is $\sqrt{0.16}$ = 0.4, so p = 0.6 for D. All babies from $D\ D$ fathers x $d\ d$ mothers will be Rhesus positive, giving p^2 x q^2 = 5.8% "at risk" pregnancies, and half the pregnancies in $D\ d$ father x $d\ d$ mother will give Rhesus positive babies, giving $1/2\ (2pq$ x $q^2) = 3.9\%$ "at risk" pregnancies, so 9.7% of pregnancies are "at risk", yet only about 0.3 to 1% of pregnancies have suffering babies.

The haemolytic disease does not usually occur in a Rhesus negative woman's first Rhesus positive baby, but can occur in second and subsequent Rhesus positive babies with increasing severity. This is because it usually takes one Rhesus positive baby pregnancy in a Rhesus negative woman to trigger the production of her anti-Rhesus antibodies. During most of the pregnancy, not enough foetal red blood cells leak across the placenta into the mother to trigger anti-Rhesus antibody production, with the main leakage occurring just before or during birth, so the mother usually only makes the antibody just after the birth of her first Rhesus positive baby. **To prevent** the mother from making such antibodies, which could harm subsequent Rhesus positive babies, she is now usually given an intramuscular injection of anti-Rhesus antibodies within 72 hours of the birth, to destroy leaked red blood cells from the baby in the mother, before they can trigger her own antibody production.

Rhesus women should only be given Rhesus negative blood in transfusions, to avoid anti-Rhesus antibody production. If a Rhesus positive baby is born with severe haemolytic disease of the newborn, it is usually given transfusions of Rhesus negative blood (without anti-Rhesus antibody) because the mother's anti-Rhesus antibody which has got into the baby could attack Rhesus positive blood if transfused in.

With Rhesus, the anti-Rhesus antibody is **IgG** (immunoglobulin G), which can cross the placenta. The ABO antibodies are IgM, which cannot cross the placenta, so mother/child incompatibility for ABO does not usually cause any problems. Mother/baby ABO incompatibility can help with Rhesus incompatibility if the ABO antibodies destroy the baby's leaked blood in the mother before it can trigger anti-Rhesus production. With a Rhesus positive mother carrying a Rhesus negative baby, if her antigens do get into the baby by leakage across the placenta, there is no danger of the mother being harmed by the baby producing anti-Rhesus antibodies, as babies do not produce the antibody (in response to a challenge) until about six months after birth.

13.10.3. Blood groups in farm animals

Haemolytic disease of the newborn from mother/baby incompatibility is known in horses and pigs, but not in sheep or cattle. Blood groups are many in cattle, sheep, horses and pigs, and have been used in animal **paternity tests**, e.g., checking whether a female was mated to an excellent sire, or to a cheaper substitute male. There is a cattle blood grouping service in Edinburgh, Scotland, but as in human forensic medicine, DNA analysis is often preferred now to blood group studies.

13.11. The major histocompatibility complex

The **major histocompatibility complex** (MHC) includes the **Human Leukocyte Antigen** (HLA) system, which encodes a number of leukocyte antigens. It is the most polymorphic genetic system in man and regulates immune responses. For most loci, no single allele is very common and heterozygosity is extremely frequent. It is a gene cluster containing about 80 genes over 4 Mb of DNA on chromosome 6 (6p21.3). The products from class I (*HLA-A, HLA-B* and *HLA-C*) and class II genes (*DP, DQ* and *DR*), which are respectively at the telomeric and centromeric ends of the cluster, control the presentation of processed antigens to T-cells, so are very important in immune responses, e.g., to virus infections. Class I products are found on the surface of most nucleated cells, while class II products are found on the surface of B-lymphocytes, activated T-lymphocytes and on antigen-presenting cells including macrophages. The class III genes, in the centre, include genes of unknown and immunological function, including tumour necrosis factors, heat-shock protein 70 and complement proteins.

Multiple alleles are known for most class I and class II genes, and definition of genotypes is normally performed by DNA tests, which have largely replaced immunological tests. Because the genes are all close, they tend to be inherited together as particular **haplotypes** (the combination of genes on one chromosome is a haplotype). If say parent 1 has haplotypes A and B on his copies of chromosome 6, and parent 2 has haplotypes C and D, then amongst their children, the chance of two particular siblings having the same haplotype combination, e.g., AD, is one quarter, because whatever haplotype the first child has, the second child has equal chances of the four possible haplotypes, AC, AD, BC, BD, unless crossovers (which have a frequency of several per cent within MHC) within the region have produced new haplotypes. This is very important in tissue (e.g., bone marrow) or organ transplants, as transplants between individuals of the same HLA type are usually accepted, not rejected, by the body's immune system. For example, the half-life of transplanted kidneys is 26 years when fully HLA matched, compared with 12.2 and 10.8 years for grafts from parents and siblings, respectively, without full matching (see Howell and Navarette, 1996).

There is enormous variation between **racial groups** in antigen frequencies for class I and class II genes. For example, *A1*, *A25*, *B37*, *B38* and *B63* are largely confined to Europeans, *A34*, *A36*, *A43*, *B42*, *B53* and *B58* to Negroes, and *B52*, *B54* and *B61* to Orientals. There is extensive linkage disequilibrium, with the frequencies of combinations of alleles often differing from those expected from their individual frequencies (Chapter 4.2.3). That means that if one HLA allele predisposes people to a particular disease, then other alleles with which that allele is often associated will often have positive correlations with the disease. The strength of the association of an HLA allele with a particular disease is measured by the **relative risk**, which is the frequency of the disease in those with the allele divided by the frequency of the disease in those without the allele. Thus HLA allele *B27* occurs in more than 90% of people with ankylosing spondylitis (severe stiffening of the spine), whereas only 8% of the general population has that allele, and the relative risk is nearly 90. The *HLA-B27* allele is used diagnostically for ankylosing spondylitis. The relative risks for various HLA-allele associated conditions are: psoriasis with *DR7*, 43; psoriasis with *B17*, 6; multiple sclerosis with *DQB1*0602*, 36. The notation with * and four or five numbers indicates an HLA DNA marker, while notation such as *DR7* indicates an antigenic marker. Autoimmune diseases such as rheumatoid arthritis (relative risk 3 to 6 with *DR4*) often have HLA allele associations, with self-antigens being treated as foreign, with immune responses against the host cells. See Connor and Ferguson-Smith (1997). The combination in *cis* or *trans* of *DQA1*0501* with *DQB1*0201*, which both encode the DQ2 antigen, confers a relative risk of about 250 of coeliac disease. See Howell and Holgate (1996) for the relation between HLA genes and allergic disease, including asthma, hay fever and eczema.

13.12. Prenatal, neonatal and adult screening

Prenatal diagnosis is done to reassure couples if the foetus is normal, and to detect various genetic and chromosomal abnormalities if present, so that the couple can be offered a termination of the pregnancy if they wish. In the UK, prenatal diagnosis is offered in about 8% of pregnancies. Most of the foetuses investigated have no defect, with selective termination being offered in about 7% of cases. Termination is only offered if there are serious defects present, such as Down syndrome. Termination of a baby solely on the grounds of unwanted gender is not allowed.

13.12.1. Amniocentesis

In **amniocentesis** at about 16 weeks pregnancy, about 20 ml of fluid is taken from the amniotic fluid surrounding the foetus through a syringe needle inserted through the woman's abdominal wall, using ultrasound monitoring to avoid damaging the foetus, umbilical cord or placenta. The fluid can be immediately analysed for some biochemical disorders, but the main use is of the foetal cells which have been shed and can be spun down. The cells are few and many are dead. They can be immediately sexed from the presence or absence of Barr bodies, or stained with fluorescent dyes to detect the Y chromosome. FISH (fluorescence in-situ hybridisation) can be used with specific DNA probes, but for most purposes there are too few living cells, so they need to be multiplied in culture for two to three weeks. Then they can be karotyped, and used for DNA analysis or metabolic tests, to detect chromosome, gene or enzyme defects in the foetus.

The **risk of miscarriage** or foetal damage is about 0.5% and the procedures are expensive, so amniocentesis is not done routinely in all pregnancies. In Britain it is offered to pregnant women of 35 or over, because of their higher risks of Down's and other age-related syndromes, or where there is a known genetic risk in the family. It is particularly useful for women who are translocation carriers, to check whether the foetus is normal, a carrier or a sufferer. The woman (or better, the couple) is usually offered a termination if a bad defect is found, but there is never any compulsion to have one. If the test is only done at 20 weeks and the test results take 3 weeks, a termination would be very close to the legal upper limit of 24 weeks gestation in the UK. However, under the 1990 Human Fertilisation and Embryology Act, termination is permitted up to term if the defect is serious.

13.12.2. Early amniocentesis

It was found in 1988 that really skilled people could take 10 ml amniotic fluid at 13-14 weeks pregnancy. There are more viable cells then than at 16-20 weeks, so karyotyping could be done within 8 days, so any termination could be offered much earlier than with normal amniocentesis. The problems are a slightly higher risk of miscarriage, and taking 10 ml reduces the amniotic fluid by 30%, which can sometimes affect lung development.

13.12.3. Chorionic villus sampling

A catheter is inserted through the vagina and cervix to take a sample of **chorionic villi**, removing about 20 mg of tissue; any maternal tissue must be dissected away, or errors could occur. The chorion is of foetal origin, anchoring the foetus to the uterine wall before full placental development. This sampling can be done at 8-12 weeks pregnancy, even before the pregnancy is visible externally. The villus cells are dividing rapidly, so can be used immediately for karyotyping, DNA analysis and metabolic tests. The advantage is early diagnosis, but there is a risk of about 2% of miscarriage, higher than for amniocentesis. The chorion is sometimes a mosaic of normal and chromosomally abnormal tissue even if the foetus is normal, which can lead to wrong diagnoses.

13.12.4. Ultrasound screening

Ultrasound screening is noninvasive, carries almost no risks, and can detect multiple births and major developmental abnormalities. At 10-12 weeks gestation, anencephaly (absence of part or all of the brain) can be detected, but 16-18 weeks is the earliest for detecting many abnormalities of head, body or limb development, including open neural tube defects such as spina bifida (open lower end of the neural tube). Males can be identified by observation of the external genitalia at about 18-20 weeks gestation.

Ultrasound is particularly useful in detecting multifactorial anatomical traits such as club foot which cannot be done by DNA or biochemical analysis. It is widely used to detect congenital heart disease, kidney and bladder problems, polydactyly, dwarfism, etc. It has led to developments in prenatal surgery, and certainly not all the conditions mentioned here would lead to offers of termination of pregnancy.

13.12.5. Maternal blood sampling

Pregnant women at 16-18 weeks gestation are usually screened for their blood **alpha-fetoprotein** (AFP). Foetuses with neural tube defects such as anencephaly or spina bifida tend to leak AFP into the mother's blood. Unfortunately the test is not very accurate as many women have raised AFP levels from other causes, so it is necessary to screen the foetus by ultrasound as well.

There is also the **triple marker** (Bart's test), done on maternal blood at 16-18 weeks gestation, measuring three serum components. It is usually only given to women of age 35 or over. It detects 2 in 3 cases of Down syndrome, 9 in 10 cases of anencephaly and 4 in 5 cases of spina bifida. It takes 10 days to get the results, and its failure to detect all cases of these conditions can cause great distress.

13.12.6. Foetal DNA screening

Foetal DNA, obtained from chorionic villus sampling (the best method) or amniocentesis is amplified in specific regions by PCR (the polymerase chain reaction) for testing with labelled probes for particular disease alleles. Some of the main diseases which can now be tested for in this way are α- and β-thalassaemia, cystic fibrosis, fragile-X syndrome, haemophilia A (factor VIII deficiency), Huntington disease, and Duchenne muscular dystrophy. For example, for Duchenne muscular dystrophy, the dystrophin gene is amplified and can be hybridised with a specific probe.

A major problem with DNA testing is that probes are normally **allele specific**, detecting only one type of mutation each. There are many possible mutations at the locus for muscular dystrophy, so one probe or even a few probes for the commonest mutations will still fail to detect the other mutations. For cystic fibrosis, where 1 in 22 "whites" is a heterozygous carrier, 70% of mutant genes carry the same mutation, a deletion of a phenylalanine codon at position 508. A probe for this will miss the other 30% of mutations, although the recent DGGE CFTR kit from Ingeny is claimed to detect all mutations. With sickle-cell anaemia, nearly all mutant genes have the same defect in the β-globin gene, a base-substitution which changes a glutamic acid to valine, and which changes a restriction site in the DNA, so the mutant gene is easily recognised by DNA analysis.

Restriction fragment length polymorphisms (RFLPs) can be used in certain favourable families if the gene for a disease has not been characterised enough for DNA analysis for the mutant allele, and if there is a closely linked RFLP. If one parent is heterozygous both for an autosomal dominant disease and for a very closely linked RFLP marker, the foetuses can be tested for the RFLP marker in samples from amniocentesis.

Suppose that the father is *D d* for the dominant disease and *m1/m2* for the RFLP marker, and that the normal mother is *d d, m2/m2*, then one needs to find out whether the father is *D,m1/d,m2* or *D,m2/d,m1*. If the first child born is a sufferer *D d* and *m1/m2*, then the father must have been *D,m1/d,m2*, as the mother provided the *m2* RFLP and *d* in the foetus. If amniocentesis of subsequent foetuses shows any to be *m1/m2*, one can predict that they will be sufferers (*D d*) and termination can be offered, while foetuses which are *m2/m2* should be non-sufferers, inheriting the father's harmless *d* allele. If the first child is normal and is *m2/m2*, then subsequent foetuses with that RFLP combination should be normal, but those which are *m1/m2* have a high risk of having the dominant disease. If the RFLP site is not very closely linked to the disease site, then errors of diagnosis may be made when recombination between the sites occurs.

13.12.7. Neonatal screening

There is no routine **neonatal** (birth to one month of age) DNA screening. The important routine screening is for **phenylketonuria** (Chapter 13.3.1) in the first week of life, introduced in 1961. The mother controls the foetus's levels of phenylalanine in the blood before birth, but recessive homozygotes for PKU have blood phenylalanine levels rising to 15 to 30-fold of normal during the first week after birth. This can soon cause brain damage. Around day three, blood is taken from the baby, usually from a heel-prick, and a dried spot is bioassayed for levels of phenylalanine - details were given in Chapter 13.3.1 - for detection, with a rapid start of dietary treatment for sufferers. PKU has an incidence of about 1 in 10,000.

Although it would not be economic to screen separately for some very rare diseases, other blood drops from the PKU sample are used to test for galactosaemia (sufferers cannot metabolise galactose, the main sugar in milk; 1 in 70,000 are affected), maple syrup urine disease (1 in 250,000) and homocystinuria (1 in 100,000). All these are treatable to some degree by controlling early diet. Congenital hypothyroidism (1 in 3,500 in UK "whites", 1 in 900 in Asians, 1 in 25,000 USA "blacks") can be detected by testing a dried blood spot for elevated levels of thyroid-stimulating hormone.

13.12.8. Adult or adolescent screening

This is usually only done where there is a high risk of certain conditions in a particular population. **Genetic counselling before marriage** can be given where such screening is carried out. It is usually uneconomic to screen the general population for inherited diseases, especially just to detect carriers. In Ashkenazi Jews, **Tay-Sachs disease** usually has frequency of about 1 sufferer in 3,000, with 1 carrier in 30 compared with 1 carrier in 300 in most other groups. It is due to a lack of hexosaminidase A enzyme, giving death before the age of three, with no real treatments available. Blood enzyme (or DNA) screening can detect carriers and they can be given genetic counselling before marriage, and sufferers can be detected from amniocentesis. Counselling plus prenatal detection and selective abortion reduced by 65% the frequency of sufferers born to Ashkenazi Jews over the period 1970-1980.

Cystic fibrosis genes can be screened for in adults, including carrier detection, but as noted above, one probe only detects one type of mutant gene, so the probe for the most common mutation detects only 70% of cystic fibrosis cases in Northern Europeans. **Sickle-cell anaemia** sufferers and carriers can easily be screened for by electrophoresis of blood haemoglobins, or by a reliable DNA analysis mentioned above.

In Cyprus, malaria resulted in about 16% of the adult population being carriers of β-thalassaemia, with 0.6% of births being sufferers who usually died within one year of birth. With such a high frequency of the deleterious gene, it was worth screening the whole population before they reached marriageable age, so that carriers could be given genetic counselling. Malaria has been eliminated from Cyprus, so the lack of heterozygote advantage should reduce the frequency of the bad gene. Cypriots (and others from malarial areas) who migrated to non-malarial areas such as Britain have a lower chance of producing sufferer babies if marrying a native of that area rather than another Cypriot. A Cypriot/Cypriot marriage has a 1 in 200 chance of an affected child because there is a chance of about 2% of such marriages being between two carriers.

With the great advances in molecular genetics, there are **worries** that employers and insurance companies might insist on DNA screening of potential employees or of people wanting insurance, to detect those with genes which might cause increased risks of early death or of illnesses such as heart disease or cancer. In Britain in about 1998, the Human Genetics Advisory Commission found only one case of employers using such screening. It was the Ministry of Defence screening potential aircrew members for sickle-cell anaemia, a very sensible precaution (Chapter 13.3.1). In America in 1998, the American Management Association found nine out a sample of 1,085 employers used genetic testing of staff, and 11 states have passed legislation limiting such genetic testing.

For Alzheimer's disease, Huntington's disease, sickle-cell anaemia, muscular dystrophy, thalassaemia, Tay-Sach's disease and some forms of bowel cancer, there are reliable tests for the relevant genes. For dyslexia, some mental illnesses, diabetes, arthritis, heart disease and many cancers there are genes which **predispose** people to having those diseases, but mere possession of the gene does not necessarily mean that that disease will develop. About 5% of women who develop breast cancer have a mutant gene, *BRCA1*, but not all who have that gene get the cancer, and most women who develop breast cancer have no known genetic predisposition to it.

In 1997 in Manchester, England, two sisters aged 26 and 28 opted to have both breasts removed in an attempt to avoid breast cancer, which ran in their family, with their maternal grandmother dying of it at 26 and their mother dying of it at 31. The sisters were told that they had a 50% chance of having the bad genetic make-up, and that if they had that, there was an 85% chance of getting breast cancer. Because there were no surviving relatives on the mother's side of the family for testing, doctors were not able to make meaningful tests to assess the genetic risk to the sisters.

13.13. Effects of human inbreeding

Inbreeding in humans has the same effect as in plants and animals (Chapter 6.2): it makes deleterious and beneficial recessives homozygous and exposes them to expression and selection, if they were previously hidden in heterozygotes. In Japan after World War II, large-scale human genetic studies were made on survivors of the atomic bombs in the Hiroshima and Nagasaki areas, which gave data on the effects of inbreeding as well as on radiation effects. Where parents were first cousins, $F = 1/16$, the frequency of congenital malformations in the children rose from 0.011 to 0.016, a 48% increase compared to children of unrelated parents; stillbirths rose by 25% and the infant death rate rose by 35%. Even the amount of inbreeding involved in first-cousins matings was therefore clearly deleterious.

In data from rural France after World War II, the frequency of deaths before adulthood, including stillbirths, was 0.12 if parents were unrelated, and 0.25 if parents were first cousins ($F = 1/16$ for their offspring). The increase in deaths from making one-sixteenth of the heterozygous loci homozygous was thus $0.25 - 0.12 = 0.13$, so making all heterozygous loci homozygous would give $16 \times 0.13 = 1.8$ deaths, so on average all children would be dead almost twice over! As a heterozygote $A\ a$ could go equally to $A\ A$ or $a\ a$, only half its homozygous offspring would die, so from the deaths we register only half the recessive lethal genes, thus the estimated **average number of recessive lethal alleles per person** from these data is 3.6. More modern data give estimates closer to 2 recessive lethals per person, obviously usually in the heterozygous condition, but the true number must be much higher as those giving death in the embryo or early foetus are probably undetected, especially in figures relating only to stillbirths and to deaths later in development.

To help study the effects of **incestuous matings**, there is a confidential **incest register** in the UK, recording the consequences of mating between very close relatives. Data from the UK and USA on 31 children from father/daughter and brother/sister matings showed that only 13 (42%) were normal and 18 were handicapped, with six dying as young children.

The **frequency of marriages between relatives** differs greatly between communities, depending on religion, customs, laws and population size. In data from the 1950's, in Baltimore, USA, the frequency of first-cousin marriages was about 0.05%, while in rural India, in Andhra Pradesh, it was 33%, with an average population F of 0.032. That same rural population also had an astonishing 9% of maternal uncle/niece marriages (F of offspring, 1/8), but a much lower incidence of paternal uncle/niece marriages, which is as if it didn't matter if the relationship was only on the female parent's side. In a very small population, or in a small, closed religious group marrying almost entirely within the group, even random mating may involve relatives such as first or second cousins.

Consider **first-cousin marriages**. If an individual is heterozygous for a rare deleterious allele, the chance of a first cousin also carrying that bad allele is 1/8 (first cousins have 1/8th of their genes in common through recent descent, the theoretical coefficient), and two heterozygotes mating will have an average of one quarter of their children homozygous recessive and suffering the deleterious phenotype. So for a person heterozygous for a deleterious allele and marrying a first cousin, the chance of a child being a sufferer is 1/8 x 1/4 = 1/32, about 3.2%. There will be a separate chance of 1/32 of being a sufferer for each recessive deleterious or lethal gene possessed, so with two to three lethal recessives on average per person, and other non-lethal deleterious recessives, it is not surprising that offspring of first-cousin matings suffer increased death rates and other problems, compared with children whose parents are not related. Offspring of first-cousin marriages tend on average to have slight reductions in height, girth and aptitudes (IQ down about 4 points), compared to non-inbred children. About 20% of human albinos come from cousin marriages.

In one sense, inbreeding in humans is eugenic as it exposes harmful recessives to selection, so reducing their frequency. Continued inbreeding in a group should eventually adapt it to inbreeding for this reason. The pharaohs of ancient Egypt practised brother-sister marriages for many generations and yet were generally successful. As usual in human genetics, one cannot be sure whether the stated parents were the actual parents in those days.

If **consanguineous marriages** are defined as those between relatives who are second cousins or closer, the frequency of consanguineous marriages is between 20% and 50% in many Muslim societies in north Africa, Asia and parts of the former Soviet Union. In the primarily Hindu southern states of India, consanguineous marriages average 20 to 45%, and when families from these Moslem and Hindu communities migrate, say to the UK or the USA, they often retain a high degree of consanguineous marriages.

In the UK the closest legal marriage is usually considered to be between double first cousins (where a pair of sibs marries a pair of sibs, so that both sets of grandparents of the children are in common, not just one set of grandparents in common as for normal first cousins), with $F = 1/8$ in the offspring. However, if a pair of identical twins marries a pair of identical twins, the offspring of one such marriage are legally first cousins of the offspring of the other marriage, but genetically they are the equivalent of full siblings, so if they intermarry, their offspring will have $F = 1/4$. Uncle-niece and aunt-nephew marriages are banned in the UK, where the Church of England's Table of Kindred and Affinity is the main basis of the marriage laws. Roman Catholics need special permission for first-cousin marriages. In more than half of the USA's states, uncle-niece, aunt-nephew and first-cousin marriages are banned, as they are in some largely non-Moslem African countries. Marriage between relatives is most common in India, Pakistan and parts of the Middle East. In Britain, first-cousin marriages are most common in the Moslem community, with the deleterious genetic consequences being offset by earlier arranged marriages, earlier first pregnancy and more children per family than in other groups, on average. Marriage within kinships may have economic benefits in keeping property together. See Bittles et al. (1991), for more information on human inbreeding.

13.14. Genetic counselling

The aim of **genetic counselling** is to give information to sufferers, carriers or their relatives, or anyone with a high risk of having or passing on a deleterious genetic condition. First, the condition must be accurately identified; for example, not all forms of muscular dystrophy are caused by X-linked recessives, and some forms are not heritable. Second, treatment must be given if possible, e.g., by drugs, hormones, special diets, blood transfusions, physiotherapy or surgery, as previously mentioned for different conditions. Third, a prognosis must be given, informing the sufferer or relatives or guardians what the likely course of the disease will be in future. Fourthly, if the sufferer might be fertile and is capable of understanding the issues, he or she (or relatives or other carers) is given information on the likelihood of having suffering or carrier children by different types of partner.

For example, a sufferer from **phenylketonuria** would be advised that all children by a fellow sufferer would be sufferers, that of children by a carrier, about half would be sufferers and half would be carriers, whereas children by a normal dominant homozygote would all be carriers. Information would be given about detecting carriers, about prenatal diagnosis and about the easy detection of sufferers in the first week of life and the effective treatment by diet. Women with PKU are advised to adhere strictly to a low phenylalanine diet during pregnancy because a normal foetus can be adversely affected when growing in an affected mother.

Counsellors need to be aware of phenomena such as incomplete penetrance (Chapter 3.2.5) through which an autosomal dominant such as inherited colon cancer can unexpectedly "skip a generation", and variable expressivity (Chapter 3.2.4). Even for conditions with full penetrance and constant expressivity, the existence of gonadal mosaics can complicate the calculations of recurrence risks. An individual might be symptomless because a dominant new mutation was confined to his or her gonads, but the recurrence risk could be a half if all the gonadal cells carried it. Under the UK Congenital Disabilities (Civil Liability) Act, 1976, a person whose breach of duty to parents results in the birth of a disabled or abnormal child can be sued, and there is increasing litigation in the USA and UK over genetic disease.

For autosomal dominants (Chapter 2.1.2), a heterozygous sufferer married to an unaffected person has a risk of a half for a child being affected, but that can be modified by incomplete penetrance, variable expressivity and variable age of onset, as with Huntington disease. For autosomal recessives (Chapter 2.1.2), the recurrence risk from two carriers is one quarter. For X-linked dominants, affected fathers will pass the condition to daughters only, and heterozygous affected mothers will pass the condition to half the sons and half the daughters. For X-linked recessives, affected males married to normal females will have all carrier daughters and normal sons: see Chapter 2.1.4 for a colour-blindness example showing all possible crosses.

Very **strong emotions** can be generated by suggestions of a genetic defect in a family, so great tact and confidentiality are required in counselling. A family in which someone is diagnosed as have a defective gene or chromosome sometimes feel that all members' marriage chances may be diminished, perhaps together with their employment prospects, reputation for mental stability, or insurability at normal rates. The affected person can feel guilty about having the condition. What is given is always information, never commands. Relatives of an affected person sometimes refuse to be tested, as they would rather not know whether they carry the bad gene or chromosome. Relatives may have died or be out of touch or living in another country.

Gene therapy is dealt with in Chapter 14.2.

Suggested reading

Baraitser, M. and R. Winter, *A Colour Atlas of Clinical Genetics*. 1988. Wolfe Medical Publications, London.

Bittles, A. H. et al., Reproductive behavior and health in consanguineous marriages. *Science* (1991), 252, 789–794.

Carter, C. O., Genetics of common disorders. *British Medical Bulletin* (1969), 25, 52–57.

Connor, J. M. and M. A. Ferguson-Smith, *Essential Medical Genetics*, 5th ed. 1997. Blackwell Scientific, Oxford.

Edwards, J. H., M. F. Lyon and E. M. Southern, *The Prevention and Avoidance of Genetic Disease*. 1988. The Royal Society, London.

Fraser-Roberts, J. A. and M. E. Pembrey, *Introduction to Medical Genetics*, 8th ed. 1985. Oxford University Press, Oxford.

Gelherter, T. D., F. S. Collins and D. Ginsburg, *Principles of Medical Genetics*, 2nd ed. 1998. Williams and Williams, Baltimore.

Hartl, D. L. and E. W. Jones, *Genetics. Principles and Analysis*, 4th ed. 1998. Jones and Bartlett, Boston.

Howell, W. M. and S. T. Holgate, Human leukocyte antigen genes and allergic disease, pp 53–70, in: Hall, I. P., ed., *Genetics of Asthma and Atopy*. Monographs in Allergy. 1996. Vol. 33. Karger, Basle.

Howell, W. M. and C. Navarette, The HLA system: an update and relevance to patient-donor matching strategies in clinical transplantation. *Vox Sanguinis* (1996), 71, 6–12.

Jacobs, P. A. and T J. Hassold, The origin of numerical chromosome abnormalities. *Advances in Genetics* (1995), 33, 101–133.

Lakich, D. et al., Inversions disrupting the factor VIII gene are a common cause of severe haemophilia A. *Nature Genetics* (1993), 5, 236–241.

Lesage, M., *Vitiligo. Understanding the Loss of Skin Colour*. 1997. The Vitiligo Society, London.

McConkey, E. H., *Human Genetics, the Molecular Revolution*. 1993. Jones and Bartlett, Boston.

Montagu, A., *Human Heredity*, 2nd rev. ed. 1963. Signet, New York.

Morgan, D. R., A guide to living with risk. *Biologist* (1989), 36, 117–124.

Page, S. L. and L. G. Shaffer, Nonhomologous Robertsonian translocations form predominantly during female meiosis. *Nature Genetics* (1997), 15, 231–232.

Pasternak, J. J., *An Introduction to Human Molecular Genetics. Mechanisms of Inherited Diseases*. 1999. Blackwell Science, Oxford.

Scriver, C. R. et al. *The Metabolic and Molecular Basis of Inherited Disease*, 7th ed. 1995. McGraw-Hill, New York.

Sherwood, L., *Fundamentals of Physiology. A Human Perspective*, 2nd ed. 1995. West Publishing Company, St. Paul/Minneapolis.

Stern, C. *Principles of Human Genetics*, 3rd ed. 1973. W. H. Freeman and Co., New York.

Strachan, T. and A. P. Read, *Human Molecular Genetics*, 2nd ed. 1999. Bios, Oxford.

Strickberger, M. W., *Genetics*, 3rd ed. 1985. Collier Macmillan, London.

Weatherall, D. J. et al. *Oxford Textbook of Medicine*. 1995. Oxford Medical Publications, Oxford.

CHAPTER 14

GENETIC ENGINEERING IN PLANTS, ANIMALS AND MICRO-ORGANISMS, AND HUMAN GENE THERAPY

14.1. Genetic engineering

14.1.1. Introduction

"Genetic engineering" is now used for *in vitro* genetic manipulations involving the artificial joining of two different DNA molecules to generate genomes with particular desired properties. It aims to replace or supplement traditional methods of plant, animal and microbial improvement from random mutation and recombination, with directed modification of the genome. In essence, two DNA molecules are cut at specific base sequences by restriction endonuclease enzymes, are rejoined in a desired way, and are reintroduced into a living cell which can sustain and multiply the cloned DNA; selective techniques are used to identify cells which carry this DNA. For expression in another organism, a **transgene** (a gene modified *in vitro* for reintroduction to an organism) is constructed to have a highly active promoter (preferably tissue-specific in higher Eukaryotes), protein-coding sequences, and the right terminator sequences to get the protein product moved to the right part of the cell, or to be excreted.

For a multifactorial or polygenic human disorder such as club foot or heart disease, influenced by many genes of small effect and by environment and chance, genetic engineering offers little in the way of diagnosis or treatment. It is not the universal "cure-all" as sometimes represented in the media. The current use of genetic engineering in farm animals for those animals' own traits was summed up by Maijala (1997): "However, there appeared to be many problems and, since the conventional selection methods with modern statistical techniques give good rates of progress, molecular-genetic techniques have not yet led to new breeds in sheep."

It must be stressed that whether new varieties are produced by genetic engineering or classical breeding, it is essential that they be **rigorously tested** in a range of environments over a number of seasons, to prove their commercial reliability and to find under what conditions they are better than existing cultivars (Chapter 12.9).

Even a very deliberate molecular change in one gene may have unpredictable and unfavourable effects on other aspects of an organism's performance: see Chapter 14.1.10, Bergelson et al. (1998). For more examples of genetic engineering than are given below, see the suggested reading list.

There are now some very sophisticated systems available to **control the activity of individual genes** in higher Eukaryotes, quantitatively and reversibly, with regulation factors of up to 10^5. For example, Clontech Laboratories UK Ltd. have introduced gene activators which are switched on by the presence of tetracycline, and transcriptional silencers where genes can be turned off by tetracycline, to different extents in different transformants.

14.1.2. Restriction endonucleases and ligases

Endonucleases cut chains of nucleotides within the chain, while **exonucleases** remove bases from the chain ends. **Ligases** can join up two ends of DNA, making an intact chain by joining a terminal 5′ phosphate to a terminal 3′ hydroxyl group.

Restriction enzymes were given that name when they were found to restrict the range of strains of bacterium *Escherichia coli* which were attacked by a particular bacteriophage such as *lambda* (λ). λ grown on *E. coli* strain C could grow well on that strain but had a very low frequency of infection on *E. coli* strain K, while λ grown on strain K had a high frequency of infection on both strains C and K. In 1970, Meselson and Yuan showed that the effect was due to strain K having a restriction endonuclease which could break up unmethylated DNA, and to strain K having a **DNA methylase**, while strain C lacked both enzymes. λ grown on C had unmethylated DNA and was able to attack strain C, but when it injected its DNA into strain K, that DNA was usually destroyed by the endonuclease. If λ DNA managed to escape this fate and was replicated in strain K, then the λ produced had methylated DNA. The methylation protects the DNA from the endonuclease, so λ grown on K could infect K and C easily.

When an organism produces an endonuclease, it usually has the corresponding DNA methylase which protects its own DNA from being attacked by its own endonuclease. If strain K had the endonuclease without the methylase (methyltransferase), it would cut up its own chromosomal DNA.

Restriction enzymes each recognise a specific DNA base sequence (usually of four to six bases), while the associated DNA methyltransferase enzymes recognise the same sequence and methylate a key base within that sequence, thus preventing the DNA from being cleaved by its own endonuclease. There are two main types of restriction endonuclease. **Type I** enzymes recognise a specific sequence in the

DNA but cleave the DNA more or less randomly after moving along the chain from the recognition sequence. They include the *E. coli* K restriction enzyme. **Type II** enzymes cut at specific sequences and so always produce the same fragments from the same DNA, with the same exposed ends. They are found in all types of organism.

Type II enzymes usually recognise rotationally-symmetrical sequences in the DNA, and either make **blunt** (flush) **end breaks**, by cleaving the two chains at the same point, or make **staggered breaks**, by cleaving the two chains at different points, leaving "sticky ends" of unpaired bases, with complementary sequences at the two ends. Thus a blunt-end-break enzyme from the bacterium *Arthrobacter luteus* recognises the sequence:

5′ AGCT 3′

3′ TCGA 5′, which has a central axis of rotational symmetry between the C's and G's, with the sequence going away from this in the top strand being the same as the sequence going away from it on the opposite side in the other strand. The enzyme cuts each strand between the C and the G, leaving blunt ends,

——AG 3′ on one fragment and 5′ CT—— on the other fragment.

——TC 5′   3′ AG——

The most useful enzymes for genetic engineering are those making staggered breaks. For example, *Eco*RI, from *E. coli*, recognises the sequence

5′ G*AATTC 3′

3′ CTTAA*G 5′, cutting the upper chain after the G and the lower chain before the G, at the positions marked by asterisks, leaving the left-hand fragment ending in

——G 3′   and the right-hand fragment ending in   5′ AATTC——

——CTTAA 5′   3′ G——

These are **sticky ends**, in that the left- and right-hand ends now have complementary single-stranded base sequences and could pair up, with DNA ligase rejoining each of the two chains. While DNA ligase from phage T4 can join blunt ends, unlike most ligases, there is no specificity as to which blunt ends join which other ones; with sticky ends from staggered breaks, the complementary base pairing between the sticky ends ensures specificity in which ends are joined.

The main reason that genetic engineering is possible is because the same type II staggered-break restriction enzyme will make the same sticky ends in any type of DNA containing its recognition sequence. If one wants to insert say a particular fragment of human DNA into a bacteriophage genome, one has to find a restriction enzyme which will cut the human DNA twice, once at each end of the desired fragment, and once in the bacteriophage genome. There are several hundred known restriction enzymes that could be tried, each with a unique recognition sequence of bases. If *Eco*RI is the .enzyme chosen, it will make the same sticky ends in the human DNA as in the bacteriophage DNA, so the two types of fragment can be brought together and allowed to base-pair at the sticky ends before DNA ligase is used to join the fragments covalently. Not all molecules will be of the desired type, as the phage DNA could circularise by pairing of its own sticky ends with no insert, or several molecules could join, or the insert could be joined in the wrong orientation. The techniques for overcoming these problems are outside the scope of this book.

14.1.3. Vectors

The **vector** is the carrier molecule into which the desired DNA is inserted. Ideally, it should have only one copy of the recognition site for a particular restriction enzyme, but it can have just one recognition site for each of a number of different restriction enzymes, so that the user has a choice of enzymes. Common vectors include *E. coli* plasmids, phages such as λ and M13, and derived plasmids such as cosmids, which are made using for example the cohesive ends of λ, with other DNA sequences.

One has to be able to get the vector plus insert back into a living cell. Cell walls may need to be enzymatically removed from plants and filamentous fungi before vectors can penetrate into the cells. With the **biolistic** method, DNA-covered gold or tungsten particles are shot from a gene gun into tissues or fungal hyphae, without needing protoplasts. With plasmid or cosmid vectors, one can use the DNA to transform *E. coli*, but as that bacterium has no active DNA uptake system, one has to use **passive DNA uptake** after treating the culture with calcium chloride to make the walls and membranes permeable to the DNA, or one can use **electroporation** (see Chapter 14.1.6), but the bigger the molecule, the less efficient the transformation. With phage DNA, one can use **transformation**, or one can add empty phage coats to package the DNA into **infective phage** particles.

Transformation efficiencies are usually low: for λ, it is about 10^{-4} to 10^{-6}; for plasmid ColE1 it is 10^{-3}, but goes down to 10^{-5} or 10^{-6} with a large insert. With such low efficiencies, one needs a selective system to detect and isolate bacteria which have taken up and expressed the vector DNA plus insert. Ideally one has an insert with an antibiotic resistance gene, say to tetracycline, and plates the bacteria on medium with the antibiotic, so the only survivors are the desired bacteria plus inserts.

A vector should also have an origin of replication, so that it can be replicated inside the bacterium or other organism used. "**Shuttle vectors**" have a bacterial origin of replication and often a yeast origin of replication, so that they can replicate in Prokaryotes and in Eukaryotes.

Plasmids are used for cloning small fragments of DNA, say five to ten kilobases. λ phage particles are of a defined size and will only package DNA which is within the range 80-105% of a normal λ genome length, which is 50 kb. That would only allow an insert length of 5% of that, but by deleting 20% of the λ genome which is not essential for infection, about 25% of the λ genome length can be used for inserts, about eight genes. 20 kb is the largest insert usable in λ, but cosmid vectors, which are plasmids, can take up to about 45 kb inserts. Deletions of phage P1 can accept up to 85 kb of insert.

An example of a commercially available cloning vector is the **pBluescript plasmid** (Stratagene Cloning Systems, La Jolla, California), which is circular, with 2961 base pairs. It has two origins of replication, one from plasmid ColE1 (a high copy-number plasmid, so that one can get about 300 copies per cell of the vector) and one from single-stranded DNA phage f1. When bacterial cells containing a recombinant plasmid are additionally infected with f1 helper phage, the f1 origin enables packaging of a single strand of the inserted fragment into phage particles, which is useful for *in vitro* mutagenesis. The vector carries resistance to the antibiotic ampicillin, for selection of transformed cells which have taken up the vector, and a **multiple cloning site** within part of the *E. coli lacZ* gene. The multiple cloning site (nucleotides 657 to 759) contains one copy of each of the recognition sites for about 23 different restriction endonucleases. Having this within part of the *E. coli lacZ* gene is very convenient for identifying cells which have an insert into the multiple cloning site, as that makes them Lac⁻ compared with cells without the insert, which are Lac⁺; with the β-galactoside X-gal, Lac⁺ colonies are deep blue, while Lac⁻ colonies are white.

14.1.4. Getting a particular piece of DNA into a vector, and recognising a clone containing it

If one cuts DNA with a particular restriction enzyme, the number of fragments will depend on the frequency of that enzyme's recognition sequence in that DNA. If one cuts a small molecule like λ DNA (about 50 kb), there might be as few as 10 fragments, while cutting *E. coli* DNA might produce thousands of fragments, and cutting the much larger human DNA might produce a million different fragments. Random cloning of the cut DNA into a vector could therefore produce anything from a few different vector + insert combinations to millions of different such combinations, each with a different insert.

The simplest cloning selection can be done if the wanted DNA insert directly confers a selectable phenotype to the organism receiving the vector + insert. One could select for the insert being a *histidine A* gene by cloning from an organism which was *hisA*$^+$, putting it into *hisA*$^-$ bacteria, and plating on minimal medium, so only *his*A$^+$ bacteria could grow. Antibiotic resistance or phage resistance genes could be selected on medium with the antibiotic or the phage.

If direct selection is not possible, it may be possible to purify the DNA before adding it to the vector. The DNA after treatment with the restriction enzyme may be subjected to electrophoresis, which can partition the fragments according to size and electrical charge. That would be useful if the size of the desired restriction fragment were known, so that only fragments of that length were added to the vector. If the DNA sequence is known at the ends of the desired fragment, and if the DNA length to be used is less than about 50 kb, the **polymerase chain reaction** (PCR) can be used to amplify the DNA of interest many thousands of times. Oligonucleotide DNA primers about 20 nucleotides long are artificially synthesised, one complementary to the left hand end sequence and one to the right hand end sequence of the target DNA. With thermal cycling to denature the target DNA into single strands (about 95°), to anneal the primer oligonucleotides (about 55°) to the target DNA, and to allow for elongation of each primer (about 70°) by DNA synthesis off the complementary single target strand, only the target DNA is multiplied exponentially over a series of cycles, while the rest is not. If a sufficient number of PCR cycles is used, with n cycles theoretically giving 2^n copies of the target DNA, most vectors should pick up the target DNA, though other fragments will also be present. The taq polymerase used in PCR is very heat-stable, coming from the heat-resistant bacterium *Thermus aquaticus*, and so survives this thermal cycling for many rounds of replication.

Most Eukaryote genes are rather long compared with Prokaryote genes, because the Eukaryote coding sequences in the **exons** of the gene are usually interrupted by non-coding **introns**. This often makes the whole gene too long to amplify by PCR and may involve there being several copies of a restriction enzyme's recognition sequence within the one gene, so that it is difficult to isolate the whole gene. When working with Eukaryote genes, especially if one wants eventually to have them expressed in Prokaryotes (which cannot excise the introns and would therefore not produce the right proteins), it is often best to isolate already processed messenger RNA, with the introns removed in the Eukaryote. This processed mRNA can then be transcribed into **complementary DNA (cDNA)** by **reverse transcriptase enzyme**. Isolating fully processed mRNAs for a particular gene is best done by using a tissue in which that gene is most strongly expressed, such as the oviduct of mature hens for chicken ovalbumin mRNA, coding for the main egg-white protein.

It is usual to use a vector to obtain a large set of different clones, referred to as a **library**, then to identify which particular clones carry the gene one is interested in. This may require having labelled DNA or RNA probes for that gene, which can be applied to DNA from each clone, and will only hybridise with those having complementary sequences. If the protein produced by the gene is known, clones having and expressing that gene can be detected by using labelled antibodies to that protein.

Other useful techniques include map-based cloning for genes originally known only from their phenotypes, e.g., cystic fibrosis, and transposon-tagging for that purpose in plants.

14.1.5. Site-directed mutagenesis

Because spontaneous and induced mutations are random, it is usually very difficult to obtain a specific mutation at a specific place by those methods. If a gene has been cloned and its base sequence is known, then there are methods for direct DNA manipulation to obtain a **particular mutation**. Suppose it is desired to change a particular GGA DNA codon (CCU in mRNA, proline) to a GAA DNA codon (CUU, leucine), then one synthesises an oligonucleotide containing DNA complementary to that region of the target gene, with CTT for the desired complementary codon.

The target gene can be cloned into a single-stranded DNA plasmid which has two antibiotic resistance markers, say ampicillin and tetracycline resistance, where the ampicillin-resistance gene has a single base substitution mutation, making the double-stranded form confer the phenotype Amp^S, Tet^R on its bacterial host. The single-stranded plasmid DNA is annealed with two synthesised oligonucleotides (20 to 50 bases), one carrying the desired mutant sequence complementary to that part of the target gene, and the other carries a sequence complementary to the part of the *amp* gene, but with the amp^R sequence, not amp^s. When the two oligomers have annealed to the plasmid, they act as primers for synthesising the second strand of the DNA by DNA polymerase plus ligase. This gives a plasmid with two base-pair mismatches, one in each oligomer region, with CTT opposite GGA in the target gene, giving mispair T/G.

The plasmid with the two mismatches is transformed into *E. coli* cells which carry *mutS* - and therefore lack a mismatch repair system - and amp^s, tet^s. At replication, the double-stranded molecule segregates, one daughter molecule having the unmutated target gene sequence and the amp^s gene. The other, derived from the synthetic oligomers, carries the mutated form of the target gene (GAA paired correctly with CTT) and the *amp^R* gene. When the bacteria are plated on medium with ampicillin and tetracycline, that selects for cells carrying DNA derived from the mutated strand, as only they will be *amp^R* and *tet^R*. Most will have the desired target gene mutation. There are many other ways of carrying out *in vitro* mutagenesis. The term **reverse genetics** is used for studies in which a known DNA change is made to see its effect on the phenotype, instead of finding a mutant phenotype and then finding what DNA change was responsible for it.

14.1.6. Gene targeting

Gene targeting is the modification of a gene in a chromosome by the use of **homologous recombination** with DNA of an introduced vector. It has been particularly used in mouse embryonic stem (ES) cell lines, established from the inner cell mass of blastocysts from preimplantation mouse embryos. If treated correctly to prevent differentiation, these ES cells can resume normal development and contribute to all cell lineages, including germ-line cells of the resulting **chimeric** (having different cells with different genotypes) mice, and so can be expressed in the next generation. In practice, insertion of the vector only occurs sometimes in the desired gene, and at other times elsewhere in the genome.

The ES cells are given the targeting vector by **micro-injection** or by the easier but less efficient **electroporation**, which is transformation aided by short electrical pulses. Depending on the vector design, homologous recombination can yield an insertion, a replacement or a deletion in the target locus. As only one ES cell in 10^2 to 10^5 may undergo the desired targeting event, selective methods are needed to detect the right cells. Often a **positive-negative** selection is used. There is positive selection for cells which have incorporated the vector anywhere in the genome, often using neomycin (*neo^R*) resistance to antibiotic G418. There is a negative selection against cells which have randomly integrated the vector into regions lacking homology to the vector, often using the hypoxanthene phosphoribosyl transferase gene (*Hprt*). In the presence of the base analogue 6-thioguanine, only the *Hprt^−* cells survive, those in which the desired homologous recombination occurred.

When the desired mutation in the targeted ES cells (say from a brown mouse) has been selected, the cells are grown on and injected into the blastocoel cavity of a pre-implantation mouse embryo (say from a black mouse), and the blastocyst is transferred to the uterus of a foster mother (any colour). If the injected cells survive, the mice born will be chimeric, composed of cells from the donor ES cells and of ES cells from the host embryo. They may be identified as chimeric if the coat shows brown and black regions. If the mutated donor cells have helped to form the germ line, then mating the chimeric mouse to a wild-type mouse can yield some mice in which all cells are heterozygous for the mutation, and further breeding can give homozygotes, giving expression even of recessive changes in the target gene.

Suppose it was desired to inactivate gene 1 in an $Hprt^-$ cell by insertion of a neomycin resistance gene within gene 1. The vector could be made with part A of gene 1, then the neo^R gene, part B of gene 1, and an $Hprt^+$ gene. If there is a double crossover of the vector with the chromosomal intact wild-type gene 1, with one crossover within part A and one crossover within part B, then the chromosomal gene will be interrupted by the neo^R gene, but the chromosomal gene will not incorporate $Hprt^+$ as that is beyond the double crossover region, in a region without homology to the target area. Cells which have had this event can be selected on the neomycin analogue, G418. Cells in which the whole of the vector was randomly inserted, without the double crossover in regions of homology, will carry the $Hprt^+$ marker from the intact vector, so will be killed when the cells are grown with 6-thioguanine.

This kind of gene targeting, where genes can be disrupted (knocked out) by insertions or deletions, or given replacement DNA sequences, is very useful in studying how genes work and what they do. Systems for gene targeting are better developed for the mouse than for most other organisms. The fact that potential **germ-line** cells can be treated is important for breeding purposes in plants and animals.

14.1.7. Genetic engineering in plants

Transgenic crop plants have been produced by using the ***Agrobacterium* Ti plasmid**, by direct uptake of DNA by protoplasts, by high-velocity DNA-coated particles, and by micro-injection of DNA into cells. Plant genetic engineering is largely based on the tumour-inducing (**Ti**) plasmid from the bacterium *Agrobacterium tumefaciens*, a soil bacterium causing crown gall disease of dicotyledonous higher plants, where tumours (galls) develop on the stem near the soil level. The plasmids are circular, double-stranded and large, about 200 kb. They have an origin of replication, a 25 kb region called T-DNA, a *virulence* region for T-DNA transfer into the plant genome (more or less at random, without a need for homology) and a region for nopaline (or octopine) utilisation. The transferred part, the T-DNA, codes for the production of tumours (by uncontrolled plant cell division) and for the synthesis of opines such as nopaline by the plant under the control of the plasmid. These opines are used by the bacterium, helped by the opine-utilising genes on the plasmid.

Natural Ti plasmids have few suitable restriction sites, are too large, and the tumour production is unwanted. For plant genetic engineering, parts of the plasmid

amyloliquefaciens protein, **barstar**, is used in the restoration of fertility, with its gene also fused to *TA29* for tapetum-specific expression. Barstar protein combines with barnase enzyme to form an inert complex. Barstar on its own is harmless and restores male fertility in the presence of barnase. When a male-sterile inbred line carrying *TA29-barnase* is crossed with a male-fertile plant, *TA29-barstar*, all seed harvested on the male-sterile plant must be crossed seed, F1 hybrid. The F1 plants will be male fertile. See Mariani et al. (1990 and 1992). The barnase method has also been applied to cabbage, chicory, cotton, maize and oilseed rape.

Attempts have been made to genetically engineer plant tissue cultures to produce **high-value compounds** such as the anti-cancer agents vinblastine and paclitaxel (Taxol®, normally obtained from yew bark), the anti-viral castanospermine and the anti-malarial artemesinin. There have been problems with genetic instability, low yields and the high costs of tissue culture.

One technique with wide applications in plants is the use of **anti-sense RNA**. Normally, the promoter ensures that only the correct "sense" strand of DNA is transcribed, but by moving the promoter in relation to the gene, the other strand, "anti-sense", can be transcribed, making RNA complementary to the "sense" mRNA and which can therefore bind to and inactivate the "sense" mRNA. Anti-sense mRNA was engineered into tomatoes to reduce the production of the enzyme polygalacturonase which is involved in fruit softening during ripening. The transgenic tomatoes have a longer shelf-life before softening, and bruise less easily than normal tomatoes. Extended shelf-life has also been obtained in tomatoes by introducing a gene for a bacterial enzyme which reduces ethylene production; ethylene is a major ripening agent in fruits (Chapter 17.1.4).

Homology-dependent gene silencing (cosuppression) is now often used for the same purpose as anti-sense mRNA, and some cases of "anti-sense" suppression may actually be gene silencing. People seeking to enhance the expression of an endogenous gene often introduce additional copies of that gene, with more active promoters. Such transgenes are not always expressed in the expected way, and sometimes actually reduce the activity of the introduced gene and of the homologous endogenous gene, hence the term "homology-dependent gene silencing". The gene silencing can affect repeats in transgenes, or transgenes with sequence homology to endogenous genes. In plants, both transcriptional and post-transcriptional silencing have been observed: see Grant (1999) and Waterhouse, Graham and Wang (1998).

According to the British Society of Plant Breeders, the **main aims** today for UK crops involve the following modifications, including the use of genetic engineering: maize - insect resistance, herbicide tolerance; oilseed rape - modified oil, herbicide tolerance; sugar beet - modified sugar content, herbicide tolerance; wheat - modified starch, disease resistance; potato - modified starch, disease resistance, insect resistance; tomato - slower ripening; apple - disease resistance, slower ripening; field vegetables, pest resistance; soft fruit, slower ripening.

Another development on which a number of firms are working is "**terminator technology**", so that cereal seed grown by farmers can be sold for food but will not germinate. That would force farmers to buy fresh seed from the producers each year. Several groups are working on "**plant vaccines**", to produce engineered foods giving protection against cholera, hepatitis B, Norwalk virus and travellers' diarrhoea, *Exterotoxigenic E. coli*. Potatoes, tomatoes and bananas are being tried and could deliver vaccine directly to the intestinal surfaces, but year 2,004 is the earliest likely time for a working system.

For more information on transgenic plants, see Galun and Breiman (1997), who have an appendix on intellectual property rights and the commercialisation of transgenic plants, and Owen and Pen (1996) for plant systems for producing industrial and pharmaceutical proteins.

14.1.8. Genetic engineering in animals

Transgenic manipulations in animals (see Bishop, 1999) are done in order to introduce selected isolated genetic material from species or varieties which could not cross with the target animal. The gene's promoter is usually altered to remove the normal one and replace it with a specially designed one to switch the gene on at a desired time in development and in the correct tissues. Germ-line transgenic animals can be produced by embryonic stem-cell blastocyst injection technology, as described in Chapter 14.1.6, or micro-injection of nuclei in one-celled embryos.

In the micro-injection method, the cloned DNA is inserted directly into fertilised eggs which are then implanted in the recipient's uterus. It is difficult to control into which chromosome the DNA will be inserted and whether the whole engineered sequence will be incorporated; there is rather variable expression in different offspring. The success rate is usually less than 5% of the treated eggs giving fully-expressing offspring, although it can reach 20% in mice.

In 1999, Transgenic Technology Services of Imperial College of Science, Technology and Medicine, London, offered a micro-injection service in which they prepared the DNA and did the micro-injection of mouse oocytes and the implantation into foster mice. They could use DNA from plasmids, phage P1 artificial chromosomes, bacterial artificial chromosomes, and yeast artificial chromosomes. They claimed 8% of offspring were transgenic with plasmid DNA and 5% with yeast artificial chromosomes, at a cost of £300 per session, each session giving about 13 mice born.

In **sheep**, many of the injected embryos die, and only about 1% become transgenic lambs (Wilmut, Campbell and Young, 1997). The site of integration of the transgene appears to be random, usually within genes which are disrupted, giving lethal mutations in more than 5% of the cases, with the lethality often only showing once the gene becomes homozygous. **Levels of expression** in mice and sheep are very variable, depending on where the transgene is inserted. The recombinant proteins from human genes in sheep are not identical to ones produced in humans, because of differences in post-translational processing. In sheep, commercial applications of genetic engineering have been for pharmaceutical proteins, not for agricultural purposes.

Somatic or germ-line transformations are possible with plasmids, and also with retrovirus vectors inserting DNA, made by **reverse transcriptase** from their RNA, into animal host chromosomes. If present, tumour-causing genes must be deleted from the retroviruses, and that deletion makes room for desired inserts.

For making proteins such as hormones or drugs for use in humans, one advantage of engineering genes into mammals rather than into bacteria is that mammals can carry out various post-translational modifications of proteins which prokaryotes cannot do, so that the final proteins are much more like normal human proteins than are ones grown say in *E coli*.

Table 14.1.8.1 shows various products being made or developed in **transgenic farm animals**, including the value of that product per animal per year. Production of products in blood is less useful than in milk, as animals need to be bled or slaughtered to obtain the product. The production of tissue plasminogen activator, for treating blood clots as in coronary thrombosis, is advantageous in the milk of sheep or goats because the yields in about five litres of milk a day are as much as in a 1,000 litre bioreactor using mammalian cell cultures. The genes are placed under the control of a β-lactoglobulin promoter which is active only in mammary tissue, and females carrying the desired genes (after micro-injection of fertilised eggs) are identified by PCR. Expression in milk is often at high levels and the required proteins are easily purified. Cell cultures were previously used to make human α1-antitrypsin, but production in milk of sheep gives large amounts of glycolsylated bioactive protein at low cost (see Wright, 1991). About 200 g are needed per patient per year, to treat people with genetic disorders causing their livers to under-produce this protein. These disorders are common in Caucasian males, giving emphysema which is often lethal if untreated. In 1998, PPL Therapeutics announced that human clotting factor IX expression levels of 300 mg/litre had been obtained in milk from transgenic sheep. The current annual value of factor IX production is about £100 million.

Table 14.1.8.1. Products produced in farm animals after genetic engineering.

Product	Animal	Value/animal/ year (in 1995), US$	Use of product
Haemoglobin	pig	3,000	Blood substitute in transfusions
Alpha-1-anti-trypsin	sheep	15,000	Treatment of inherited emphysema
Lactoferrin	cow	20,000	Infant formula feed additive
Factor VIII	sheep	20,000	Treatment of haemophilia A
Factor IX	sheep	37,000	Treatment of haemophilia B
Tissue plasminogen activator	goat	75,000	Treatment of blood clots
Cystic fibrosis transmembrane conductance regulator	sheep	75,000	Treatment of cystic fibrosis
Human protein C	pig	1,000,000	Anticoagulant to treat blood clots

Unfortunately, some products from engineered human genes have caused problems when used on humans. In 1991 it was reported that 400 diabetics in Britain wanted to sue for damages when they suffered serious side-effects from human-gene insulin. They and their families claimed that they suffered a range of side-effects including paralysis, permanent memory damage and even death as a result of being transferred from natural animal insulin to the genetically-engineered "human" insulin when that became available in the mid-1980's.

Farm animals may also be modified by genetic engineering to affect their own normal products, not just therapeutic proteins for humans. This includes casein for cheese-making and milk with low levels of lactose for people with lactose intolerance. Sheep can have improved wool production from bacterial genes for improved dietary cysteine utilisation.

The gene for **human growth hormone** has been fused with regulatory sequences from the mouse metallothionein-I gene and put into pigs and sheep where it is expressed in cell types which usually produce metallothionein-I, which includes the liver. The hormone increased the growth of pigs and reduced their fat content, but overproduction of the hormone gave problems from premature aging, sterility and arthritis in pigs, and diabetes in sheep. Hormone balances are so crucial and so difficult to control for engineered products, that use of such inserted genes to improve production of farm animals has largely been abandoned. Some success has been achieved with the sockeye salmon growth-hormone gene in the coho salmon, giving faster growth and maturity.

Attempts are also being made to incorporate human genes into pigs, to lower the rejection rate of pig organs when used in **transplants** into man, as pig organs are of a suitable size and physiology, and there is a serious shortage of human donors. Transplant rejection is not the only problem, though, as pig diseases might be transmitted during transplanting, especially viruses.

Transgenic chickens have been produced which express a gene for a protein from avian leukosis virus, giving protection from that virus.

14.1.9. Genetic engineering in micro-organisms

Genetic engineering has been very successful in bacteria, for production of bacterial drugs and some human proteins, with a huge variety of vectors and methods available. For fungi, see Anke (1997) and references in Chapter 18. The fungus baker's yeast (*Saccharomyces cerevisiae*) is also very suitable as there is a convenient 2 μm length plasmid, 6.3 kb, which is transmitted through mitosis and meiosis. It can be transformed into the cell. Bacterial plasmids can also be transformed into yeast cells. As bacterial plasmids do not have a suitable origin of replication to multiply autonomously in yeasts, they are used after integration by crossing-over into a yeast chromosome. The yeast **2-micron plasmid** can replicate autonomously in the nucleus, so does not need to integrate. It can be given a bacterial origin of replication too, making it a shuttle vector which can grow in bacteria and yeasts. This plasmid can also be given a yeast centromere, ensuring that both daughter cells inherit it after cell division. In yeast, "**ars**" vectors containing a replication origin, or autonomously replicating sequence (ars), are also used. Various industrial yeasts have been engineered to change the flavours in beer or the CO_2 production in baker's yeast.

Yeast artificial chromosomes (YACs) can be constructed from a linearised plasmid, a centromere, yeast replication origins and yeast telomere sequences at both ends. Heterozygosity for a locus on two homologous YACs can show proper Mendelian 2:2 segregation ratios at meiosis. YACs have been widely used as cloning vectors for large segments of Eukaryote DNA, such as genes or regions up to 1000 kb long. The longest human gene is that for Duchenne muscular dystrophy (3 sufferers in 10,000 males), which is 2.4 Mb long and takes about 16 h to be transcribed, but exons make up only 0.6% (about 14 kb) of this gene. See also Chapter 13.3.2.

Genetic engineering is often the method of choice for altering highly adapted microbial genomes if information is available on what changes might be beneficial. If it is desired to knock out a particular yeast gene for brewing or baking, or to change the base sequence, then site-directed mutagenesis will change only that gene, whereas random mutation with radiation or chemicals could affect many genes and has a low chance of making the desired change to the target gene. Many characters such as yield, however, are polygenic, with many of the genes unidentified, when genetic engineering is often of little use.

Of all known natural antibiotics, over 60% come from Streptomycetes. DNA manipulation has been used extensively in them to improve titres of antibiotics, and to generate new antibiotics.

Many **vaccines** against viral diseases consist of inactivated virulent strains or live attenuated strains. Attenuated strains must be non-revertible (see Chapter 7.1.1) to wild-type when replicating in the recipient, to avoid chance infections as has happened rarely with live polio vaccine. Engineered large non-reverting deletions are often used. A recombinant vaccinia-based vaccine against rabies is now being employed, distributed on chicken heads, to vaccinate wild foxes in Europe against rabies.

14.1.10. Some dangers of genetic engineering; the amount of genetically engineered crops grown

There are many other engineered plants, animals and micro-organisms under production or testing, including ones for the production of vaccines, pharmaceuticals and high-value biochemicals. There is, however, a great deal of **public opposition** to genetic engineering and the use of genetically engineered products such as foods. Polls in the UK in 1999 showed that only one per cent of the public was keen to see development of the biotechnology industry, and 96% did not want GM foods. In England in 1999, 87,000 packets of organic maize tortilla chips were destroyed as they were found to contain genetic contamination from cross-pollination of the organic maize by genetically modified maize carrying a promoter from cauliflower mosaic virus. Maize pollen from genetically modified (GM) crops ("Bt-corn") has been shown in the USA and Switzerland to kill caterpillars of the monarch butterfly and aphid-eating beneficial lacewings. This led to Professor Beringer, a government advisor and chairman of the Advisory Committee on Releases to the Environment, to call for the withdrawal of licences for maize with Bt toxin, if the research was substantiated.

The position in Britain in 1999 is a voluntary three-year embargo on commercial plantings of GM crops, while their effects are studied in further field trials. Some supermarket chains and national retailers have banned from their shelves all products from genetically modified organisms (GMOs). The Local Government Association has advised members to ban GM foods from schools and hospitals, because of health worries. In May 1999, the British Medical Association's report, *The Impact of Genetic Modification on Agriculture, Food and Health*, stated that: "As we cannot yet know whether there are any serious risks to the environment or human health, the precautionary principle should apply." The BMA called for an indefinite moratorium on the planting of GM crops. Government advisors from English Nature expressed worries about the effects on wildlife of growing herbicide-resistant crops, and asked the biotechnology industry to produce crops which could not cross-contaminate wild plants. Putting herbicide-resistance genes into chloroplast genomes, instead of the nuclear genome, would help to restrict contamination by cross-pollination as chloroplast genes in most flowering plants are maternally transmitted, not through pollen.

In the European Union, EC Directive 90/220 (as amended) controls the deliberate release into the environment and the marketing of GMOs. According to a MAFF Joint Food Safety and Standards Group document in February 1999, *Genetic Modification of Crops and Food*, no GM crops have yet been approved in the UK for commercial cultivation. The document states that there will be a programme of managed development of herbicide-tolerant GM crops, with very limited farm-scale plantings at first (in spring 1999) for monitoring their ecological effects. It states that herbicide-resistant GMOs are the only type of GMO likely to receive market approval and to be marketed in the UK in the next three years, and that evidence from the USA shows less overall use of pesticides where herbicide tolerant crops are used. Evaluations will be carried out using herbicide-tolerant spring and winter oilseed rape, maize and sugar beet. The document also states that the GM foods and ingredients currently on sale in the UK include tomato paste, maize and soya as ingredients in a variety of foods, and chymosin (an enzyme from a GM micro-organism) used in cheese manufacture as a "vegetarian" alternative to rennet from calves' stomachs.

In 1998, 2,000 ha of GM maize were grown in France, and 30 million hectares of GM crops were grown in the world, mainly in the USA, China, Canada, Argentina, Australia and Mexico. The French government reportedly wants to ban GMOs but would face legal action from the European Commission if it did. The first GMO approved in the USA by the Food and Drug Administration, in 1992, was the Flavr Savr tomato, with a gene for softening switched off, but it was not a commercial success for other reasons. In the USA, there is not much opposition now to GM crops; half the soya beans grown there are GM and that proportion is increasing. GM soya is resistant to Monsanto's "Roundup" herbicide which can be used on soya seedlings instead of at pre-planting time, so farmers need to spray less often: yields are higher and costs are up to 30% lower. Monsanto is a major firm involved in GM crop production as well as in herbicide production. It is producing herbicide-resistant ("Roundup Ready") potatoes, wheat and rice, and has produced pest-resistant potatoes (which kill Colorado beetles but not other insects) and cotton.

There are worries about the spread of **antibiotic resistance** from plants to gut bacteria, from antibiotic-resistance genes used for selection during the engineering, and about the creation of "**superweeds**", such as herbicide-resistant weed species, from accidental pollination of wild plants by genetically engineered ones, or the engineered crops themselves becoming weeds of other crops, especially if crops grown in rotation are resistant to the same weedkiller. The vast use of herbicides would reduce biodiversity of plants and animals in the agricultural environment. It would be very easy for alien genes to escape from sugar beet, oilseed rape and cereals into wild relatives, making herbicide or insect-resistant "superweeds", disrupting the natural ecology. Genetic modification of grasses such as cereals or forage crops is especially risky because grasses tend to cross not just with closely-related species, but also with grasses of entirely different genera, so alien genes could become widely distributed (Pain, 1999). Biodiversity can be threatened by any bred crop, whether GM or not. For example in Switzerland the wild sickle medic, *Medicago falcata*, has largely been replaced by escaped forage alfalfa, *M. sativa*, and by hybrids between the two species.

The possible **unexpected effects** of genetic modification were clearly demonstrated by Bergelson et al. (1998) with *Arabidopsis thaliana*. This is inbreeding and normally self-pollinates, but the incorporation of the dominant GM herbicide-resistance gene *Csr1-1* changed the breeding system, making the plants twenty-times more outbreeding (from 0.3 % outbreeding to 6.0 %) and promiscuous, and so much more likely to spread the alien gene to wild relatives. The same chlorsulphoron-resistance gene had no effect on breeding when not in the pBin vector. The Royal Botanic Gardens, Kew, will not store GM plants in its Millennial Seed Bank (Prance, 1999).

People allergic to Brazil nuts are also allergic to soya beans engineered to have certain Brazil nut proteins. It is possible that **new viruses** could form from transgenic RNAs combining with natural viral RNAs. The spread of Bt toxin genes to wild plants could upset ecological balances depending on herbivorous insects and their predators. Replacement of many traditional cultivars by a few genetically-engineered ones would reduce the genetic base available for future breeding (see Chapter 15.3, gene conservation). There are also **ethical problems**, such as vegetarians objecting to plant crops containing animal-derived genes giving frost-resistance.

One interesting ethical question arose from the production of blood clotting factor IX in sheep milk (for extracting to treat haemophilia B; Chapter 13.3.2) by injecting the human gene into fertilised sheep blastocysts. About 90% of the resulting female sheep had no human gene detectable; 1% had the gene but did not express it, and 9% gave factor IX in the milk. The question was whether the non-producing sheep, including all the males, could be used for meat production when they might contain human genes. The Advisory Committee for Novel Foods and Processes (which carries out the assessment of novel foods in the UK) declared that there was no health risk, but several pressure groups did not want to eat human genes, on ethical grounds.

A recent report (Laboratory News, December 1999) listed some opinions from an FT Conference on Genetically Modified Organisms. They included: "There is now no perceived benefit of being in the business of GMOs. It is a meaningless commercial lecture, but it is meaningful on a bad PR level." "GMOs are dead."

14.2. Human gene therapy

14.2.1. Introduction

Gene therapy is the treatment of disease by the transfer of therapeutic DNA or RNA into the patient. In theory, one can use wild-type genes to correct mutant genes causing human diseases, if one can get them to the right places at the right time. If the mutation is dominant, it would be very difficult to get added wild-type recessive genes expressed unless the dominant alleles could be removed or silenced, which is generally impractical once a defective baby has been born. With recessive mutations, added dominant wild-type genes should be expressed. Only somatic tissue changes are permitted, not germ-line therapy.

The introduced DNA is not usually found at the homologous locus, occurring instead in widely diverse places in the genome. It does not replace the defective gene, but it can mask its effect if the transgene is dominant. Normally only some of the treated cells receive the transgene, and even fewer actually express it.

The **criteria for effective human gene therapy** were laid down in 1984 by one of the pioneers, French Anderson:

- A cloned healthy gene must be available.
- Transfer of this gene to diseased recipient cells must be possible, e.g., by retrovirus vectors or electroporation.
- The target tissue must be accessible, e.g., you can remove bone marrow cells (for blood-disease treatments) for manipulation, then transfuse them back in blood into the body, or you can give viral aerosols into the lungs to treat cystic fibrosis.
- Treatment must not harm the patient. For example, if treating β-thalassaemia with β-globin genes in bone marrow cells, one would not want white blood cells expressing the globin genes intended for red blood cells. Random integration of retroviruses into host DNA usually gives insertional mutagenesis, disrupting a gene; this would not matter in some cells, but it would matter if it caused tumour formation.
- Treatment must lead eventually to a large improvement in the patient's health. If only a few cells were transformed, or if there was poor expression of the therapeutic gene, the procedure would not be worthwhile.

14.2.2. Methods for somatic gene therapy

Micro-injection of somatic cells is impractical because far too many cells would need treating, and many are inaccessible, e.g., lung cells. **Electroporation** is the use of short pulses of high voltage to open up membrane pores to DNA for short periods. It can be used on cells outside the body, such as bone marrow cells, but is difficult to use *in situ*.

Retroviruses, with RNA genetic material and reverse transcriptase to form DNA for integrating into host cells, can be effective vectors for therapeutic genes, with a high rate of random integration into host chromosomes. They are the usual choice, once oncogenic sequences have been deleted. One does not want the retroviruses replicating in the host, so replication-defective strains (lacking *gag, pol*, and *evn* but possessing the packaging region) are grown up with the aid of "helper" strains (with *gag, pol*, and *evn* but lacking the RNA signal for packaging its genome in protein coats) in packaging cell cultures to provide the engineered viral particles for infection of the host. Unfortunately retroviruses can only carry inserts up to about 8 kb long, too small for some genes, and they might recombine with endogenous retrovirus strains already in the host to give infective replicative forms, causing disease. A serious problem for treatment of fully differentiated cells is that integration of the therapeutic gene from retroviruses normally only occurs when the host cell replicates.

Integration within a host gene will usually knock out that gene and with some types of gene may occasionally cause tumour formation.

Other possible virus vectors include **adenovirus** which does not need host cell division for integration and has a low risk of insertional mutagenesis, but gene expression is transient as the viral DNA is an unintegrated episome, and hosts may have or develop immunity to such viruses. Adeno-associated virus, herpes, vaccinia and influenza viruses are also being investigated as possible vectors.

Of non-viral vectors, **direct injection of DNA** into muscle can be effective, with good expression of the therapeutic gene after transformation of the recipient cells. DNA can be coated with cationic lipids to make **liposomes**, which can fuse with cell membranes, releasing the DNA, e.g., after the particles have been blown into the lung. Experiments are also taking place using **Human Artificial Chromosomes** as vectors, with human telomeric DNA, genomic DNA and arrays of α-satellite DNA (which has centromeric functions). They would need to be transformed into the target cells.

14.2.3. Progress

The first human gene therapy attempt was made in 1990 by French Anderson and Michael Blaise in the USA on a four-year-old girl with **adenosine deaminase deficiency** (ADA). This very rare autosomal recessive disease gives severe combined immunodeficiency disease (SCID), making sufferers very susceptible to infectious diseases, usually causing death in early childhood. Such children used to have to live in a germ-free protective chamber, but now they can live more normally if given weekly injections of polymer-coated enzyme (PEG-ADA), which slowly releases the enzyme into the blood.

This treatment was kept up during the gene therapy. Large samples of T-lymphocytes were taken from the girl's blood and were multiplied in cell culture. They were then transfected with mouse leukaemia virus carrying a normal human ADA gene. The transfected T-lymphocytes were then returned to the body, in eight blood transfusions over 11 months. There followed six months without transfusions, then transfusions every four months. The inserted gene was expressed well and the girl recovered well from ADA deficiency and went to a normal school.

Gene therapy for **cystic fibrosis** (Chapter 13.3.1) is under trial, using an aerosol with treated adenovirus or liposomes into the lungs. Unfortunately, the treated cells are continually shed, requiring repeated treatments, perhaps every three weeks. Although the treatment agents reach the lung epithelium, they do not usually reach the submucosal cells in which the cystic fibrosis transmembrane conductance regulator protein gene is most strongly expressed. Treatment does not transform all the lung cells, so some continue to secrete viscid mucus, and immunity to the treatment agent, especially viral coat proteins, may build up during successive treatments. Although liposomes are easily absorbed by endocytosis of the cell membrane, only about one DNA particle in 1,000 reaches the nucleus. Furthermore, treating the lungs does not help the sweat gland or pancreas symptoms. Therapy for cystic fibrosis has not been nearly as successful as with ADA deficiency.

Trials are under way for gene therapy for other diseases, but there are doubts as to whether gene therapy for some diseases will ever be cost-effective. One treatment with promising initial results has been for **malignant melanoma**, a skin cancer. Skin tumours were directly injected with plasmid DNA engineered to express the HLA class I gene B7, with a mismatch to the patient's own HLA type. This is to stimulate the patient's immune system to attack the melanoma tumour cells.

For some diseases, exact regulation of the amount of product is not crucial, e.g., factor VIII for treating haemophilia A (Chapter 13.3.2), but if one were treating β-thalassaemia by introducing the normal β-globin gene into bone marrow cells, one would want accurate transcriptional control of the level of expression. An excess or a deficit of β-chains compared to α-chains could cause precipitation of one type of chain, and result in anaemia. Gene therapy might work well in future, but progress has so far been slow and at very high financial cost.

Suggested reading

Anke, T., *Fungal Biotechnology*. 1997. Chapman and Hall, London.

Bergelson, J., C. B. Purrington and G. Wichmann, Promiscuity in transgenic plants. *Nature* (1998) 395, 25.

Bishop, J., *Transgenic mammals*. 1999. Longman, Harlow.

Culver, K. J. W., *Gene Therapy - a Handbook for Physicians*. 1994. Mary Anne Liebert Inc., New York.

Galun, E. and A. Breiman, eds., *Transgenic Plants*. 1997. Imperial College Press, London.

Grant, S. R., Dissecting the mechanisms of posttranslational gene silencing: divide and conquer. *Cell* (1999), 96, 303–306.

Griffiths, A. J. F., W. M. Gelbart, J. H. Miller and R. C. Lewontin, *Modern Genetic Analysis*. 1999. W. H. Freeman and Co., New York.

Hartl, D. L. and E. W. Jones, *Genetics. Principles and Analysis,* 4th ed. 1998. Jones and Bartlett, Boston.

Klein, T. M., E. D. Wolf, R. Wu and J. C. Sandford, High velocity microprojectiles for delivering nucleic acids into living cells. *Nature* (1987), 327, 70–73.

Maijala, K., Genetic aspects of domestication, common breeds and their origin. pp 13–50, in: Piper, L. and A. Ruvinsky, eds., *The Genetics of Sheep.* 1997. CAB International, Wallingford.

Mariani, C. et al., Induction of male sterility in plants by a chimeric ribonuclease gene. *Nature* (1990), 347, 737–741.

Mariani, C. et al., A chimaeric ribonucleotide inhibitor gene restores fertility to male sterile plants. *Nature* (1992), 357, 384–387.

Old, R. W. and S. B. Primrose, *Principles of Gene Manipulation*, 5th ed. 1994. Blackwell Scientific Publications, Oxford.

Owen, M. R. L. and J. Pen, *Transgenic Plants: a Production System for Industrial and Pharmaceutical Proteins*. 1996. John Wiley and Sons, Chichester.

Pain, S., Selfish genes warrant caution. *Kew* (1999), Spring, 56.

Prance, G., Genetic modification and Kew. *Kew* (1999), Spring, 3.

Primrose, S. B., *Molecular Biotechnology*, 2nd ed. 1991. Blackwell Scientific Publications, Oxford.

Sedivy, J. M. and A. L. Joyner, *Gene targeting.* 1992. W. H. Freeman and Co., New York.

Waterhouse, P. M., M. W. Graham and M-B. Wang, Virus resistance and gene silencing in plants can be induced by simultaneous expression of sense and antisense RNA. *Proc. Natl. Acad. Sci. USA* (1998), 95, 13959–13964.

Weatherall, D. J et al., eds., *Oxford Textbook of Medicine*, 3rd ed. 1995. Oxford University Press, Oxford.

Wilmut, I., K., H. S. Campbell and L. Young, Modern reproductive technologies and transgenics, pp 395–411 in: Piper, L. and A. Ruvinsky, *The Genetics of Sheep.* 1997. CAB International, Wallingford.

Wright, G. et al., High-level expression of active human alpha-1-antitrypsin in the milk of transgenic sheep. *Biotechnology* (1991), 830–834.

Xoconostle-Càzares, B., E. Lozoya-Gloria and H. Herrera-Estrella, Gene cloning and identification, pp 107–125, in: Haywood, M. D., N. O. Bosemark and I. Romagosa, eds., *Plant Breeding. Principles and Prospects.* 1993. Chapman and Hall, London.

CHAPTER 15

GENETIC VARIATION IN WILD AND AGRICULTURAL POPULATIONS; GENETIC CONSERVATION

15.1. The forces controlling the amounts of variation in a population

Chapter 1 dealt with definitions of populations (Chapter 1.3.11) and polymorphisms (Chapter 1.3.12), which can be for alleles at a locus or chromosome aberrations. Chapter 4 covered basic population genetics, including genetic drift and the effects of population size, and mutation, migration and selection, including the interactions of mutation and selection. Chapter 5 covered various types of selection, Chapter 7 covered mutation, Chapter 8 covered recombination, and Chapters 9 and 10 covered chromosome aberrations of structure and number, respectively. We now look at the factors affecting the amount of genetic variation in a population, both in the wild and in agriculture.

15.1.1. The forces or processes which increase or maintain genetic variation within a population

The forces or processes which generally **increase** the amount of genetic variation within a population are:

- **gene mutation.** Mutation frequencies are controlled by the amount of exposure to external and internal mutagens; by the rate of spontaneous tautomerisation of the DNA bases; by the efficiency of various methods of protection from mutagens and of repair processes, such as proof-reading during DNA replication, and excision-repair of pre-mutational lesions such as thymine-thymine dimers. Mutation frequencies are affected by environmental factors such as temperature. Various mutagenic processes can cause single base-pair changes, such as frame shifts and base substitutions, and larger changes varying from deletions or additions of a few bases to changes in many bases, especially large deletions.
- **production of chromosome aberrations,** by agents which break chromosomes, and where broken ends may re-anneal in different patterns. Chromosome aberrations can also be produced by unequal crossing-over (Chapter 9, Fig.

269

9.4.1). Unequal crossing-over could occur at meiosis, mitosis (more rarely), at DNA replication, or at other times by chance breakage or from agents (including radiation and enzymes) causing double-strand breaks.

- **changes in the number of copies of all chromosomes. Increases** in the number of copies of whole genomes can come about (Chapter 10.2) by the failure of a cell to cleave after the chromosomes have replicated, when a diploid could give an autotetraploid, or by double fertilisations giving triploids, or unreduced gametes giving triploids or occasionally autotetraploids. Series of events can give higher ploidies. In haploid fungi, nuclei may rarely fuse by chance in vegetative hyphae, giving diploid nuclei, capable of giving diploid individuals (Chapter 18.4). **Reductions** in number can come about by development of a female gamete without fertilisation, giving a monoploid from a diploid, or a diploid from a tetraploid. This occasionally happens spontaneously and sometimes may be triggered by pollination with pollen of related varieties which do not fertilise the egg cells (Chapter 10.2). Fungi with a parasexual cycle can have spontaneous haploidisation by progressive loss of chromosomes from the diploid until a stable haploid condition (with one of each type of chromosome) is reached (Chapter 18.4).

- **changes in the number of copies of one type of chromosome. Non-disjunction** for a single chromosome at meiosis or mitosis in a diploid can cause monosomy and trisomy, generating aneuploid chromosome numbers such as 2X - 1 and 2X + 1. Non-disjunction for a pair of homologous chromosomes in a diploid can produce a 2X + 2 cell and a 2X - 2 cell, but the latter will almost certainly die if the chromosome contains any essential genes, as for a haploid which loses one chromosome. Aneuploids from polyploids are less affected by the abnormality than are aneuploids of diploids, and may be vigorous. Animals are more sensitive than plants to abnormal chromosome numbers, and in plants and animals, certain aberrations of number or of structure are more often passed through female gametes than male gametes. Non-disjunction frequencies are influenced by environment, with extremes of temperature and some heavy metals increasing them. In fungi, parafluorophenylalanine can be used to increase the frequency of non-disjunction, to get haploids from diploids (Chapter 18.4).

- **immigration.** Immigration increases the number of individuals in the population and if the immigrants have different alleles or allele frequencies from the resident population, they will increase the amount of genetic variation and/or change the allele frequencies (Chapter 4.5.2). Migrants from different environments will often have different allele frequencies from the resident population because of selection, and even if from similar habitats they will often have different allele frequencies by chance. In a farming environment, immigration may be accidental, as with some ewes escaping from a field and mating with another farmer's ram, or deliberate, say by introduction of animals or semen from another farm.

* **recombination**. If there is genetic variation giving heterozygosity at two or more loci, then recombination will produce new combinations of existing genetic variation. Recombination can occur by independent assortment for non-syntenic loci, by meiotic crossing-over for syntenic loci, by mitotic crossing-over, by gene conversion at meiosis or mitosis, or by haploidisation in some fungi (Chapters 8 and 18.4). Crossing-over and gene conversion frequencies are under genetic control and are also influenced by environment, especially by temperature.

The forces or processes which tend to **maintain** the amount of genetic variation within a population are:

* **heterozygote advantage**. By favouring individuals in which the frequency of two alleles is equal, heterozygote advantage helps to maintain existing genetic variation, even if other forces tend to favour one allele (Chapter 15.3).
* **cyclic selection**. With cyclic selection, different forms or alleles are favoured at different times, tending to maintain both in the population.
* having **patchy or diverse habitats** tends to maintain genetic variation, as some types are favoured in some habitats and others in other habitats.
* **selection for rarity**. If rare forms are favoured because they are rare, their rare genes will tend to be maintained in the population, and may even increase until they become relatively common (Chapter 15.3).

15.1.2. The forces or processes which reduce genetic variation within a population

The forces or processes which generally **reduce** the amount of genetic variation within a population are:

* **directional selection**. If the breeder or nature or the opposite sex selects for organisms at one extreme for breeding, then genes giving average phenotypes or phenotypes towards the other extreme will be lost, whether major genes or polygenes.
* **stabilising selection**. If the breeder or nature or the opposite sex selects for phenotypes in the middle of the range, then genes giving extreme phenotypes at either end of the range will be lost.
* **genetic drift**. Genetic drift will decrease the amount of genetic variation purely by chance, with the effect being much greater in small populations than in large

ones (Chapter 4.3): The closer an allele's frequency is to zero, the more likely that that allele will be lost by chance, even if it is selectively neutral or mildly advantageous.

- **emigration**. Emigration reduces population size and therefore makes drift more likely. The loss of some individuals through emigration leads to the loss or the reduction in frequency of certain alleles if they are more frequent in the emigrants than in the resident population.

15.1.3. The interactions of forces or processes affecting the amount of variation within a population

We have seen that there are some forces or processes creating new genetic variation, some tending to maintain existing variation, and some tending to reduce the amount of genetic variation in a population. These **forces** will **interact** in different ways in different organisms. For example, in haploid organisms deleterious recessives cannot be hidden from selection in heterozygotes, and the loss of a chromosome by non-disjunction will probably be lethal. Polyploids are less sensitive than haploids to aneuploidy and to the effect of a deleterious recessive. New recessive mutations could easily show in a haploid, but would need several to many generations to become homozygous and expressed in an autotetraploid.

The amount of phenotype variation in a population depends on its amount of genetic variation, on the amount of environmental variation, and on the heritability of different characters. The amount of phenotype variation for particular types of genetic variation also depends on dominance, additive action, epistasis, etc. For example, if a diploid maize population is homozygous for *a a* (colourless grain) and that allele shows recessive epistasis to the *P/p* locus (purple versus recessive red grain), then genetic variation at the *P/p* locus will not show in the phenotype (Chapter 2.2.2 and Plate 7). Similarly, inducible and repressible genes and conditional mutations will only be subjected to selection in environments in which they are expressed. Selection only acts on phenotypes, not genotypes.

The amount of migration depends on population structure and the mechanisms for gene flow in an organism (Chapter 4.4). The amount of genetic drift will be affected by whether population sizes are fairly stable or whether they undergo big seasonal or annual fluctuations. Periodic events such as droughts, fires or floods, or farmers selling off all their stock for meat, can wipe out whole populations, but migration from surviving neighbouring populations can give recolonisation later, perhaps with different allele frequencies.

The most frequent type of variation encountered is for alleles of major genes and polygenes. One kind of variation which is of little consequence for phenotypes and performance, but which can be very useful for research, is in repeated DNA sequences in "**minisatellites**". These are segments of DNA, typically a few thousand base pairs long, consisting of tandem repeats of sequences perhaps 30 bases long. There is frequent variation in the number of repeats, perhaps from incorrect alignments in replication or crossing-over, as well as variation within the sequence of a repeated unit. By using the Polymerase Chain Reaction to amplify the appropriate region many times, the sequences can be studied from very small amounts of DNA, e.g., from mouth cells, hair roots or semen. They have been used extensively by Professor Jeffreys of Leicester University for "**DNA fingerprinting**" of humans, with forensic applications. The method can be used on humans, animals or plants to trace ancestries and relationships between individuals or breeds. The high rate of change in the number of repeats means that most individuals are different from one another for a particular sequence, so there is an enormous amount of genetic variation for minisatellite DNA, but with little phenotypic or ecological consequence. Unusually, most minisatellite mutations are generated in the male line, at spermatogenesis. For one very unstable minisatellite sequence, about one sperm in seven carries a new mutation. Variation for such DNA is largely lost through drift, not selection.

15.2. Using a knowledge of the origins of genetic variation to solve a practical problem

Suppose we wanted to use some land with **salt-pollution** for growing an annual seed crop. How could we utilise our knowledge of genetic variation to solve the problem? A first approach would be to use artificial migration, selecting possibly pre-adapted seeds from existing varieties of this crop already growing on salt-polluted land elsewhere. If that did not work, or if one wanted to adapt an existing non-tolerant variety to grow in this habitat, one could try mutation, recombination, chromosome aberrations, or changes in chromosome number, in existing non-tolerant strains, using spontaneous or induced changes as convenient. One could make polyploids such as autotetraploids or allotetraploids to test in the new environment.

One could try inbreeding existing lines to expose any beneficial recessive alleles, and could make F1 hybrids (Chapter 12.1.3) to exploit the power of hybrid vigour in this difficult environment. One could also try a composite cross, with agricultural mass selection (Chapter 12.3), of different varieties in a semi-polluted salty habitat, trying a sample of seed each year on the fully salt-polluted land. One could also try gradual adaptation, growing seeds from a composite cross in successively more salty environments each season, or try interspecific crosses.

Once a salt-pollution-resistant strain had been found, the safest policy for keeping it would be to reproduce it vegetatively (Chapter 17.1.5), in case favourable gene combinations were lost at meiosis and fertilisation. Recurrent backcrossing to high-yielding non-tolerant lines might be used to improve yield (Chapter 12.4). Various selfings and crosses to non-tolerant plants could be made to find out whether the character is qualitative or quantitative, and to study segregation ratios and the genetics of tolerance, such as the number of loci involved, dominance, epistasis, etc. The genes for tolerance could be crossed into other varieties if required, or could be incorporated in polyploids or F1 hybrids, or used in DNA manipulations into other species.

15.3. The maintenance of polymorphism in populations

For adaptation and evolution in plants, animals, micro-organisms and humans, it is useful for populations to have a variety of genetic polymorphisms. These will however tend to disappear because of selection if any alleles have any phenotypic advantage over the others. Even for selectively neutral alleles, genetic drift will tend to eliminate polymorphisms, especially from small populations. With selection often being strong and mutations being fairly rare per gene per generation, it might seem strange that polymorphisms for many characters are so common in most organisms. There are several mechanisms which favour their retention.

The first is **selection for rarity**, that is, an allele is favoured by frequency-dependent selection because it is rare. A plant example is self-incompatibility alleles in the evening primrose, *Oenothera*. If self-sterility allele (*s*) type s^1 has a frequency of 1% in a particular population, then plants carrying it can cross with approximately 99% of the population (subject to assumptions about dominance in pollen or style). If allele s^2 has a frequency of 40%, existing in heterozygotes, then plants carrying it can only cross with about 60% of the population, so much s^2 pollen is wasted in incompatible pollinations. Rare alleles for this character are therefore favoured until they become common, when they are at a disadvantage to rarer alleles. This leads to situations with many alleles at a locus, all of them fairly rare yet still preventing self-pollination. *Trifolium* (clovers) and *Brassica* populations typically have many self-incompatibility alleles, with up to 200 different self-incompatibility alleles in a single field of clover.

Selection for rarity can also occur for prey/predator relations and host/parasite relations. A bird might eat the common form of a butterfly, getting used to the taste and visual form of that common type, but might ignore a rare form, not associating it with the common form as the same favoured food source. Similarly, a fungal parasite might adapt to attacking the common form of some plant species, but not adapt to attacking a rarer form. Even a relatively low frequency of disease- or pest-resistant polymorphisms helps to curtail the spread of diseases or pests. A build-up of different alleles giving resistance to challenges by different races of a pathogen helps to control epidemics, but those alleles' frequencies only rise in response to such challenges.

A second possible mechanism for preserving polymorphisms is **an equilibrium between mutation, selection and gene conversion**. See Chapter 4.5.3. Mutation pressures might favour one direction of change of allele frequency over the other (typically favouring wild-type to mutant changes over mutant back to wild-type changes). Selection might favour a different direction of change, and there might well be disparity in the direction of gene conversion (say with conversion from mutant to wild-type being more frequent than conversions from wild-type to mutant). Depending on the parameter values for mutation, selection, gene conversion frequency and the disparity in the two directions of correction for a pair of alleles, an equilibrium may be set up between the three forces, tending to preserve the polymorphism. Gene conversion disparity can also remove polymorphisms. The relevant equations and data on conversion frequencies and typical amounts of conversion direction disparity were given by Lamb and Helmi (1982) and Lamb (1985, 1998).

A third and very important mechanism favouring polymorphism retention is **heterozygote advantage** (Chapter 12.1). The heterozygote is sometimes fitter than either homozygote, showing overdominance for fitness. In Mendel's pea plants one usually describes height as showing complete dominance, but $t\,t$ is about 30 cm high, $T\,T$ is about 182 cm and $T\,t$ is about 213 cm, showing some overdominance. Heterozygote advantage favours the retention of polymorphism because it favours the genotype with equal numbers of the two alleles.

The classic human case of heterozygote advantage is **sickle-cell anaemia** (Chapter 13.3.1) in a malarial area. The recessive homozygote, $a\,a$, has anaemia which is often lethal in childhood because at low oxygen concentrations the red blood cells go sickle-shaped and are removed by the spleen. The bad allele, however, is maintained at much too high a frequency to be due to recurrent mutation. The explanation is that the heterozygote, $A\,a$, is more resistant to a common form of malaria than is the dominant homozygote, $A\,A$. In a malarial region, $A\,a$ is fittest, so let its fitness be 1.0; let the fitness of $a\,a$ be reduced by t from anaemia, and let the fitness of $A\,A$ be reduced by s by malaria, so that the relative fitnesses are $(1 - s)$ for $A\,A$, 1.0 for $A\,a$, and $(1 - t)$ for $a\,a$. The genotype frequencies before selection in a generation are the Hardy-Weinberg ones of p^2, $2\,pq$, q^2, respectively, so what is eliminated by selection in a generation is $p^2 s$ by malaria and $q^2 t$ by anaemia. As the frequency of A alleles is p and p^2s are eliminated, the proportion of A alleles eliminated is p^2s/p, which is ps; similarly, the proportion of the q of a alleles eliminated is q^2t/q, which is qt. At equilibrium, the proportions of the two alleles being eliminated must be equal, which is why the allele frequencies remain constant, so at equilibrium $ps = qt$ and because $p + q = 1$,

equilibrium $p = \dfrac{t}{s+t}$ and equilibrium $q = \dfrac{s}{s+t}$.

In Africa, $a\ a$ is usually lethal, so $t = 1$, and in malarial regions, $A\ A$ has about a 10% fitness disadvantage compared with $A\ a$, so $s = 0.1$. At equilibrium we therefore get $p = \frac{1}{0.1+1} = 0.91$, and $q = \frac{0.1}{0.1+1} = 0.09$. The genotype frequencies expected at birth, before selection, are therefore $A\ A$, 0.826; $A\ a$, 0.165, $a\ a$, 0.008, in a malarial area. If however, by population migration to a non-malarial area or the elimination of mosquitoes, malaria disappears, then t stays 1.0 for anaemia, but s becomes 0.0, so at equilibrium we get $p = 1.0$, $q = 0.0$, with complete elimination of the deleterious sickle-cell gene once the heterozygote advantage due to malaria goes. In accordance with these predictions, this polymorphism and sickle-cell anaemia are slowly being eliminated where swamp drainage and/or insecticide sprays have killed off the mosquitoes and also where Africans have migrated to non-malarial countries such as Britain.

In malarial regions, it is wasteful to have some $A\ A$ dying of malaria and nearly all $a\ a$ dying of anaemia. The ideal population there would be all $A\ a$, but without vegetative reproduction (cloning), that is not possible in humans because of Mendelian segregation, $A\ a$ x $A\ a$ giving 1/4 $A\ A$, 1/2 $A\ a$, 1/4 $a\ a$. In plants with vegetative reproduction, or yeasts reproducing by budding, or animals with apomixis or parthenogenesis, successful heterozygous genotypes can be maintained once they have arisen by recombination or mutation. The loss of fitness due to segregation and recombination is called the **segregation load**, in parallel to the mutation load. The segregation load reduces immediate fitness, but like mutation, segregation and recombination, it permits adaptation, long-term fitness and evolution by providing variation on which selection can act.

15.4. The need for genetic conservation; methods of conservation

15.4.1. The need for genetic conservation and the value of some old varieties

It has been estimated by staff at Kew Gardens that in 50 years on present trends, one quarter of the world's **wild species** will have been lost, for plants and their dependent animals. Staff at Marwell Zoo, Hampshire, estimate that 41% of all mammalian species are at risk of extinction. In **agriculture**, as large numbers of mediocre varieties of crops and farm animals are replaced by a few superior modern varieties, the **gene pools available for future breeding** and for genetic engineering are being seriously diminished. To counteract this, large collections of older stocks and their wild relatives should be maintained for possible future use, even from inferior material which might contain some very useful genes which have not yet been recognised and exploited: see Brush (1999). Some genes might become useful only in the future, through changes in climate, agricultural practices, medical drug discoveries or consumer tastes. Some might just be polygenes which would only show to a very small degree individually, but which collectively might have a large impact on performance.

In Afghanistan, a mediocre **local wheat** strain which was being replaced by modern varieties was found to have a valuable rust-resistance, which was then bred into the replacement wheats. A local wheat collected in Turkey in 1948 had poor agricultural characteristics but has been the source of several fungal-resistance genes for modern wheats. It has genes for resistance to *Puccinia striiformis* rust, to 35 strains of *Tilletia caries* (bunt) and to *T. foetida*, to 10 strains of *T. controversa*, and tolerance to several species of *Urocystis*, *Fusarium* and *Typhula*. In **rice**, *Oryza sativa*, resistance to the important Asian "Grassy Stunt" virus was found in just one population of the wild relative, *Oryza nivara*; this resistance was incorporated into the high-yielding variety *IR36*.

In some less developed parts of the world, **human malnutrition** is often more common than starvation. The problem is usually one of insufficient protein, especially of the amino acid lysine in cereals, and to a lesser extent, a deficiency of methionine and cysteine in legumes. In 1964, people at Purdue University, USA, reported that maize homozygous for the mutations *opaque-2* or *floury-2* had lysine at 3.7 g/100 g protein instead of the normal 1.6 g/100 g. *Opaque-2* has a grain yield 99.9% of normal varieties, so its flour is very useful in combating lysine-deficiency in human diets. Both these **high-lysine mutants** had been known as morphological mutants since 1935, but their nutritional properties had not be investigated. Some endangered animal species have known **dietary problems**, such as the great sensitivity to vitamin E deficiency in Przewalski's Horse and Black Rhino, so conservation or captive breeding programmes can take that into account.

We have seen that within a variety, a **need for uniformity** (for marketing or ease of mechanisation) results in stabilising selection being used. In Chapter 15.1.2, we saw that stabilising selection, directional selection and genetic drift all reduced genetic variation. Although mutation and recombination could restore some variation, most mutations are harmful and would be eliminated by natural or artificial selection; many recombinants would be less good than their parents, and would be eliminated.

If genetic stocks are kept in small populations because their performance is mediocre and they cost money to maintain, then **genetic drift** will be severe and reduce genetic variation within each stock. It is particularly expensive to maintain big herds of large farm animals if they are inferior to current commercial stocks, as individual animals cost a lot more to maintain than do most plants. In rare breed cattle, sheep and pigs, most males are sold off for meat as only a few are needed for breeding with the females, but the fewer the males that are used for breeding, the lower the effective population size and the more severe the genetic drift (Chapter 4.3).

If the world really is undergoing a **climatic change**, with global warming, then even more species will become endangered. There are already worries that the hole in the ozone layer, by allowing more UV light to reach the earth's surface, will reduce the viability of pollen, especially in wind-pollinated plants. Plants particularly affected in experiments were maize, rye, pears, cherries, pistachios and poppies ("Cuttings", 1998).

Sometimes ideal types have been lost by **indiscriminate breeding** with other types. In South America, the guanacos are the wild, undomesticated ancestors of the llamas, alpacas and vicunas. **Llamas** were selected over 5,000 years ago by man as pack animals for use the Andes mountains, but were also used for meat and wool. **Alpacas** and **vicunas** were selected for fine wool production, and the Incas are said to have had advanced breeding systems for improving them. The Spanish conquests of South America in the 16th century led to the mass slaughter of alpacas in favour of sheep and cattle, and indiscriminate breeding of the remaining animals led to the loss of the fine lines. There is now a British Alpaca Society, with a stud at Arunvale, Sussex, selecting for soundness of frame and wool quality.

15.4.2. Conservation programmes and methods of genetic conservation

Ideally, one should conserve **wild habitats** which contain *in situ* wild relatives of species used by man in case they are needed in future (see Maxted et al., 1997), but the pressures of human populations often make that difficult. There is a strong case for *ex situ* regional, national or **international centres** for plant, animal and microbial conservation, with public funding and national conservation bodies, especially as one cannot always predict what will be useful in future. For methods of plant conservation, see Given (1994), Brush (1999), Maxted et al. (1997), and Razdan and Cocking (1997 and 1999).

The Food and Agriculture Organisation of the United Nations (**FAO**) promotes cooperation, and various International Agricultural Research Centres specialise in particular crop collections. The **numbers of accessions of plants** at international research centres, for *ex situ* conservation, research and further breeding can be enormous. They include 79,500 for rice at the International Rice Research Institute, in the Philippines, about 12,350 for maize, 25,500 for wheat and 8,200 for *Triticale* at the International Center for Maize and Wheat Development, in Mexico, about 54,000 for grasses at the International Center for Agricultural Research in Dry Areas, in the Lebanon, 31,500 for pearl millet at the International Crops Research Institute for the Semi-Arid Tropics, in India, and about 880 for potatoes and 6,300 for sweet potatoes at the International Potato Center, Peru (information from ASSINSEL, 1997).

The FAO has a **Global Project for the Maintenance of Domestic Animal Genetic Diversity** for 14 species, including sheep. One aim is to conserve genetic uniqueness, using microsatellite markers to calculate the genetic distance between breeds. The FAO has a "world watch list" for domestic animal diversity, which is a catalogue of breeds at risk for 28 mammalian and avian species (Sherf, 1995). For the United Nations Environmental Programme, Heywood (1995) has listed the indigenous breeds of the main mammalian livestock species being conserved in various countries.

In Denmark, Finland, Iceland, Norway and Sweden, the "**Nordic Gene Bank Cooperation - Farm Animals**" was started in 1984, covering 103 breeds over 13 animal species. Farmers can receive a subsidy for loss of profit on maintaining a conserved breed. Flocks and herds are also kept by teaching organisations and agricultural museums. The loss of some cryopreserved stocks from power failure in this Nordic programme shows the importance of keeping live animals as well as frozen stocks. See Ponzoni (1997) for other animal conservation programmes.

In the UK there is the non-governmental, privately funded, **Rare Breeds Survival Trust**. Since its foundation in 1973, no farm animal breed has become extinct in Britain. About 1,000 owners maintain small groups of rare-breed animals. It owns Linga Holm island off the north coast of Scotland where **North Ronaldsay sheep** can survive on kelp (seaweeds). This breed is unusually salt tolerant, with a tidal, not diurnal, grazing pattern, and can survive on a diet so low in copper as to kill other breeds.

The **Gloucester Old Spot pig** (Plate 39) is an example of what was an endangered breed, but enthusiasts such as Dave Overton of Exfold Farm, Dorking, Surrey, have led to their numbers reaching about 700 in Britain. The pigs are very hardy and were traditionally fattened on whey from cheese, and windfall apples. When crossed to white breeds, all offspring are white, giving commercial F1 pigs with hybrid vigour, suitable for pork or bacon. Pedigree pigs are a bit fatty for most commercial processors but they have a superior flavour and sell through specialist "rare breed butchers". Gloucester Old Spot pigs became better known through mention on a popular radio farming programme, *The Archers*. Gilts have 5-10 piglets and sows have 9-15; up to 18 is not uncommon but is undesirable as mortality is then high. Piglets weigh about 1.6 kg, weaning at 11-27 kg at eight weeks, with slaughter at 18-24 weeks, with about 64 kg liveweight giving a 45 kg carcase.

Most **zoos** are involved in animal conservation projects and in breeding programmes to return endangered species to the wild, but these are mainly not for agricultural animals. For example, the Jersey Wildlife Preservation Trust (now the Durrell Conservation Trust) has bred Antiguan racer snakes and the Mauritius pink pigeon; Marwell Zoo has bred Przwalski's horse, the scimitar-horned oryx, red pandas, peccaries, reddish buff moth and other species.

In Britain, the **National Council for the Conservation of Plants and Gardens** (NCCPG) aims to retrieve plants believed lost from commercial cultivation but which may still be preserved in private gardens, and to conserve as many varieties as possible. For example, a double-flowered bramble, *Rubus rosifolius* "Coronarius", first described in 1816, was believed to have been lost for 100 years, but was spotted in a garden in Virginia and is now listed by about 30 nurseries. The NCCPG has established about 600 National Collections of separate genera, containing as many types as can be traced, with more than 50,000 types kept by institutions or enthusiastic amateurs. For example, amateur Veronica Read is the Holder of the National Collection of Hippeastrums (Knight's star lily), with more than 300 *Hippeastrum* plants covering 74 named cultivars and 20 species, including varieties no longer commercially available. Similar organisations have been set up in America, Australia, New Zealand and several European countries.

The £74 million **Eden Project** in Cornwall, England, will have two "biomes", climate-controlled environments, one for plants - especially economic plants - from the humid tropics (e.g., Amazonia, West Africa, Malaysia), and one for Mediterranean-climate plants (southwestern USA, South Africa, and Mediterranean areas). One greenhouse for tropical rainforest trees will be 1,000 m long by 120 m wide, and 60 m high. The project will be for conservation and education.

Unfortunately, there can be severe **legal and financial problems** about ownership, exchange and use of conserved stocks. The Convention on Biodiversity agreement, from the 1992 Earth Summit in Rio de Janeiro, has been ratified by 174 countries, including the UK and European Union, but the sharing of the benefits from the use of biodiversity is the most controversial part (see Prance, 1998).

As inferior stocks will not usually be commercially viable on their own, there is an excellent case for **"added value" activities** with them, whereby members of the public will pay admission charges to see rarities, or will sponsor individual animals. Thus zoos, safari parks, game reserves, rare-breeds centres, botanical gardens, show farms, etc., are all valuable places for conserving non-commercial varieties. While some rare breeds have public appeal in their appearance or behaviour, others do not. To minimise the problems of genetic drift and inbreeding in small populations, many zoos and safari parks have scientifically planned programmes of exchange of animals between institutions, either permanently or just for mating. Sale of surplus seed, fruit or animals is one way of helping to fund genetic conservation. Any restrictions on the range of plant or animal varieties which can be sold commercially within the EU are extremely unhelpful to those trying to preserve older or less productive types.

Micro-organisms are the easiest and cheapest to conserve. One can preserve most of them dried down on silica gel, where say yeasts are suspended in non-fat milk and then poured into small chilled glass vials containing dehydrated sterile silica gel, before storing at 4° in an airtight jar containing moisture-indicating silica gel. Bacteria can be stored at room temperature in high-salt "stab" nutrient media. Phage can be stored in phage buffer at 4°. Many micro-organisms can be stored in freezers after suitable treatment, say at liquid nitrogen temperature, as can animal sperm and mammalian embryos. All these treatments can result in stocks being stored in suspended animation for ten to many years, though viability should be checked periodically.

With **plants**, one can store pollen, seed, cell cultures, organ or tissue cultures, or whole plants, in different ways. Living plants at the National Fruit Collections at the Brogdale Horticultural Trust, Faversham, Kent, include at least two plants of each variety of 2,009 dessert and culinary apples, 75 cider apples, 272 cherries, 495 dessert and culinary pears, 20 perry pears, 4 medlars, 41 hazelnuts, 120 blackcurrants, 55 grape vines, 60 strawberries, etc. (1995 figures).

Seed banks are now being built to conserve seeds of many of the world's flora. For example, the Royal Botanic Gardens, Kew, are building a Millennial Seed Bank in Sussex at Wakehurst Place. The project has a cost of £80 million, with £30 million from the National Lottery Millennium Project. The construction costs are £13 million, with the buildings due to open in 2,000. The project will soon safeguard all the British flora and will eventually include 10% of the world's flora, including endangered species, with some specialisation in plants from arid regions of Africa, India and Latin America. The collected seeds are dried for about a month at 10-15% relative humidity and are frozen at -20°. With storage in airtight 3-litre preserving jars, the **estimated survival time** is 200 years for about 80% of the species, and more than 1,000 years for some legumes.

There is a very large seed bank in Colorado, USA, and another very large one, the International Seedbank at Svalbard, buried in the permafrost in Norway, so seeds will be preserved even if electricity fails for very long periods. In seed banks, one wants to conserve variation within species, so a typical collection would be of 20,000 seeds of one species, taken from different plants in different areas, regions or countries.

Near Gannoruwa in Sri Lanka, there is the modern **Plant Genetic Resource Centre**, including a seed bank for Asian food plants. Outdoors in soil there are "**living germplasm**" collections consisting of many small beds of different strains of local food plants such as **gotukola**, a low-growing plant (Plate 40) whose small leaves are used like spinach. Inside the building there are **tissue cultures** of commercial plant cultivars including pineapples, bananas and jack fruit. The main facility consists of cold stores for the **seed bank**. For this, many seeds of a crop plant are collected from different sources, the identification is checked, and the seeds are tested for pests and diseases, and for germinability, before drying. The Base Collection can store 25,000 species, at 1° and 30-35% relative humidity, with typical projected seed longevities of 35-50 years. There is also an Active Collection for sending out as required, with a capacity of 25,000 species, kept at 5°, 35-40% relative humidity, with a projected longevity of about 20-25 years. The seeds are kept in large screw-topped plastic jars in filing drawers inside large walk-in stainless steel chambers: see Plates 41 and 42.

"**Recalcitrant seeds**" are difficult to store for long periods, remaining viable for only two weeks to several months, as with important Asian crops, cocoa, coconut and rubber. Some plants can be **cryopreserved** *in vitro* at -196° as zygotic embryos, tissue cultures or single-cell cultures: see Withers (1991), and Razdan and Cocking (1997 and 1999). Some species are highly heterozygous and their genotypes are maintained by vegetative propagation, as for many fruit trees, strawberries, potatoes, artichokes, bananas and cassava. They can be kept in field gene banks as whole plants, or as cuttings, bulbs, tubers, etc., at low temperatures and controlled humidity for relatively short periods. They can also be stored as tissue cultures, meristems, or shoot tips in media giving very slow growth, with low light and low temperatures. This is used for potatoes and sweet potatoes at the Centro Internacional de la Pipa, Peru.

With **animals**, one can conserve them as living populations, or by cryopreservation of gametes (especially as sperm for AI; Chapter 17.2.4) or embryos (Chapter 17.2.5). DNA can be stored, but one cannot regenerate whole animals from isolated DNA at present. Semen and embryos freeze and thaw well, but not ova. Cryopreservation avoids genetic drift and disease, but power failures can ruin stocks. For sheep over a 20 year period, the costs of keeping live populations and the same number of cryopreserved animals are very similar.

Suggested reading

ASSINSEL. *Feeding the 8 Billion and Preserving the Planet.* (undated, but 1997) ASSINSEL (International Association of Plant Breeders), Nyon, Switzerland.

Brush, S. B., *Genes in the Field: Conserving Plant Diversity on Farms.* 1999. Lewis Publishers, Boca Raton, Florida.

Collins, W. W. and C. O. Qualset, eds., *Biodiversity in Agroecosystems.* 1998. Lewis Publishers, Boca Raton, Florida.

"Cuttings", Ozone loss may damage pollen. *Kew* (1998), Autumn, 7–9.

Ford-Lloyd, B. and M. Jackson, *Plant Genetic Resources: an Introduction to their Conservation and Use.* 1986. Edward Arnold, London.

Given, D. R., *Principles and Practice of Plant Conservation.* 1994. Chapman and Hall, London.

Heywood, V. H., ed., *Global Biodiversity Assessment.* 1995. Cambridge University Press, Cambridge.

Lamb, B. C., The relative importance of meiotic gene conversion, selection and mutation pressure, in populations and evolution. *Genetica* (1985), 67, 39–49.

Lamb, B. C. Gene conversion disparity in yeast: its extent, multiple origins, and effects on allele frequencies. *Heredity* (1998), 80, 538–552.

Lamb, B. C. and S. Helmi, The extent to which gene conversion can change allele frequencies in populations. *Genetical Research* (1982), 39, 199–217.

Marwell: *Your Guide to Marwell and the World of Animals.* Undated, available in 1999. Marwell Zoological Park, Hampshire.

Maxted, N., B. V. Ford-Lloyd and J. G. Hawkes, eds., *Plant Genetic Conservation: the in situ Approach.* 1997. Chapman and Hall, London.

Ponzoni, R. W., Genetic resources and conservation. pp 437–469, in: Piper, L. and A. Ruvinsky, eds., *The Genetics of Sheep.* 1997. CAB International, Wallingford.

Prance, G. Questions of ownership and access. *Kew* (1998), Autumn, 3.

Prasad, B. N., ed., *Biotechnology and Biodiversity in Agriculture/Forestry.* 1999. Science Publishers, Inc., Enfield, USA.

Razdan, M. K. and E. C. Cocking, *Conservation of Plant Genetic Resources* in vitro. Vol. 1: *General Aspects.* 1997. Science Publishers, Inc., Enfield, USA.

Razdan, M. K. and E. C. Cocking, *Conservation of Plant Genetic Resources* in vitro. Vol. 2: *Applications and Limitations.* 1999. Science Publishers, Inc., Enfield, USA.

Sherf, B. D., ed., *World Watch List for Domestic Animal Diversity*, 2nd. ed. 1995. FAO, Rome.

Withers, L. A., Maintenance of plant tissue cultures, in: Kirksop, B. E. and A. Doyle, eds. *Maintenance of Microorganisms.* 1991. Academic Press, London.

CHAPTER 16

GENETIC METHODS OF INSECT PEST CONTROL

16.1. Introduction

The control of insect pests of plants, animals and man is of enormous economic and medical importance. Insects attack animals, as in fly-strike of sheep and screw worm of cattle; they destroy or damage plants, as with locusts, or sawfly of apple, and may be vectors of disease, as with aphids transmitting viruses to plants, and mosquitoes transmitting malaria to man.

The main methods of controlling insect damage (see ADAS, 1998) are **insecticides** to kill the insects, **chemosterilants** that sterilise but do not kill the insects, **biological control agents**, the **release of sterile insects** into the agricultural and natural environment, and breeding **genetically resistant** plants or animals.

Insecticides can be very effective, as shown by the use of DDT in Sri Lanka to kill the **mosquitoes** which cause malaria. With a human population of about 15 million, there were about 2.8 million cases a year of malaria in Sri Lanka in 1948. After the widespread use of DDT under a World Health Organisation programme, the number of cases fell to only 17 a year in 1963. When the use of DDT was banned on environmental grounds, the number of cases of malaria rose within five years to 2.5 million a year. This WHO programme also involved the use of the antimalarial drug chloroquin, and the development of chloroquin-resistant strains of malaria was another factor promoting the rise in malaria cases. Hormone-mimicking insecticides can be fairly specific, e.g., to Lepidoptera, but some are broad-spectrum (Dhadialla, Carlson and Le, 1998).

Over time, many insect pests develop **genetic resistance** to particular insecticides. For example, the housefly, *Musca domestica*, developed resistance to DDT, breaking it down with a dehydrohalogenase enzyme controlled by a major gene, with polygenic modifiers controlling the rate of uptake of the DDT. Insecticides are generally sprayed over the target areas, but can be used in **traps** with attractants such as pheromones or lights.

In theory, **chemosterilants** can be more efficient than insecticides. A chemosterilant spray which made 90% of an insect population sterile would be more effective than an insecticide giving 90% kill. Providing the chemosterilant does not affect mating fitness, many of the matings of the remaining 10% of fertile insects will be with sterile insects, and therefore be infertile, leaving only 1% of fertile mating combinations (10% x 10%). Using insecticides and chemosterilants does however mean loading the environment with potentially harmful chemicals, which is why other methods are used when they are suitable.

Biological controls of insects, e.g., using fungi, nematodes, bacteria and viruses, vary greatly in effectiveness and specificity. For example, a baculovirus, nuclear polyhedrosis, is used on one million hectares in Brazil to control soya bean caterpillar, *Anticarsi gemmatalis* (Moscardi, 1999), and fungi can be used to control roaches and termites (Gunner, 1999).

16.2. The release of sterile insects

The **release of sterile insects** has been extremely successful with some pests but is unsuitable for others. Typically one cultures vast numbers of the pest one wishes to eliminate, e.g., rearing codling moth of apples on apples or a cheaper artificial diet. One then sterilises them with the minimum necessary dose of gamma rays from a cobalt-60 source (or X-rays, but they are more expensive to use), and releases them over the pest's natural habitat from the air or the ground, depending on economic and biological factors. In the natural habitat, the sterilised insects mate with natural populations, reducing those populations because the matings are sterile. One advantage of using sterile insects rather than insecticides or chemosterilants is that the sterile insects can move and actively seek out the natural insects, even in niches which sprays might not reach. One could use chemosterilants to sterilise the insects before release, instead of radiation.

The **success** of such programmes depends on many factors. They work best if the females mate only once rather than many times, since a combination of mating with sterile and fertile insects is usually fertile. The ratio of sterile insects to natural ones is crucial and should be as high as possible, at least 10 sterile insects to one natural, with 40 sterile codling moths being used for every natural one, which means high rearing costs. There must be a good geographical spread of the sterile insects, and the sterile insects must be reliably sterile, otherwise one is spreading the pest itself.

With **codling moth**, the best results are from giving males a gamma ray dose which permits 10-15% egg hatch on mating to natural females, as those hatchlings are too feeble to do much damage. A dose sufficient to give complete male sterility reduces the mating competitiveness of those irradiated males. In codling moths, sterile males disperse further and more rapidly in search of wild females if they are released separately from sterile females than if they are released with both sexes together, when sterile males tend to mate with the sterile females.

In cattle screw worm control, one would like to - but cannot - kill off all females from the sterilised insects because their laying of sterile eggs still damages the cows' hides and they lessen dispersal of the males. Even worse, females need a dose of 7,500 rads, while the males only need 2,500 rads to sterilise them, and giving the males the higher dose reduces their competitiveness. Another problem with mass culture is that the reared insects adapt to the culture conditions, and may be less competitive than wild insects.

Despite these difficulties, the classic success story of release of sterile males is **screw worm of cattle,** *Cochliomyia hominivorax.* Just before 1957, it did about $120 million damage a year to cattle in the USA. The female lays eggs in skin lesions, where larvae feed for five to six days before falling off and pupating in the soil. In 1957, a campaign was begun with sterile insects to eliminate screw worm from Florida and southeastern USA. A mass-rearing factory was started, producing 50 million flies a week. Pupae were given 7,500 rads of gamma rays, which stops female oviposition. Because the sexes are not easily separated, mixed sex releases were made, with 800 flies per square mile (per 259 ha) per week dropped from planes, with additional releases in badly infested areas. After 17 months, with the release of 3,500 million sterile insects, complete eradication was achieved in the Florida area. The cost was $10.6 million, giving estimated savings of $20 million a year. The Florida peninsula is fairly isolated from sources of reinfection; one must eliminate the pest from a whole zone, not just from isolated pockets. The same method was used successfully in 1990/91, at a cost of $82 million, to eliminate screw worm from Libya after its accidental introduction in 1988 in a sheep shipment from South America; up to 40 million sterile males a week were released.

Because of the cost of mass rearing of insects for such programmes, and the risk of accidental escape of unsterile insects, e.g., of 100,000 fertile medflies in the 1980 California medfly sterile release programme, **alternatives to radiation** for sterilisation have been explored, and methods of getting rid of females. Heterozygous translocations reduce fertility, but releasing homozygotes for translocations to mate with natural non-translocated ones would be pointless if the translocation homozygotes could breed amongst themselves. Tetraploids give sterile triploids on mating with diploids, but tetraploid males usually have low fertility and are difficult to keep in culture. Other possibilities include cytoplasmic male sterility, or the release of conditional lethal mutations with stocks raised under conditions where the mutations were not expressed. So far, the use of gamma-ray sterilised individuals has been the most successful genetic method of insect control.

Myers, Savoie and van Randen (1998) considered the **cost-benefit assessment** of eradication programmes, pointing out that it costs as much to eliminate the last 1-10% of a population as it does the first 90-99%. Some eradication programmes have succeeded, such as the screw worm ones in Florida, 1958-1960, in USA and Mexico (at a cost of $750 million) by 1991, and Honduras by 1995. In Nigeria, traps and insecticide-impregnated targets were used to reduce the **tsetse fly** population, before the release of 10 sterile males to each wild male, giving success by 1985 (Oladunmade, Dengwat and Feldman, 1986). A sterile male release programme was tried in the Okanagan Valley, British Columbia, in 1995-96, for **codling moth** (*Cydia pomonella*) of apple, but the sterile males were not active enough in early spring to compete with wild males. The wild population was reduced but not eradicated. The Californian State Medfly Project was unsuccessful and resulted in 14,000 claims for damage to car paintwork. In 1997, millions of sterile male mosquitoes were released from aircraft and from the ground in the southern Indian state of Tamil Nadu in an attempt to eradicate malaria.

16.3. The breeding of insect-resistant varieties

The **breeding of pest-resistant** plants (see Maxwell, 1990) and animals is an ideal method of control. See Niks, Ellis and Parlevliet (1993) for plant defence **mechanisms of resistance**, including avoidance, resistance and tolerance, gene-for-gene interaction and specificity of resistance. See Chapter 14.1.7 for genetic engineering with *Bacillus thuringensis* Bt toxin genes, against specific insect pests of plants. One worry about breeding pest-resistant plants and animals is that success could lead to a big reduction in the natural predators of the pest species, and that mutations in the pest which might eventually enable it to overcome the resistance could lead to a huge outbreak of the pest, with little control by its predators. There are already insect pests which have developed resistance to Bt toxins. In general, insect-resistance breeding has been much more successful in plants than in animals.

In sheep, fly-strike has a heritability of 0.1 to 0.58 in Merinos, while fleece-rot has heritabilities of 0.05 to 0.8 (Raadsma, Gray and Woolaston, 1995). Fly-strike increases with rainfall and if dags (a build-up of faeces in wool around the anus) are not removed by jetting with water. Increased genetic resistance to fly-strike would reduce treatment costs. Because fleece-rot is easier to measure than fly-strike and has higher levels of expression and a good genetic correlation with fly-strike, Australian Merino breeders have selected just for resistance to fleece-rot, to improve that character and resistance to fly-strike: see Atkins (1987).

Lindley, in 1831, found that the apple variety Winter Majetin was much more resistant to **woolly aphid** than were other varieties, so it provided breeding material for transfer of the resistance to other cultivars. This resistance has proved to be stable ever since. Although it has been known for many years that American grape vine species are largely resistant to the **phylloxera** root pest (a relative of the aphids), the resistance has not yet been bred into European vines such as *Vitis vinifera*. Instead, most *vinifera* grape vines for wine are grafted onto American rootstocks to minimise phylloxera, which devastated the vineyards of Europe and other places in the second half of the 19th century. Even if genetic resistance is incomplete, as for carrots to carrot fly (*Psila rosae*) and potatoes to the blight fungus, *Phytophthora infestans*, partial resistance often permits significant reductions in the amount of pesticide or fungicide required for control.

After the breeding of wheats resistant to **Hessian fly** (*Mayetiola destructor*), infestation levels for this important pest of wheat in the midwest of the USA have dropped from over 90% to under 10%. Similarly, wheat stem sawfly (*Cephus cinctus*) has been controlled by breeding resistant varieties. The costs over 10 years of producing wheat cultivars resistant to Hessian fly, European corn borer (*Ostrinia nubialis*) and wheat stem sawfly was US$ 9.3 million, giving estimated savings of US$ 308 million a year, which include reduced pesticide and labour costs (see Niks, Ellis and Parlevliet, 1993).

In 1997, scientists at Horticultural Research International, at East Malling, Kent, reported finding an apple gene, the dominant Sd_1, which gave resistance to two local biotypes of the damaging pest, the rosy leaf-curling aphid, *Dysaphis devecta*, which causes severe leaf curl and conspicuous red galling (HRI, 1996). It was found in Cox's Orange Pippin and can be crossed into other varieties. See also Chapter 8.7.

Unlike the action of most pesticides, insect resistance in plants is nearly always **species-specific**, but resistance to more than one pest can be incorporated, as with alfalfa resistant to spotted alfalfa aphid and to pea aphid. Insects sometimes form **different biotypes**, so that a plant may be resistant to some biotypes but not others, as with Hessian fly of wheat. Insects can mutate and/or have recombination, so that a plant's resistance may be overcome. Gene-for-gene systems controlling host-pest resistance and virulence have been found, parallel to the ones which Flor found for rust-resistance in flax. Resistance genes are often found in primitive cultivars or related species, and can be bred into cultivars: an example is the transfer of greenbug resistance from rye to wheat.

Suggested reading

ADAS. *The ADAS Pest Control Manual: a Reference Manual for the Management of Pests*. 1998. ADAS, Guildford.

Atkins, K. D., Resistance to fleece rot and body strike: its role in a breeding objective for Merino sheep, pp 3.1–3.7, in: *Proceedings of the Sheep and Blowfly and Flystrike Management Workshop*. 1987. Department of Agriculture, Trangie, New South Wales.

Dhadialla, T. S., G. R. Carlson and D. P. Le, New insecticides with ecdysteroidal and juvenile hormone activity. *Annual Review of Entomology* (1998), 43, 545–569.

Gunner, H., Mycopesticides: killing roaches and termites the natural way. *Annual Review of Microbiology* (1999), 53, in press.

HRI (1996). New molecular markers and their use in apple breeding. [Work of P. Roche et al.] *Annual Report 1995-96*. pp 64–65. Horticulture Research International, Wellesbourne, Warwick.

Maxwell. F. G. ed., *Breeding Plants Resistant to Insects*. 1990. Wiley, New York.

Meara, T. J. and H. W. Raadsma, Phenotypic and genetic indicators of resistance to ectopathogens, pp 187–218, in: Gray, G. D., R. R. Woolaston and B. T. Eaton, eds., *Breeding for Resistance to Infectious Diseases in Small Ruminants.*. 1995. ACIAR Monograph No. 34, ACIAR, Canberra, Australia.

Moscardi, F., Assessment of the application of baculoviruses for control of Lepidoptera. *Annual Review of Entomology* (1999), 44, 257–289.

Myers, J. H., A. Savoie and E. van Randen, Eradication and pest management. *Annual Review of Entomology* (1998), 43, 471–491.

Niks, R. E., P. R. Ellis and J. E. Parlevliet, Resistance to parasites, pp 422–447, in: Hayward, M. D., N. O. Bosemark and I. Romagosa, eds., *Plant Breeding. Principles and Prospects*. 1993. Chapman and Hall, London.

Oladunmade, M. A., L. Dengwat and H. U. Feldman, The eradicating of *Glossina palpalispalpalis* using traps, insecticide impregnated targets and sterile insect technique in Central Nigeria. *Bulletin of Entomological Research* (1986), 76, 2775–2786.

Raadsma, H. W., G. D. Gray and R. R. Woolaston, Genetics of disease and vaccine response, pp 199–224, in: Piper, L. and A. Ruvinsky, eds. *The Genetics of Sheep*. 1997. CAB International, Wallingford.

CHAPTER 17

REPRODUCTIVE PHYSIOLOGY IN PLANTS, ANIMALS AND HUMANS; CROSSING METHODS

17.1. Plants

17.1.1. Plant sexual reproduction

Plants normally reproduced **sexually,** by seed, in commerce include: cauliflower, Brussels sprouts, pea, bean, beetroot, sugar beet, celery, leek, radish, oilseed rape, cucumber, marrow, onion, carrot, turnip, tomato, lettuce, cereal grasses (including wheat, maize, barley and rice), poppies, most trees including oaks, and most garden flowers. See North (1979) for details of individual plants' reproduction, including agricultural crops, and for relevant anatomy, physiology and development. For plant breeding systems, see Richards (1997). For plant propagation by seeds or vegetatively, see Hartmann et al. (1997).

Asparagus is one crop which can be reproduced sexually or vegetatively: seeds give a sex ratio of about 1 male (XY chromosomes) : 1 female (XX), with the females usually being rogued out as they are inferior to males in spear size and quality. Rare supermales (YY) can be propagated vegetatively or crossed to females to produce only males. Excellent males can be clonally propagated *in vitro*, vegetatively.

To cross different varieties, one needs them to flower at the same time, unless one stores pollen. *Solanum* species have long **flowering periods,** so that if one species flowers from January until June and another flowers from April until December, one could cross them between April and June. Cut-flower crops such as daffodils have much shorter flowering periods, and cereals average about 14 days. In cereals, normal spacing between plants allows only one or two tillers (offshoots) to develop. To get a longer flowering period, the breeder can plant the seeds further apart, allowing many tillers to develop, some of which may flower 7 to 21 days later than others. Some fruit trees have a juvenile period of about six years before they flower, so grafting buds of a newly-bred seedling onto a mature tree can speed the breeding process.

Flowering times can be manipulated by various treatments, including temperature, soil conditions and **light** intensity and duration. **Short day plants**, such as rice, soya beans and many tropical crops, have flowering promoted by day-lengths shorter than some critical value. Thus with less than six hours light per day it might take 11 days to induce flowering, but with 10 hours light per day it might take 21 days to induce flowering, and with more than 12 hours a day the plants might never flower. **Long day plants**, such as wheat or sugar beet, are induced to flower by light of longer duration than some critical value, and **day-neutral plants** such as groundnuts, tomatoes and some dwarf wheats need light for growth, but are largely indifferent for flowering to the length of the daily light period. Plate 43 shows artificial lights and rolled-back black plastic sheeting to control day-length in a commercial greenhouse for the production of flowering **chrysanthemums** for a much longer period than is possible with natural daylight alone.

The induction of flowering in a given plant is sometimes permanent, sometimes reversible. The leaves are the main sensors of day-length, but often go through a non-sensitive juvenile period, i.e., rice leaves are insensitive to daylight length for two to five weeks after emergence.

Higher **temperatures**, unless too high, usually speed development, including flowering. Some plants, such as groundnuts and peas, need a certain **accumulated temperature** for flowering; this is measured in "**degree days**", i.e., total days x degrees C, say needing 1,000 degree days to initiate flowering. **California's wine regions** are classified into five zones, based on the number of degree days experienced in each zone in a typical growing season. Those degree days are measured, in Fahrenheit, as the length of time the average temperature remains over 50°F (10°C) between 1st April and 31st October, and by how much the average temperature exceeds this base value. Thus if the mean temperature over a five-day period in summer was 70°F (21°C), the number of degree-days accumulated over that period would be 5 days x (70 - 50)°F = 100 degree days. The five wine zones are: zone I, 2,500 degree days or fewer (e.g., the cool Napa Valley); II, 2,501 - 3,000; III, 3,001 - 3,500; IV, 3,501 - 4,000 (e.g., San Diego), and V, more than 4,000 (e.g., Fresno in the hot Central Valley).

The introduction of stone fruits to warmer areas was made possible by developing varieties without a cold requirement for fruiting. Pollination at 40° has been used in *Trifolium* (clover) to overcome self-incompatibility and between-species incompatibility.

Some plants need particular temperature and day-length sequences. A principal example is **vernalisation**, where plants need a period at low temperatures (e.g., 2 to 8°) to change the physiological condition of the growing points, so that the resulting leaves become perceptive to day-length. In barley, wheat and rye, there are separate **spring and winter varieties**, where the spring varieties do not need vernalisation but winter varieties do. The winter varieties need low temperatures at the seedling stage in order to be able to respond to long days later. Winter varieties are therefore planted in late autumn or early winter, with increasing daylight the following spring initiating flowering. As the spring varieties do not need the cold winter temperatures, they can be planted after the winter.

Brussels sprouts and pyrethrum plants (*Tanacetum* - also called *Chrysanthemum* or *Pyrethrum* - *cinerariifolia*, and other species used for insecticide from the flowers) need **low temperature** to induce flowering. Pyrethrum is often grown in the tropics, but the need for low-temperature induction of flowering means that the plants have to be grown at high altitude. Three varieties of pyrethrum were grown at altitudes of 1,830, 2,130 and 2,440 metres in Kenya. Variety A yielded **1,000** units of flowers at the lowest altitude, 1,100 units at the middle altitude, and 1,200 units at the highest altitude. Variety B yielded 400, **1,200** and 1,400 units at these altitudes, and variety C yielded 50, 1,100 and **1,600** units, respectively. At 1,830 m, one would therefore grow variety A; at 2,130 m one would grow B, and at 2,440 m one would grow C. These data show **genotype/environment interactions**, with different varieties behaving differently in different environments.

Other tropical crops which are subject to long dry periods may need **rain** to get flower opening. With coffee, long dry periods give ripe flower buds, then one shower of rain causes the flowers to open, with excellent synchronisation for pollination. The breeder can use the same method for synchronising different varieties for crossing.

Some **tuber crops** are difficult to get into flower for the breeder to make crosses, as in yams and some potato varieties. In potatoes, one can graft the stems onto tomato plants, which suppresses tuber formation and diverts the energy into flowers. Breeders can also manipulate flowering times by controlling glasshouse temperatures, or by using hormones such as gibberellins or antigibberellins.

With fruit trees, seedlings may take many years before first flowering. Flowering may be greatly speeded up by **grafting** seedling shoots or buds onto adult trees, say in apples or pears. To increase flowering in perennials, one can spray with anti-gibberellins to reduce vegetative growth, or girdle the stems with incisions into the layers below the bark to retard the downflow of photosynthetic products to the roots. In general a **high carbon/low nitrogen ratio** encourages flowering rather than vegetative growth, so some fruit trees are encouraged to flower by partial root-pruning, or planting in poor soil, or even over concrete to restrict root growth, as with fig trees.

The duration of **stigma receptivity** is very variable between species, with 14 days for the "silks" on a maize plant and 10 days for grape vines, but only three to four hours for sweet potato and mango. The duration of **pollen viability** is often only one to three days, with cereals under dry conditions losing pollen viability after six hours. Pollen grains may be **binucleate**, with a vegetative nucleus and a generative nucleus which only divides in the pollen tube during germination on the stigma, or **trinucleate**, when the generative nucleus divides while the pollen is still inside the anther. This binucleate/trinucleate difference correlates with many other pollen properties, such as the incompatibility system, storage life and ease of *in vitro* germination of pollen on artificial media. The **number of pollen germ pores** is usually specific, usually two or three in pollen from diploids, then gradually increasing with ploidy, as in clover and potato. In potato, pollen from diploids generally has three germ pores but pollen from the tetraploid mainly has four pores.

Normally only one pollen tube grows per grain, with germination on a compatible stigma taking only a few minutes. The time from pollination to egg fertilisation is typically 12–48 hours, but is much longer in some shrubs and trees, with 3–4 months in hazel (*Corylus avellana*). It takes 12–14 months in oaks (*Quercus* spp.) and in *Hamamelis*, where part-grown pollen tubes overwinter.

Pollen size and **surface stickiness** are correlated with whether the plants are **wind-pollinated** (grains are often 20–60 μm diameter, with a dry surface) or **insect-pollinated** (grains are larger and sticky). For example, wind-pollinated *Myosotis* (forget-me-not) has grains four to five μm across, and dry, while insect-pollinated *Curcurbita* (marrow, squash) species often have grains 200 μm across, and sticky. Pollen development typically includes about three days for meiosis, about eight days for microspore interphase, and about four days for mitosis 1 and 2. One maize plant may produce several million pollen grains, so many grains will carry spontaneous mutations (see Chapter 7.2.1 for mutation frequencies).

Breeders should find out whether their crops are wind- or insect-pollinated when considering how to avoid **bastardisation**, the contamination of their seed lines with stray pollen from other crop or wild varieties. Fig. 17.1.1 shows two arrangements of a breeder's plots in relation to the prevailing wind, with arrangement (ii) being much better than (i) for avoiding bastardisation between those plots.

Seed production in open plots should not be done near to other varieties of the crop, nor to wild relatives which can cross with them. Bastardisation can be reduced if there is a screen of tall trees (e.g., windbreak poplars), a hill or some buildings between an upwind source of contaminating pollen and the seed fields, so that the prevailing wind either takes the contaminating pollen right over and beyond the seed plots, or the obstruction intercepts most of the pollen. If there is an unavoidable

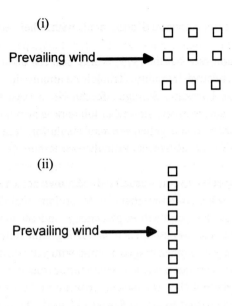

Figure 17.1.1. Plot arrangements affecting bastardisation of seed lines by each other in a wind-pollinated species. Arrangement (ii) is much better than arrangement (i) for avoiding bastardisation.

upwind source of contaminating pollen, the breeder can discard seed from plants nearest that source, which may be pollinated by the contaminant, but the breeder can use seed from plants further downwind, as these will largely have been pollinated by the desired plants. Depending on flight patterns, insect and plant biology, and winds, the breeder of insect-pollinated plants may take a similar approach, not harvesting seed from the outer rows of plants on the seed plots, as these "guard rows" might have been pollinated from insects flying in from contaminating plants elsewhere, but using seed from more central areas of the seed plot. These plants will probably be pollinated by insects which have picked up pollen from the seed plot itself. In Holland, there are legal minimum distances between particular seed crops and the nearest known source of contaminating pollen, e.g., 50 m for *Chrysanthemum leucanthemum* and 600 m for sugar beet (see North, 1979).

Pollen viability can be tested by staining the grains with acetocarmine, or with iodine in dilute potassium iodide solution, to show the percentage of grains with cytoplasm, but this may overestimate viability. It is better to stain with fluorescein diacetate and examine under a fluorescence microscope, as that tests for integrity of the plasmalemma. Especially with binucleate pollen grains, one can try *in vitro* germination in liquid or solid media (solid media for cereals). The medium usually includes sucrose, calcium nitrate and boric acid, with other requirements varying between species.

Pollen preservation (see North, 1979) works best for binucleate grains, using 0–10° and 40–50% relative humidity. Trinucleate grains do not store well and need a higher relative humidity. Storage life can be increased by raised carbon dioxide levels, lower air pressures, storing in the dark, and mixing the pollen with casein or talcum powder. Some pollen remains viable for decades at -80°. Pollen preservation works well with Rosaceae, Primulaceae, Ranunculaceae, Liliaceae and Iridaceae.

Seeds for commercial use are usually dried down to a moisture content of 16% or less and are often stored at about 5° to prolong viability. Seeds for sale must usually pass tests for germination percentage and seedling vigour. Various post-harvest treatments may be used, such as coating with fungicide, e.g., against "damping off" (*Pythium* and other fungi), or temperature treatments related to seed dormancy - some stone fruits need low-temperature treatment ("stratification"). Older seeds tend to have higher spontaneous **mutation frequencies**, with 1% in fresh *Antirrhinum* seeds rising to 5% in 9-year-old seeds. Four-year-old pea seeds have about five times the mutation frequency of fresh seeds. If one seeks new spontaneous mutations, using old seed would therefore help.

Whether **hops** are **seeded or not** is important commercially. Hops are dioecious perennials lasting about 25 years and are normally grown from cuttings from female plants. Only two or three male plants are needed per hectare to get seeded hops. The seeds add weight but have little effect on flavour, and lagers and bitters can be made from seeded or unseeded hops. Pollination is not needed to get the bitter bracts with hop oils. British hops are often seeded, but in Germany, growing seeded hops and having male plants are illegal. Wild male hops can pollinate commercial crops.

Early ripening peaches may not form mature embryos, so need embryo dissection and aseptic culture. Some orchid seeds are extremely small, with inadequate nutrients, and need cultivation on nutrient agar, or application of a symbiotic fungus to the roots.

17.1.2. Incompatibility in higher plants

The haploid pollen tube usually needs to grow down a diploid style for fertilisation, but in many species there are physiological incompatibility mechanisms which prevent self-fertilisation. About half the flowering plant species have self-incompatibility. Two major types are **gametophytic incompatibility**, where the pollen's own gametophytic genotype controls the pollen properties, and **sporophytic incompatibility**, where the pollen properties are determined by the genotype of the sporophyte, that is, by proteins from the diploid anther tissue being deposited in the grain's outer layers. There can be dominance and gene interactions in the diploid stigma and style in either system, and in the pollen in the sporophytic system. Sporophytic incompatibility occurs frequently in the Cruciferae and the Compositae, often in species with trinuclear pollen. Gametophytic incompatibility is common in the Leguminosae and Solanaceae (monofactorial), Gramineae (bifactorial), Cheniopodiaceae and Ranunculaceae (polyfactorial), often in species with binucleate pollen.

An example of **sporophytic control** is the **primrose**, *Primula vulgaris*. This has two alleles at one locus giving two morphological flower types, pin eyed (with a tall style and low-set anthers) and thrum eyed (with a short style and anthers set high in the corolla tube). The different style and anther heights provide a morphological system that promotes cross-pollination by long-tongued insects. There is also an incompletely effective physiological incompatibility. Thrum is $S s$, giving 3 thrum: 1 pin on selfing, and pin is $s s$, giving all pin on selfing, with S dominant to s. Because of this dominance and the sporophytic control, all pollen from thrum has thrum phenotype, although the genotypes are 1 S: 1 s. The $S s$ thrum style is also thrum phenotype, so if one hand-self-pollinates thrum, the thrum phenotype pollen should not be able to grow down the thrum style, but about seven seeds per flower are set. Similarly, in a selfed pin x pin cross, the s pollen should not grow down the $s s$ pin style, but about 35 seeds are set per flower. In pin female x thrum male and thrum female x pin male crosses, the pollen has a different phenotype from the style and fertilisation is good, producing about 61 and 57 seeds per flower respectively.

By detecting rare recombination of the different elements, Ernst showed in *Primula viscosa* that the S/s locus is a **super-gene**, made up of different parts. It contains G for style length, S for stigmatic surface form, A for anther height, P for pollen size, and I for pollen incompatibility type. All these traits differ between pin and thrum, and most recombinants are at a disadvantage to the two standard types because they are less efficient at preventing either self-pollination or self-fertilisation.

An example of **gametophytic** control is the **evening primrose**, *Oenothera organensis*, which has multiple alleles and no dominance at the self-incompatibility locus, so both alleles are expressed in the style, and are always different because pollen with an allele the same as one allele in the style cannot grow. If we had (i), the cross $s_1 s_2$ female x $s_1 s_2$ male, the pollen genotypes and phenotypes would be 1 s_1: 1 s_2, and neither type could grow down the style expressing s_1 and s_2, so no seed is set. If we had (ii), the cross $s_1 \, s_2$ female x $s_3 \, s_4$ male, the pollen genotypes and phenotypes would be 1 s_3: 1 s_4, so both types could grow down the style expressing s_1 and s_2, so good seed is set. If we had (iii), the cross $s_1 \, s_2$ female x $s_1 \, s_3$ male, the pollen genotypes and phenotypes would be 1 s_1: 1 s_3, and only type s_3 could grow down the style expressing s_1 and s_2, so seed is set, giving a 1: 1 ratio of $s_1 \, s_3$ and $s_2 \, s_3$ seed.

As we saw in Chapter 15.2, selection for rarity can lead to there being very large numbers of alleles at such self-incompatibility loci, such as 45 alleles at this locus in evening primrose, and 200 alleles in clover, *Trifolium repens*. In sweet cherry, apples and potatoes, such systems prevent self-fertilisation. Good breeding lines can then be propagated vegetatively. In apples and cherries, fruit does not develop without fertilisation, so **compatible pollinator varieties** must be planted in the orchards. Thus a minority of English Golden Delicious trees in an orchard of Cox's Orange Pippin apples ensures fruit formation after cross-pollination of both varieties. Some self-fertile sweet cherries have now been bred. It is clearly essential for fruit growers and breeders to know about any self-incompatibility systems in their plants.

Breeders can sometimes overcome self-incompatibility by pollination at the **bud-stage**, by high temperature, by radiation, or by chemicals which prevent RNA or protein synthesis. In some species such as sweet cherry, one can use radiation to induce mutation in self-incompatibility loci, with easy selection in gametophytic systems.

In poplar (*Populus*), Knox in 1972 obtained large quantities of interspecific hybrids. He used **mentor pollen**, which was killed compatible pollen mixed with live incompatible pollen, when the stigma allowed the incompatible pollen to germinate and fertilise. Even just proteins from the walls of compatible pollen can sometimes act as "mentor" in sporophytic systems.

17.1.3. Crossing methods

For a **controlled cross**, the stigmas of one parent must be protected from all other pollen, including its own, and pollen must be applied to the style from the anthers of the other parent. If the flowers are hermaphroditic, with male and female parts, one usually has to remove the anthers before they are ripe from the flowers to be used as females. Protection from stray pollen is provided by putting cellophane or plastic bags over individual flowers or whole inflorescences, but they must allow moisture to escape, to prevent rotting. With insect-pollinated flowers, one needs special plastic, muslin or wire cages of appropriate mesh size to exclude unwanted insects. Waterproof coverings are needed for plants grown outdoors. Details of crosses can be recorded on plastic or embossed metal tags, tied near each flower, if used outdoors, or on card tags if used indoors. Plate 44 shows two Brussels sprouts lines being crossed using blowflies within a pollination bag. **Blowflies** (readily available as larvae from fishing shops) are often preferred to honey bees for making breeders' pollinations because blowflies are more active than bees in dull weather and are more likely to cross-pollinate different lines, while bees often only visit one of the two lines to be crossed.

Hand-pollination methods vary with the species involved. Pollen can be transferred from anthers to stigmas with a small paint brush, which is sterilised between pollinations with high concentrations of ethanol, industrial alcohol or methylated spirits. If they are a convenient size, whole anthers can be detached - with alcohol-sterilised fingers or forceps - when ripe and rubbed on the stigmas. In maize, pollen is collected in bags from the dehiscing terminal tassels (stamens) and the pollen is then shaken over the appropriate silks (stigmas). Maize pollen is only viable for about one day under normal conditions, less if it is very dry. See North (1979) for details of hand pollination and emasculation in various species.

Clearly the male and female parts to be crossed must be ripe simultaneously. If the male parts are ready too early, part of the plant can be cut off and put in a fridge, with the stem in water, to delay ripening. Barriers to self-pollination, which is sometimes needed, include **protandry**, where the male parts ripen before the female ones, as in maize, carrots and parsnips, **protogyny**, where the female parts ripen first as in walnut, **dioecism**, where there are separate male and female plants, as in hops and asparagus, and **genetically determined self-incompatibility**, as in many brassicas such as cabbage and sprouts, and in apples, cherries and clover.

To prevent unwanted selfing during breeding programmes, it is sometimes convenient to turn a monoecious, hermaphroditic plant into a dioecious one by obtaining male-sterile mutants and female-sterile mutants. In maize, the mutant "tassel seed" has the anthers converted into female parts, so is male sterile, and "barren-stalk" is female sterile.

Emasculation (removal of stamens) and hand pollination are slow processes and costly in labour, so are only used commercially if there is no satisfactory alternative. They are used in making **F1 hybrid tomato seed**, because hybrids outyield selfed lines considerably, overcoming the disadvantage of more expensive seed. Tomato flowers are easy to emasculate and each flower pollinated yields a lot of seed. In lettuce and peas, which are naturally inbreeding, one can get higher yields from F1 hybrids, but the labour costs usually make the seed too expensive. Breeders could even consider breeding varieties with flowers more convenient for emasculation and hand-pollination. Attempts have been made to use chemicals, such as 0.2% aqueous FW 450 (2,3-dichloroisobutyrate), as **selective gametocides** to give male sterility, e.g., in cabbage, lettuce and tomato, with varying success.

17.1.4. Getting uniform fruit, seed or bud ripening

The **ever-increasing mechanisation of agriculture** makes getting **uniform ripening** more important. A combine harvester going through a field of wheat, barley or rye will harvest all plants whatever their degree of ripeness. The seed-buyer does not want a mixture of ripe and unripe grains. When **Brussels sprouts** were picked by hand, the pickers picked the first-ripening, lowest on the stem, sprouts first, then picked the later-ripening upper ones a few days later. For mechanical plucking, the sprouts need to ripen simultaneously all over the stem, and the breeders have produced suitable varieties. In catalogues of sprouts, some varieties are marked as suitable for mechanical harvesting and others are not, depending on uniformity of ripening.

Uniform ripening is increasingly wanted for trees such as apples and bushes such as raspberries, where fruit can be harvested by mechanically shaking the whole plant and collecting the falling fruit on sheets. **Size uniformity** is also desirable with more pre-packing of fruit, eggs and vegetables graded for size. Farmers who produce a whole range of sizes of a particular fruit have had them rejected by supermarkets because the extremes of size, too large or too small, were unwanted. In 1999, one UK supermarket started selling packets of mixed-size strawberries, as this reduced waste by 30% compared to Class One uniform fruit, enabling it to undercut the price of uniform fruit.

Breeders can collaborate with physiologists, perhaps using physiological methods to **impose phenotypic uniformity**, say for ripening time, on a genetically diverse population. Especially with outbreeders or random maters, one might breed a high-yielding variety which lacked uniform ripening, then use a spray to induce uniform ripening. **Ethylene** gas, and compounds which break down to ethylene within the plant, are very useful for this: see Abeles, Morgan and Saltveit (1992). It is interesting that growers used bonfires (which produce ethylene) and kicking plants (ethylene is produced as a wound response) to help mangoes and pineapples to ripen. One can now use ethylene-producing sprays to trigger ripening, e.g., to synchronise apple harvests, as ethylene starts formation of the abscission layer in the stalks.

17.1.5. Somaclonal variation and vegetative propagation

Somaclonal variation is the genetic variation arising during the vegetative propagation of plants. **Vegetative propagation** can involve specialised organs such as bulbs, corms, runners, rhizomes, stolons and tubers, or taking stem or leaf cuttings, or grafting buds or stems onto rootstocks or stemstocks. It can also involve more technological methods with meristem multiplication or callus culture, which can regenerate whole plants from single cells (which are dedifferentiated) in tissue culture.

Some ornamental plants can be propagated **sexually or vegetatively**, such as begonias, delphiniums, lilies, and some lupins, geraniums, primulas and violets. Plants normally propagated vegetatively include blackberries (*Rubus*), blackcurrants, blueberries, gooseberries, raspberries, red and white currants, strawberries, apples, pears, plums, potatoes, carnations, chrysanthemums, dahlias, fuchsias, gladioli, hyacinths, irises, paeonies, rhododendrons, roses and tulips. See North (1979) for details of breeding new cultivars of such plants.

Vegetative reproduction allows the production of a large number of plants without waiting for completion of the sexual cycle, and without the risks of losing favourable genotypes through crossing, and segregation and recombination at meiosis. It avoids losing heterozygosity and avoids the segregation load (Chapter 15.3). Viruses are usually unable to replicate in meristem tissue, so regeneration of plants from isolated meristems has been used to get **virus-free** rhubarb and strawberry plants, for example.

Although one might think that plants produced by mitosis from the same starting material should all be identical, one often gets **somaclonal variation**, which is called this as the origin of the variation is somatic, not from the germ-line. Tissue-culture-produced variability has been found in many species, including sugar cane, tobacco, oil palms, rice, barley, oats, alfalfa, maize, wheat and soya bean. It can arise in the nuclear genome or in the mitochondrial and chloroplast genomes.

Somaclonal variation can be useful when it produces superior plants, perhaps with a higher yield of secondary metabolites or improved disease resistance, but often it results in poorer plants than the parental line. Any favourable mutations should not be accompanied by the loss of good genetic backgrounds, as there is no sexual reassortment.

Somaclonal variation also gives **heterogeneity** in what might be expected to be uniform plants. In geraniums, plants derived from stem callus were fairly stable genetically, but ones from root and leaf callus showed much more variation - perhaps there has been natural selection for more stability in those parts which normally produce the germ-line tissue for sexual reproduction than in those which do not. While some somaclonal variation comes from changes in genes or chromosomes during tissue culture, it can also come from existing variation between cells in the material used in the explant. Thus if some cells in the parental plant are diploid and some are polyploid, plantlets derived from regenerated protoplasts will include some diploids and some polyploids. Many roots cells become polyploid during normal development in diploids.

Meristem-derived cultures which have had no real dedifferentiation tend to produce normal plantlets, but regeneration from callus, after dedifferentiation, gives more somaclonal variants. Different cultivars produce different proportions of variants. The amount of variation obtained tends to rise with the age or number of cell divisions of the culture, as gene mutations and chromosomal changes accumulate and have longer for expression. The hormones used in tissue culture can affect the rate of variation production. In tissue culture, there is no need for cells to photosynthesise to survive and the production of albino plantlets is quite common, as in rye-grass.

The **causes of somaclonal variation** include base changes in DNA, changes in the numbers of copies of genes within a chromosome, changes in methylation patterns of DNA, activation of transposable elements, mitotic crossing-over between homologous chromosomes, sister-chromatid exchange at non-homologous points, failure of cell cleavage after DNA replication giving autopolyploidy, aneuploidy from non-disjunction of chromosomes at mitosis, and chromosome breakage is frequent (especially in late-replicating chromosomal regions), giving translocations, inversions, deletions and duplications (see Chapter 9 and Lee and Phillips, 1988). Increases or decreases in methylation were observed in 16% of regenerated maize plants, persisting through several generations of selfing.

Some **beneficial quantitative changes** from somaclonal variation have been early maturity in maize and sorghum, increased dry matter in potatoes, increased submergence-tolerance in rice, and improved yield in oats. Because most changes are unfavourable, a very large number of plants must be screened to find the few beneficial changes, as in screening sugar cane for disease resistance. **New commercial varieties** produced in this way include rice with improved chilling tolerance and potatoes with improved resistance to *Verticillium dahliae*, a serious fungal vascular wilt disease. One can also use tissue culture conditions to select for types of stress-resistance at the single-cell level, e.g., tolerance to metals, low temperatures, herbicides and pathogen toxins, although the properties of the whole plants regenerated may not parallel those of single cells. For example, see Bertin, Kinet and Bouharmont (1996).

Asexual reproduction is widely used commercially in many species, giving a series of clones ("ramets") from one individual (the "ortet"). It is essential in triploid bananas, as the fruits do not set seed. Vegetative propagation is correctly called "cloning", e.g., layering of carnations, but confusion can arise because the term is also used for *in vitro* DNA work.

In **apomixis**, seeds develop an embryo without fertilisation. A somatic cell in the ovule develops parthenogenically into an embryo without having undergone meiosis. These seeds are viable and have been used to reproduce various citrus species, giving identical clones from the same plant, which can have heterozygosity perpetuated as the reproduction is essentially vegetative although involving seeds. Apomixis occurs mainly in polyploid species of the Gramineae, Rosaceae (e.g., *Rubus*) and Asteraceae families, and in *Citrus*. In some plants, such as bluegrass, *Poa pratensis*, both apomictic and sexual seeds are produced.

Apomicts are excellent for maintaining existing good lines, but for breeding better lines one needs to find sexual plants, or use good pollen from an apomict on a female-fertile plant. Some hybrids which would be sterile in normal sexual reproduction are enabled by apomixis to set viable seed. Apomixis can occur in wheat, maize, sugar beet, apples, *Panicum* (millet) and *Poa*.

Propagation by cuttings is very widely employed, where the part taken must have a terminal or lateral bud to provide a growing point, and hormones are usually applied to stimulate new growth (see Hartmann et al., 1997). It is often used for ornamental shrubs but is more labour intensive than seed production. Stem cuttings of deciduous hardwood trees are taken from matured wood when it is dormant in winter. Cuttings from semi-hardwood *Rhododendron* and *Camellia* plants are taken in summer, and cuttings from herbaceous chrysanthemums and geraniums are taken in spring. With leaf cuttings, one uses the leaf blade or the leaf blade plus petiole, e.g., in tea (Plate 45) and African Violet (*Saintpaulia*). One can use root cuttings in some plants, including the conifer *Picea abies* (Norway spruce).

Grafting is used in many woody species including grape vines, roses and apples; the rootstock chosen is very important, determining things such as phylloxera-resistance in vines and how much the apple tree is dwarfed. In grape vines, using American rootstocks such as *Vitis labrusca* gives protection to the European scion, *Vitis vinifera*, against the phylloxera root pest insect. The **scion** (top part) may be a shoot or a bud, and it must be compatible with the **rootstock**. Occasionally one has to graft the scion onto an intermediate "**interstock**" and then that onto the rootstock if the scion and rootstock are not directly compatible. Grafting is very skilled work and there are many forms of graft, such as the "omega" grape vine graft, where the stock end is cut by machine into a rounded omega-shaped protrusion, and fits into a corresponding omega-shaped hole cut in the rootstock (Plate 46), with wax protection being used over the graft to minimise drying out and infection during fusion of parts and rooting.

It is sometimes advantageous to remove the scion's mature growth and replace it with a different scion on the existing rootstock if the most desirable variety of fruit has changed, or if pathogens have developed which can attack existing scions. In South Africa, at Klein Constantia, white Sauvignon Blanc vine scions have been field-grafted onto rootstocks previously used for the red Shiraz grape. In California, changing the wine grape variety by grafting new scions onto existing rootstocks in the vineyard has been done extensively, and even changing the rootstock (by planting a new rootstock alongside the existing one, then changing the scion over; see Plate 47) when some rootstock varieties developed phylloxera in the mid-1990's.

Wild relatives of cultivated plants are often hardier than cultivars and have sometimes been used as rootstocks to extend a crop's range to areas with poorer soils or climates. For example, *Poncirus trifoliata* (trifoliate orange) is used as a cold-resistant rootstock for several citrus fruits, as well as having been crossed with *Citrus sinensis* (sweet orange) to make the hardy "citrange".

Propagation by **bulbs, corms, tubers, rhizomes** and **stolons** is common in species having these reproductive or resting-stage structures, e.g., growing new plantlets from strawberry stolons or blackberry runners, after rooting. For example, many potato plants can be obtained from one seed potato, where each "eye" axillary bud can grow into a new plant. Bananas are propagated from cutting sections of rhizome, while bamboos can be propagated by laying whole aerial shoots ("culms") horizontally in trenches, when new plants grow from the nodes. In bulbs, the basal plate can be scooped out, with adventitious bulblets developing from the base of exposed bulb scales when put on culture media.

Micropropagation uses sterilised explants from apical meristems, stems, leaves, roots, seeds or pollen grains or single cells; the explants are grown in culture with nutrients and hormones. It is particularly used on plants of high individual value, such as orchids and asparagus. The most common form is meristem culture to produce shoots, which are then subcultured by cutting them into sections, each bearing an axillary bud. These buds come from the germ-line and are genetically stable. Adventitious buds, which come from outside the germ-line, are often less stable and are used less frequently. Shoots derived from callus cultures are even less stable due to somaclonal variation, although they may still be used in cases where the callus genome is known to be relatively stable. Valuable lines of oil palm have been successfully multiplied from callus. Callus cultures sometimes produce **somatic embryos,** which are morphologically similar to zygotic embryos in form and development, with separate root and shoot meristems.

Tissue-culture cloning of ideal date palm plants is done by Date Palm Developments of Baltonborough, Somerset, exporting about 130,000 plants a year to the Middle East. Plants are grown in Somerset greenhouses and are about 61 cm high when exported for transplanting into commercial plantations, e.g. in Kuwait.

They begin to fruit at six years of age. The most popular varieties are "barhee" for fresh dates and "khalas" for stoning and exporting. Decorative palms are also supplied. No genetic engineering is involved, just tissue culture and regeneration.

Major agricultural crops for which micropropagation is used include sugar cane and potatoes, and it is used for the conifer, *Pinus radiata*. Only the very tip of the meristem is normally totally free of viruses or mycoplasmas, so meristem tips taken for culture are usually less than 1 mm long. About eight viruses have been eliminated from potatoes by micropropagation. Axillary shoot cultivation is used for *Chrysanthemum, Dianthus, Anthurium, Fuchsia, Allium, Asparagus* and *Fragaria* (strawberry).

Anther culture of Brussels sprouts can give about three diploid plants per anther, on average, as well as sometimes being used for getting monoploid plants for F1 hybrid production (Chapters 10.3 and 12.1.5).

Chimeras and **variegated plants** need special consideration. Chimeras have different genotypes in different layers of the meristems and hence in the vegetative parts, and must be propagated vegetatively. For example, *Laburnocytisus adamii* is a graft chimera between *Laburnum anagyroides* and *Cytisus purpureus*, with a periclinal arrangement of tissues. There is a single outer layer of broom (*Cytisus*) cells, giving a brown-purple colour to the flowers, overlying a core of *Laburnum* cells which give a laburnum-shaped inflorescence. This garden curiosity will not breed true from seed and must be maintained by cuttings. Reversion to one or other parental type occasionally occurs spontaneously by vegetative segregation, such as a pure laburnum shoot with yellow flowers.

Variegated plants have a mixture of green and white or yellow areas in their leaves, with occasional **vegetative segregation** (Plate 48) into pure green shoots or pure paler shoots which would die on their own from a lack of photosynthesis. Variegation normally results from meristem cells having two types of chloroplast genome, one wild-type, giving green chlorophyll pigments, and one mutant, giving no colour. Usually, daughter cells in variegated plants get both types of chloroplast after cell division, giving green cells, but occasionally a daughter cell may get only green chloroplasts, when its daughters will no longer give variegated areas, or only colourless chloroplasts, when its daughters will give a white or yellow leaf area. Variegated plants are therefore reproduced by cuttings from variegated areas. If reproduced by seed, variegation usually shows maternal inheritance, with most progeny variegated but with some all green progeny and some all white, which die. The proplastids thus go through the ovule but not usually through the pollen (with some exceptions), with occasional segregation into pure types, by chance at cell division.

17.1.6. Plant protoplast fusion

One can use **protoplast fusion** and plant regeneration to overcome sexual incompatibility and to form new allopolyploids. Enzyme mixtures are used to break down the middle lamella and cell walls, liberating protoplasts into a medium with a suitable osmotic pressure and nutrients. Polyethylene glycol or other agents are used to promote protoplast fusion, and selective systems, such as complementation between different albino mutants or non-allelic nitrate reductase mutants, can be used to select hybrid cells. In tobacco, *Nicotiana langsdorfii* (2n = 18) leaf mesophyll protoplasts were fused with ones from *N. glauca* (2n = 24), with a selective medium for growth of fused cells. The hybrids grew and developed stems and leaves, and were grafted onto one parent. Fertile flowers gave seeds which germinated to give intact allopolyploid plants, but these two species can cross sexually anyway. Fusion of protoplasts from leaf mesophyll cells has been used to get a "pomato" from potato and tomato plants. Protoplast fusion can also be used to transfer cytoplasmic male sterility to male fertile varieties. For more information, see Pelletier (1993).

17.1.7. Gene expression and natural and artificial selection at the haploid stage

In animals, sperm express some sperm-specific genes, but neither sperm nor eggs have much in the way of haploid development, nor does the plant egg cell. In contrast, pollen needs to germinate, penetrate the stigma, grow down the style, locate the ovule and penetrate its micropyle, and then fertilise the egg nucleus, to form the embryo, and fertilise the polar body fusion nucleus to form the endosperm. In maize, the pollen tube from a grain 90 µm diameter may have to grow down a silk 20 cm long. The male haploid generation in plants therefore needs extensive metabolic and growth capabilities, using its own stored nutrients and those from the style. It has been estimated that about 10% of genes are pollen-specific, 10–40% are sporophyte specific, and 50–80% are expressed in both gametophyte and sporophyte.

There can be hundreds of pollen grains deposited on one stigma, with intense **competition between pollen grains** for growth and fertilisation, at a haploid stage with no dominance. As well as **sexual selection** by incompatibility systems, there will be **natural selection** for fitness both on pollen-specific genes and on ones expressed in both gametophyte and sporophyte, including genes for growth, wall formation, mitosis and general metabolism. Any recombinants from male meiosis which prove better or worse than the parental type can be selected for or against at the haploid pollen stage, with far less resources committed to them than at the seed stage.

The breeder can try **artificial selection** at the haploid stage with *in vitro* pollen germination. This has been used to select maize and sugar beet for herbicide tolerance, and *Silene* (campion) for copper and zinc tolerance. Although the method sometimes works, with the surviving pollen tubes being used to fertilise stigmas, genes selected in the gametophyte stage often do not have reliably similar effects in the seedlings produced. See Ottaviano and Sari-Gorla (1993).

17.2. Animals

17.2.1. Sex ratios

The **sex ratio** is of commercial importance, and is measured as the number of males born per 100 females born. It can change within a species by selection, season, maternal age and the frequency of ejaculation. It tends to be high in dogs, 110 – 124, but is usually less than 100 in horses, sheep and chickens. In humans, cattle, sheep and pigs, more males than females abort before birth, with more male conceptions than female ones. For humans, the approximate figures are 130 males conceived per 100 females, but only 105 males per 100 females at birth. The sex ratio is very important economically in dairy cattle and in chickens for egg-laying, where most males are unwanted.

Much effort has therefore been put into experiments to separate X-bearing (female-producing) and Y-bearing (male-producing) sperm, or at least in getting **selective enrichment** for one type. Mechanical methods such as centrifugation, and chemicals and electrophoresis have all been tried, with mice, rabbits, sheep, cattle, pigs and humans. There have been many claims of success, including a commercial system for humans, but most have been unrepeatable by others.

In **chickens for egg-laying**, one can use a sex-linked dominant gene for barred feathers to identify males at hatching (very young chicks are difficult to sex from their anatomy), and kill off the males at hatching to reduce rearing expenses. One could cross females from a pure-breeding barred line (Z^BW - females are heterogametic in birds, with ZW rather than XY sex chromosomes) to males from a pure-breeding non-barred line (Z^bZ^b), when all males will be barred (Z^BZ^b) and all females will be non-barred (Z^bW). In mammals, one can sex embryos even when only 1 mm long, from the presence or absence of Barr bodies (Chapter 13.5), but it is uneconomic to do embryonic or foetal sex-determination and abortion of the unwanted sex in farm animals.

17.2.2. Anatomy, progeny per pregnancy, and temperature effects

The **anatomy of the sex organs** and accessory glands in males and females of different farm animals differs astonishingly widely - see Nalbandov (1976). A major reproductive difference (with anatomical and physiological consequences) between mammalian females is whether the species is **monotocous**, usually bearing one or two offspring per pregnancy, as in women, ewes, mares and deer, or **polytocous**, litter-bearing, as in sows, cats, rabbits, bitches and mice. Within a species, the number of offspring born per pregnancy affects their average and total weights, and survival. In sheep, for example, average individual weights at birth are 4 to 5 kg for a single lamb, 3 to 4 kg for twins, 2 to 3 kg for triplets, and 1.5 to 2.5 kg for quadruplets (information from the Seven Sisters Sheep Centre, East Dean, Sussex, UK, in 1999).

Fertility shows low heritabilities in all farm animals, but in sheep a major gene (or a series of closely linked genes) for **increased lambing rate** was found in Booroola Merino sheep in 1980 (Chapter 1.3.4). In breeds of low to moderate prolificacy, one copy of this allele with additive action, Fec^B, increases the number of ova by 1.0 to 1.5, on average, giving an additional 0.75 to 1.0 lambs, on average, per pregnancy. Two copies of the gene double that effect. This allele has been bred into a number of other breeds world-wide, using recurrent backcrossing (Chapter 12.4) to restore other breed characteristics. An increase of two lambs per pregnancy, from homozygous $Fec^B Fec^B$, is excessive in many breeds, as the ewe usually cannot feed more than three lambs. Major genes for increased lambing have been found in other breeds, with additive effects, but have usually been lethal when homozygous.

In seasonally breeding males, the **descent of the testicles** from the body cavity into the scrotum is very important for fertility, reducing the **testicle temperature** by one to four °. If the descent does not occur before the breeding season, infertility usually results. Rams can be made temporarily sterile, when not needed for breeding and when it is inconvenient to keep them separated from the ewes, by tying their testicles up against the body wall, to raise their temperature. Rams may suffer reduced fertility in very hot weather, when shearing their bodies and scrotums to reduce internal temperatures may help. Fertility-restoration can be very slow, because spermatogenesis takes 10 days in mouse, 48 days in bulls, and 50 days in rams. Fever in man can cause temporary sterility, and infertile men are often recommended to wear loose-fitting underwear and to spray their testicles with cold water.

17.2.3. Breeding seasons and oestrous cycles

In man and most domestic animals, both sexes can breed (mate) throughout the year, though there may be seasonal peaks of fecundity. Thus in humans in the Northern Hemisphere, conceptions occur most frequently in May and June, with peaks of births in February and March. In North America, horses mate mainly in spring and early summer, though conceptions can occur at any season. With cattle, the farmer controls conception times to take advantage of factors such as pasture conditions and milk prices. See Nalbandov (1976), Hammond et al. (1983) and the various books by Gordon (1996 and 1997) for accounts of reproduction in particular farm animals.

Some animals naturally have restricted **breeding seasons**, usually controlled by day-length, or sometimes by food availability. Sheep and deer are short-day breeders, mating in autumn or early winter as the days shorten, while ferrets, poultry and horses are stimulated to breed by long days. Artificial lighting in poultry sheds at night can be used to simulate long days. In sheep there has been selection for breeding at times of year giving the **best winter survival** of the lambs. Breeds from northerly latitudes, such as Iceland and Scotland, have a short breeding season: lambs born too early die from cold, and those born too late do not grow enough to survive the following winter. Breeds from nearer the Equator, e.g., Merinos in Spain, have longer breeding seasons, with winter survival being less difficult, and one Asiatic breed can breed all the year round. Suffolk sheep are highly seasonal breeders in England, with an anoestrous phase each year started by photoperiod, while the Romanov, Finnish Landrace and D'Man breeds are almost completely non-seasonal in their breeding. The international meat market needs lambs all the year round.

Seasonally breeding females go through an anoestrous period, with no oestrous cycles, when they are sexually inactive or less active. In continuous breeders, there is a succession of **oestrous cycles** (from the Latin, *oestrus*, a gadfly) throughout the year. Except in higher primates, females only allow copulation during definite periods of the oestrous cycle called "**heats**", when they are physiologically and psychologically ready for mating. Males usually show little interest in females which are not on heat and are repelled if they do. "Heat" occurs at the time of greatest development of the oestrogen-producing ovarian follicles, and this female sex hormone (the steroid 17–β oestradiol) brings about the physical and mental changes needed for optimum mating. Sometimes mares and cows have "quiet heats", with ovulation and the physiological changes, but are unwilling to mate, when they can be given extra oestrogen to overcome this. Heats can be induced by external oestrogen, sometimes with a little progesterone.

During heat, there are changes in vaginal mucus viscosity and pH, changes in body temperature and increases in general physical activity: when in heat, sows and cows walk about five times as much as when not in heat. These changes can be used to detect heats, and so the farmer can introduce males at the best time for producing offspring.

The length of the oestrous cycle and of "heat", and the timing of ovulation, differ greatly between species, as shown in Table 17.2.3.1 (see also Nalbandov, 1976). Thus the cycle takes 21 days in the cow and the sow, and 16 days in the ewe. These cycles are slightly shorter in heifers and gilts (i.e., young cows and young sows), and in ewes, they depend somewhat on the sex drive of the ewe or ram. In the mare, the cycle is rather variable, taking 19 to 23 days, with 4 to 7 days of heat, and ovulation varying from 1 day before to one day after heat. This vague timing of ovulation in mares is one reason why horse breeders usually use hormone treatments to induce heat and ovulation, to allow introduction of the stallion at the best time for fertilisation. In rabbits and ferrets, there is no set length to the oestrous cycle and no obvious "heat". Instead, **ovulation is induced** by mating, taking 10.5 hours after mating in rabbits and 30 hours in ferrets, so presumably the sperm stay viable for at least that period after mating. In elephants, the female is only in heat for four days every four years, and then makes low frequency (15-35 Hz) calls. Pregnancy lasts two years, followed by two years of nursing the offspring.

Table 17.2.3.1. The timing of oestrous cycles, heat and ovulation in different animals. Data from Nalbandov (1976), Table 4-4.

Species	Oestrous cycle length, days	Length of heat or sexual receptivity	Time of ovulation
Cow	21	13-17 hours	12-15 h after end of heat
Ferret	-	-	30 h after mating (induced), but continuous heat from March to August if unmated
Goat	19	Around 39 hours	9-19 h after start of heat
Horse	19-23	4-7 days	1 day before to 1 day after heat
Man	28	Continuous?	Days 12-15 of cycle
Pig	21	2-3 days	Usually 30-40 h after start of heat
Rabbit	-	-	10½ h after mating (induced)
Sheep	16	30-36 hours	18-26 h after start of heat

For introducing the male at the best time, it is very important to know when a female is on heat. The changes in body temperature, vaginal mucus and motor activity can all be used, or seeing if the female will mate with a vasectomised male. In cats and rats, there is a "Lordosis" reflex from a female on heat: when touched on the back she stands still, and when touched in the pudendal region, she adopts a mating position, especially if she smells a male of the species. In pigs, a sow in heat stands rigid if someone sits on her back or presses his or her hands on the pig's back. An experienced stockman does not need chemical tests to tell whether a cow "is bulling", but hormone tests can be used.

Control of oestrous in farm animals such as dairy cattle is often needed to get all of a herd reproducing at the same time, e.g., for a visit by an artificial inseminator or putting males with the females, or for the convenience of all births roughly coinciding, or to take maximum advantage of the best pasture or to boost milk production at a particular time, or to avoid exceeding milk quotas in a particular period. For a small dairy herd, say fewer than 50 cows, the farmer might want an unsynchronised herd, with calves produced throughout the year and a steady small income from sales of dairy bull calves or beef calves, while for a large herd, synchronisation could mean that extra farm labour was only needed at peak times, such as birth.

Oestrous cycles in cattle, sheep and pigs involve very similar fluctuation cycles in **reproductive hormones**, although the cycle length is shorter in ewes (see above). In pro-oestrous, a decline in progesterone levels from the previous cycle causes a rise in the reproductive hormone, follicle-stimulating hormone (FSH). This induces the growth of a follicle in the ovaries, and the follicle secretes increasing amounts of 17−β oestradiol (oestrogen), which acts on the hypothalamus in the brain, causing physiological changes such as increased blood flow to the genitals, and enlargement of the cervix due to swelling mucus cells, which secrete large amounts of mucus which aid the passage of sperm to an ovum.

During oestrous, the ovarian follicle releases an unfertilised egg into the fallopian tubes because of a shift from FSH to a surge in lutenising hormone (LH), which brings about metoestrous and the development of a corpus luteum from the broken ovisac cavity. The final stage, dioestrous, has a fully developed corpus luteum secreting progesterone which prepares the uterine wall for implantation of a fertilised egg. If there is no fertilisation, the corpus luteum remains functional (19 days in cows), then degenerates, with progesterone levels falling, which stimulates the next pro-oestrous. If fertilisation occurs, the corpus luteum continues to secrete progesterone until the developing placenta takes over this function.

One way to **synchronise oestrous cycles** is to end the existing cycle by the injection of a synthetic prostaglandin (PG) such as Estrumate® (cloprosterol) on arbitrary Day 0 to terminate the corpus luteum. In cattle, a second PG injection is given 12 days later, as animals in the early pro-oestrous cycle will not respond to the first PG, but they will to the second. With the termination of the corpus luteum, and decreased progesterone levels, the onset of oestrous occurs within 2 to 3 days and the females can be inseminated, usually by **Artificial Insemination** (AI). Another way to synchronise females is to lengthen the cycle and delay oestrous by a controlled intravaginal drug-releasing device (CIDR) or a subcutaneous ear implant, to release progesterone or a synthetic progesterone such as Norgestomet. This delays the ripening of a new follicle until the corpus luteum has degenerated. PG is then administered as before, to induce luteal breakdown and start pro-oestrous.

Hormone treatments can be used in sheep to increase the number of fertilised ova. Mare serum gonadotrophin is injected, combining FSH and LH activity. Follicular growth is induced, but ovulation does not occur until the corpus luteum has regressed. Prostaglandins can be used in pigs, too.

Non-chemical treatments have been used to increase the number of lambs per pregnancy in sheep, where twinning is very desirable economically, but the ewe cannot usually suckle more than two or three lambs adequately. Feeding the ewes intensively (called "flushing") before the onset of breeding increases the ovulation rate and is widely used. Twinning is unwanted in cattle.

17.2.4. Sperm; natural and artificial insemination

The volume of **ejaculate** and the number of sperm per ejaculation vary enormously between species, as shown in Table 17.2.4.1, which has implications for artificial insemination. An average bull produces enough sperm a week to inseminate 4,000 cows, because the approximately 11,400 million sperm in one ejaculate can be diluted into many doses, as only about 5 million motile sperm are needed per insemination. Thus while the 5 ml ejaculate from a bull is enough for 500 to several thousand cows, the much lower density of sperm in the 250 ml ejaculation from a boar is only enough for about 25 sows as a dose of 2×10^9 sperm is used.

Ejaculated sperm only survive typically for 20-30 hours in the female reproductive tract, although in a bull they may live for 60 days stored in the epididymis of his testicles. Although sperm are motile, with active tails, it is mainly contractions in the female reproductive tract, especially the uterus wall, which transport sperm from the vagina to the oviducts where fertilisation normally occurs. In rats, sheep and cows, it only takes a few minutes for sperm to reach the oviducts from the vagina, and even dead sperm and Indian ink particles travel as fast as live sperm, because of the wall contractions. Most sperm do not actually reach the oviducts. In rabbits, of the many million sperm introduced at ejaculation, only about 5,000 reach the oviducts and perhaps only 10 to 20 reach the ovum. Sperm need a maturation

Table 17.2.4.1. Average values for the volume of sperm per ejaculation and the concentration of sperm. Data adapted from Nalbandov (1976). The actual figures differ quite a lot in different studies.

Animal	Ejaculate volume, ml	Millions of sperm per ml
Cock	0.2–1.5	4,000
Turkey	0.2–0.8	7,000
Boar	150–500	100
Bull	2–10	1,000-1,800
Ram	0.7–2	3,000
Stallion	30–300	100
Rabbit	0.4–6	700
Dog	2–14	3,000
Man	2–6	100

process (capacitation) in the oviduct before being able to fertilise an egg, which has implications for *in vitro* fertilisation.

Artificial insemination (AI) in farm animals is the process by which sperm are deposited in the female reproductive tract by mechanical means. It was first used on a large scale in Russia in 1931, on 20,000 cows. It is widely used today in dairy cattle, sheep, turkeys, goats, pigs and horses. It can be used to minimise the dangers of having aggressive bulls on a farm, and to avoid damage to small females from mating with large males, and to help control **sexually transmitted diseases**, such as vibrosis in cattle which may be transmitted by symptomless bulls. Bulls for AI are tested before first use and annually thereafter for vibrosis, are given antibiotics even if they test negative, and sperm diluents contain antibiotics. Although using one male many times a year through AI reduces the effective population size (Chapter 4.3), and could cause inbreeding if used for more than one generation on a given herd, AI enables genetic merit in highly tested males to be spread rapidly.

There is a range of **artificial vaginas** available for different species, designed to provide the correct temperature (from a water jacket), pressure and lubrication (e.g., Vaseline or K-Y® jelly) to evoke normal ejaculation from the chosen male. Sperm are collected in a graduated glass tube at the end of the collecting tube. The artificial vagina may be held alongside a female in heat, or the male may mount a dummy containing it. A bull given a smell of a suitable cow will even mount a dummy made simply of a piece of coconut matting over a wooden trestle. Males unable to mount a dummy to serve into an artificial vagina, e.g., after injury or arthritis, or which are too small for such devices (e.g., hamsters) can be electrically stimulated to ejaculate, often by using a rectal probe, with rhythmic applications of current to achieve arousal and ejaculation. In turkeys and ganders, semen is obtained through manual massage of the ejaculatory ducts through the abdominal wall. In boars, semen may have a low sperm concentration when a dummy and artificial vagina are used, when manually deviating the penis into a flask as the boar mounts a sow gives better yields.

Bull sperm can be stored unfrozen for up to seven days, but chicken, boar and stallion sperm should be used within two days unless frozen. **Sperm for freezing** are diluted with extenders containing egg yolk or milk, buffers such as sodium citrate, antibiotics such as penicillin and streptomycin, glycerol as a protectant in freezing, and sometimes vegetable dyes to differentiate between breeds. The treated semen is cooled to just above freezing, then is packed into individual doses, usually in straws, and is frozen in liquid nitrogen (-196°) or dry ice and alcohol (-79°). Bull semen remains viable frozen for over 15 years. Rapid freezing and thawing give best viability. The transport of frozen semen by air, e.g., from the UK to New Zealand and Australia, has allowed the introduction of Charolais genes when quarantine rules prohibit import of live cattle.

For use, the straws, labelled with sire details, can be taken out, frozen, to the site of use, thawed, brought to body temperature, and put in an AI "gun". In cows, a dose of one ml is injected through the cervix, guiding the tip of the gun's tube by a hand inserted into the rectum and feeling downwards.

Females need inseminating at the **optimum time in the oestrous cycle**. In cows, ovulation occurs about 14 hours after the end of heat. Conception rates from insemination of cows at different times in the cycle have been found to be about 44% at the start of heat, 83% in the middle of heat (the optimum time), 75% at the end of heat, declining to 32% 12 hours after heat, 12% 24 hours after heat, and 0% 48 hours after heat.

In cattle, AI is widely practised in dairy herds, but less so for beef cattle, where "normal service" is usual. With semen dilution, one bull can supply 40,000 to 50,000 doses a year, so one would need fewer than 100 mature bulls for the whole of the United Kingdom. In sheep, about 1 ml is collected per ejaculate, and up to 11 collections per day can be made over short periods, with each ejaculate enough for 25 to 40 ewes. In pigs, the semen comes in three fractions, a clear pre-sperm portion, then a larger sperm-rich portion, then a gelatinous fraction amounting to half the total volume. After natural matings, the sperm is concentrated by absorption in the sow of accessory fluids. Conception rates increase from about 60% to 80% if a boar is present when the sow has AI, as the sight, sound and especially the scent of a boar helps to condition the female for pregnancy. The scent is now available in aerosol cans, and tapes can be played of boar mating cries. In turkeys, natural mating gives only 50–60% fertile eggs, but AI can give 85% fertility. Over 90% of turkey hens have AI in Britain. It has been used in chickens, quail, pheasants, ducks and geese, but with less success. In horses, AI is mainly used in pedigree stock, especially Thoroughbreds and Arabs; it saves transporting mares long distances to the top sires.

Natural insemination in farm animals is easy to control. To prevent mating, one just keeps the sexes apart. To get mating, one puts the sexes in the same enclosure, or puts them together when the female is on heat.

17.2.5. Egg transplantation, embryo freezing and animal cloning

As females of high genetic merit can naturally only have a few offspring in their lifetime, especially in monotocous animals, people have tried to find ways of increasing their reproductive potential. Many procedures for the culture and manipulation of ruminant embryos result in unusually large offspring at birth, and a high perinatal death rate.

Fertilised ova from an excellent cow can be **transplanted** into genetically inferior cows at the same stage of the oestrous cycle as the donor, i.e., post-ovulation, as the uterine lining must be ready for implantation. Unfortunately, it is difficult to extract the fertilised eggs from the donor because two to three days after fertilisation the zona pellucida has been shed and the egg is easily damaged. A flexible two-channel plastic catheter is passed through the cervix to the tip of each uterine horn, and an air cuff is inflated around it to seal off the tip area. Fluid is passed up through one channel and collected through the other. Embryos settle out in a dish and can them be implanted in the new recipients. The method works but is expensive, so it is only used on a small scale, for the top cows. In 1983, one "super-cow" had 35 calves in a year by egg transplants. See Chapter 12.7.5 for modern Ayrshire cow figures. Frozen embryos can be transported very long distances, so founder herds have occasionally been taken to other countries by embryo transport, transplanting the embryos into local cattle.

Sheep were the first domestic animals used for surgical recovery of embryos, from superior donor ewes which had been induced hormonally to super-ovulate. This is used on a small scale commercially for exporting embryos but it is expensive and involves major surgery, with a mid-ventral incision. Six to ten embryos can be recovered, giving four to six lambs as the conception rate is about 60%. The operation can cause complications in the ewe from internal adhesions. Laparoscopy ("keyhole surgery") is less traumatic for the sheep, but fewer embryos are recovered. The non-surgical trans-cervical method used in cows does not work in sheep as the cervical passage is convoluted and the sheep rectum is too small to allow an arm up it for guiding tubes through the cervix.

Embryo freezing for storage or transport works well in farm animals. In sheep, the embryo is put in 10% glycerol at 20° for a few minutes before cooling to three or four degrees below the freezing point of the freezing fluid. Rapid ice-crystal formation occurs when forceps at liquid nitrogen temperature are placed against the vessel's side. Cooling is then carried out at 0.3° per minute to -60°, with transfer to -196° for long-term storage.

In 1996, sheep were **cloned** by nuclear transfer, using a micromanipulator. Sheep embryo cells were tissue-cultured for 6 to 13 passages, then were fused to enucleated oocytes, using electric charges for activation and fusion. In 1997, a sheep named Dolly was cloned from an adult sheep udder cell nucleus, using similar technology, so the udder cell nucleus must have dedifferentiated and acted like a fertilised egg nucleus. Such procedures are interesting but very expensive, and are unlikely to become widespread in farm life of the near future, where costs are crucial.

17.3. Humans

For basic **human reproductive physiology**, see Sherwood (1995). Human reproductive advances are mainly in the field of treating infertility, where about one couple in seven has difficulty in achieving pregnancy. *In vitro* **fertilisation**, introduced in 1978, works fairly well, but collecting unfertilised eggs from the would-be mother or from an egg-donor involves hormone treatments and can be uncomfortable. By 1997, about 150,000 babies world-wide had been born by this method. Several fertilised eggs are usually placed in the recipient because most are lost through spontaneous abortion, but high survival rates can cause multiple births, even of eight babies in one pregnancy. Selective abortion is sometimes used to prevent excessive multiple births. The first baby resulting from a donated egg, fertilised *in vitro* and transplanted into a different woman's womb, was born in 1984, in Australia.

Where a couple carry a known serious genetic defect for which there is an appropriate DNA probe, it is possible to do *in vitro* fertilisation and use DNA analysis (from a single cell taken from each embryo after a very few cell divisions - removing one cell does little harm) by PCR, to find which embryos carry the defect, so that they are not implanted. This is **pre-implantation embryo selection**.

Most reproductive technology in humans has a fairly low success rate and is expensive, so natural reproduction is preferred if couples are fertile. In 1999 in

Britain, a single *in vitro* fertilisation treatment cost £2,000, with a success rate of only 15%. The National Health Service funded, on average, 11 single courses per 100,000 people in 1998. In Britain in the year to March 1998, 26,685 women received *in vitro* fertility treatment, resulting in 6,864 pregnancies, including 176 triplet births, 1,441 twin births and 4,136 singleton births.

Intracytoplasmic sperm injection was pioneered in Belgium in 1988. It enables men to reproduce even if they have no functioning sperm at all. Immature sperm are taken by needle directly from the testicles and are injected *in vitro* directly into the cytoplasm of an egg. It has improved the chance of treating male infertility from 15% to nearly 90%. About 2,000 children were conceived by this method in Britain between 1993 and 1997. If genes for an inability to produce sperm are passed to the offspring, sons will be unable to reproduce naturally.

As noted in Chapter 13.8, the frequency of meiotic non-disjunction in human females increases with age; it also increases with maternal hypothyroidism and sometimes after viral infections and irradiation. The mutation rate increases with paternal age for several autosomal dominant traits, including Apert syndrome, progressive myosotis ossificans, Marfan syndrome and achondroplastic dwarfism (see Connor and Ferguson-Smith, 1997).

Suggested reading

Abeles, F. B., P. W. Morgan and M. E. Saltveit, *Ethylene in Plant Biology*, 2nd ed., 1992. Academic Press, London.

Bertin, P., J. M. Kinet and J. Bouharmont, Heritable chilling tolerance improvement in rice through somaclonal variation and cell-line selection. *Australian Journal of Botany* (1996), 44, 91–105.

Chopra, V. L. et al., eds., *Applied Plant Biotechnology*. 1999. Science Publishers, Inc. Enfield, New Hampshire.

Connor, J. M. and M. A. Ferguson-Smith, *Essential Medical Genetics*, 5th ed. 1997. Blackwell Scientific, Oxford.

Gordon, I., *Controlled Reproduction in Cattle and Buffaloes*. 1996. CAB International, Wallingford.

Gordon, I., *Controlled Reproduction in Sheep and Goats*. 1996. CAB International, Wallingford.

Gordon, I., *Controlled Reproduction in Pigs*. 1996. CAB International, Wallingford.

Gordon, I., *Controlled Reproduction in Horses, Deer and Camelids*. 1997. CAB International, Wallingford.

Hammond, J. , J. C. Bowman and T. J. Robinson, *Hammond's Farm Animals*, 5th ed. 1983. Edward Arnold, London.

Hartmann, H. T. et al., *Plant Propagation: Principles and Practices*, 6th ed. 1997. Prentice-Hall International (UK) Ltd., London.

Ignacimuthu, S., *Plant Biotechnology*. 1997. Science Publishers, Inc., Enfield, New Hampshire.

Islam, A. S., ed., *Plant Tissue Culture*. 1996. Science Publishers, Inc. Enfield, New Hampshire.

Lee M. and R. L. Phillips, The chromosomal basis of somaclonal variation. *Annual Review of Plant Physiology and Plant Molecular Biology* (1988), 39, 413–437.

Nalbandov, A. V., *Reproductive Physiology of Mammals and Birds*, 3rd ed., 1976. W. H. Freeman and Co., San Francisco.

North, C., *Plant Breeding and Genetics in Horticulture*. 1979. Macmillan, London.

Ottaviano, E. and M. Sari-Gorla, Gametophytic and sporophyic selection, pp 332–352, in: Haywood, M. D., N. O. Bosemark and I. Romagosa, eds., *Plant Breeding. Principles and Prospects*. 1993. Chapman & Hall, London.

Pelletier, G., Somatic hybridisation, pp 93–106, in: Haywood, M. D., N. O. Bosemark and I. Romagosa, eds., *Plant Breeding. Principles and Prospects*. 1993. Chapman & Hall, London.

Prakash and R. L. M. Pierik, eds., *Plant Biotechnology. Commercial Prospects and Problems*. 1993. Science Publishers, Inc. Enfield, New Hampshire.

Prasad, B. N., ed., *Biotechnology and Biodiversity in Agriculture/Forestry*. 1999. Science Publishers, Inc., Enfield, New Hampshire.

Richards, A. J., *Plant Breeding Systems*, 2nd ed. 1997. Chapman and Hall, London.

Sherwood, L., *Fundamentals of Physiology. A Human Perspective*, 2nd ed. 1995. West Publishing Company, St. Paul/Minneapolis.

CHAPTER 18

APPLIED FUNGAL GENETICS

18.1. General fungal genetics: life cycles; wild-types and mutants; spore types; control of sexual and vegetative fusions

18.1.1. Life cycles

The **life cycles** of fungi, including industrial fungi, are extremely varied. Some fungi have no sexual cycle, a condition which has probably arisen independently in a number of formerly sexual species or groups. Fungi with no known sexual cycle include *Aspergillus niger, Penicillium chrysogenum*, and some polyploid brewing yeasts. Most fungi have cell fusion between haploid cells (which may or may not be specialised gametic cells), followed sooner or later by nuclear fusion to give a diploid nucleus. There may or may not be diploid mitosis before meiosis restores the haploid condition, completing the sexual cycle which can recombine syntenic and non-syntenic loci.

Many fungi are **predominantly haploid**, including the mycelial Ascomycetes such as *Aspergillus, Penicillium, Neurospora* and *Sordaria*. In these, mitosis is mainly or exclusively in the haploid vegetative stage, although there is often mitosis in a **dikaryotic** (n + n) stage within a reproductive structure after fusion of haploids in sexual reproduction, with meiosis occurring very soon after the fusion of haploid nuclei to give a diploid nucleus from the dikaryon. There may sometimes be a dormant diploid structure, especially the zygospore in Zygomycetes such as *Mucor*, where meiosis is delayed until long after the formation of the diploid nucleus.

In the Basidiomycetes, there is a **haploid-dikaryotic** cycle in the Hymenomycetes and Gasteromycetes, which are usually macrofungi with large fruiting bodies, such as the mushrooms. The haploid mycelium grows extensively, but if it meets a compatible haploid, fusion occurs to give a dikaryon, which can grow and can produce fruiting bodies. In these, nuclear fusion to give a diploid nucleus occurs in basidia, with meiosis rapidly occurring to give four nuclei which pass into the haploid and dispersive basidiospores. The parasitic Uredinales (rusts) also have a haploid and a dikaryotic mycelium, but usually on different hosts, e.g., barberry and wheat for wheat stem rust.

Some fungi can have prolonged **haploid and diploid stages**, both stages having vegetative reproduction after mitosis. Examples include the yeasts such as *Saccharomyces cerevisiae*, and the Chytridiomycete *Allomyces*. In both these types of organism, the haploid cells and diploid cells are very similar in appearance, but the diploid cells have roughly twice the volume of the haploid ones. In Myxomycetes, there is an amoeboid haploid phase and a plasmodial diploid phase.

In the Oomyces, such as *Saprolegnia* and the parasitic *Pythium* and *Phytophthora*, the mycelium is **diploid**. Meiosis occurs in the male and female sex organs, the antheridia and oogonia respectively, with gametic fusion restoring the diploid state.

Two examples of life cycles will be given, both from Ascomycetes, but with very different amounts of morphological complexity; both fungi have unordered asci. *Saccharomyces cerevisiae*, yeast, is of enormous importance in the wine, beer, spirits and baking industries, as well as being useful for food and for producing vitamins and other biochemicals, including industrial alcohol. The asci are formed singly, not in fruit bodies, and there are only single cells, not a mycelium. The diploid vegetative cell grows then undergoes mitosis, budding off a smaller diploid bud from the mother cell. Both cells then grow and can bud, continuing the diploid vegetative cycle indefinitely under suitable conditions. Adverse growing conditions such as starvation can induce meiosis, giving an ascus, a sack containing four haploid ascospores, two of mating type *a* and two of mating type α. The ascal sack eventually breaks down, releasing the four ascospores. They each give rise to a haploid vegetative cell which can reproduce indefinitely by budding after mitosis. There are no specialised gametic cells but a vegetative α cell can mate with a vegetative *a* cell, usually but not necessarily derived from a different ascus. Cell fusion is followed quickly by nuclear fusion, and a diploid bud gives rise to a new diploid vegetative cell. There is thus indefinite haploid and diploid mitosis. See Plate 3 for a yeast cross made for a *cis/trans* test.

Aspergillus nidulans is not of industrial importance but is closely related to industrial fungi such as *Aspergillus niger* (citric acid production), to the Aspergilli used in producing soya sauce and sake, and to *Penicillium chrysogenum* used in penicillin production. *A. nidulans* has both a sexual and a parasexual cycle (Chapter 18.4). The vegetative mycelium is haploid, septate and multinucleate, growing indefinitely by hyphal extension. It can fuse vegetatively with other mycelia, giving a heterokaryon. It reproduces asexually by producing aerial conidial heads, with chains of uninucleate haploid dispersive green conidia (Plate 49), which form new mycelia if they land on suitable substrates.

The mycelia can also reproduce sexually. Hyphal sexual initials give rise to closed spherical fruiting bodies called cleistothecia. Haploid nuclei pair up to give dikaryotic hyphae within the fruit bodies. In terminal croziers, nuclear fusion gives a diploid nucleus which quickly undergoes meiosis. The four haploid nuclei in the resulting ascus undergo two mitoses, producing eight sculptured binucleate haploid ascospores per ascus. There is no discharge mechanism for the ascospores, unlike in *Neurospora, Sordaria* and *Ascobolus*, and the ascospores are liberated by breakdown of the cleistothecium and of the ascus. An ascospore can germinate on a suitable medium to give a new mycelium. There are no mating types in this homothallic species, so any strain can be self-fertile and can usually cross, by vegetative fusion, with any other strain. A selfed cleistothecium from a homokaryon will produce identical ascospores, but a heterokaryon from vegetative fusion between two genetically different haploids can - depending on whether similar or different nuclei form the dikaryotic hyphae - produce selfed cleistothecia and crossed cleistothecia, where the latter will produce a range of genotypes in the ascospores.

18.1.2. Wild-types and mutants

The **wild-type** of a fungus is the typical form found in the wild. The term is also used for the wild-type allele at a locus, with other alleles being considered mutant. Wild-types of many, but by no means all, fungi are **prototrophic**, growing on a simple defined minimal medium, e.g., water, glucose, biotin and inorganic salts (including a nitrate) for *Neurospora crassa*. Other fungi may require organic nitrogen or other vitamins. The wild-type of a given fungus has a given morphology for its vegetative and sexual parts.

One can get **spontaneous or induced mutants** where each mutant has an altered morphology (growth rate, hyphal growth pattern, fruit body form, asexual spore forms, colony form, etc.), or altered pigmentation (fruit body, sexual or asexual spores, mycelium or colony), or is **auxotrophic**, requiring an additional nutrient to minimal medium, say an amino acid, vitamin or nucleic acid base, or is resistant to a drug, metabolic inhibitor or antibiotic. One can therefore obtain a range of mutations to use in basic genetic studies of the organism, for mapping or for use in breeding better strains of useful fungi (Chapter 18.6). Mutations can be nuclear or cytoplasmic; e.g., yeast with deletions of mitochondrial DNA can give *petites*, which grow slowly by fermentation but cannot carry out aerobic respiration, giving very small colonies (Plate 50), while *petites* can also be due to mutations at one of at least nine nuclear loci.

Mutagens (see Chapter 7) used with fungi include:
- UV light, especially on conidia or hyphae, giving all kinds of gene mutation (base substitutions, frame shifts, deletions) and chromosome aberrations;

- X-rays and gamma rays, on hyphae, sexual or asexual spores, giving all types of gene mutation and many chromosome aberrations;
- EMS (see Chapter 7.2), giving mainly base substitutions but some frame shifts;
- NMG (see Chapter 7.2), which is fairly specific for base substitutions;
- ICR-170, mainly giving frame shifts, but also some base substitutions;
- intercalating agents such as proflavin, which are fairly specific for frame shifts.

18.1.3. Spores

When working with any fungus, it is important to know what kinds of spore it produces, and what their properties are, including how to germinate them. **Sexual spores** include ascospores and basidiospores. They are usually haploid but may be uninucleate, binucleate or occasionally multinucleate, depending on the group or species. They may be tough, resistant, resting and survival bodies, as well as dispersive reproductive bodies.

Asexual spores are absent from yeasts, but there may be several different types as in stem rust of wheat, *Puccinia graminis tritici*, which has aecidiospores, uredospores and teleutospores, as well as pycniospores involved in mating, and sexual basidiospores. Conidia are very small and efficient dispersal agents in many Ascomycetes. They are haploid and uninucleate in *Aspergillus nidulans* (Plate 49) and *Penicillium chrysogenum*, but have three to eight haploid nuclei in *Neurospora crassa*.

18.1.4. The control of vegetative and sexual fusions

In some fungi there are no genes which normally control **vegetative fusions**, so all mycelia can usually fuse with all other mycelia, as in *Aspergillus nidulans*. If the two strains fusing are genetically different, then **heterokaryons** are formed containing two types of nuclei. The ratio of the two types of nuclei may change drastically or slightly during growth, from natural selection for the optimum nuclear ratio. In yeast, there are no vegetative fusions, only a sexual one between haploids of opposite mating type.

In *Neurospora crassa*, only opposite mating types *A* and *a* can fuse sexually, e.g., a conidium or hypha with the trichogyne of the protoperithecium of the opposite mating type, to give a large black perithecium containing about 100 asci, each with eight black haploid ascospores in a linear order. Opposite mating types in this species will not fuse vegetatively, say two hyphae, one *A* and one *a*. The only compatible vegetative fusions are of *A* with *A*, and *a* with *a*. There are also at least

10 non-mating-type loci controlling vegetative fusions, where the fusing hyphae must carry identical alleles, so that C,D will fuse with C,D but not with C,d, c,D or c,d. In *Ascobolus immersus*, however, hyphae will fuse whether they are of the same or opposite mating types, (+) with (+), or (+) with (-), or (-) with (-) (Lamb and Chan, 1996). See Glass and Kuldau (1992) and Carlile and Watkinson (1994).

The **control of sexual fusions** has different degrees of genetic complexity in different fungi, starting with self-fertile strains with no mating types in **homothallic** species such as *Aspergillus nidulans* and *Sordaria fimicola*. In **heterothallic species** (having more than one type of body with respect to sex), one can have two alleles at one locus as in yeast, *Neurospora crassa*, *Ascobolus immersus* and *Sordaria brevicollis*. *Coprinus comatus* has multiple alleles at one locus, a_1, a_2, a_3 ... a_{100}, where alleles must differ for fertility, and *Coprinus fimeterius* has multiple alleles at two loci, so that a_4, b_1 x a_2, b_1 is sterile because of one common allele, but a_4, b_8 x a_{23}, b_{98} is fertile. There are even more complex systems.

18.2. The commercial importance of fungi

Carlile and Watkinson (1994) give the following figures from the 1980's for the annual value in billions of US$ for fungal products, but modern values would be considerably higher: alcoholic beverages, 37; cheese (only some use fungi, especially blue-veined cheeses and some soft ones), 22; fungal antibiotics, 5 (penicillins, 3; cephalosporins, 2); mushrooms, 8; industrial alcohol from yeast, 4; fungal enzymes, 1; high fructose syrup, 2; yeast biomass, 1; citric acid, 0.5; steroids, 0.5.

Different strains of **yeast**, usually *Saccharomyces cerevisiae*, are used for baking, wine, beer, whisky, gin, industrial alcohol, and food yeast, and for the production of vitamins, amino acids and other biochemicals. Both *Saccharomyces cerevisiae* and *Torulopsis utilis* are used as food yeasts, being added to many foods to improve flavour and nutritional quality.

Penicillium chrysogenum is used for **penicillin** production and other species such as *P. camembertii* and *P. roquefortii* are used for soft cheeses (e.g., Camembert) and blue-veined cheeses (e.g., Roquefort). Antibiotics are produced in many fungi, including *Penicillium*, *Aspergillus*, *Trichoderma viride* and *Cephalosporium*.

Aspergillus niger is important in producing citric acid from molasses, starch hydrolysates and other cheap substrates. *A. terreus* produces itaconic acid, used for pharmaceuticals, synthetic fibres, resins and surfactants, from beet and cane molasses. *A. oryzae* is used on a large scale in Japan for making soy sauce and rice wine (sake). Aspergilli also produce amyloglucosidases (for converting starch to glucose), amylases (starch and dextrins to maltose etc.), pectinases, proteases, glucose oxidase, catalases, tannases (to break down tannins), lipases to break down fats, lactases, cellulases, etc.

Various Basidiomycetes are well known for **food**, e.g., *Agaricus bisporus*, the cultivated mushroom, *Lentinus edodes*, the shiitake, and some Ascomycetes, such as *Tuber*, the truffle. *Fusarium graminearum* mycelium (with thin-walled mutants to reduce wall material in relation to protein) is used to make fungal protein products, such as the meat-substitute Quorn®.

Fungi are used for **organic acids** such as citric, gallic, gluconic and itaconic acids, for amino acids such as lysine and threonine, a vast range of enzymes, vitamins such as riboflavin (many are from yeast hydrolysates), antibiotics, alkaloids, and many other biochemicals. Fungi can be used to transform metabolites in very specific ways, e.g., *Rhizopus nigricans* does one stereospecific step in steroid biosynthesis.

Fungi are also extremely important as **parasites** on plants, animals and humans, e.g., rusts and mildews on crop plants, as **saprophytes** rotting down plant remains and animal dung, and as **spoilage organisms** on stored foods and drinks, timber, leather, optical surfaces, fuels, etc. See the textbooks at the end of the chapter for more details.

Another use of fungi is for **biological control**, e.g., of roaches and termites (Gunner, 1999), and of parasitic fungi on plants, such as applying *Trichoderma harzianum* spores to seeds to control seed rots and damping off of seedlings caused by the fungus *Pythium*. Fungi are used to control plant nematodes either by fungal products, e.g., DiTera from Abbott Labs, or directly, as with nematode-trapping fungi such as *Dactylaria*. *Paecilomyces lilacinus* destroys the egg masses of the nematode *Meloidogyne arenaria*, increasing yield in oats. Fungi can also be used as **mycoherbicides**, with host-specific pathogens of weeds of crops. For example, sprays of spores of *Phytophthera palmivora* have been used to control milkweed vine, *Morrenia odorata* (see Carlile and Watkinson, 1994).

18.3. Recombination and sexual mapping

For reasons of space, not all the background theory will be given in this section (see Fincham, Day and Radford, 1979, for details). Chapter 8 dealt with general aspects of mapping and recombination, including the physical DNA mapping methods now widely used. For a summary of recombination mechanisms in fungi, see Lamb (1996). In fungi with different mating types, such as yeast, it is easy to make defined crosses for mapping and for getting desired recombinants, using sexual spores from such a cross. If the fungus is homothallic, with no mating types, as in

Aspergillus nidulans, one makes a heterokaryon between the strains to be crossed, or inoculates the two strains on opposite sides of a petri dish, but the fruit bodies formed could be selfed or crossed. Suppose in *A. nidulans* one crosses a multiply-marked strain 1, with white conidia, with a multiply-marked strain 2, with yellow conidia (Plate 49), with *w* and *y* being unlinked. One plates only some of the ascospores from specific cleistothecia onto complete medium. If a cleistothecium produces colonies only with white conidia, it must be from strain 1 selfed. A cleistothecium with ascospores giving only yellow conidia is from strain 2 selfed, while a cleistothecium giving some colonies with white conidia, some with yellow conidia, and some with recombinant green conidia, must be a crossed one, and recombination frequencies can be obtained for pairs of markers by scoring the colonies growing from its remaining ascospores when their colonies are tested on suitable media.

Sexual spores such as ascospores or basidiospores can be used as isolated, separate spores for mapping, using recombination frequencies for pairs of loci, or for obtaining desired recombinant types, e.g., for industrial or agricultural use. This is **random spore mapping**. One can do more sophisticated mapping when the spores can be isolated as intact groups from a single meiosis, such as the unordered tetrads of ascospores in yeast, or the linear octads of ascospores in *Neurospora crassa*.

With **ordered tetrads or octads**, one can find the distance in cM between a locus and its centromere from the formula,

$$\text{centromere distance} = \frac{\text{\% second division segregation}}{2}.$$

Ordered asci with **first division segregation** for the locus *m*, in a $m^+ \times m^-$ cross of two haploid strains, have the two different spore-types confined to half the ascus, e.g., $m^+ m^+ m^+ m^+$ in one half and $m^- m^- m^- m^-$ in the other half. Asci with **second division segregation** for that locus would have both different types of spore in each half ascus, e.g. $m^+ m^+ m^- m^- m^+ m^+ m^- m^-$ (see Lamb, 1996, for diagrams, photos and theory). If the mutation is for a character scorable in ascospores, such as for black versus mutant white pigmentation, the scoring of intact asci can be done directly down a microscope after displaying the asci from say a dissected perithecium. If the character is not scorable visually from ascospores, say being a requirement for some metabolite, then analysis is much more lengthy: ascospores have to be dissected out of each ascus in their original order, then have to be germinated, then the resulting colonies need biochemical tests, e.g., for growth on minimal or complete medium, or on minimal medium with various supplements.

Centromeres can also be mapped from unordered tetrads, as was done extensively in yeast using the **centromere marker technique**. If one has a cross of two haploids, *A, B* x *a, b*, then one can identify three **types of unordered tetrad**. A **parental ditype** (PD) has two types of spore, with parental arrangement of alleles, which here would be two spores *A, B* and two spores *a, b*. A **non-parental ditype** (NDP) has two types of spores, with a non-parental (recombinant) arrangement of alleles, two spores *A, b* and two spores *a, B*. A **tetratype tetrad** (T) has four different types of spore, two parental and two recombinant, one spore each of *A, B*; *a, b*; *A, b*; *a, B*. For non-syntenic loci, one expects half the asci to be PD and half to be NPD when both loci show first division segregation, and all the asci to be T when one locus shows first division segregation and the other locus shows second division segregation. When both loci show second division segregation, one expects a ratio of one PD to one NPD to two T (see Fincham, Day and Radford, 1979, for the theory). For linked loci, one expects PDs (from no crossover between the loci) to be more frequent than twice the number of Ts (from a single crossover) which in turn will be more frequent than NPDs (from a four-strand double crossover). A simple test is that loci are linked if PDs are significantly more frequent than NPDs.

The **centromere marker test** is based on the fact just described, that to get a tetratype, one or both loci must show second division segregation. One chooses a marker (mutant) which is very closely linked to its centromere, so that it hardly ever shows second division segregation. In yeast, *trp1* (linkage group IV) is such a centromere marker, so one crosses it to an unlinked marker whose centromere distance one wishes to determine, say *ade1* (linkage group I), getting a tetratype frequency of say 24%. As *trp1* rarely shows second division segregation, and tetratypes show that at least one locus has segregated at the second division, the tetratype frequency will represent the second division segregation frequency of the other marker. One can therefore deduce that *ade1* has a second division segregation frequency of 24%, which we halve, using the above formula, to get a centromere distance of 12 cM for *ade1*.

If a fungus does not have ordered asci, one needs a method to identify centromere markers for the above technique. This can be done using **three unlinked markers**, in a method put into equation form by Whitehouse (1957). Let the loci be *a, b* and *c*, respectively with second division fractions of *x, y* and *z*, and first division fractions of $(1-x)$, $(1-y)$ and $(1-z)$. For unlinked loci, tetratype frequencies depend solely on the second division segregation frequencies of the two loci.

The first way to get tetratypes is with first division segregation at one locus and second division segregation at the other, which for loci *a* and *b* has a frequency of $x(1-y) + y(1-x)$. The second way to get tetratypes is with second division at both loci, when half the tetrads are Ts, which will have a frequency of $1/2 \, xy$.

We therefore get a tetratype frequency for loci a and b of
$T_{ab} = x(1-y) + y(1-x) + 1/2\ xy$, which simplifies to $x + y - 3/2\ xy$.
With T_{ab} obtained as data from a cross, one cannot solve for the two unknowns,
x and y. However, we have parallel equations for the other tetratype frequencies,
determined from the crosses: $T_{ac} = x + z - 3/2\ xz$ and $T_{bc} = y + z - 3/2\ yz$.

With three values and three unknowns, one can solve for x, y and z by
simultaneous equations. Alternatively, one can solve the following equation,

$$x = \frac{2}{3}\left(1 \pm \sqrt{\frac{4-6Tab-6Tac+9Tab.Tac}{4-6Tbc}}\ \right)$$

and substitute the value of x obtained from it in the simpler equations above. In
practice it is easy to choose whether to use the positive or negative value of the
square root, as one gives a sensible answer and the other gives an impossible
answer, such as more than 100% second division segregation.

Because all spores in an NPD are recombinant and half the spores are recombinant
in a T ascus, we can calculate recombination frequencies from unordered tetrads
from RF% = $\dfrac{1/2\ T + NPD}{PD + T + NPD}$.

Considerations of the effects of single, double and multiple crossovers give a
maximum expected value of 50% recombination between two syntenic loci, however
far apart they are, if there is no interference between crossovers (Chapter 8.2).
Similarly, with no interference there is a maximum expected value of 66.67%
tetratypes for two loci, and of 66.67% second division segregation frequency for
one locus, however far it is from its centromere.

There are two major types of interference, chromosome and chromatid
interference. **Chromosome interference** (chiasma position interference) is concerned
with whether one crossover in an interval tends to inhibit (positive chromosome
interference), tends to promote (negative chromosome interference), or has no effect
on the chance of a further crossover in that interval. In some organisms, e.g.,
Drosophila, there is localised positive interference, with a crossover tending to
inhibit another one from occurring close by, but with no effect on crossovers further
away, while the fungus *Aspergillus nidulans* generally has no chromosome
interference. With no chromosome interference the chance of getting x crossovers
in an interval can be calculated from the Poisson distribution, $(m^x.e^{-m})/x!$, where m
is the average number of crossovers in the interval. The main departure from this
expectation is that almost all chromosomes have at least one crossover, which is
probably essential for regular chromosome segregation in meiosis. Short
chromosomes tend to have one crossover, with more in longer chromosomes, but
there are seldom more than two or three visible chiasmata per chromosome arm.

Foss et al. (1993) stated that chromosome interference extends over different physical distances in different organisms: 10^4 kb in *Drosophila*, and in fungi, 10^3 kb in *Neurospora* and 10^2 kb in yeast. For the different findings of negative, positive and no interference in different fungi, see Lamb (1996).

Chromatid interference (strand interference) is concerned with whether different chromatids are involved at random in double crossovers. Random involvement would give a ratio of one two-strand double to two three-strand doubles to one four-strand double. This ratio has been found, approximately, showing no or little chromatid interference in *Neurospora*, *Aspergillus* and yeast (see Lamb, 1996). Fungi are particularly suited for studying chromatid interference because it requires tetrad analysis.

Chapter 8.2 mentioned **mapping functions** for allowing for undetected crossovers in multiple crossovers, e.g., from the two crossovers cancelling each other out in a two-strand double crossover. The mapping functions needed for tetrad mapping are different from those used in random-spore mapping: see Barratt et al. (1954). They depend on the amounts of chromosome and chromatid interference.

One further use of tetrad analysis in fungi is that in favourable circumstances it enables one to tell whether two loci giving about 50% recombination are non-syntenic or are syntenic but far apart, which cannot be done just from the recombination frequency. For example, in *Neurospora crassa*, tetrad analysis of a cross of two mutants gave 7 PD, 11 NPD and 2 T. With the recombination, by chance, being a little over 50%, and with PDs approximately equal to NPDs, we deduce that the loci are unlinked and either non-syntenic or syntenic but far apart. The low tetratype frequency tells us that both loci usually segregate at the first division (because second division segregations often give tetratypes), and so both loci must be near their centromeres. They cannot both be near to the same centromere and simultaneously be far away from each other, so they must be close to different centromeres, and be non-syntenic.

18.4. The parasexual cycle and parasexual mapping

A **parasexual cycle** is a series of events which make possible the recombination of non-syntenic loci and/or of syntenic loci, independently of the sexual cycle and meiosis. A parasexual cycle can recombine non-syntenic loci during **haploidisation**, by chance chromosome losses as a diploid nucleus becomes haploid by a series of non-disjunctions, and it can recombine syntenic loci in a diploid nucleus by **mitotic crossovers** or, less often, by **mitotic gene conversion**.

Full parasexual cycles have been demonstrated in a range of fungi, including ones of industrial importance and parasites of major crops, in Ascomycetes, Basidiomycetes and Fungi Imperfecti. They include *Penicillium chrysogenum* (penicillin production), *Aspergillus niger* (citric acid), *Cephalosporium* (the antibiotic cephalosporin C), *Coprinus lagopus, Verticillium* and *Fusarium. Saccharomyces cerevisiae* (yeast) and *Ustilago maydis* (maize smut) have partial parasexual cycles, with mitotic crossovers but no haploidisation by non-disjunction. Parasexual cycles are particularly important in nature and for the breeder where the fungus has no sexual cycle, as in *A. niger* and *P. chrysogenum.*

Aspergillus nidulans has a full sexual cycle and a full parasexual cycle, so is excellent for comparing the two cycles. It is the organism in which Roper, Käfer, Pontecorvo and others worked out the parasexual cycle during the 1950's. This account includes practical details of how it is used. The wild-type has green conidia (y^+, w^+), is haploid and prototrophic, growing on a defined minimal medium. To map the markers involved and to get recombinants, one can take two strains with complementary auxotrophic markers and with a conidial colour marker each, for example, strain 1 with $a, b, c, d, e^+, f^+, y, w^+$, which is auxotrophic for a, b, c and d, and has yellow conidia, and strain 2, $a^+, b^+, c^+, d^+, e, f, y^+, w$, which is auxotrophic for e and f, and has white conidia (*white* is epistatic to *yellow*).

Conidia of the two strains are mixed in water over complete medium, so that conidia can germinate and hyphae can fuse, giving a strain 1 + strain 2 heterokaryon which is prototrophic and can outgrow both homokaryons as the nutrients get depleted. The mycelial mat which grows overnight is allowed to drain and is washed free of nutrients. Small bits of mat are placed well apart on minimal medium, so that only the heterokaryon can grow. It produces a mixture of white and yellow conidia, because the conidia are haploid and uninucleate, so each conidium is either of strain 1 or of strain 2 type. Within the growing 1 + 2 heterokaryon, one gets rare chance nuclear fusions, 1 + 1, 2 + 2, and 1 + 2. The first two types of diploid nuclei give only auxotrophs. The 1 + 2 diploid nuclei, however, give prototrophic green conidia. As the nuclear fusions are rare (being non-sexual), one plates millions of conidia from the heterokaryon on minimal medium, to select the very few 1 + 2 diploid nuclei in conidia.

These particular nuclei give green prototrophic colonies which are isolated as the diploids, and their conidial size is checked against that of the parental haploids: diploid conidia have about twice the volume of the haploid conidia. The diploids are then grown on complete medium, so that any segregants can grow, even if auxotrophic. Well-spaced inocula are used, and after about three days, colonies are checked visually for the presence of **sectors** with white or yellow conidia. The sectors will be large if they arose early in colony development, or very small if they arose late. Conidia from the sectors are isolated onto complete medium, with subculturing to get rid of any wild-type green contaminants.

The purified segregants are checked for **conidial size**, to find which are diploids and which are haploids. They (and the two parental strains and the diploid as controls) are then inoculated onto particular positions on a master plate of complete medium: see Plate 49. After the growth of new conidia, they are then replica-plated (using a pin replicator or a sterile velvet pad) onto a series of plates, e.g., minimal medium, complete medium, minimal supplemented with all requirements except that required by *a*, minimal supplemented with all requirements except that required by *b*, etc. That enables the genotypes of the yellow haploid segregants to be worked out, and that of the white haploids except for the *y* locus (because *white* is epistatic to *yellow*). In the diploid segregants, those loci which show a recessive phenotype have become homozygous - one cannot directly tell whether loci showing the dominant phenotype are still heterozygous or whether they have become homozygous dominant.

Suppose that maps for the two parental strains are as shown in Fig. 18.4.1, with *a, b, c, y* and *d* syntenic, and the other three loci non-syntenic with any other of these loci. The diploid between strains 1 and 2 is heterozygous for all these loci. The three causes of yellow or white segregants from that diploid are haploidisation, giving haploid segregants; mitotic crossovers, giving diploid segregants, and non-disjunction diploids, also giving diploid segregants. The commonest cause of colour segregants is mitotic crossovers, then haploidisation, with non-disjunction diploids the rarest.

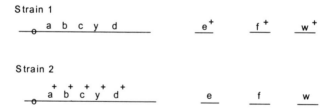

Figure 18.4.1. Possible maps for strains 1 and 2, with syntenic and non-syntenic loci.

In **haploidisation**, a diploid nucleus occasionally has chance mitotic non-disjunction, giving one 2X + 1 product and one 2X - 1 product. The latter monosomic nucleus is often less stable than the diploid nucleus and tends to lose more chromosomes at later mitoses, only reaching stability when it has the haploid chromosome number, with a complete genome, one of each type of chromosome. The diploid only shows the dominant markers in the phenotype, but in haploids any retained recessive markers show in the phenotype, e.g., *y* or *w*, which respectively give yellow or white conidia, in contrast to the green conidia of the diploid. Because whole chromosomes are retained or lost, syntenic loci stay in their parental arrangements, with 0% recombination, but non-syntenic loci recombine freely, with 50% recombination. One can see from Fig. 18.4.1 that if the chromosome carrying *y* is retained and the homologue carrying y^+ is lost, then the haploid yellow segregant will carry and express those markers on the *y* chromosome, *a, b, c, d*, but it is a matter of chance whether this yellow haploid carries *e* or e^+, *f* or f^+; because of the epistasis of white over yellow, it must carry w^+, not *w*, or it would be a white segregant, mutant at both conidial colour loci. Syntenic loci *y, a, b, c, d* therefore show 0% recombination with any of each other, but 50 % recombination with non-syntenic loci e/e^+ and f/f^+; we cannot map *y* relative to *w* because of the epistasis. Looking for such a very big difference as 0% versus 50% is very easy. Using haploids is therefore excellent for finding out whether loci are syntenic or not, but does not give any indication of distances between syntenic loci. About one in 200 haploids has had mitotic crossovers before haploidisation, so in practice there is very rare recombination for syntenic loci in the haploid products, rather than none.

For distances, we use the **diploids from mitotic crossing-over**, for loci which the haploids have shown to be syntenic, which in this case are *y, a, b, c, d*. Mitotic crossovers occur by chance during mitosis, after replication, when each chromosome is present as two chromatids, with no zygotene pairing or synaptonemal complex holding non-sister homologues together, unlike in meiosis. Double crossovers are so rare in mitosis that they can be ignored.

Consider one crossover at the four-stand stage of mitosis in the *y*-bearing chromosomes, between *a* and *b*, as in Fig. 18.4.2, with chromatids 1 and 3 going to one pole and chromatids 2 and 4 going to the other pole at anaphase. One daughter nucleus, the lower left-hand one in the figure, will express the alleles a^+, *b, c, y, d*, so it will be a yellow diploid not requiring the compound needed by *a*, but needing the compounds required by *b, c* and *d*. The other nucleus, the lower right-hand one in the figure, will give rise to a phenotypically normal colony which is not isolated as a colour segregant.

The important point to note is that, given the right centromere segregations, any loci **proximal** to the crossover (between the crossover and the centromere) stay heterozygous and express their dominant allele, like locus *a* in Fig. 18.4.2. Loci **distal** to the crossover (further away from the centromere than the crossover) become homozygous and the recessive alleles can show in the phenotype, like *b, c, y* and *d* in the figure.

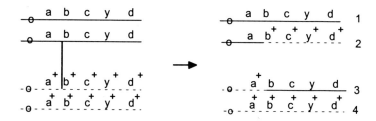

If chromatids 1 and 3 segregate to one pole and chromatids
2 and 4 go to the other pole, it then gives:

The alleles expressed in the phenotypes will be:
left set,   right set,
a^+ b c y d a^+ b^+ c^+ y^+ d^+

Figure 18.4.2. The effect of a mitotic crossover on loci proximal and distal to the crossover.

We could have crossovers in other places along the chromosome. One between
the centromere and *a* would make all the loci in that chromosome arm homozygous,
so *a* would be expressed as well as the other loci. A crossover between *c* and *y*
would make only *y* and *d* homozygous, so the colony would express a^+, b^+, c^+, *y*,
d. Suppose our yellow diploid segregants were phenotypically 20 *a, b, c, y, d*; 10
a^+, *b, c, y, d*; 30 a^+, b^+, *c, y, d; 15 a^+, b^+, c^+, y, d*, and no others for these syntenic
loci. The fact that *d* is always homozygous whenever *y* is homozygous recessive
shows that *d* is **distal** to *y* (or extremely closely linked and proximal to *y*). Alleles
a, b and *c* must be proximal to *y* because they can stay heterozygous, expressing
the dominant allele, while *y* and *d* are homozygous and expressed. The example in
Fig. 18.4.2 shows that *a* must be proximal to *b* and *c*, and the a^+, b^+, *c, y, d* class
shows that *b* is proximal to *c*. We can therefore deduce the order of the syntenic
loci within one arm of a chromosome: centromere, *a, b, c*, y and *d*. The frequencies
of the different diploid yellow segregants give the relative distances, since the
larger physical distances between loci are expected to have the most crossovers.
We then get the order and relative distances of: centromere - 20/65 - *a* - 10/65 -
b - 30/65 - *c* - 15/65 - *y*, and an unknown distance to *d*.

We only detect half the mitotic crossovers because the alternative segregation of
centromeres, as in Fig. 18.4.2, of chromatid 1 with 4, and chromatid 2 with 3,

results in daughter nuclei heterozygous at all loci, so they will not be detected as colour segregants.

In **non-disjunction diploids**, the diploid has a first non-disjunction, giving 2X + 1 and 2X - 1. Suppose the extra chromosome in the 2X + 1 is the one carrying the *y* allele, so the trisomic has y^+, *y*, *y*. It may later have a second non-disjunction in which the y^+ chromosome is lost, giving a stable diploid homozygous for *a*, *b*, *c*, *y* and *d*, and expressing all those alleles in the phenotype as a yellow sector. It will differ from mitotic crossover diploids in that it will be homozygous for all markers in **both arms of the chromosome**, not just those distal to a crossover in one arm.

The analysis has been explained for sectors for conidial colour mutants. One can also use some other types of markers, such as **acriflavine resistance** and sensitivity. On medium with the right concentration of acriflavin, the heterozygous diploid acr^R/acr^S grows slowly. By eye, one can identify fast-growing sectors which have become acr^R/acr^R by mitotic crossovers or acr^R by haploidisation. If one needs to increase the frequency of haploidisation, say to analyse a diploid fully, one can use 60 µg/ml of parafluorophenylalanine to increase non-disjunction.

Mitotic crossovers are much rarer than meiotic crossovers, with roughly 0.1 to 0.3% of crossovers per chromosome arm per mitosis, compared with 100 to 300% of crossovers per arm in meiosis. In *Aspergillus nidulans* we can compare the maps from meiotic and parasexual mapping. The order of loci is the same in both types of map, but the relative distances are sometimes different, showing different "hot spots" for recombination at meiosis and mitosis (see Pritchard, 1963, quoted in Rédei, 1982).

The **complete parasexual cycle** in *Aspergillus nidulans* can be summed up as a fusion of unlike haploid hyphae to give an n + n heterokaryon, which can reproduce by growth but not from conidia which only contain a haploid single nucleus; in the heterokaryon one gets rare nuclear fusion to give rare diploid nuclei, 2n, which may sometimes be fusions of the unlike nuclei; the diploid nuclei may get into uninucleate conidia to give diploid colonies, capable of indefinite asexual reproduction by conidia and vegetative growth. Rare mitotic crossovers within diploid nuclei give recombination of syntenic loci, and haploidisation by progressive non-disjunctions gives haploid nuclei, which can get into conidia, be dispersed, and start normal homokaryotic haploid colonies, reproducing by growth and conidia. Haploidisation can recombine non-syntenic loci.

One of the fascinating developments in the 1970's was the application by Pontecorvo and others of ideas and methods from fungal parasexual analysis to **human genetics**, especially the fusion of say a human skin fibroblast or a peripheral blood leukocyte with a tissue-culture adapted mouse cell to give a heterokaryon, then with nuclear fusion giving a synkaryon. Studies of gene and chromosome losses in proliferating hybrid cell populations greatly helped with mapping and with assigning linkage groups to visible chromosomes (Chapter 8.6.2). Cell fusion methods in humans also helped with diagnosis of genetic defects in hereditary diseases through locus assignment of mutations by *cis/trans* tests (Chapter 1.3.8) in tissue culture.

18.5. The induction and isolation of mutants, including auxotrophs

If a fungus has asexual spores such as conidia, they are the best material for mutation, although ascospores, basidiospores and hyphae can be used if necessary. If the conidium is haploid and uninucleate, as in *Penicillium* and *Aspergillus*, a typical procedure would be to **assess survival** at various dosages of mutagen, and to chose a dose giving 99% kill, 1% survival. That is a good compromise between getting enough mutants and not making all survivors too weak to be of use. If the conidia are multinucleate, a higher percentage kill is used, say 99.9% for *Neurospora crassa*. That fungus has three to eight haploid nuclei per conidium, so if one were isolating *his⁻* auxotrophs, one would want the mutagen to destroy all but one nucleus in a conidium, and to mutate that remaining nucleus, because a conidium with a mixture of mutant and wild-type nuclei has the wild-type phenotype, prototrophic, so could not be used for selecting *his⁻* mutants. One could grow the conidia on media giving low numbers of nuclei per conidium, or use microconidial mutants of *Neurospora*, which usually have only one nucleus per conidium.

Some types of mutant can be identified by **direct selection**. One can select conidial colour mutants by eye, subculturing any growth with mutant colour. Using the naked eye or the microscope, one can identify non-conidiating mutants, morphological mutants with altered colony form or colour, growth rate, growth pattern (e.g., growth in waves, or clock mutants with periodic growth patterns), branching habit, failure to produce fruit bodies, etc. One can select directly for resistance to drugs or metabolic inhibitors by plating spores on medium containing the appropriate chemical at a suitable concentration, e.g., acriflavin.

Although **selection for auxotrophs** is not often needed directly for industrial strains, it is very useful for getting mutants for mapping and general genetic studies of a fungus. Auxotrophs are invaluable for many selective techniques, such as using two non-allelic adenine-requiring (red) mutations in repulsion crosses in yeast, so that the prototrophic zygotes (white) can be selected on minimal medium, to separate them from the two auxotrophic parental strains, with a colour confirmation.

For selecting auxotrophs, one can use the **filtration enrichment** method with filamentous fungi and highly flocculent yeasts. The suitably mutagenised fungus, say conidia or yeast cells, is put into liquid minimal medium and allowed to grow say for four to six hours, with mechanical agitation to reduce aggregation and possible fusions. Prototrophs can grow hyphae, or yeast cells can form flocculating groups, but auxotrophs cannot grow in the minimal medium, so remain as conidia or separate yeast cells. The suspension is poured through a filter with an appropriate pore size, so that auxotrophs can pass through, but grown prototrophs are retained. About four to six periods of growth and filtration are used, so that nearly all protrophs are removed. The number of auxotrophs should remain nearly constant, but become a much higher proportion of the remaining organisms, hence the term "filtration enrichment". Finally, the filtered suspension is plated out on complete medium, and all colonies growing are isolated into individual tubes of complete medium. They are then tested on minimal and complete medium to check which are auxotrophs, and then their particular requirements are investigated.

If one wants **particular types of auxotroph**, say histidine-requirers, the method can be made selective. For *his⁻* mutants, one would replace the liquid minimal medium by liquid minimal medium supplemented with all the common vitamins, DNA bases and amino acids except histidine. All auxotrophs except *his⁻* ones should therefore be able to grow and would be filtered out, along with the prototrophs. The final plating out should be on minimal medium plus histidine, not complete medium.

Another method to get auxotrophs in fungi and bacteria is **killing enrichment**. It can be used with filamentous or single-cell fungi such as yeasts. For fungi, one uses an anti-fungal antibiotic such as nystatin, but for bacteria one uses an appropriate anti-bacterial antibiotic such as penicillin. The mutagenised organism is placed in liquid minimal medium plus the antibiotic. Suitable antibiotics kill growing cells much more quickly than non-growing cells, so prototrophs grow in the minimal medium and are rapidly killed, while auxotrophs do not grow, and they survive. After a period of growth, the suspension is centrifuged down; the organisms are washed free of the antibiotic and are plated at a number of dilutions on complete medium without antibiotic, for the isolation of individual survivors, which can then be tested for auxotrophy. This method can also be made specific for particular types of auxotrophs by using liquid minimal medium plus all except one nutrient.

A third method is **double-mutant survival**, using the fact that certain double auxotrophs survive longer in minimal medium than do single auxotrophs, which tend to starve to death more quickly. For example in *Aspergillus nidulans*, biotin-requiring conidia were mutagenised with UV and were spread on minimal medium, and covered with additional solid minimal medium. After 96 hours at 37°, 99% of *bi⁻* conidia had died. A layer of complete medium was poured over the top, with nutrients diffusing down to the conidia. Colonies which grew were isolated and tested, with up to 60% having a second auxotrophic mutation, e.g., being *bi⁻*, *arg⁻*. In *Neurospora crassa*, an inositol-requiring mutation was used, with minimal medium plus sorbose to slow down growth.

Any auxotrophs obtained by any method can be tested on minimal medium with a series of different supplements to find what compound is required. A quicker method is to use a **combinatorial approach** developed by Holliday in 1957 for 36 common requirements (see Clowes and Hayes, 1968). This involves having 12 pools of nutrients, each pool containing six different nutrients, with each nutrient occurring in only two different pools. Pool 7 might contain adenine, biotin, phenyl alanine, alanine, arginine and leucine, while pool 1 might contain adenine, uracil, and four other compounds; pool 8 might contain serine, glycine and four other compounds, and pool 12 might contain uracil, valine and four other compounds.

A mutant which grew only on medium supplemented with pool 7 and on medium supplemented with pool 1 could then be identified as an adenine-requirer, because adenine only occurs in pools 7 and 1. A mutant growing on only one pool could have a double requirement, so that a uracil/valine double mutant could grow on pool 12, which contains both nutrients, but could not grow on pool 1, containing uracil but not valine, nor on pool 6, containing valine but not uracil. Growth on more than two pools indicates an alternative requirement. For example, a mutant requiring either serine or glycine could grow on pools 3, 6 or 8.

The next section contains information on obtaining other types of mutant, especially overproducers of extracellular enzymes and antibiotics.

18.6. Obtaining improved strains for industry

18.6.1. Aims and methods

The aims are to increase the yield and quality of desirable products and the rate of production, to decrease the yield of unwanted products, to reduce the costs of nutrients, equipment, fuel, labour, transport and purification, and to increase the reliability of the yield. Depending on the fungus, the main techniques to use are selection, mutation (spontaneous or induced), recombination through the sexual or parasexual cycles, genetic engineering, or trying different ploidies, chromosome aberrations, heterokaryons to utilise complementation, hybrid vigour in diploids or polyploids, and even mixtures of genotypes or of species. For details of fungal genetic engineering, see Chapter 14.1.9, and for fungal biotechnology, see Anke (1996), Peberdy et al. (1991), Bennett and Lasure (1991) and Bos (1996).

Selection only works if there is heritable variation present, which can come from mutation, recombination, immigration, or genetic engineering. Selection procedures can be very simple, such as selecting wine yeasts with higher alcohol tolerance by setting the yeasts to grow at increasing alcohol concentrations. Selection for high yields of diffusible excreted products, such as amino acids, extracellular enzymes or antibiotics, can be done very easily by using the **overlay method**. Suppose one wants overproducers of tryptophan from a fungus. In a petri dish one puts fungal minimal medium, then the mutagenised fungus at low concentration, then an overlay of *try⁻* bacteria in a bacterial minimal medium. After a period of growth at temperatures suiting both organisms, one examines the size of the bacterial colony over each fungal colony. Fungal colonies which overproduce tryptophan will have larger bacterial colonies above them than other fungal colonies, and under-producers will have smaller bacterial colonies.

For **antibiotic production**, say of penicillin by a fungus, one would have fungal medium (not minimal), then over that the mutagenised fungus, then a penicillin-sensitive bacterium in bacterial medium. The size of the zone of killing of the bacterium would indicate the amount of penicillin excreted and diffusing through the bacterial medium.

For **extracellular enzymes**, one would have fungal medium, the mutagenised fungus, then a layer containing the enzyme's substrate. After growth, one would look for breakdown of the substrate, or for the presence of the breakdown products. This could be used for proteases, cellulases, amylases, lipases, pectinases, tannases, etc. For enzymes breaking down starch or cellulose, one could look for visible zones of clearing of the starch or cellulose powder, or one could stain for starch with dilute iodine in potassium iodide solution.

For bacterial antibiotics, one can use **indirect selection**, and the method could be applied in fungi. Where the antibiotic has amino acid precursors, one can select for overproduction of those amino acids, and this often results in higher antibiotic titres, e.g., penicillin has cysteine and valine precursors (see Fig. 18.6.3.1). One uses a chemical analogue of the amino acid to act as a competitive inhibitor of the normal amino acid in protein synthesis. By selecting for resistance to the analogue, one can often get overproducers of the normal amino acid, although some mutations might have other modes of action, such as permease changes affecting access of the analogue to the cells.

Types of mutation which one might get or want include the following. Mutations in **structural genes** coding for enzymes might give no enzyme or an enzyme with altered catalytic properties, giving faster or slower rates of action or altered substrate specificity or different end products, or altered regulation. Many biochemical pathways in synthetic metabolism have end-product inhibition of the first enzyme by the final product, as shown in Fig. 18.6.1.1 where the end-product might be an amino acid such as alanine. If a mutation made enzyme 1 no longer sensitive to allosteric binding and end-product inhibition by alanine, one could get overproduction of alanine because all enzymes in pathway could work even in the presence of a high concentration of the end-product.

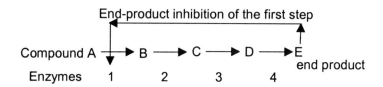

Figure 18.6.1.1. End-product inhibition of the first enzyme in an anabolic pathway.

One could get mutations in **regulator genes** controlling transcription, if there were bacterial-type repressor proteins and operator genes controlling transcription, changing inducible enzyme production to constitutive production. Stronger promoters or ones with altered regulation could be induced by mutation or introduced by genetic engineering (Chapter 14).

Mutations in **permease genes** could affect the rate at which substrates or products got into or out of cells, especially where active transport is involved.

Mutation can be nuclear or cytoplasmic, where nuclear genes show more reliable segregation. Mitochondrial DNA mutations can cause "petites", with slow growth (Plate 50) and an inability to respire aerobically. Such mutations occur spontaneously in beer yeasts and can give off-flavours. Mutations can also be in plasmids, which are used in fungal genetic engineering (Chapter 14) and in bacterial starter cultures for dairy products.

Mutations can result in metabolic gains in ability, or more often, in losses of ability. An example of a **desirable loss mutation** occurred in *Penicillium chrysogenum* strain Q176, when a UV-induced mutation gave strain BL3-D10, which lacks chrysogenin, the wild-type's unwanted yellow pigment which has to be purified away in penicillin production, so the mutation reduced purification costs.

18.6.2. Different aims in different yeasts

As an example of a very valuable industrial fungus, let us take **yeast**, for which we have already considered the life history. The most common yeast used is *Saccharomyces cerevisiae*, as a vegetative diploid, which can therefore show hybrid vigour. Most strains are sexually fertile and can form ascospores, but some cannot. One can use the sexual cycle for getting recombinants, and can select recessives using haploid cells. Yeast species sometimes hybridise naturally, e.g., *S. cerevisiae* and *S. bayanus* (Masneuf et al., 1998). The breeding aims are quite different in different industries.

For **baker's yeast**, one wants a rapid production of CO_2 to get dough to rise, a good flavour and aroma, alcohol tolerance only up to 5%, and an ability to work at 37° during leavening and at even higher temperatures during the early stages of baking. Bread yeasts can grow up to 42° and ferment up to 55°.

Wine yeasts must produce 8 to 15% alcohol, depending on the type of wine; they must work at pH 2.9 to 3.6, give a good flavour and bouquet, and must not autolyse rapidly as that can give off-flavours. They must resist the SO_2 used to control bacteria and wild yeasts, and should settle out well at the end of fermentation. Many wineries use carefully selected cultivated yeasts, and those which use "natural" yeasts are usually using ones which have been through the winery many times, with "natural oenological selection" for the right properties. One very widely used wine yeast is Prisse de Mousse, an Institute Pasteur *bayanus* race, ideal for crisp, dry wines, including sparkling and still white wines and fruit wines. It has low foaming, is excellent for barrel fermentation, has good flocculation and ferments well at low temperatures (13-15°). It can also be used for high-alcohol beers, such as barley wine.

Beer yeasts: for bitters and stouts, one uses a top-fermenting yeast, *S. cerevisiae*, but a bottom-fermenting yeast, *S. carlsbergensis*, is used for lagers. These two species confer quite different flavours. Beer yeasts are selected for flavour and aroma characteristics, flocculence and speed of fermentation. They work at higher pHs, 4.5 - 5.5 (but the pH is usually below 4.2 in the finished beer), than do wine yeasts and do not need such high alcohol tolerance or SO_2 tolerance. They must resist hop oils. Different beer yeasts are needed for different types of beer, and these descriptions modified from a WYEAST Laboratories (Mt. Hood, Oregon, USA) Brewer's Choice™ leaflet illustrate that. Note the ability of the lager yeast to ferment at the low temperatures traditionally used for producing and maturing lagers.

Ale yeasts: (i), German ale yeast. Ferments dry and crisp, leaving a complex but mild flavour. Produces an extremely rocky head and ferments well down to 13°. Flocculation - low. (ii), London ale yeast. Rich, minerally profile, bold and crisp, with some diacetyl production. Flocculation - medium. 16-22°. (iii), London ESB ale yeast. Highly flocculent top-fermenting strain with rich, malty character and balanced fruitiness. This strain is so flocculent that additional aeration and agitation are needed. An excellent strain for cask-conditioned ales. 18-22°. **Lager yeast**: Danish lager yeast II. Clean dry flavour profile often used in aggressively hopped pilsner. Clean, very mild flavour, slight sulphur production, dry finish. Flocculation low. 8-13°.

Beer yeast **flocculence** is controlled by three main loci, with alleles *Flo-1*, *Flo-2* (these are 8 cM apart) and *flo-3* (unlinked to the other two) increasing flocculence. Of these three alleles, one recessive and two dominant, only *Flo-1* is common in commercial strains. Even without knowing genetics, brewers have controlled flocculence for centuries by the time in fermentation when they skim off yeast (from the top, for top-fermenters, from the bottom for lager yeasts) to use in the next fermentation. The most flocculent yeasts come to the surface early and less flocculent ones rise later, for ale yeasts. Some breweries reuse yeasts indefinitely, but most keep lab stocks of highly selected yeasts which they grow up, test, use sequentially for four to six fermentations, then they sell off the surplus yeast (e.g., for food after de-bittering) because mutations tend to build up and give off-flavours. Adnam's Brewery in Suffolk, England, uses a mixture of four different yeast strains in their bitter, as they find that this gives better results than any single yeast.

For a **food yeast**, one wants maximum production from very cheap media, good protein and vitamin content, and no toxins or off-flavours. For details of the preparation of **microbial starters** for the food industry, including mixtures of bacteria for dairy-products, and the use of plasmids in bacteria, see Frazier and Westhoff (1988). For details of fermented beverages, see Lea and Piggott (1995).

If one has a highly selected yeast, sending it through the sexual cycle to get better recombinants often loses good gene combinations, so mutation or genetic engineering (e.g., using the 2 μm plasmid, Chapter 14) is often preferred.

18.6.3. Penicillin production

Another extremely valuable industrial fungus is *Penicillium chrysogenum*, for the production of penicillin, an antibiotic effective against gram positive bacteria. Sir Alexander Fleming's original strain (in 1928) produced 2 units/ml, compared with more than 6,000 units/ml for modern strains. The antibiotic was in use during the Second World War, after the work of Lord Florey, Chain, Abraham and others.

A **1944 survey** of 241 wild isolates showed penicillin yields varying from zero to 100+ units/ml in surface culture and from one to 80 in submerged culture. NRRL 1951 B25, a culture from a single haploid uninucleate conidium from a culture on a rotten cantaloup melon, gave yields of 100-200 units/ml in submerged culture and is the ancestor of all industrial strains, which therefore had no genetic variability. Yields were increased purely by mutations, including spontaneous ones, and ones induced by X-rays, UV, nitrogen mustard, NMG, diepoxybutane, and other agents. Interestingly, the best dose of UV for improving penicillin yield was one giving 25-30% survivors, rather than a higher dose. The mutation giving a loss of chrysogenin pigment was mentioned in Chapter 18.6.1.

Selection was made for adaptation to particular culture conditions, and for the production of different penicillins, such as I, F, G, K or X, which have different effectiveness against different bacteria, e.g., penicillin X is less effective against *Staphylococcus aureus* than against *Bacillus subtilis*.

The **genetics of penicillin production** were studied using the parasexual cycle, e.g., by Ball, Normansell and others in the 1970's. It was often difficult to obtain heterokaryons, diploids were often unstable, and some high-yielding strains had poor conidiation. Ball found from haploidisation that there were three linkage groups, while wild-type *Penicillium notatum* (*P. chrysogenum* is in this group) has n = 5. The many mutagenic treatments may have caused translocations. High yield was generally recessive to low yield in diploids. Many genes not directly connected with penicillin acted as modifiers of penicillin yield, e.g., various auxotrophs and conidial colour mutations.

Normansell and others studied the genetics by using *npe* mutations, giving **no penicillin** or very low levels. Complementation *cis/trans* tests, using protoplast fusion, showed that the 12 mutants were in five complementation groups, A to E, suggesting that there were five main loci for penicillin production. Work was also done on penicillin production in *Aspergillus nidulans*, because it was easier technically, and where yields were increased by recombination as well as by repeated mutation. 52 different wild-type strains had yields of 0 to 14 units per ml.

Fig. 18.6.3.1 shows a modern **pathway for penicillin synthesis**, with identical steps in *Aspergillus nidulans* and *Penicillium chrysogenum*, and with the first steps also in common to the production of penicillin N and cephalosporin C in *Cephalosporium acremonium*. The genes *pcbAB* and *pcbC* are common to both pathways; *penD* and *penE* are specific to penicillin G synthesis, and *cefD*, *cefEF* and *cefG* are specific to the cephalosporin C pathway. When one looks at the structure of the genes and transcripts in *P. chrysogenum*, *A. nidulans* and *C. acremonium*, there are great similarities, e.g., with *pcbAB* and *pcbC* being adjacent but transcribed in opposite directions, with *penDE* next to *pcbC* in the first two fungi (see Martin and Gutierrey, 1995).

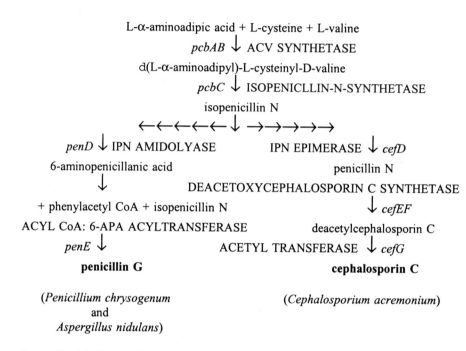

L-α-aminoadipic acid + L-cysteine + L-valine

pcbAB ↓ ACV SYNTHETASE

d(L-α-aminoadipyl)-L-cysteinyl-D-valine

pcbC ↓ ISOPENICLLIN-N-SYNTHETASE

isopenicillin N

←←←←←←↓→→→→→→

penD ↓ IPN AMIDOLYASE   IPN EPIMERASE ↓ *cefD*

6-aminopenicillanic acid   penicillin N

↓   DEACETOXYCEPHALOSPORIN C SYNTHETASE

+ phenylacetyl CoA + isopenicillin N   ↓ *cefEF*

ACYL CoA: 6-APA ACYLTRANSFERASE   deacetylcephalosporin C

penE ↓   ACETYL TRANSFERASE ↓ *cefG*

penicillin G   **cephalosporin C**

(*Penicillium chrysogenum*   (*Cephalosporium acremonium*)
and
Aspergillus nidulans)

Figure 18.6.3.1. Compounds, enzymes and genes in the penicillin and cephalosporin C pathways. Simplified from Chiang and Elander (1992).

The penicillin story is remarkable for the starting material with no genetic variation, and for the major improvements coming from repeated mutagenesis with a variety of mutagens. Improvements are possible from genetic engineering (see Chiang and Elander, 1992), but there are worries that physiological constraints may mean that yields cannot be improved dramatically any further.

Suggested reading

Anke, T., ed., *Fungal Biotechnology.* 1996. Chapman and Hall, London.

Ball, C., ed., *Genetics and Breeding of Industrial Microorganisms.* 1984. CRC Press, Boca Raton, Florida.

Barratt, R. W. et al., Map construction in *Neurospora crassa. Advances in Genetics* (1954), 6, 1–93.

Bennett, J. W. and L. L. Lasure, eds., *More Genetic Manipulation in Fungi.* 1991. Academic Press, New York.

Bos, C. J. ed., *Fungal Genetics, Principles and Practice*. 1996. Marcel Dekker, New York.

Carlile, M. J. and S. C. Watkinson, *The Fungi*. 1994. Academic Press, London.

Chiang, S. D. and R. P. Elander, The application of genetic engineering to strain improvement in b-lactam-producing filamentous fungi, pp 197-211, in: Akora, D. K. et al., eds., *Handbook of Applied Mycology* Vol. 4, *Fungal Biotechnology*. 1992. M. Dekker, New York.

Clowes, R. C. and W. Hayes, eds., *Experiments in Microbial Genetics*. 1968. Blackwell Scientific, Oxford.

Elliot, C. G., *Reproduction in the Fungi: Genetical and Physiological Aspects*. 1993. Edward Arnold, London.

Fincham, J. R. S., P. R. Day and A. Radford, *Fungal Genetics*, 4th ed. 1979. Blackwell, Oxford.

Foss, E. et al., Chiasma interference as a function of gene distance. *Genetics* (1993), 133, 681–691.

Frazier, W. C. and D. C. Westhoff, *Food Microbiology*, 4th ed. 1988. McGraw-Hill, New York.

Glass, N. L. and G. A. Kuldau, Mating type: vegetative incompatibility in filamentous ascomycetes. *Ann. Rev. Phytopathology* (1992) 30, 201–224.

Gow, N. A. R. and G. M. Gadd, eds., *The Growing Fungus*. 1995. Chapman and Hall, London.

Gunner, H., Mycopesticides: killing roaches and termites the natural way. *Annual Review of Microbiology* (1999), 53, in press.

Kirsop, B. E. and A. Doyle, eds., *Maintenance of Microorganisms and Cultured Cells*, 2nd ed. 1991. Academic Press, New York.

Lamb, B. C., Ascomycete genetics: the part played by ascus segregation phenomena in our understanding of the mechanisms of recombination. *Mycological Research* (1996), 100, 1025–1059.

Lamb, B. C. and W. M. Chan, Heterokaryon formation in *Ascobolus immersus* is not affected by mating type. *Fungal Genetics Newsletter* (1996), 43, 33–34.

Lea, A. G. H. and J. R. Piggott, *Fermented Beverage Production*. 1995. Blackie Academic and Professional, London.

Manners, J. G., *Principles of Plant Pathology*, 2nd ed. 1993. Cambridge University Press, Cambridge.

Martin, J. F. and S. Gutierrey, Genes for b-lactam antibiotic biosynthesis. *Antonie van Leeuwenhoek International Journal of General and Molecular Microbiology* (1995), 67, 181–200.

Masneuf, I. et al., New hybrids between *Saccharomyces* sensu stricto yeast species found among wine and cider production strains. *Applied Environmental Microbiology* (1998), 64, 3887–3892.

Peberdy, J. F. et al., eds., *Applied Molecular Genetics of Fungi*. 1991. *18th Symposium of the British Mycological Society*. Cambridge University Press, Cambridge.

Rédei, G. P., *Genetics*. 1982. Macmillan, New York.

Webster, J., *Introduction to Fungi*. 1980. Cambridge University Press, Cambridge.

Whitehouse, H. L. K., Mapping chromosome centromeres from tetratype frequencies. *Journal of Genetics* (1957), 348–360.

CHAPTER 19

THE ECONOMICS OF AGRICULTURAL PRODUCTS AND BREEDING PROGRAMMES

19.1. Basic economics: economic systems; price theory; factors affecting supply and demand; perfect and imperfect competition; monopolies; inflation

19.1.1. Factors of production; types of economic system

The **production of most goods and services** requires labour for manpower; enterprise to organise and bear the risks of production; land, e.g., for crops, shops, factories or offices; capital equipment, such as machines, factories, vehicles and roads; and working capital in the financial sense. There is a limited willingness of people to provide labour, land and finance, but an almost unlimited potential demand for goods and services. An **economic system** determines who produces what and who consumes what. The two main systems are the command system and the money-price system, and most countries have a mixture of the two, though often in very different proportions.

The **command system** may be autocratic, where the rulers or planners do not consult the public as to their wishes, but the rulers dictate who will produce what and who can consume what. Alternatively, it can be consultative, where the planners do consult the public about their willingness to supply labour, land, etc., and what they want to consume. A complete consultative system for Britain or America for every commodity would be administratively impossible, and most people are keener to consume than to produce.

Command systems are difficult to operate, slow to respond to changes in demand, and may involve a large loss of individual freedom of choice in jobs and products. They characteristically have occasional large gluts of certain products, and, more frequently, severe shortages, often with a private "black market" supplying more choice than the state system, at higher prices.

The **money-price system** has individual firms producing what they think the consumers want; if the firms produce goods or services at a price people are prepared to pay, their products sell. The **preferences of the public** in their own buying, and in their willingness to hire out their own labour, land and savings, determine what is produced and at what prices. There is a circular functioning, as shown in Fig. 19.1.1.1. The households' expenditure becomes the firms' income; what the firms pay out in wages and rents become the households' income, etc. If people decide to spend more on vegetables and less on meat, then vegetable suppliers can expand, but meat producers would have to cut back. The sum of decisions of individual purchasers, and people's willingness to do different jobs or give others use of their land or savings, determine such things as what is produced and the relative wage rates for different jobs. The money-price system is highly democratic in that each purchase is a vote for the continued production of that item or service.

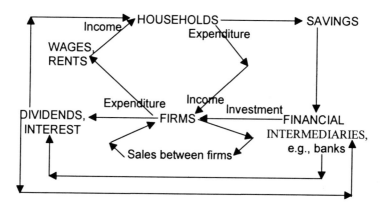

Figure 19.1.1.1. The circular functioning of a money-price economic system.

One factor missing from Fig. 19.1.1.1 is the **role of governments**, local and national. For administration, legal services, armed services, education, health, roads, etc., governments consume resources which could otherwise have gone to consumer spending. The allocation of resources between governments and citizens is sometimes called the "guns or butter" problem. Governments curtail the spending power of citizens by direct and indirect taxation, and alter the distribution of incomes by differential taxation, taxing higher incomes at proportionately higher rates. During and after wars, or at other times of shortages, governments may regulate consumption by **rationing**, issuing ration coupon books. Governments also affect consumption by **differential taxes** on different items, such as different rates of tax on fruit juices, beer, cider, wine and spirits. Governments also work through the command system, e.g., banning cyclamate sweeteners or the sale of beef on the bone. National and local governments together consume a large proportion of the gross national product in Britain. The 1998 estimates from the Office for National Statistics were that taxes, at £300 billion, were 37% of the Gross Domestic Product, rising to 53.5% if tax reliefs are included.

19.1.2. Price theory

A **price** is a value agreed upon between producers and consumers for the exchange of goods or services. The main **factors affecting demand** are: income; the prices of substitutes and complements; tastes, and the availability and convenience of use of products.

Income: for most goods, demand rises as incomes rise. A 19th century economist, Engels (not the Engels who co-wrote the Communist Manifesto) made an empirical finding that as incomes go up, proportionately less is spent on food. In Britain between 1955 and 1992, average real earnings (i.e., earnings in relation to the index of retail price) rose by 210%; expenditure on food fell from 35% to 15% of household spending, while expenditure on cars and transport went up from 7% to 16%. There is a different **elasticity of demand** for different products. "**Inferior goods**" are those on which spending goes down in absolute terms as incomes rise; this happens when cheap items are replaced by more expensive ones as people get richer, so potatoes and margarine have sometimes had reduced sales when people have been able to afford more varied vegetables or butter, as incomes have risen.

Availability and price of substitutes: demand for beef, for example, is affected by the price and availability of lamb, pork, chicken and fish, as people can substitute one product for another, depending on price, availability and quality. The demand for one type of product might be fairly consistent, say for green vegetables, but the demand for a particular type of green vegetable might be elastic, depending on price and quality. In the 1990's, health scares over beef from BSE (bovine spongiform encephalopathy) increased the demand for other meats as beef substitutes in the UK.

Availability and price of complements: the demand for cranberry jelly, for example, is influenced by the price of turkey and of game birds, as it is usually bought as a complement to those meats.

Tastes: people's tastes can change. Certain foods and drinks come into or go out of fashion, irrespective of price. For example, in 1999, a major UK supermarket chain asked its melon growers in Spain to produce small melons (up to 540 g) rather than large ones (around 965 g), as the small ones had started selling very much better than the large ones. When breeders first hybridised different species of *Hippeastrum* lily, they aimed for large showy flowers, but gardening tastes have changed, with smaller-flowered types coming back into fashion (Mathew, 1999). Tastes are influenced by advertising, by press comments, by consumer organisations, by trend-setters, and by people's upbringing.

Availability and convenience of use: people usually prefer to buy what is locally available, rather than to go far to obtain something. Convenience foods are very popular as they take less time to prepare, even though they cost more than starting with the basic ingredients. There is a trend for more seedless varieties of fruit, such as seedless oranges and grapes, while fruits such as Russett apples, with thick corky skins which usually need to be peeled off, are less popular than thinner skinned varieties which do not need to be peeled. "Easy-peel" citrus fruits are favoured.

A **demand curve** shows the relation between price and demand for a particular product, and the term is used even if the relation is linear. Fig. 19.1.2.1 shows two possible demand curves for one product. In curve 2, more is demanded at the same price than in curve 1; for example, at price OA, amount OC is demanded for curve 2, compared with only OB in curve 1. Demand curve 1 could change to curve 2 if incomes rise, or if the price of substitutes rises, or if the price of complements falls, or with favourable changes in tastes, or wider availability or increased convenience. For almost all types of goods or service, the demand curves slope upwards to the left (less demand at higher prices), but the degree of slope varies with the **price-elasticity of demand** for that product, as shown in Fig. 19.1.2.2. Some "**snob goods**" actually have increasing demand as their prices rise, as some people value exclusivity of possession, e.g., for very expensive cars.

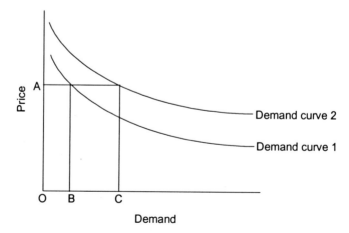

Figure 19.1.2.1. Different demand curves in relation to price.

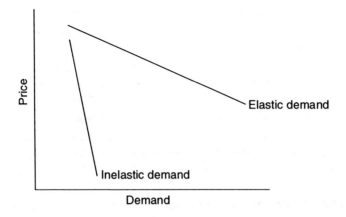

Figure 19.1.2.2. Differences in price-elasticity of demand.

With elastic demand, one gets large changes in demand with changing price, which is true of many manufactured goods such as cameras. With inelastic demand, one only gets relatively small changes in demand with changing price, which is true of most agricultural produce and other primary products such as metal ores.

The price-elasticity of demand =

the change (e.g., rise) in the quantity demanded .
the opposite change (e.g., fall) in price

For example, if demand goes down 20% when the price rises 10%, the price-elasticity of demand is 2.0. Elastic demand has values of 1.0 or over, and values less than that are inelastic demand.

One must distinguish between price and **total expenditure**, the relation between which depends on the shape of the demand curve. In Fig. 19.1.2.3, the total expenditure at the higher price OA is price x demand = OA x OB, which - depending on the exact shape of the curve - may be greater than, equal to, or less than the total expenditure at the lower price OD, which is OD x OE. In agriculture, one must also distinguish between expenditure on food and **farmers' personal income**, because there are many costs in the food chain. It has been estimated that a Danish farmer in the 1990's receives on average only 27% of the market price for his or her produce. Of that 27%, only 2% is for the farmer personally, on average, and 98% goes to employees, suppliers, government, transport, debt-servicing, and other items.

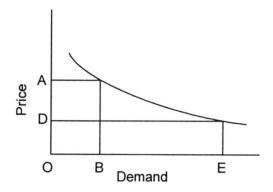

Figure 19.1.2.3. The relation between total expenditure and price.

One of the main problems with agriculture is inelastic demand: falling prices only expand demand a little, so falling prices greatly diminish the total expenditure on food. If there is a food surplus, cutting prices will not help much in getting rid of the surplus, but it will cut farmers' income greatly.

The determination of supply. Profit = revenue - costs. Most firms aim to maximise long-term profits, and this may well mean not taking the maximum profit per item, if a smaller profit per item results in proportionately larger sales and a greater total profit. Let P be the market price, then the supply curve for one firm might be roughly as shown in Fig. 19.1.2.4.

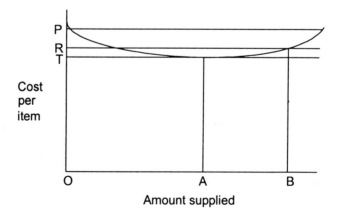

Figure 19.1.2.4. Curve for cost per item for different amounts supplied. See text for details of symbols.

As supply expands, between O and A in Fig. 19.1.2.4, costs per item usually go down at first, with economies of scale, but eventually costs per item rise again, as in the region to the right of A, due to limits of managerial efficiency in very large firms; inefficiencies which might be spotted in a small firm may go unnoticed in large ones. The biggest firms are not always the most efficient nor the most profitable. The firm's total profit, at the maximum profit per item, would be PT x OA, but by taking a smaller profit per item, PR at greater output OB, the total profit, PR x OB, may be greater than PT x OA.

In general, firms will want to supply more when market prices are high but firms differ in efficiency, so one gets different supply curves for different firms, as shown in Fig. 19.1.2.5. At market price OA, only firms 2 and 3 could operate.

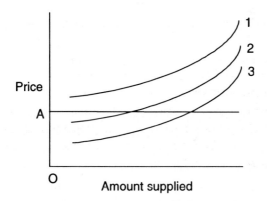

Figure 19.1.2.5. Different supply curves for firms of different efficiency.

Putting the demand and supply curves together, we get an equilibrium point where the demand and supply curves intersect, as shown in Fig. 19.1.2.6. Quantity OB would be exchanged, at price OA. The intersection point is actually an equilibrium point and might take time to reach. For example, firms might have to build or sell factories, hire or shed labour, to cope with increased or decreased demand.

Speculators are routinely damned as harmful by political parties of the left and of the right, but they are useful if their operations are not large in relation to the market. A successful speculator buys when prices are low and sells when there is a shortage, with high prices. Speculators therefore improve demand when it is slack, and increase supplies when they are short, driving down high prices. Speculators can therefore help to reduce peaks and troughs in demand and supply.

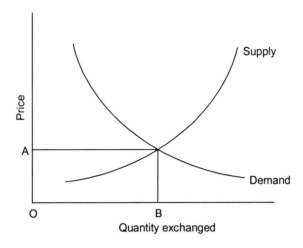

Figure 19.1.2.6. The determination of price and quantity exchanged by the intersection of supply and demand curves.

19.1.3. Types of competition

The commonest types of economic competition are called "perfect", "imperfect", oligopoly, and monopoly, in order of decreasing numbers of competing firms. With **"perfect competition"**, no firm is large in relation to the total market; no one acts as a price leader. There is free entry into the market and free exit from it, with no great costs of entering or leaving the market. Many firms make almost identical products, and the price is determined by market conditions of supply and demand, not by individual firms. "Perfection" is rarely found but is approximated by the Chicago wheat, maize and meat markets, where many farmers supply almost identical products. If a farmer raised his prices, no one would buy his produce, while if he lowered his prices, he could expand sales but would not get an adequate return.

With **"imperfect competition"**, the products of different firms are not identical. They may differ in quality, advertisement, packaging, service, availability, etc., say in the pleasantness of sales staff or in the location of sales outlets. For example, someone might go 50 metres for better bread, but not 50 km. With this difference between products, people are willing to pay different prices for them. Each firm can fix its own prices and try to maximise long-term profits, but usually they do not know what the demand curve is. Many firms charge a fixed "mark-up"; a trendy dress shop in London charged a 300% mark up, so that if the shop paid £200 for a dress, they sold it for £200 + (£200 x 3) = £800. A firm may charge different mark-ups for different items. For example in 1999, a computer firm charged £19 retail each for a keyboard and a mouse, with cost prices of £5-91 and £3-54 respectively.

With an **oligopoly**, there are few sellers, mostly large in relation to the market, as with petrol companies in the UK. The action of one firm affects all the others. For example, if Shell lowered its petrol prices, it could expand sales, but that would diminish sales by Texaco, BP and Esso. The other firms could regain sales by lowering their prices, but then all firms' profit margins would be lower, including Shell's. **Price wars** can harm all firms in a market, as happened to freight shipping rates in the Indian Ocean in the 1950's. Firms kept cutting their prices in order to retain custom, hoping that other firms would abandon the market as unprofitable, so that rates could rise again. Prices actually fell to covering only 10% of costs, so all firms made losses and many went bankrupt. Because of the dangers to all firms from a price war, there is often an informal but usually illegal "gentlemen's agreement" between the firms not to start price wars, to compete by advertising and services rather than by price, but even a big advertising war can be harmful by putting up costs, unless it expands the total market. In 1998, the European Commission fined four British sugar companies, including British Sugar and Tate & Lyle, £36 million for attempting to fix sugar prices in the late 1980's. The Commission criticised British Sugar for instigating the price-fixing cartel in 1986 at the end of a price war.

With a complete **monopoly**, one firm has all the market, with no competitors. In some countries, a monopoly may be defined legally as having some lesser share of the market, say 1/4 or 1/3 for a particular product. In a complete monopoly, there is no need for a firm to worry about the action of competitors, or of being undercut on price. The monopoly firm can usually make more profits than under competition, making excess monopoly profits. In Britain, supermarkets control about 70% of the fruit and vegetable market, but with no single chain having a monopoly position.

In Britain, the Fair Trading Act 1973 defines a monopoly as a situation where a company - or a linked group of companies - supplies or purchases 25% or more of all the goods or services of a particular type in the UK or in a defined part of it, so there can be national or local monopolies. There is no assumption that monopolies are wrong in themselves: the invention of a new device will make the inventor a monopolist, or a firm may gain a monopoly from its efficiency. The Fair Trade Act just defines situations where it is possible that excessive market power could exist and could be misused contrary to the public interest. The Monopolies and Mergers Commission (MMC), the Director General of Fair Trading and the Restrictive Practices Court all have responsibilities roughly corresponding to their names. Under the Competition Act 1980, an anti-competitive practice is defined as one which restricts, distorts or prevents competition in some market in the UK. Practices which are acceptable in a market where competition is strong may not be acceptable in a market where competition is weak. Companies are excluded from provisions of the act if they have a turnover of less than £10 million a year, or if they have less than 25% of a relevant UK market. In 1992 the MMC found that Bryant and May had 78% of the UK market in matches and disposable lighters, and were making monopoly profits, so price controls were imposed. In 1991, the MMC found that Nestlé had 56% of the UK soluble coffee market, and higher profits than other firms, but MMC attributed this to a good performance in the face of strong competition, and no action was taken. Rules on monopolies vary widely between countries.

There are several origins of monopolies:

• Legal. This is conferred by a government licence or legal charter, as for the Royal Mail having a monopoly in Britain for most types of letter delivery.

• Too high a cost of entering a market. For example, few firms could afford to lay out a competing London underground railway system, or a national railway system.

• One firm may "corner the market", e.g., buying up all sources of supply of a commodity, say nutmeg.

• Patents on new processes and inventions. This applies to plant breeders' rights on a new variety (Chapter 19.2.6), or to a new drug. After World War II, Mr Biro had a monopoly on the sale of his invention, of ball-point pens. Such monopolies are usually broken eventually by better new inventions.

Monopolies are often contrary to the consumers' interests, as there is no other source to buy from if they do not like that firm's prices or products. Monopolies can have some benefits, such as big economies of scale of production, and more efficient planning in the absence of uncertainty about competitors' plans. Big firms often do more research and development than small firms. Sometimes there has been no reduction in prices on breaking up a monopoly, but the break-up of many state monopolies (nationalised industries) in Britain under Mrs Thatcher's Conservative Government's privatisation programme from 1979 into the 1990's generally resulted in lower prices (e.g., for telephones and gas) and better services. For example, by 1999, household electricity bills had fallen by 26% since privatisation in 1990. Many governments abroad also instituted programmes of privatisation.

It is not only firms which try for monopolies. The **Trades Unions** try to get a monopoly control over the supply of labour in particular firms or industries, to try to drive wages up above the free-market price of labour.

19.1.4. Inflation

Inflation is the reduction with time in the purchasing power of a unit of money, which is the same thing as a general rise in prices. For example, the retail price index rose 24% in 1975 under Mr Wilson's Labour Government in Britain. A London suburban semi-detached house which cost £6,500 in 1967 was, through inflation, worth £320,000 in 1999. Inflation is most damaging to the standard of living of people on fixed incomes, or who live on savings, or who are in weak bargaining positions, such as university lecturers.

Demand inflation occurs when spending power is greater than the availability of goods, so that excess demand drives up prices. This is fairly rare except during and after wars when people have not had much to spend money on, as resources have gone into war-related products and overseas supplies may have been cut off by blockades. At the end of a war, consumers want to buy more than the suppliers can provide, until war factories have changed to producing consumer goods and imports have been resumed. Possible remedies include increased production, reduced government consumption, or for governments to increase taxes "to mop up surplus spending power". The British Government restricted consumption during and after World War II by **rationing** many items, so that one had to exchange ration-book coupons as well as money to buy meat, sugar, butter, margarine and cooking fats, tea, cheese, chocolate, sweets, coal, clothes, petrol, furniture and other items. For example, the weekly allowances per person in 1941 were: butter, 113g; sugar 340g; tea 57g; cooking fats and margarine, 57g; bacon and ham, 113g; other meats were rationed by price, not weight, to 1s 10d (9p) a week. Meat rationing only ended in 1954, long after the war finished in 1945.

Cost inflation is much more common and more serious than demand inflation. After rises in costs, firms raise their prices; higher prices lead to demands for higher wages, leading to cost increases if these demands are met. These increased costs lead to price rises, then to demands for increased wages, etc., in an inflationary spiral. This is usually associated with wages rising faster than productivity, often under Trades Unions' use of monopoly control over the labour supply. Inflation can continue until high unemployment weakens the Trades Unions' bargaining position. There are no easy, painless remedies for cost inflation, although there are many different actions that governments can take. They are too complicated and controversial to discuss here; they include control of the money supply, interest rates, and international exchange rates, and government-imposed price and/or wage controls.

19.2. Economics applied to agriculture

19.2.1. Gluts and shortages; how governments intervene in agriculture; European Union policies

Although plant and animal breeders breed to increase yields, if all farmers have a high yield in one season, it can lead to a **glut** and actually decrease farmers' incomes because of inelastic demand. With a glut, farmers try to reduce the amount of surplus produce coming onto the market, to stop it driving down prices. Farmers' organisations agree to plough crops under, or dump them at sea, or render them inedible by sprays, or leave them to rot. This arouses great journalistic fury and public resentment about waste, but it may be the only way for farmers to stay in business. In 1995, the European Union (EU) taxpayers paid for 720,000 tons of surplus peaches to be destroyed.

Consider the difference between a year of glut and a year of shortage, as shown in Fig. 19.2.1.1. Because of inelastic demand, the farmers' income is lower with supply curve 2, with a glut, than in a poor harvest year, supply curve 1, because reducing prices substantially only results in a slightly increased demand: the quantity demanded at low price OC, is OD, which is only slightly more than demand OB obtained at substantially higher price OA in the poor year.

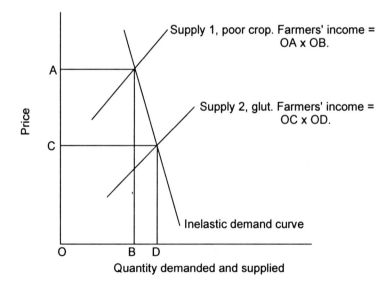

Figure 19.2.1.1. Farmers' income in a free market, in a glut year and in a shortage year.

It is not in anyone's interests to have the farmers ruined, so most **governments intervene** in farming and agricultural supply. As Devon farmer Richard Yeomans put it in a letter to the Freedom Association (*Freedom Today*, 1999, February/ March, p15): "Agriculture cannot be left to the market. The goods are perishable, production lines can take years to stop, and over-production of one per cent can drive down prices 10 per cent. So over-production has to be controlled in some way, because agriculture must be kept alert, modern, prosperous, vibrant and 'ready to go' ".

Agriculture is heavily supported by governments in all the major economies, including the USA, Japan and the EU. A study of seven major countries recently found that government assistance to agriculture exceeded one third of the value of the agricultural produce, with about half of the money coming from taxpayers (from subsidies) and about half from consumers (from higher-than-free-market prices).

A common government policy is to try to avoid overproduction by restricting supply, by **licensing**. A government agency issues licences for a particular crop, restricting the area planted, as with potatoes in the UK, or restricting yields by means of quotas as with milk yield in the EU, where there are fines for exceeding one's quota. Spotter planes are sent out in Britain to look for unlicensed commercial-scale potato growing. This method of restricting supply has the obvious and serious drawback that one cannot predict yields in advance. If farmers in 1999 are licensed to grow only a restricted area of potatoes in the following year, and that year turns out to be a bad year, then shortages will result. The EU's method of controlling fish stocks and sales, by making fishermen throw back into the sea those species for which they do not have a quota, is the worst possible conservation strategy, as those fish die. They amount to 40% of the catch.

The second type of government policy to control production is to adopt **financial measures**, such as subsidies or intervention buying. The system of **subsidies** (deficiency payments) was used in Britain before Britain adopted EU methods. It ensured a good supply of food at fairly low prices to the consumer, but with a high cost to the taxpayer. It is illustrated in Fig. 19.2.1.2.

OA is the price with no intervention, with supply OB, but the government (Ministry of Food) agreed price OC after discussions with farmers' representatives (the National Farmers' Union). With agreed price OC, the farmers supplied OD, but with supply OD, the market price obtained was only OE. The government (the taxpayer) then had to pay for the amount by which the actual market price fell short of the agreed price, (OC - OE = CE), with total payment CE x OD. The amount of subsidy varied widely but was often about one third of the shop price for many food products.

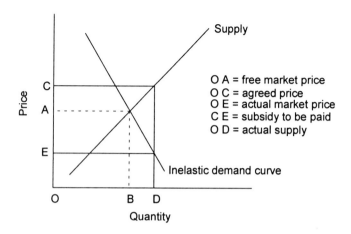

Figure 19.2.1.2. The former British system of subsidies.

The American system of **intervention buying** is different. There the farmers and the government agree on a market price, with government agencies buying up the surplus in order to maintain that agreed market price. This is shown in Fig. 19.2.1.3. With the agreed price OC, farmers supply OD, but at that price consumers only buy OF, so the difference, FD, has to be bought up by the intervention agency. This gives higher prices than free-market prices, a bill OC x FD for the taxpayer, and leads to **food mountains** from the intervention buying. In America, the produce

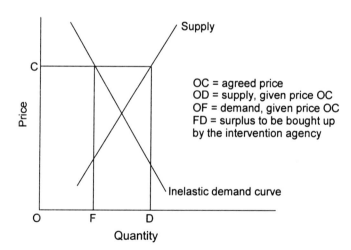

Figure 19.2.1.3. The traditional American intervention-buying system.

bought has been used for strategic stockpiles (e.g., for use in the event of war), for foreign aid programmes, and for selling abroad. In the EU, intervention buying has at various times led to so-called butter mountains, beef mountains, wine lakes, etc.

The EU method is similar to the USA system. It aims to protect EU farmers from overproduction within the EU and from competition from overseas countries. A **target price** is agreed for various foods but it is not guaranteed. If EU production is large enough to cause the market price to drop much below the target price, intervention agencies buy up enough to restore the market price to the target level. Imports from outside the EU which reach the EU at prices below the target price are subject to **import levies** to bring their price up to EU levels, so the EU consumer does not benefit from cheaper world prices.

In order to get rid of the mountains of intervention surpluses, the EU pays a **subsidy** to lower the price of exports to the world price, although the World Trade Organisation tries to limit the amounts involved. For example, the EU placed a massive subsidy on the sale of butter to Russia to help get rid of a butter mountain, so that Russians were able to buy EU butter at much lower prices than EU consumers, with a large cost to the EU citizens. The cost of the **EU's Common Agriculture Policy** is about half the EU total budget, with high bills to EU members and distorted competition with other countries because of the import levies and export subsidies. To avoid surpluses of major crops, the EU often subsidises minor crops, and even subsidises the growing of no crops on agricultural land, under the **set-aside scheme**: see Ansell and Tranter (1992). In 1998, the EU decreed that 10% of arable land in Britain, about 650,000 ha, must be "set-aside" to reduce grain production by 1.6 million tons. See Chapter 12.9, end, for a linseed flax example of the absurdities to which EU policies can lead. The EU suffers from over-regulation, even issuing a directive on the amount of permitted curvature in bananas, which is surely a matter for consumer purchasing choice, not bureaucrats.

The EU policies, like those of many areas, are a political compromise between producers and consumers, politicians and voters. The various EU countries agree that the present system is cumbersome, expensive and unfair, but they cannot agree on how it should be reformed, especially those countries making massive gains from the existing system. The EU Common Agricultural Policy costs an average of £550 a year per taxpayer in Britain, plus £300 extra per head on food prices (1998 figures), with about 48% of farmers' incomes coming from various subsidies. In recent years, New Zealand has greatly reduced government intervention in agriculture, reducing subsidies from 60% to about 4%, with promising results for farmers and consumers, and less overgrazing. 1998 figures (*The Daily Telegraph*, 16/10/98) for EU subsidies per hectare in England included £242 for cereals, £349 for proteins,

£427 for oilseeds, £568 for non-cannabis-producing mutant hemp for fibre, £305 for set-aside, £112 for suckler cows, and £105 bull premium.

Even with all these subsidies, there is a crisis in many branches of agriculture, especially livestock farming. The Meat and Livestock Commission found in 1998 that the UK shop price of lamb had risen slightly over a year to £4-58 a kilo, while the price paid to farmers had dropped from £2-42 to £1-69 a kilo. Of the shop price of £4-58 a kilo, the farmer typically gets £1-69, slaughter costs 22p, cutting and packing cost 33p, inspection and offal disposal cost 18p, transport costs 11p, and shops' overheads were £2-05.

19.2.2. Seasonal and perishable produce

In a free market, prices for the same commodity can show sharp **seasonal variations**, with higher prices at times of scarcity and the lowest prices during peak supplies. Many agricultural products are seasonal, with the lowest prices just when most farmers are selling their crops. There is therefore an economic advantage in breeding early-maturing varieties, with peak production when prices are high, as with early tomatoes and strawberries, and new potatoes.

One can also supply goods out of season by **cold-storage**, as for many British apples harvested in autumn for sale during the winter and spring. One grower in Kent, England, puts his apples into cold store at 2° for just 10 to 14 days, to avoid supplying the market just when the other growers are selling their crop. There are losses from the costs of moving goods into and out of store, from pests and diseases, hiring storage space, and from bruising from extra handling, which have to be offset against any gain in price obtained.

Many foods which used to be seasonal in the UK are now available year-round, from growing early varieties, from cold storage (e.g., frozen peas and beans), and from imports (e.g., of South African grapes, apples and citrus fruits, Plate 51; Spanish and American strawberries; Kenyan beans). The prices and qualities may still show strong seasonal variation, e.g., for strawberries, where UK-grown ones in summer have the best flavour. Many imported fruits are picked before optimum ripeness and flavour, to reduce losses during transport and handling, and to improve shelf-life.

The **perishable** nature of many foodstuffs means that breeders, farmers and food technologists try to produce varieties, foods or processed food products with a longer shelf-life, to reduce wastage. Chilled storage and preservatives such as sulphur dioxide are widely used. Waxes and medicinal paraffin are often used on citrus fruits and partly dried grape products such as raisins to delay spoilage by micro-organisms.

19.2.3. The value of rarities

Some **rarities** are valuable, some are not, depending to a large extent on whether they catch the public's interest and how well marketed they are. **Orchids** are generally perceived as exotic, mysterious, rare and interesting, with a £6 billion a year international trade, including rarities fetching more than £16,000 a plant.

A classic case of rarities fetching high prices was that of **tulips** in Holland, where the first bulbs were planted in 1593. They became extremely popular, and in 1610-1637, and especially in 1634-1637, Holland suffered from "Tulip Mania", with a huge rise in the price of bulbs, particularly for rarities even though only the rich could afford them. At the peak of demand, the variety Semper Augustus fetched from 5,000 to 13,000 florins a bulb, equivalent to the price of a grand house on the Amsterdam canals, when the average annual income was 150 florins: see Pavord (1999). In April 1637, the Dutch government intervened, cancelled all speculative agreements, and fixed the maximum price at 50 florins a bulb. There are now more than 3,000 tulip cultivars, classified as early-flowering, medium-early flowering, and late-flowering, and subdivided into types such as single, double, Darwin, lily-flowered, fringed (Plate 52), parrot, viridiflora, Rembrant and Botanical tulips.

An example of a well-documented animal rarity is the **white tiger**. In 1951, a white tiger aged about nine months, Mohan, was caught in India from the wild. He was later mated to a normal tigress, Begum, and all 10 cubs were normal. Mohan (Plate 53) was mated to one of his daughters, Radha, and this eventually produced 11 white cubs and three normal ones. All white x white matings give all white offspring, because the condition is due to an autosomal recessive allele. White tigers are not albinos. They have icy-blue eyes, not pink, and the off-white coat has brown or chocolate stripes, and the animals are larger and heavier than normal tigers. Selection within zoos has now made the coat a brighter white than Mohan's. In 1983, there were 65 white tigers in the world, 21 in India, 40 in USA, 4 in the UK. Their extreme attractiveness to zoo visitors and their rarity then made them worth about 1.1 million Indian rupees (at about 17 rupees to the pound sterling) a breeding pair. When Bristol Zoo, in England, first had white tiger cubs, the number of paying visitors increased dramatically.

19.2.4. "Health foods"

Many people in developed countries are becoming more **health conscious**, helped by an avalanche of articles and programmes in the media, so that it is worth breeding to cater for their wishes, whether soundly based or not. Meat with reduced fat and cholesterol commands a good premium, and there is a huge market for reduced-fat milk, yoghurt, spreads, sauces and manufactured foods. **"Organic"** foods are increasingly popular, commanding higher prices than normal food, e.g., 29p a litre for organic milk against 18p for normal milk (1999 figures), and see Murphy (1992). See Table 5.13.1, Chapter 5, for the calories, protein, fat and cholesterol content of different meats, where ostrich meat has the least fat and by far the least cholesterol. Ostrich meat is regarded as healthy and is increasing sales in Europe and elsewhere.

An example which combines animal conservation (Chapter 15.4) with health food is the **bison** (buffalo). In North America in 1830, there were about 60 million wild bison grazing on the prairies. By 1880, hunting had reduced them to a few hundreds. A small number of ranchers saved them from extinction and some established breeding stock for meat production. The meat is high in flavour, has a quarter the fat of most other red meats, and is low in calories, so it can command prices up to twice those of beef. Bison-rearing is therefore growing in importance in Canada, the USA and Scotland. The animals need no shelter even in Edmonton, Canada, where winter temperatures fall to -40°, and the animals tend to be healthy and easy to manage.

As well as good sales for fat-reduced milk products on possible health grounds, there is also a market for non-dairy milk-substitutes for vegetarians and people unable to take dairy products, say because of lactose intolerance. A Swedish company markets "Mill Milk", made from rapeseed oil and oats, in natural and mango flavours. "Milk" is also produced from rice and from soya beans.

19.2.5. Discounted cashflow assessment of breeding programmes

Different breeding programmes give different rates of response and cost different amounts to run. Some result in an immediate improvement, while others may take years, as with the progeny testing of bulls for milk yield genes (Chapter 5.7). Breeders are interested in the rate of response per unit time, and the times of investment and benefit. To make the best decisions in the long term, between alternative breeding programmes, one needs to work out their economics over time, say a 20-year period. The **discounted cashflow** method is the standard approach.

One predicts some realistic annual rate of compound interest that an investment might receive, say 8%, in order to compare the effects of investment in the breeding programmes at different times. The earlier the money is invested in the breeding programme, rather than in a bank, the greater its relative cost. £1 invested in year 1 would have produced £1.08 by year 2 if invested at 8% interest, and £$(1.08)^2 =$ £1.166 by year 3. The cost of investing £1 in the breeding programme in year 1 is

£1, compared with £1/1.08 = £0.92 in year 2, £1/1.166 = £0.857 in year 3, while investing £1 in year 20 only costs £0.232, in relation to year 1. Thus we can calculate a discount factor for each year, multiplying costs and returns in each year by the discount factor, to assess the net benefits or losses at different stages of the programmes.

This is shown in Table 19.2.5.1 for two possible breeding programmes. Programme 1 has consistent fairly heavy costs in each year, and no commercial benefit until year six, perhaps from progeny testing of bulls. Programme 2 has a very heavy initial cost, then low costs thereafter, with benefits from year seven. Programme 1 shows a net profit after 17 years, compared with only 14 years for

Table 19.2.5.1. A discounted cashflow assessment of alternative cattle breeding programmes.

		Programme 1					Programme 2				
		Actual			Discounted		Actual			Discounted	
Year	Dis-count factor	Costs that year, £k	Returns that year, £k	Net balance that year, £k	Net balance that year, £k	Sum over all years, £k	Costs that year, £k	Returns that year, £k	Net balance that year, £k	Net balance that year, £k	Sum over all years, £k
1	1	250	0	-250	-250	-250	600	0	-600	-600	-600
2	0.926	250	0	-250	-232	-482	100	0	-100	-93	-693
3	0.857	250	0	-250	-214	-696	100	0	-100	-86	-779
4	0.794	250	0	-250	-199	-895	100	0	-100	-79	-858
5	0.735	250	0	-250	-184	-1,079	100	0	-100	-74	-932
6	0.681	250	100	-150	-102	-1,181	100	0	-100	-68	-1,000
7	0.630	250	150	-100	-63	-1,244	100	50	-50	-32	-1,032
8	0.584	250	200	-50	-29	-1,273	100	100	0	0	-1,032
9	0.540	250	300	50	27	-1,246	100	200	100	54	-978
10	0.500	250	400	150	75	-1,171	100	300	200	100	-878
11	0.463	250	500	250	116	-1,055	100	400	300	139	-739
12	0.429	250	600	350	150	-905	100	600	500	215	-524
13	0.397	250	700	450	179	-726	80	800	720	286	-238
14	0.368	250	800	550	202	-524	80	1,000	920	339	**101**
15	0.340	250	900	650	221	-303	80	1,200	1,120	381	482
16	0.315	250	1,000	750	236	-67	80	1,400	1,320	416	898
17	0.292	250	1,100	850	248	**181**	80	1,500	1,420	415	1,313
18	0.270	250	1,200	950	257	438	80	1,600	1,520	410	1,723
19	0.250	250	1,300	1,050	263	701	80	1,700	1,620	405	2,128
20	0.232	250	1,500	1,250	290	991	80	1,800	1,720	399	2,527

programme 2. Programme 2 has a net discounted sum advantage over programme 1 after only four years, so with this interest rate, programme 2 would be preferred to programme 1. One could repeat the calculations with different interest rates. Inflation would affect both costs and returns, though perhaps not equally, so the calculations are partly balanced against the effects of inflation. Even a very well-planned breeding programme can have its success seriously affected by changes in government policies, regulations and price-support, or by unforeseen circumstances such as the BSE beef crisis in the 1990's in the UK.

19.2.6. Breeders' rights

In 1987, the **British Society of Plant Breeders** (BSPB) was established by combining the Plant Royalty Bureau and the British Association of Plant Breeders. Its members comprise nearly all public and private sector plant breeders in the UK. The BSPB represents the political, commercial and technical interests of the plant breeding industry. It promotes plant breeding's role in maintaining the competitiveness of UK agriculture, and encourages the education of plant breeders.

Its other function is to collect and distribute **royalties** on behalf of the breeders, originally under the 1964 UK Plant Varieties and Seeds Act. That act gave plant breeders the exclusive right to charge a royalty for a licence to produce and sell a variety. The Plant Varieties Act 1997 brings UK law into line with the EU legislation agreed in 1994. A major change was to enable the BSPB to charge royalties (at a reduced rate compared to seed sales) on **farm-saved** seed, for reuse on the same farm, which had been exempt from previous legislation which only referred to sale of seed. Popular old varieties are often free of this charge, but all recent varieties come under it. In 1997/98, the BSPB received about £2.3 million from farm-saved seed royalties. Payments for particular crops were agreed between the BSPB and the farmers' unions. Unlike the royalties of certified seed, farm-saved seed payments are fixed for each crop species, regardless of variety. They are charged on an area-sown basis, e.g., for 1998/99 per ha, £4.76 for wheat, £8.57 for oilseed rape, or per tonne if cleaned by a BSPB-registered processor, e.g., for 1998/99 per tonne, £25.03 for wheat, £27.42 for beans.

Licensing schemes now operate in the UK for cereals, fodder plants, tree fruits, linseed, potatoes, cane fruits, rootstocks, oilseed rape, field peas and beans, bush fruits and *Triticale*. The breeder grants a Head Licence to the BSPB which can issue sub-licences to seed merchants for the production and sale of seeds of that variety. The BSPB collects the royalties and passes them on to the breeder, after deductions for the BSPB running costs. The royalty is about £8 a tonne on seed potatoes, equivalent to less than 50p on each tonne of potatoes sold. On cereals it is just over £6 a hectare. For certified seed for sale (i.e., not farm-saved), the royalties differ between varieties of a crop, and are fixed by the breeding firm owning the rights. Thus in 1999, royalty rates per tonne on Otira spring barley were £70 for first generation seed and £50 for second generation seed, while for Saloon spring barley they were £91 and £53.50 respectively. Breeders' rights last 30 years for potatoes, raspberries, apples and rootstock crops, and 20 years for other crops such as cereals. Wheat charges average about £42 a ton, perhaps £70 when first released, going down to perhaps £30 later. This information was provided by the BSPB in 1999. Similar schemes operate for plant crops in the rest of the EU, but elsewhere arrangements are very variable.

There is no equivalent organisation or system of breeders' rights for **animals** in the UK. For beef cattle, sheep and pigs, there is a **slaughter levy** to fund the Meat and Livestock Commission, which does recording of animal performance. In recent years, that Commission has formed an organisation called Signet to provide performance records and genetic evaluations on beef cattle and sheep. For dairy cattle, there is a **milk levy** for the Milk Development Council, which funds an Animal Data Centre for genetic evaluation. Animal breeders just get payment for the animals or semen they have developed, but with no continuing royalties on subsequent sales.

Suggested reading

An Outline of United Kingdom Competition Policy. 1998. The Office of Fair Trading/ The Stationery Office, London.

Ansell, D. J. and R. B. Tranter, *Set-aside: in Theory and in Practice*. 1992. University of Reading, Reading.

Epp, D. J. and J. W. Malone, *Introduction to Agricultural Economics*. 1981. Macmillan, New York.

Mathew, B., The secret of the knights star lily. *Kew* (1999), Spring, 34–37.

Monopolies and Anti-competitive Practices. 1995. The Office of Fair Trading/The Stationery Office, London.

Murphy, M. C., *Organic Farming as a Business in Great Britain*. 1992. Agricultural Economics Unit, University of Cambridge, Cambridge.

Pavord, A., *The Tulip*. 1999. Bloomsbury, London.

Turner, J. and M. Taylor, *Applied Farm Management*. 1988. Blackwell Scientific, Oxford.

Whitaker's Almanack. 1999. J. Whitaker & Sons Ltd, London.

COLOUR PLATES

Plate 1. Variation in sorghum panicles, displayed at the International Congress of Genetics, New Delhi. There is a wide range of colour and form variants available for breeding and selection. Breeders are concerned with grain yield, nutritional value, whether the strain is for feeding animals or humans, and whether the plants are grown for the grain or for grain plus foliage. It is often planted on poor soils, with drought-resistance required in some areas. Chapter 1.2.

Plate 2. A spontaneous leaf-shape mutation in beech (*Fagus sylvatica*). The cut-leaf mutant branch on the left was on the same tree as the normal leaves on the right. Chapters 1.3.6 and 7.2.

367

368

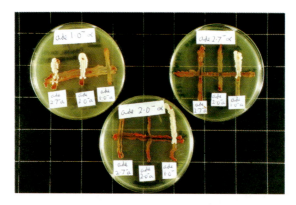

Plate 3. The *trans* part of a *cis/trans* test for allelism between three haploid, red, adenine-requiring mutants in yeast, *ade-1.0⁻*, *ade-2.0⁻* and *ade-2.7⁻*, where the wild-type is non-adenine requiring and white. The horizontal streaks are the α mating types of these mutations, and the three *a* strains of the same mutations have been streaked upwards through them. The non-mated areas such as below the α strains are controls for spontaneous reversion to wild-type, which would give white colonies. In the mated areas above the α strains, complementation shows as white patches of diploid, while non-complementing mutants give red diploids. The *ade-1* strains complement the *ade-2* strains. No strain complements its own mutation, and the two *ade-2* mutants, *ade-2.0* and *ade-2.7*, do not complement each other because they are at the same locus. The plates are 9 cm in diameter. Chapters 1.3.8 and 18.1.1.

Plate 4. An autosomal recessive mutation, albinism. The parents are unaffected heterozygotes, *A a*, while the twin boys are albinos, *a a*, with white hair, pale skins, poor eyesight, and eyes which are pink in direct light, but look blue here. Blue eyes in humans are due to light-scattering, not pigments, and albinos cannot make normal skin, eye or hair melanin pigments. The two unaffected children could be heterozygotes or normal *A A*. Chapters 2.1.2, 13.2.2 and 13.3.1.

Plate 5. Complete linkage in maize grain between coloured/colourless, *C/c*, and full/shrunken, *Sh/sh*. The parents were coloured full and colourless shrunken, giving 0% recombination in the F2 after test-crossing the F1 (coloured full) to the double recessive, *c c, sh sh*. A 1:1:1:1 ratio is expected if there is no linkage. Chapter 2.2.1. The grid is of one inch (2.54 cm) squares.

Plate 6. Recessive epistasis of *a* to the *Pr/pr* locus in maize grain. Selfing the F1 gave a modified dihybrid ratio of 9 purple (*A -, Pr -*) : 3 red (*A -, pr pr*) : 4 colourless (3 *a a, Pr -* : 1 *a a, pr pr*). Allele *A* gives red, which *Pr* modifies to purple. Chapter 2.2.2.

370

Plate 7. A modified dihybrid ratio in maize from selfing the colourless F1, *A a, I i*. Dominant inhibitor *I* prevents expression of colour from *A*, giving 3 coloured (*A -, i i*) : 13 colourless (9 *A -, I -*; 3 *a a, I -*; 1 *a a, i i*). Chapter 2.2.2.

Plate 8. Two unlinked complementary dominant genes in maize, *A* and *C*, together giving grain colour. The modified dihybrid ratio from selfing the F1 is 9 coloured (*A -, C -*) : 7 colourless (3 *A -, c c*; 3 *a a, C -*; 1 *a a, c c*). Chapter 2.2.2.

Plate 9. A polled (hornless) Hereford bull, under two years of age. Chapter 2.3.

Plate 10. A lodging-resistant rice strain under test at the Agriculture Research Station, Bentota, Sri Lanka. This strain can raise the upper stem if beaten down, by bending upwards from the nodes. Chapter 5.3.

Plate 11. The Ames test for mutagenicity. *Salmonella typhimurium* histidine-requiring bacteria are plated on minimal medium with a trace of histidine, which enables a background lawn to grow slightly, but revertants from *his⁻* to *his⁺* can form individual visible colonies. There is a cleared zone of killing close to the disk with NMG (left), but as the concentration of NMG decreases away from the disk, there are many individual revertant colonies, showing that this particular base substitution amber mutation, 881, is reverted by NMG at sub-lethal concentrations. The disk with streptomycin (right) shows a clear halo of growth outside the cleared zone of killing, but no individual revertants. In the halo, misreading of the amber stop codon is induced by streptomycin at low concentration, so that some complete protein is formed, allowing histidine

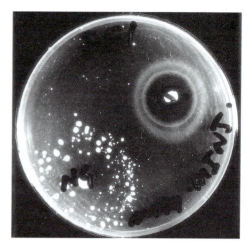

synthesis: that is phenotype suppression, not reversion, and the growth in the halo would not grow again if subcultured on minimal alone, unlike the revertants from NMG. This plate shows that NMG is mutagenic for base substitutions, but streptomycin is not. There are also some spontaneous revertants away from both disks. The dish is 9 cm in diameter. Chapters 7.2.2 and 13.7.2.

Plate 12. The effects in tomato (*Lycopersicon esculentum*) of gamma ray irradiation of seeds by a cobalt-60 source. From the front to the rear, the trays are those with 60,000 rads, 55,000 rads, 45,000 rads, and no irradiation. The seeds were from selfed *Xa-2 xa-2* plants, where there is incomplete dominance of xanthophyllic, *Xa-2*, so one expects a ratio of 1/4 green: 1/2 yellow-green: 1/4 yellow seedlings, where the latter die after germination because they cannot photosynthesise. Irradiation causes stunted growth, morphological deformities and some leaf-colour mutations (Plate 13). The seedlings had been planted for five weeks. Chapters 7.2.2 and 1.3.4.

Plate 13. A seedling from irradiated tomato seed (see Plate 12 for details). This *Xa xa* yellow-green plant has some distortion and possibly a small yellow-white sector on the right side of the upper left leaflet, which would be mutation from wild-type *xa* to mutant *Xa*, giving an *Xa Xa* sector. The upper right leaflet shows some distortion and a clear green sector, from an *Xa* to *xa* mutation, giving *xa xa*. An advantage of using a locus with incomplete dominance is that mutations in either direction in the heterozygote are visible. One can use induced mutations for fate-mapping: this seedling shows what area of leaf is determined by a single mutated cell in the seed. Chapters 7.2.2 and 1.3.4.

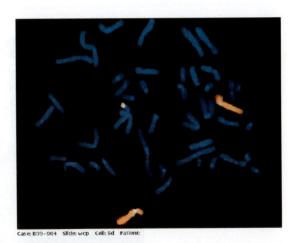

Case: B99–904 Slide: wcp Cell: 5d Patient:

Plate 14. A human cell at metaphase, with a "whole chromosome 5" FISH paint. There are two normal chromosome 5's, showing pink, and a 21 with part of 5, showing pink for its left end, with the translocated part of 5 almost as long as the 21, which does not physically join the two adjacent longer chromosomes. The 21 was identified as such from G-banding in other preparations. This is an unbalanced non-reciprocal translocation. Chapters 9.5 and 13.8.1.

374

Plate 15. Plates 15 and 16 are from the work of Dr D. Ockendon when at the National Vegetable Research Station, Wellesbourne, Warwickshire. Monoploid Brussels sprouts plants produced from anther culture, showing differences in appearance from each other. The foreground plant has had colchicine applied to the growing point, to induce some diploid side-shoots. Chapters 10.3 and 12.1.5.

Plate 16. Flowers on monoploid (left, from anther culture) and diploid Brussels sprouts plants. Chapters 10.3 and 12.1.5.

Plate 17. A segmental allopolyploid, *Primula kewensis*, from a natural crossing between two different primula species at Kew Gardens. See text. Chapter 10.6.

Plate 18. An F1 hybrid coconut plant, from crossing tall x dwarf trees, at the Coconut Research Station, Lunuwila, Sri Lanka. Chapter 12.1.3.

Plate 19. Lord Lambourne dessert apples, produced from a James Grieve x Worcester Pearmain cross by Laxton Brothers in 1907. Photographed at the Brogdale Horticultural Trust, Faversham, Kent. Chapter 12.3.

376

Plate 20. Apple seedlings for molecular-marker screening at Horticultural Research International, at East Malling, to decide which are worth growing to fruiting and later determination of commercial characters. Chapters 12.3 and 8.7.

Plate 21. An intergeneric hybrid, the Leyland Cypress, x *Cupressocyparis leylandii*, which arose in 1888 by natural cross-pollination between its two parents, *Chamaecyparis nootkatensis* and *Cupressus macrocarpa*. Chapter 12.5.2.

Plate 22. Ayrshire cows from the Pylon herd, at the South of England Show, Ardingly, Sussex. Chapter 12.7.5 includes a description of show points for this dairy breed.

Plate 23. Showing bulls in the Sainsbury Super-Bull Championships for beef bulls under two years of age, at the South of England Show. A Charolais is on the left and a polled Hereford on the right. Chapters 12.7.5 and 12.8.

378

Plate 24. Limousin cows, a beef breed, at the South of England Show. The cow with the crinkly coat was first in her class and the foreground smooth-coated cow was third. Chapters 12.7.5 and 12.8.

Plate 25. Southdown lambs at the South of England Show. The many awards shown by the cards increase the animals' value. Chapters 12.8 and 12.7.4.

Plate 26. Vitiligo causing white patches on the skin of a lady from Jamaica. It is multifactorial and definitely not infectious. Her white skin patches started when she was 12, and now 45% of her skin is white, with the main spreading occurring during her forties. She was chairman of the Vitiligo Society in Britain from 1993 to 1997. Sometimes she uses make-up to cover the white patches, which are very susceptible to sunburn. Vitiligo can cause embarrassment, low self-esteem and social problems, but is not life-threatening. Chapter 13.2.4.

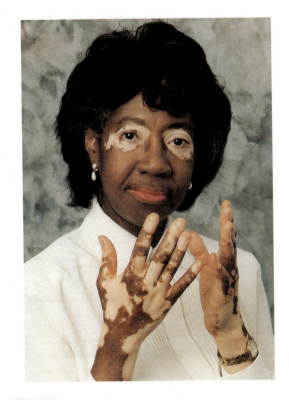

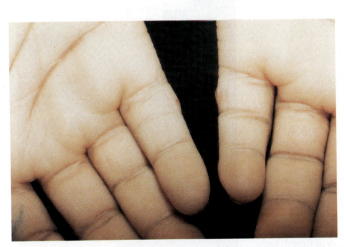

Plate 27. The little fingers each bear a small, slightly itchy bump, all that remains of dominant autosomal isolated polydactyly, which was treated on the day of birth by tying off the small extra finger stubs. Chapter 13.3.1.

380

Plate 28. Achondroplastic dwarfism, showing that the limbs are very short, especially the proximal parts. The condition is caused by an autosomal dominant allele but which acts as a recessive lethal. Seneb was Chief of the Royal Wardrobe at Gizeh, Egypt, in the 5th Dynasty. He lived from 2563-2423 B. C., and had two normal children by his normal wife. The males are shown darker than the females. Funerary image photographed at the Egyptian Museum, Cairo. Chapter 13.3.1.

Plate 29. A tortoiseshell female cat, showing X-inactivation. She is heterozygous for X-linked alleles O (orange) and o (black), with different X's inactivated in different body regions. The white areas are controlled by a different system. Chapter 13.5.

381

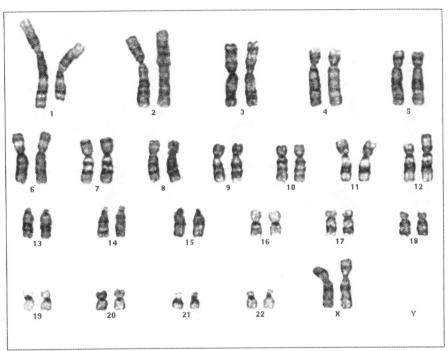

Case: B98-1826   Slide: G-Banding   Cell: 3   Patient:

Plate 30. A G-banded human karyotype of a girl who was slightly dysmorphic and developmentally delayed. 46,XX, t(1;12)(q41;p13.33). The end of the long arm of one chromosome 1 (q41 onwards) has become detached from 1 and attached to the top of the short arm of chromosome 12, at p13.33. See Plate 31 for more details. This is a balanced non-reciprocal translocation. Chapters 13.8.1 and 9.5.

382

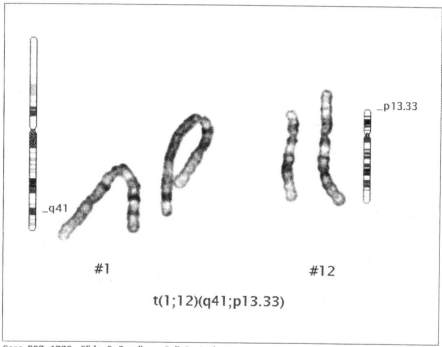

#1   #12

t(1;12)(q41;p13.33)

Case: B98-1826   Slide: G-Banding   Cell: 3   Patient:

Plate 31. See Plate 30 for details. This shows enlargements of chromosomes 1 and 12, with normal and translocated ones, and a representation of the standard Giemsa banding for the normal versions of those two chromosomes. Chapters 13.8.1 and 9.5.

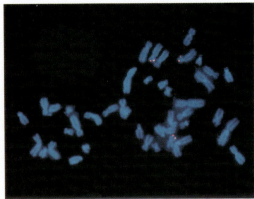

Plate 32. A FISH preparation with a sub-telomeric probe for chromosome 5. There are two normal 5's, near the top, and near the bottom a 17 with terminal extra material on the short arm, from chromosome 5. This is an unbalanced non-reciprocal translocation. Chapter 13.8.1 and 9.5.

Plate 33. An interphase cell, an uncultured white blood cell from a newborn baby with ambiguous genitalia. FISH with dual-coloured centromeric probes shows the baby to be chromosomally X (red centromere) Y (green-blue centromere). Chapter 13.8.1.

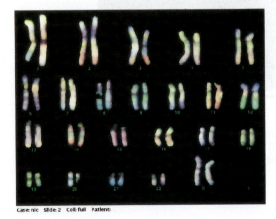

Plate 34. A karyotype with R$_x$FISH (Applied Imaging) from a normal human female cell, viewed with special imaging systems. Each type of chromosome has a different colour pattern. Chapter 13.8.1.

384

Plate 35. A Down syndrome girl, photographed at 10 months. Like many Down's children, she had heart defects and needed a major operation, which set back her development still further. Chapters 13.8.2, 1.3. and 10.8.

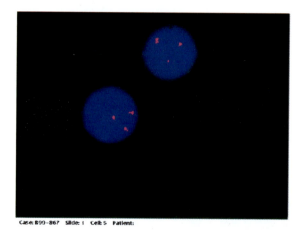

Case: R99–867   Slide: 1   Cell: 5   Patient:

Plate 36. Uncultured white blood cells from a newborn baby with some Down syndrome features. The "Quintessential" probe for chromosome 21 shows that there are three copies of 21, but it cannot distinguish between a standard Down's and a translocation Down's. Chapters 13.8.2 and 13.8.1.

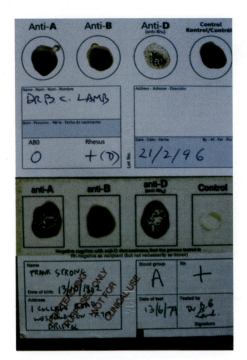

Plate 37. Testing ABO and Rhesus blood groups using a card with dried antibodies. The upper card shows an O Rh positive grouping, with only agglutination on the anti-D (anti-Rh$_o$) panel. The lower card shows strong reactions in the anti-A and anti-D panels, showing an A Rh positive grouping. Chapters 13.10.1 and 13.10.2.

Plate 38. Testing ABO and Rhesus blood groups using a centrifuge and monoclonal antibodies. Collected blood (the right-most tube) is diluted in phosphate buffer, as in the next tube along, then small samples are mixed separately with the monoclonal antibodies for anti-A (liquid coloured blue), anti-B (liquid coloured yellow) and anti-D (colourless), in different tubes, needing three tubes per blood sample. After mixing and centrifuging the sample, one shakes each tube gently. A positive reaction (agglutination) gives a pellet which stays intact, but having no reaction allows the pellet of red blood cells to be resuspended. For each antibody, one tube here has a pellet and one has resuspended cells; e.g., the left-most capped centrifuge tube has no B antigen but the next tube along has B antigen, which has agglutinated with the anti-B antibody to give a firm pellet and clear yellow liquid. Chapters 13.10.1 and 13.10.2.

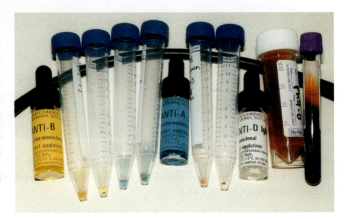

386

Plate 39. A Gloucester Old Spot pig, probably from Exfold Farm. She is a gilt, not having had her first litter, hence the underdeveloped udders. This breed was endangered, but the danger has lessened. Chapter 15.4.2.

Plate 40. Plates 40, 41 and 42 were taken at the Plant Genetic Resource Centre, Gannoruwa, Sri Lanka. This shows living plant germplasm collections, especially different collections of gotukola (foreground and middle ground). Chapter 15.4.2.

Plate 41. The stainless steel refrigerated chambers for the seed banks, especially of Asian food plants. See text for details. Chapter 15.4.2.

Plate 42. Inside a refrigerated chamber shown in Plate 41. The seeds are in screw-top plastic jars, filed in drawers. Chapter 15.4.2.

388

Plate 43. Glasshouse growing of chrysanthemums at Headcorn, Kent. The lights and black plastic curtains are used to control day-length and so get flowering for a longer part of the year than would happen naturally. The plants are vegetatively propagated, with good uniformity for flowering time. Chapter 17.1.1.

Plate 44. Blowflies pollinating two Brassica strains, within a pollination bag. Fruits (siliquas) can be seen above. Chapter 17.1.3.

Plate 45. A rooted leaf-cutting of tea, with a strong new axillary shoot. The fern leaves are for shade and moisture-retention. Taken at Eildon Hall Estate, Sri Lanka.

Plate 46. Top: an omega graft in grape vines, near the left-hand end below the buds. Middle: *Vitis vinifera* scion before shaping to an omega protrusion about one third of the way down from the left. Bottom: American rootstock before having an omega-shaped hole cut in the left-hand end below the bud. The grafted pairs are waxed over the graft before being incubated to get rooting. Taken at Geisenheim Wine School, Germany. Chapter 17.1.5.

390

Plate 47. Inarch grafting to replace a rootstock which had become phylloxera-susceptible. The new rootstock was planted behind and to the left of the existing one, then when it was big enough it was grafted into the existing large scion. When the graft had taken, the first rootstock was sawn off, with its upper end visible coming down to the right of the supporting pole. Taken at Domaine Carneros, Napa Valley, California. Chapter 17.1.5.

Plate 48. Vegetative segregation of an all-white branch (lower left) and some all-green leaves (centre) from a variegated garden *Euonymus*. Chapter 17.1.5.

Plate 49. *Aspergillus nidulans* colonies inoculated onto a master plate of complete medium for replica-plating with a 27-pin pin-inoculator onto a series of diagnostic plates to determine which auxotrophic mutants are expressed by which colonies. This is an experiment on the parasexual cycle and mapping. The two haploid parental complementing auxotrophs are at the top, one with yellow conidia and one with white conidia. The green colonies include the prototrophic diploid obtained from them, and a haploid wild-type. The other yellow and white colonies are diploid and haploid segregants from the green diploid; a few are not pure, having white and yellow conidia. Chapters 18.4 and 18.1.1.

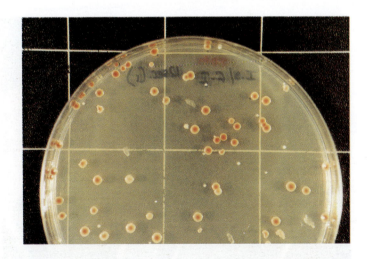

Plate 50. A red *ade-1* yeast which has been exposed to UV light. Most of the colonies are normal-sized red ones, but induced "petites" show as small white colonies. "Petites" cannot respire aerobically, but can grow slowly by fermentation as the medium here includes glucose. They have mutations at one of at least nine nuclear loci, or in mitochondrial DNA. Spontaneous "petite" mutations may accumulate in beer yeasts which are used in several successive fermentations, and may cause off-flavours. The dish is 9 cm diameter. Chapter 18.6.1.

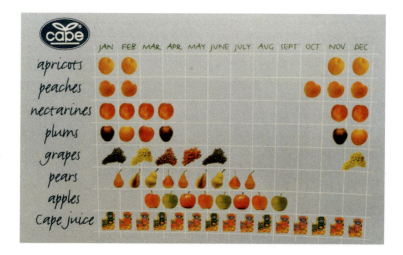

Plate 51. A chart showing the seasonal availability of fruit from the Cape, South Africa, for export to the UK. Within one type of fruit, say grapes, different varieties are available at different times. Chapter 19.2.2.

Plate 52. Fringed tulips, variety Johann Gutenberg, photographed at Pashley Manor, Ticehurst, Sussex. There is an enormous range of different types of tulip, for flower form and colour. Chapter 19.2.3.

Plate 53. Mohan, a male Indian white tiger from the wild, homozygous for an autosomal recessive allele. He is not an albino and has blue eyes and some coloured stripes. Such rarities can have very high values: see text. Photographed at the International Congress of Genetics, New Delhi. Chapter 19.2.3.

INDEX

Entries in italics refer to the plates.

Anaesthesia: Review 4

Squire's Ether Inhaler
The anesthetic apparatus used for the demonstration by
Robert Liston at University College Hospital, London,
on 21 December 1846 of the use of ether in a surgical
operation. (Courtesy of the Wellcome Institute Library
London).

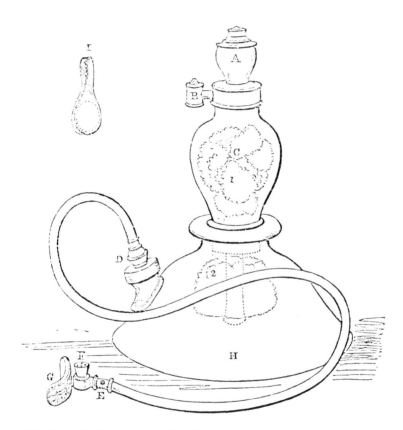

A. The Urn with its stopper, into which the ether is poured.
B. Valve which admits the air.
C. Contains sponge saturated with ether.
D. Valve which opens at each inspiration, and closes at each expiration.
E. Ferule for regulating the quantity of atmospheric air admitted.
F. Valve for the escape of expired air.
G. Mouth-piece.
H. Lower vase.
I. Spring for closing the nose.

Anaesthesia:
Review 4

Edited by

Leon Kaufman

MD, FFARCS
Consultant Anaesthetist, University College Hospital, London;
Consultant Anaesthetist, St Mark's Hospital, London;
Honorary Senior Lecturer, Faculty of Clinical Sciences,
University College London, UK

CHURCHILL LIVINGSTONE
EDINBURGH LONDON MELBOURNE AND NEW YORK 1987

CHURCHILL LIVINGSTONE
Medical Division of Longman Group UK Limited

Distributed in the United States of America by
Churchill Livingstone Inc., 1560 Broadway, New York,
N.Y. 10036, and by associated companies, branches and
representatives throughout the world.

First published 1987

ISBN 0-443-03450-8
ISSN 0263-1512

British Library Cataloguing in Publication Data
Anaesthesia. —— Review 4.
  1. Anaesthesia —— Periodicals
  617′.96′05   RD78.3

Printed in Great Britain by Bell and Bain Ltd., Glasgow

Preface

The publication of Review 4 has been planned to coincide with the 140th anniversary of the first public demonstration of ether anaesthesia, which took place at University College Hospital, London, on 21st December 1846. To commemorate the use of ether by Morton in Boston two months earlier, we have invited contributions for the first time from our transatlantic colleagues.

The details of the first surgical operation under anaesthesia are set out in the first chapter and it is worth recalling that on this historic occasion the anaesthetist was a medical student. An early anaesthetic text of the 19th century sounded warnings about possible perils and stressed the need for having more than one anaesthetist present at operation. The hazards of anaesthesia are emphasised in the medico-legal section, noting the need to maintain high standards of safety and care as well as the necessity for keeping accurate records.

As in previous volumes there is a distinct emphasis on medicine which could be applied to anaesthetic practice. There are chapters on the asthmatic patient, followed by an account of high frequency ventilation. Many detailed aspects of the cardiovascular system are included, with sections on drug therapy in hypertension, the production of induced hypotension, the management of cardiac arrhythmias, and anaesthesia for cardiac surgery and post-operative care.

We have the benefit of one of our transatlantic colleagues of his experience in trauma emergency care, while another section deals with many of the problems involved in anaesthesia for laser surgery. The section on paediatric anaesthesia and intensive care has been updated and now occupies two chapters. The use of narcotics in general anaesthesia is considered, as well as studies on the recovery from neuromuscular blockade.

Early anaesthesia was concerned with providing insensibility during operation, whereas at present we are also concerned with the management of post-operative and chronic pain relief as well as debating the benefits of supressing the endocrine response to surgery. The productivity of the 'literature' continues unabated so that the updated material now occupies two chapters. This is indeed a far cry from the medical articles that were available 140 years ago.

My colleagues have presented their chapters with enthusiasm and punctuality. Miss S. Wiggins has again borne the task of coordinating the whole volume with cheerful efficiency, assisted by Miss J. Rowan and Mr T. Wells.

London, 1986   L.K.

Contributors

Beverley Astley MB BS, FFARCS
Consultant Anaesthetist, University College Hospital and The Middlesex Hospital, London, UK

P. M. Bailey MB BS, FFARCS
Consultant Anaesthetist, Bloomsbury Hospital Group, London, UK

W. F. Bynum MD, PhD
Reader in History of Medicine, University of London; Head, Academic Unit, Wellcome Institute for the History of Medicine, London, UK

A. John Camm BSc, MD, FRCP
Professor of Cardiovascular Medicine, St Bartholomew's Medical College, London, UK

Christopher S. B. Child MB BS, FFARCS
Consultant Anaesthetist, The Royal Sussex County Hospital, Eastern Road, Brighton, UK

D. Wyn Davies MB, MRCP
Honorary Senior Registrar in Cardiology, St Bartholomew's Hospital, London, UK

G. Farnsworth MB, ChB, DRCOG, FFARCS
Consultant Anaesthetist, St George's Hospital, London, UK

Stanley P. Galant MD
Clinical Professor of Pediatrics and Acting Director, Allegy/Immunology Division, Department of Pediatrics, University of California, Irvine Medical Center, Orange County, California, USA

Ronald Green MA, MB, BChir, FFARCS
Honorary Consultant Anaesthetist, The Royal Free Hospital and St George's Hospital, London, UK

E. M. Grundy RD, BSc, MRCP, FFARCS
Consultant Anaesthetist, The Middlesex Hospital and University College Hospital, London, UK

Charles R. K. Hind BSc, MD, MRCP
Consultant Physician, Royal Liverpool Hospital, Prescott Street, Liverpool, UK

C. W. Howell MB, FFARCS
Consultant Anaesthetist, King's College Hospital, London, UK

Ian G. James MB, Ch, FFARCS
Consultant Anaesthetist, The Hospital for Sick Children, Great Ormond Street, London, UK

Leon Kaufman MD, FFARCS
Consultant Anaesthetist, University College Hospital and St Mark's Hospital, London; Honorary Senior Lecturer, Faculty of Clinical Sciences, University College London, UK

William A. Littler MD, FRCP
Professor of Cardiovascular Medicine, The University of Birmingham, UK

Colin F. Mackenzie MB, ChB, FFARCS
Associate Professor of Anesthesiology, University of Maryland School of Medicine, Baltimore, Maryland; Formerly Senior Attending Anesthesiologist, Maryland Institute for Emergency Medical Services Systems, Maryland, USA

Margaret J. S. Robertson MB, ChB, FRCP(Glas), FFARCS
Consultant Anaesthetist, Islington District Hospital, London, UK

Stephen Andrew Smith MB, ChB, MRCP (UK)
Research Fellow and Honorary Registrar, Department of Cardiology, University of Birmingham and East Birmingham District General Hospital, Birmingham, UK

Edward Sumner MA, BM, BCh, FFARCS
Consultant in Paediatric Anaesthesia and Intensive Care, The Hospital for Sick Children, Great Ormond Street, London, UK

A. C. Wainwright MB BS, FFARCS
Consultant Anesthetist, Southampton General Hospital, Southampton, UK

David White MB BS, FFARCS
Consultant Anaesthetist, Northwick Park Hospital, Harrow, Middlesex, and Division of Anaesthesia, Clinical Research Centre, Harrow, Middlesex, UK

Contents

Ether at University College Hospital, 1846

INTRODUCTION

The difficulty of controlling pain adequately during both minor and major surgical procedures had long influenced the way these operations were performed.

Speed was at a premium and the conscious patient was generally a witness to whatever was done to his body, although his subsequent testimony might well have been clouded by the pain itself, as well as by the alcohol, opium, cold or other measures often employed to deaden the discomfort or blunt the memory of the traumatic event. The production of unconsciousness by these traditional means could not be achieved without serious risk to the patient, and as long as consciousness remained, pain control was not complete. The capacity of nitrous oxide ('laughing gas') to block the memory of pain in the conscious subject was known since the very beginning of the nineteenth century (Cartwright 1952), but its medical implications were not exploited for several decades.

Mesmerism was occasionally helpful but often unreliable, and Henry Hill Hickman's (1800–1830) experiments in the 1820s with carbon dioxide provided a vision but no practicable results. Like nitrous oxide, ether, too, had a social history ('ether frolics') before being appropriated into medicine and surgery as an anaesthetic (the root word being coined by Oliver Wendell Holmes (1809–1894)). During the 1840s, however, both nitrous oxide and ether were employed by dentists and surgeons deliberately to alter consciousness and hence to control pain during operations (Cartwright 1967). Beyond their obvious significance, these developments have two noteworthy resonances: they derived from the United States and thus represent the first major American contribution to medicine (Davison 1965), and both substances were simple, laboratory-manufactured compounds, thus providing an undeniable chemical triumph and so stimulating the growth of experimental pharmacology and the systematic application of chemistry to medicine (Bynum 1970). Despite earlier instances, as far as the public was concerned, the first major operation using an anaesthetic occurred in the Massachusetts General Hospital on 16 October 1844, with John Collins Warren (1778–1856) operating on a patient rendered unconscious by William Thomas Green Morton (1819–1868), administering ether. Warren is reported to have said at the conclusion of the painless operation, 'Gentlemen, this is no humbug. We have seen something today that will go round the whole world'.

UNIVERSITY COLLEGE HOSPITAL, 21 DECEMBER 1846

Warren was correct. The news did travel, not only throughout the United States, where it provoked various acrimonious priority disputes, but also to Europe, where in London the anaesthetic properties of ether were demonstrated at a meeting of the Medico-Chirurgical Society in London on 11 November 1846. During the following weeks, ether was used more than once in Britain, in the house of an American doctor, Francis Boott, practising at 24 Gower Street, London, and at the Dumfries and Galloway Royal Infirmary. Nevertheless, the analogous operation to Warren's at the Massachusetts General Hospital took place at the University College Hospital on 21 December. It was public, attended by numerous important medical men and widely reported and discussed, for example, in the pages of the Lancet and the London Medical Gazette. The episode is almost as famous as the one that had taken place two months earlier on the other side of the Atlantic.

The surgeon was Robert Liston (1794–1847), professor of clinical surgery at the medical school and senior surgeon to the hospital. Liston knew Boott, who had received a letter describing the Boston operation, and between Saturday 19 December and Monday 21 December, Liston discussed the matter with Boott, with Mr Robinson, the dentist who had used ether for a tooth extraction in Boott's house, with William Squire, a medical student at UCH, and with Squire's uncle, Peter Squire (1798–1884), a successful Oxford Street chemist (Todd 1984). The uncle prepared the ether which his nephew administered to a Harley Street butler named Frederick Churchill, whose chronic disease of the knee joint made amputation necessary. Liston had already operated on him on 25 November in an attempt to save the leg, but post-operative infection had supervened and Churchill agreed to amputation (Cock 1911). Liston's dresser recorded the following note in the case book, describing the historic occasion:

'Dec. 21. It having been decided to remove the limb today, at 25 minutes past 2 p.m., the patient was taken into the operating theatre. Prior to the operation ether vapour was given to breathe for between 2 and 3 minutes, the effect of this was so far to stupefy as to cause complete insensibility to pain although consciousness was retained and questions were answered. Professor Liston finished the complete removal of the limb in 25 seconds — not the slightest groan was heard from the patient nor was the countenance at all expressive of pain. This is the first capital operation which has been performed in this country under the narcotizing influence of ether vapour, and it was perfectly successful. The patient did not know that the limb was removed ... ' (Merrington 1976).

Churchill had a relatively uneventful course and was discharged from the hospital as 'cured' on 11 February 1847. It had taken Liston under half a minute to remove his leg. Forty years later, Squire remembered the patient as saying, when regaining consciousness, 'Take me away. I can't have it off, I must die as I am.' (Squire 1888). Liston was supposed to have said, 'This Yankee dodge, Gentlemen, beats Mesmerism hollow'. To prove his point, he evulsed both sides of a great toe-nail of another etherized patient.

The apparatus used by Peter Squire to administer the ether had been devised by his uncle and was described in an eye witness account:

'In these cases the ether vapour was administered by means of an ingenious apparatus extemporaneously contrived by Mr Squire of Oxford Street. It consisted of the bottom part of a Nooth's Apparatus, having a glass funnel filled with sponge soaked in pure washed ether, in the upper orifice, and one of Read's flexible inhaling tubes in the lower. [This tube was fitted with an expiratory valve and a mouthpiece. By turning a ferrule placed behind the mouthpiece the supply of fresh air which the patient received could be regulated.] As the ether fell through the neck of the funnel it became vaporized, and the vapour being heavy descended to the bottom of the vase, and was thence inspired through the flexible tube. No heat was applied to the apparatus or the ether.' (Duncum 1947).

Liston and all others present were conscious of the historical significance of the occasion and on the same day the surgeon wrote to Dr Boott, who had been unable to attend, 'It is a very great matter thus to destroy sensibility to such an extent, and without, apparently, any bad result'(Duncum 1947). At supper that night, Liston's excitement was such that he could hardly contain himself. As he had reason to believe, the medical press picked up the story immediately, and the journals for 1847 are filled with discussions of the new possibilities. Within months, the use of anaesthesia was no longer a novelty.

AFTERMATH

The operating theatre where Liston performed the surgery had been crowded and we know the names of a number of witnesses. There is, however, no conclusive evidence that two medical students popularly believed to have been there were actually present: Joseph Lister (1827–1912) and Joseph Thomas Clover (1825–1882) (Calverley 1985). Clover did witness an unsuccessful attempt by Squire ten days later (1 January 1847) to anaesthetize two patients, and he went on to become the leading anaesthetist of his generation (Thomas 1972). Curiously, Squire did not keep up his early interest in the field, becoming instead a consultant physician and a Fellow of the Royal College of Physicians. Liston did not live long to enjoy his acclaim: he died suddenly of a dissecting aortic aneurysm less than a year later, on 7 December 1847. The butler Frederick Churchill hobbled off the stage of history when he left UCH in February 1847.

REFERENCES

Bynum W F 1970 Chemical structure and pharmacological action: a chapter in the history of 19th century molecular pharmacology. Bulletin of the History of Medicine 44: 518–538
Calverley R K 1985 J. T. Clover: A giant of Victorian anaesthesia. In: Ruprecht J, Lieburg M J V, Lee J A, Erdmann W (eds) Anaesthesia, essays on its history. Springer Verlag, Berlin
Cartwright F F 1952 The English pioneers of anaesthesia. John Wright, Bristol
Cartwright F F 1967 The development of modern surgery from 1830. Arthur Barker Ltd, London
Cock F W 1911 The first operation under ether in Europe — the story of three days. University College Hospital Magazine 1: 127–144
Davison M H Armstrong 1965 The evolution of anaesthesia. John Sherratt, Altrincham
Duncum B M 1974 The development of inhalation anaesthesia. Oxford University Press for Wellcome Historical Medical Museum, London
Merrington W R 1976 University College Hospital and its medical school: a history. Heinemann, London
Squire W 1888 On the introduction of ether inhalation as an anaesthetic in London. Lancet ii: 1220–1221

Thomas K B 1972 The Clover/Snow Collection. Papers of Joseph Clover and John Snow in the Woodward Biomedical Library, University of British Columbia, Vancouver. Anaesthesia 27: 436–449
Todd R 1984 Peter Squire: 1798–1884. The Pharmaceutical Journal April 7: 419–424

Medicine relevant to anaesthesia

RESPIRATION

Pulmonary oedema

Pulmonary oedema is known to occur in climbers at high altitude and may be fatal. Treatment has consisted of oxygen therapy by mask until the climbers descend to lower altitudes. Larson (1985) reported on the use of a tightly fitting mask with spring loaded valves to obtain a positive expiratory pressure of 10 cm of water: oxygenation improved by 20%.

Sudden infant death syndrome

In a prospective study Wilson et al (1985) studied the respiratory and heart rate patterns in infants destined to be victims of sudden death syndrome. They found that the respiratory variables were of little value although the heart rate was higher in the affected group.

An analysis of the carotid-bodies from infants dying of sudden infant death syndrome showed that the dopamine content had a 10-fold increase and a three-fold increase for noradrenaline compared with controls (Perrin et al 1984). They suggested that these amines inhibited respiration and that infants were unable to respond to hypoxia.

Another mechanism for sudden unexplained infant death is that there is defective surfactant during lung development so that large areas of the lungs may collapse producing a right to left shunt, affecting control of breathing. The situation may be reversed with large changes in intrathoracic pressure. These episodes may be repeated but sudden death may occur if the infant fails to respond (Talbert & Southall 1985).

Respiratory distress syndrome

In a retrospective study it has been shown that the ventilation of neonates has improved the outcome in respiratory distress syndrome. The commonest cause of death was intraventricular haemorrhage. Pneumothorax developed in a third of the babies irrespective of age. The presence of a patent ductus arteriosus did not affect the survival rate but increased the need for prolonged artificial ventilation (Greenough & Roberton 1985).

5

Laryngeal obstruction

In acute epiglottitis complete laryngeal obstruction may occur during procedures which precipitate crying in children. Tarnow-Mordi et al (1985) described such an episode when the setting up of an intravenous infusion caused a child to cry resulting in respiratory arrest from laryngeal obstruction. Intubation was unsuccessful and oxygen had to be administered via a cannula inserted through the cricothyroid membrane.

Scoliosis

Respiratory function has been studied by Cooper et al (1984) showing that scoliosis of less than 60° had a decreased maximum inspiratory pressure but the expiratory pressure was normal suggesting that the inspiratory muscles work at a mechanical disadvantage. The vital capacity is still a useful guide to assessment as well as the arterial PO_2 during sleep. Guilleminault et al (1981) found that respiratory failure occurred at night during REM sleep.

Quadriplegia

Quadriplegia with respiratory paralysis due to cervical trauma was treated with phrenic nerve continuous electrical pacing. The hemidiaphragms were stimulated simultaneously. The pulse frequency of stimulation ranged from 7.1–8.3 Hz and the respiratory rate from 5–9 per min in the supine position (Glenn et al 1984).

Haemophilia

Pulmonary function has been studied in haemophiliac patients as it was thought that repeated infusions of commercial factor VII concentrates might affect lung function. It had been recommended that factor VII concentrate should only be infused through a filter (40 μm) but Chediak et al (1984) were unable to confirm that lung diffusing capacity for carbon monoxide was affected in non-smokers. As expected diffusing capacity was reduced in smokers.

Heroin addicts

Respiratory function is impaired in heroin addicts. Recently Itkonen et al (1984) demonstrated a decrease in lung diffusing capacity for carbon monoxide following the use of pentazocine and tripelennamine, an antihistamine (known as 'Ts and Blues'). The mechanisms for this abnormality may be due to arteriolar-capillary obstruction granuloma leading to fibrosis. There may also be a decrease in ventilation–perfusion ratio.

Post-operative pulmonary function

Simonneau et al (1983) suggested that the reduction in pulmonary function following upper abdominal surgery was due to diaphragmatic dysfunction. They found that there was pulmonary collapse with elevation of the diaphragm with a reduction in vital capacity (VC) and functional residual capacity (FRC). This persisted for 7 days and was not relieved by post-operative analgesia provided by epidural fentanyl. The use of dry gases may be a factor as studies on the hyperpnoea with fully saturated air at 37°C did not cause heat or water loss or coughing (Blanner et al 1984). Attempts to prevent pulmonary collapse by deep breathing may result in a significant rise in blood lactic acid (Freedman et al 1983).

The effect of acute changes in arterial CO_2 have been studied in normal men post-operatively and when the end-tidal CO_2 reached 7.5%, diaphragmatic contractility was reduced which predisposed to fatigue. This might also explain the acute episodes of respiratory failure seen in patients with chronic obstructive respiratory disease (Juan et al 1984).

Theophylline has a potent and long lasting effect on diaphragmatic strength and fatigue in patients with fixed airway obstruction (Merciano et al 1984). Theophylline was administered, 30 mg per kg body weight in two doses per day, and transdiaphragmatic pressure increased by 16% after 7 days and persisted even after 30 days. The benefits of theophylline are not only due to the bronchodilation but possibly due to an action on skeletal muscle or possibly by increasing calcium influx through the slow calcium channel of the sarcolemma. This could be mediated by adenosine receptors.

Extrapyramidal disorders

Vicken et al (1984) studied lung function (in particular airflow) in patients with extrapyramidal disorders and found evidence of upper airway obstruction due to the involvement of the upper airway musculature. This may cause severe limitation of airflow although it has always been supposed that the respiratory muscles have been compromised. Laryngeal stridor was a feature in patients with Parkinson's disease and emergency intubation may be necessary to prevent the retention of secretions, lung collapse and associated respiratory infection.

Asthma

Circulating catecholamines in acute asthma have been studied by Ind et al (1985). Plasma noradrenaline concentration was increased two to three times normal but plasma adrenaline surprisingly did not increase even during the acute attack. They suggested that there was impairment of adrenaline secretion though the mechanism was still unknown. β_2-Agonists cause hypokalaemia. The inhalation of fenoterol was found to reduce the plasma potassium level to 2.9 mmol/l. (Haalboom et al 1985)

Severe asthmatics may fail to respond to conventional treatment with hydrocortisone, salbutamol and mechanical ventilation. Ether has recently been shown again to have a place in life-threatening acute severe asthma (Robertson et al 1985). They reported two cases in whom ether produced a dramatic fall in airway pressure and an improvement in clinical condition as well as in blood gases.

The view that adrenoceptor agonists act by increasing cyclic AMP and decreasing calcium ions has suggested that calcium-channel blocking agents may have a place in the treatment of asthma. The effect of drugs, such as verapamil and nifedipine on airway smooth muscle has recently been reviewed by Barnes (1983), and although results seemed initially disappointing it seems that they may inhibit bronchioconstriction without causing bronchodilation.

Eschenbacher et al (1984) have shown that morphine can inhibit bronchoconstriction in mild asthmatics and this effect can be reversed by naloxone. There is a positive correlation between the inhibiting effects of morphine and atropine. 0.15 ml per kg of morphine does not appear to have a bronchoconstrictive effect in mild asthmatics, although there may still be histamine release and respiratory depression. Gross & Skorodin (1984) suggested that the parasympathetic system played a major part in the bronchoconstriction in emphysema. Bronchodilatation was obtained using either an adrenergic agent such as salbutamol or atropine mathonitrate (1.5 mg by inhalation). The dominant reversible component of airway obstruction in emphysema is parasympathetic activity and this is reversed by atropine.

Shock lung — adult respiratory distress syndrome

Wardle (1984) has reviewed the aetiology, pathology and treatment of adult respiratory distress syndrome (ARDS) but the type of intravenous fluid for therapy has always remained controversial in view of the problems associated with extravasation of fluids from the pulmonary capillaries. Metildi et al (1984) in a study of 46 patients with an intrapulmonary shunt greater than 20% found that clinically there was no advantage in promoting colloid solutions (albumin) in preference to crystalloid (a Ringer lactate solution).

CARDIOVASCULAR SYSTEM

Cardiac Disorders

In asymptomatic healthy patients with ventricular ectopic beats a longterm follow-up study failed to find an increased risk of sudden cardiac death compared with a control group (Kennedy et al 1985).

The mechanism underlying multifocal atrial tachycardia is often unknown and may be associated with hypoxaemia, beta adrenergic stimulation, hypokalaemia, acidaemia or due to an overdose of digoxin. Verapamil given intravenously in the dose of 1 mg followed by 4 mg over a 5 minute period reduced atrial and ventricular rates and even converted the atrial tachycardia to normal sinus rhythm (Levine et al 1985a).

Multifocal atrial tachycardia may be caused by theophylline, the mechanism being thought to be due to abnormal atrial automaticity and calcium-channel blockers have also been recommended for treatment (Levine et al 1985b).

Maisel et al (1985) described three patients who became hypotensive after receiving oral quinidine and intravenous verapamil. They suggested that the hypotension was due to alpha-receptor blockade as well as the other known actions of quinidine and verapamil.

The pathology in sudden ischemic cardiac death has been investigated by Davies & Thomas (1984). In a study of 100 deaths the incidence of coronary thrombi was 74%: the right coronary artery which supplies the conducting system was commonly affected. 'Plaque fissuring' was present in most of the remaining cases post-operatively. Plaque Fissuring is described as the development of an opening of the lumen of the vessel into the intima (dissection haemorrhage: plaque haemorrhage).

Heart block following the onset of bacterial endocarditis is often an indication for surgical intervention. Dinubile (1984) successfully treated a patient with aortic and mitral endocarditis with first degree heart block with medical therapy alone which prompted him to suggest the following guidelines for selecting patients for operation. These are: progressive heart block; aortic valve involvement; no improvement in heart block, despite a week of antibiotics; no obvious other causes of the conduction defect.

Wolfson & Swartz (1985) commented that despite many years of measurement of serum bactericidal activity there was very little data on whether it assisted in the management of patients with bacterial endocarditis. In an addendum they reported that the test could predict bacteriological cure but not clinical outcome.

Coriat et al (1982) felt they were able to predict the incidence of myocardial ischemia in patients with known coronary artery disease about to undergo vascular surgery. If the pre-operative resting ECG revealed ST abnormalities irrespective of whether the patients had angina or not, the chances of developing perioperative myocardial ischaemia were increased. Induction of anaesthesia in particular appeared to be hazardous.

A recent study of patients who have had myocardial revascularization suggests that patients who have had significant coronary artery disease and have had myocardial surgery can tolerate general surgical procedures satisfactorily. In a study of 73 patients cardiac complications including signs of myocardial infarct and pulmonary embolus, developed in one patient while another had complete heart block with congestive heart failure, requiring a permanent cardiac pacemaker (Prorok & Trostle 1984).

Sick sinus syndrome

This syndrome in which there are alternating periods of bradycardia and tachycardia is thought to be due to a defect not only in the sinoatrial node but also in other pacemakers in the heart. Watt (1985) postulated that there was increased sensitivity to adenosine and that treatment might include phosphodiesterase inhibitors such as theophylline. As the bradycardia has been regarded as an impending Stokes-Adams attack, treatment with a cardiac pacemaker has been recommended. Mazuz & Friedman (1983) found that the survival rate was not improved by the insertion of a permanent pacemaker. The severity of the bradycardia was not related to the degree of symptoms and a pacemaker was only indicated when symptoms occurred.

Belic & Talano (1985) advocated the use of intravenous hydralazine before resorting to pacemaker therapy. The complications of permanent transvenous pacing are discussed by Phibbs & Marriott (1985).

Anaesthesia and the circulation

There have been some recent studies on the effect of halogenated anaesthetics on the heart. Trigt et al (1984a) proposed the use of an index to quantify the amount of anaesthesia to depress the inotropic state of the heart by 20% (ID_{20}). Halothane appears to have a greater myocardial depressant effect than enflurane but when the ID_{20} is related to the minimum alveolar concentrations (MAC of each agent) the myocardial depressant effects are of the same degree. The addition of nitrous oxide did not further depress the myocardium. Halothane caused no change in diastolic compliance (Trigt et al 1984b). In further animal experiments Manohar & Parks (1984) noted that the addition of nitrous oxide to enflurane anaesthesia increased cerebral blood flow.

The mechanism of arrhythmias produced by hydrocarbon anaesthetic agents have been studied by Reynolds (1984) showing that superventricular arrhythmias occur when the pacemaker moves from S–A node to the A–V node which is more resistant to anaesthetic agents. Ventricular arrhythmias occur because of impaired conduction across the A–V node leading to ventricular ectopic foci or reentrent excitation. Hydrocarbon anaesthetics and adrenaline create conditions favourable to both these mechanisms. Increased intraventricular pressure may trigger off latent pacemakers leading to ectopic beats. Ectopic beats may also arise if there is a reduction in duration of the Purkinje fibre action potentials. The study also reiterates some practical applications from previous literature. Three times more adrenaline is required to produce ventricular ectopic beats when isoflurane is administered compared with halothane. Lignocaine when administered with adrenaline increases the safety dose of adrenaline during halothane anaesthesia.

Antiarrhythmic drugs

The importance of metabolites of drugs is often forgotten and recently Kates et al (1984) have drawn attention to the problems associated with some of the newer anti-arrhythmic drugs. One of the metabolites of encainide has a half life of 5 hours in contrast to the 1.5–3.5 hours of the parent drug. It is nine times more potent than encainide. Lorcainide is dealkylated to norlorcainide which is just as active as the original drug.

The degree of metabolism of these drugs may influence the duration of action when encainide patients metabolize the drug poorly. The half life will vary from 7–22 hours while in those in whom metabolism is rapid the half life will vary from 1.5–4 hours but as the metabolites are active the drug action may persist.

Amiodarone, a class III anti-arrhythmic drug may give rise to interactions with other agents such as warfarin (see Marcus 1983). The plasma concentrations of warfarin rise after even 3 to 4 days of amiodarone therapy but the onset may be delayed for as long as 3 weeks. It affects other drugs used in the management of cardiac disease such as digoxin. The elimination of digoxin is decreased. There are also interactions with quinidine, disopyramide, mexiletine and procainamide. Cardiac arrest has been reported following the use of amiodarone with β-blockers such as propranolol and calcium antagonists such as verapamil. Hypotension has also been reported during anaesthesia (Gallagher et al 1981). (See Aronson 1985.)

MYOTONIA CONGENITA

In dystrophia myotonia there is wasting of muscles as well as the inability to relax. There is an increase in morbidity and mortality following operations associated with respiratory failure (Kaufman 1960). In myotonia congenita, myotonia is usually elicited in the limb muscles and wasting is absent. Lung function tests may be within normal limits. Myotonia may occur in the respiratory muscles including the diaphragm leading to the inability of respiratory muscles to relax during expiration (Estenne et al 1984).

DYSTROPHIA MYOTONICA

A survey of patients with this dystrophy revealed that the mortality was often as high as 25% and the cause of death was respiratory failure due to muscle weakness and associated with respiratory infection (Kaufman 1960). Perusal of case notes of a recent death suggested that the cause of death was pulmonary oedema. Perloff et al (1984) in a detailed study of 25 patients found that the dystrophic process affected the His-Purkinje system rather than the cardiac muscle and although cardiac failure was not often present there was a high incidence of atrial and ventricular arrhythmias. Damage to the conduction system may result in death due to AV block or ventricular tachycardia. Cardiac arrhythmias have deleterious effects in the presence of pre-existing cardiac disease (Naito et al 1983).

DIABETES

Diabetes should be regarded as a multi-system disease. There is even an increased sensitivity to platelet aggregation when the blood sugar is elevated (Mustard & Packham 1984).

The incidence of bundle branch block is significantly higher in diabetics when compared with hypertensives in whom left bundle branch block is not an uncommon feature. The myocardial damage is believed to be due to a microangiopathic cause rather than arteriosclerotic disease (Blandford & Burden 1984).

Williams et al (1984) studied 52 patients and found that 25% had impaired sensitivity to hypoxia with a reduction in ventilation in response to hypercapnia. Seven per cent of the patients had an abnormal pattern of respiration during exercise, some of whom had evidence of autonomic neuropathy as judged by the small pupil size. Abnormal patterns of breathing with central or obstructive apnea occurred during sleep in type 1 diabetes with autonomic neuropathy (Mondini & Guilleminault 1985).

Autonomic neuropathy is known to affect cardiac function and oesophageal emptying may also be affected. In a study of insulin dependent diabetics with autonomic neuropathy it has been shown that domperidone, a potent peripheral dopamine antagonist did not improve oesophageal emptying of solids and will therefore be ineffectual in these patients (Maddern et al 1985).

There are hazards associated with the addition of insulin to intravenous infusion bags. It has been suggested that the insulin may adhere to the plastic of the infusion bag. Talbot (1984) has shown that in many instances when the insulin is being injected into the infusion bag the needle may fail to penetrate the inner seal of the injection port and be made into the port dead space. This may explain some of the instances of so called insulin resistance.

Wiseman et al (1985) found that in early insulin dependent diabetes there was an increase in glomerular filtration rate and an increase in kidney size. When the blood glucose level was brought under control the glomerular filtration rate fell but the kidney did not decrease in size.

The problems associated with continuous subcutaneous insulin infusion were reviewed by Watkins (1985) who found that ketoacidosis occurred more frequently than in those treated by conventional means and the method did not appear to be suitable for the 'brittle' diabetic. Other mishaps occurring during insulin pump therapy include infection, technical problems with the siting of the needle, pump failure, blocked infusion lines, loose connections and pump malfunction. There also appears to be an increased incidence of hyperkalaemia. In addition hypoglycaemia may occur following decreased food intake, vigorous exercise or the pump delivering too much insulin ('pump runaway') (Leading Article, Lancet 1985a).

Continuous infusion therapy is not recommended during pregnancy as pump failure may lead to ketoacidosis with a high incidence of intrauterine death. The reduction in perinatal mortality appears to be associated with good blood glucose monitoring maintaining the levels between 5–7 mmol/l (Leading Article, Lancet 1985b). Intranasal insulin has been tried by Salzman et al (1985) using an aerosol preparation which also contained surfactant. Their results suggested that intranasal insulin might be used in association with conventional treatment. Schade and Eaton (1985) were critical of the study in that subcutaneous insulin appeared to give better glycosylated haemoglobin concentrations than with intranasal insulin. There are also possible problems as to whether nasal mucosal absorption would be effective during influenza or even the common cold.

In type 1 diabetes being treated with insulin propranolol may cause hypoglycaemia. However in type II diabetes it had no effect when administered alone but it potentiated the hyperglycaemic action of hydrochlorothiazide. This effect was unrelated to the serum potassium level or insulin secretion. The diuretic may act directly on hepatic glucose production whereas the propranolol may inhibit glucose uptake (Dornhorst et al 1985).

Alberti & Thomas (1979) introduced a regime for the management of diabetics during surgery which involved the infusion of a solution containing 10% dextrose, potassium and insulin. This had subsequently been modified by Bowen et al (1984) but in their series of 27 patients 2 developed hypoglycaemia and the surgical procedures were mostly of a minor nature. Thomas et al (1984) claimed good results in the management of diabetics at operation using their original regime but this failed to convince Hall (1984) who suggested that there were too many variables in the study including the endocrine and metabolic response to the surgical stimulus. Regimes involving intravenous insulin also require separate infusion lines.

NEUROLEPT MALIGNANT SYNDROME

Malignant hyperthermia is a hazard which may occur during the course of anaesthesia and may be triggered by various anaesthetic agents. There is an increase of the calcium ion concentration in the sarcoplasm in muscle. The condition has recently been reviewed by Nelson & Flewellen (1983) and Ellis (1985). The earliest physiological change that might be detected is a rapid increase in the end-tidal CO_2 (Verburg et al 1984). Treatment is with intravenous dantrolene.

There have been suggestions that the malignant hyperthermia syndrome might be related to the neurolept malignant syndrome (NMS) which is characterized by muscle rigidity, hyperthermia, autonomic instability and stupor or coma. NMS is associated with the major tranquillizers particularly butyrophenones and pheonthiazines suggesting there is dopamine receptor blockade in the basal ganglia and hypothalamus. The onset may be gradual but may develop within 24–72 hours with muscle rigidity and increased temperature up to 40°C. There is tachycardia, fluctuations in blood pressure and sweating. There may be rigidity with opisthotonos or involuntary movement such as tremor and even oculogyric crisis. There may be dyspnoea due to rigidity of the chest wall. Damage of skeletal muscle may lead to myoglobin release with acute renal failure. The condition may be precipitated by drugs such as haloperidol, chlorpromazine or prochlorperazine. Mortality is high due to hypoventilation, cardiovascular collapse, renal failure and aspiration pneumonia.

The major difference between NMS and malignant hyperthermia is that the muscular rigidity in the former is central (presynaptic) whereas in the latter it is peripheral (postsynaptic). In NMS neuromuscular blocking drugs can produce flaccid paralysis whereas in malignant hyperthermia they are ineffectual. (See Leading Article, Lancet 1984, Szabadi 1984.) However this has been disputed by Downey et al (1984) who reported one case of NMS in whom the muscle biopsy response was similar to that seen in malignant hyperthermia.

Treatment is to stop the offending drug and to treat the symptoms such as hyperthermia as they arise. Intravenous dantrolene appears to be effective (Ritchie 1983) while oral doses appear to be effective in less acute cases. Dantrolene was ineffectual in the case described by Downey (1984). Bromocriptine, a dopamine agonist has been successfully used to treat NMS (Mueller et al 1983).

DIURETICS

The clinical pharmacology and therapeutic use of diuretics have been reviewed by Lant (1985a,b). The 'loop' diuretics may be administered during anaesthesia and these include frusemide or bumetanide which is 40–60 times more potent. Renal blood flow is temporarily reduced and then increased following intravenous frusemide, effects which can be blocked by indomethacin.

There have also been studies postulating that diuretics including frusemide and the less efficient thiazides reduce blood pressure not only by their diuretic effect but also by a direct vasodilator action. Longterm treatment does lead to a decrease in total peripheral vascular resistance (Freis 1983). Materson (1983) also concluded that the mechanism of action of diuretics in reducing blood pressure was unknown. However Schuster et al (1984) were able to demonstrate that frusemide was able to expand blood volume irrespective of its diuretic properties. They suggested that frusemide decreased the resistance of the capacitance vessels leading to a lowering of capillary hydrostatic pressure while there was also an increase in osmotic pressure. Both these factors assist in the reabsorption of oedema fluid and may account for the effectiveness of the drug in the treatment of pulmonary oedema.

Other studies have shown that in congestive cardiac failure frusemide causes venous dilatation and venous pooling before the onset of diuresis (Dikshit et al 1973). This effect is related to renin release and angiotension formation. These vascular effects are blocked by captopril (Johnson et al 1983) and by propranolol which also prevents an increase in plasma aldosterone (Johnson et al 1985).

Aldosterone secretion is stimulated by ACTH, renin, a rise in potassium or a fall in plasma sodium. Renin acts on circulating globulins to form angiotension I which is converted into angiotension II which acts on the adrenal cortex causing secretion of aldosterone. In animal experiments plasma aldosterone initially increased and then returned to base levels despite increased plasma renin during repeated injections of frusemide. The plasma aldosterone levels closely followed those of the serum potassium. The addition of potassium supplement to prevent hypokalaemia still resulted in elevated levels of renin and aldosterone. It appeared that serum potassium modulates the synthesis of aldosterone induced by angiotension II and ACTH (Fujimaki et al 1984). Captopril also abolished the rise in plasma aldosterone induced by frusemide.

The effect of frusemide on potassium loss depends on fluid and salt intake and that potassium balance is maintained if there is a liberal sodium intake (Wilcox et al 1984). Frusemide also has an effect on the cerebrospinal fluid composition resulting in a fall in CSF sodium in normocapnia and a reduction in CSF chloride during hypercapnia (Johnson et al 1984).

Kaplan et al (1985) treated hypertensive patients with diuretic-induced hypokalaemia. The diuretic was not specified but they showed that when patients whose serum potassium was below 3.5 mmol/l the addition of 60 mmol/d of potassium chloride led to a fall in blood pressure which correlated with the fall in plasma renin activity (PRA) but was not related to plasma aldosterone. Kassirer & Harrington (1985) however urged caution in their interpretation of these results.

An attempt has been made to evaluate the possible role of adrenergic mechanisms in the release of renin following frusemide. Cannella et al (1983) found that in patients with mild essential hypertension frusemide (1 mg/kg intravenously) produced a prompt and long-lasting increase in PRA. The increase within the first half-hour was closely related to the peak increase in urinary output and sodium loss. There was a significant decrease in plasma volume and central venous pressure. Plasma catecholamines were unchanged initially but the noradrenaline level increased significantly after 1 hour and the adrenaline level at 1.5 hours — times at which there was a marked loss of body fluid. The excretion of water and sodium returned to normal after 2.5 hours. There were no changes in blood pressure but the pulse rate rose significantly after 1.5 hours. Their findings suggested that adrenergic mechanisms are only involved in response to loss of body fluid.

Fieldman et al (1985) found that plasma arginine vasopressin (AVP) rose markedly within an hour of the onset of surgery with levels well above that necessary to produce a maximum antidiuresis. Kaufman & Bailey (1986) confirmed that ADH levels rose rapidly during the first half-hour of surgery and continued to rise throughout the surgical procedure. If bumetanide was given early during the operation the rise in ADH levels were suppressed and a diuresis initiated.

Cardiac receptors and vagal afferent pathways may be involved in the ADH response to volume depletion (Grimaldi et al 1985). In a study of diabetic patients 50% were considered to have cardiac autonomic neuropathy. Volume depletion was assessed after intravenous administration of frusemide. In those with autonomic dysfunction the mean blood pressure fell and there was no increase in ADH. PRA and plasma aldosterone were elevated in both the patients with autonomic dysfunction and the controls.

The factors affecting plasma levels of aldosterone have been already outlined but Gilchrist et al (1984) have shown that metoclopramide (10 mg i.v.), a dopamine blocking agent, caused a significant increase in plasma aldosterone within 10 minutes without any effect on plasma ACTH, plasma renin, angiotension II or potassium. At 20 minutes the level had increased to a peak of three-fold. Metoclopramide also caused a rise in plasma vasopressin after 15 minutes with further peaks at 90 and 120 minutes accompanied by a slight but significant drop in blood pressure (Nomura et al 1984).

Thus it can be seen that diuretics and the interaction between the various hormones and electrolytes produce results which are complex and difficult to interpret.

REFERENCES

Alberti K G M M, Thomas D J B 1979 The management of diabetes during surgery. British Journal of Anaesthesia 51: 693–710
Aronson J K 1985 Cardiac arrhythmias: theory and practice. British Medical Journal 290: 487–488
Banner A S, Green J, O'Conner M 1984 Relation of respiratory water loss to coughing after exercise. New England Journal of Medicine 311: 883
Barnes P J 1983 Calcium-channels blockers in asthma. Thorax 38: 481–485
Belic N, Talano J V 1985 Current concepts in sick sinus syndrome. II. ECG manifestation and diagnostic and therapeutic approaches. Archives of Internal Medicine 145: 722–726
Blandford R L, Burden A C 1984 Abnormalities of cardiac conduction in diabetes. British Medical Journal 289: 1659
Bowen D J, Daykin A P, Nancekieveill M L, Norman J 1984 Insulin-dependent diabetic patients during surgery and labour. Use of continuous intravenous insulin-glucose-potassium infusions. Anaesthesia 39: 407–411
Cannella G, Galva M D, Campanini M, Cesura A M, De Marinis S, Picotti G B 1983 Sequential changes in plasma renin activity and plasma catecholamines in mildly hypertensive patients during acute, furosemide-induced body-fluid loss. European Journal of Pharmacology 25: 299–302
Chediak J, Chausow A, Solarski A, Telfer M C 1984 Pulmonary function in hemophiliac patients treated with commerical factor VIII concentrates. The American Journal of Medicine 77: 293–295
Cooper D M, Rojas J V, Mellins R B, Keim H A, Mansell A L 1984 Respiratory mechanics in adolescents with idiopathic scoliosis. American Review of Respiratory Disease 130: 16–22
Coriat P, Harari A, Daloz M, Viars P 1982 Clinical predictors of intraoperative myocardial ischemia in patients with coronary artery disease undergoing non-cardiac surgery. Acta Anaesthesiologica Scandinavica 26: 287–290
Davies M J, Thomas A 1984 Thrombosis and acute coronary-artery lesions in sudden cardiac ischemic death. New England Journal of Medicine 310: 1137–1140
Dikshit K, Vyden J K, Forrester J S, Chatterjee K, Prakash R, Swan H J C 1973 Renal and extra renal haemodynamic effects of frusemide in congestive heart failure after acute myocardial infarction. New England Journal of Medicine 228: 1087–1090
Dinubile M J 1984 Heart block during bacterial endocarditis: a review of the literature and guidelines for surgical intervention. American Journal of the Medical Sciences 287: 30–32
Dornhorst A, Powell S H, Pensky J 1985 Aggravation by propranolol of hyperglycaemic efect of hydrochlorothiazide in type II diabetics without alteration of insulin secretion. Lancet 1: 123–126
Downey G P, Rosenberg M, Caroff S et al 1984 Neuroleptic maligant syndrome. Patient with unique clinical and physiologic features. American Journal of Medicine 77: 338–340
Ellis F R, Heffron J J A 1985 Clinical and biochemical aspects of malignant hyperpyrexia. Recent Advances in Anaesthesia. In: Atkinson R S, Adams A P (eds) Churchill Livingstone, Edinburgh, 173–207

Eschenbacher W L, Bethel R A, Boushey H A, Sheppard D 1984 Morphine sulfate inhibits bronchoconstriction in subjects with mild asthma whose responses are inhibited by atropine. American Review of Respiratory Disease 130: 363–367

Estenne M, Borenstein S, De Troyer A 1984 Respiratory muscle dysfunction in myotonia congenita. American Review of Respiratory Disorders 130: 681–684

Fieldman N R, Forsling M L, Le Quesne L P 1985 The effect of vasopressin on solute and water excretion during and after surgical operations. Annals of Surgery 201: 383–390

Freedman S, Cooke N T, Moxham J 1983 Production of lactic acid by respiratory muscles. Thorax 38: 50–54

Freis E D 1983 How diuretics lower blood pressure. American Heart Journal 106: 185–187

Fujimaki M, Nagahama S, Suzuki H, Saito I, Saruta T 1984 Modulation of aldosterone secretion in frusemide-induced hypokalaemia. Acta Endocrinologica 107: 91–96

Gallagher J D, Lieberman R W, Meranze J, Spielman S R, Ellison N 1981 Amiodarone-induced complications during coronary artery surgery. Anesthesiology 55: 186

Gilchrist N L, Espiner E A, Nicholls M G, Donald R A 1984 Effect of metoclopramide on aldosterone and regulatory factors in man. Clinical Endocrinology 21: 1–7

Glenn W W L, Hogan J F, Loke J S O, Ciesielski T E, Phelps M L, Rowedder R 1984 Ventilatory support by pacing of the conditioned diaphragm in quadriplegia. New England Journal of Medicine 311: 1150–1155

Greenough A, Roberton N R C 1985 Morbidity and survival in neonates ventilated for respiratory distress syndrome. British Medical Journal 290: 597–600

Grimaldi A, Pruszczynski W, Thervet F, Ardaillou R 1985 Antidiuretic hormone response to volume depletion in diabetic patients with cardiac autonomic dysfunction. Clinical Science 68: 545–552

Gross N J, Skorodin M S 1984 Role of the parasympathetic system in airway obstruction due to emphysema. New England Journal of Medicine 311: 421

Guilleminault C, Kurland G, Winkle R, Miles L E 1981 Severe kyphoscoliosis, breathing, and sleep. 'Quasimodo' syndrome during sleep. Chest 79: 626–630

Haalboom J R E, Deenstra M, Struyvenberg A 1985 Hypokalaemia induced by inhalation of fenoterol. Lancet 1: 1125–1127

Hall G M 1984 Diabetes in Anaesthesia — a promise unfulfilled. Anaesthesia 39: 627–628

Heatley R V, Thomas P, Prokipchuk E J, Gauldie J, Sieniewicz D J, Bienenstock J 1982 Pulmonary function abnormalities in patients with inflammatory bowel disease. Quarterly Journal of Medicine, new series 203: 241–250

Ind P W, Causon R C, Brown M J, Barnes P J 1985 Circulating catecholamines in acute asthma. British Medical Journal 290: 267–269

Itkonen J, Schnoll S, Daghestani A, Glassroth J 1984 Accelerated development of pulmonry complications due to illict intravenous use of pentazocine and tripelennamine. The American Journal of Medicine 76: 617–622

Johnson D C, Frankel H M, Kazemi H 1984 Effect of furosemide on cerebrospinal fluid composition. Respiration Physiology 56: 301–308

Johnson G D, Nicholls D P, Leahey W J, Finch M B 1983 The effects of captopril on the acute vascular responses to frusemide in man. Clinical Science 65: 359–363

Johnson G D, O'Connor P C, Nicholls D P, Leahey W J, Finch M B 1985 The effects of propranolol and digoxin on the acute vascular responses to frusemide in normal man. British Journal of Clinical Pharmacology 19: 417–421

Juan G, Calverley P, Talamo C, Schnader J, Roussos C 1984 Effect of carbon dioxide on diaphragmatic function in human beings. New England Journal of Medicine 310: 874–879

Kaplan N M, Carnegie A, Raskin P, Heller J A, Simmons M 1985 Potassium supplementation in hypertensive patients with diuretic-induced hypokalemia. New England Journal of Medicine 312. 746–749

Kassirer J P, Harrington J T 1985 Fending off the potassium pushers. New England Journal of Medicine 312: 785–787

Kates R E, Woosley R L, Harrison D C 1984 Clinical importance of metabolites of antiarrhythmic drugs. The American Journal of Cardiology 53: 248–251

Kaufman L 1960 Anaesthesia in dystrophia myotonica. A review of the hazards of anaesthesia. Proceedings of the Royal Society of Medicine 53: 183–187

Kaufman L, Bailey P M 1986 (in press)

Lant A 1985 Diuretics. Clinical pharmacology and therapeutic use (pt I). Drugs 29: 57–87

Lant A 1985 Diuretics. Clinical, pharmacology and therapeutic use (Pt II), Drugs 29: 162–88

Larson E B 1985 Positive airway pressure for high-altitude pulmonary oedema. Lancet 1: 371–373

Leading Article 1984 Neuroleptic Maligant Syndrome. Lancet 1: 545–546

Leading Article 1985a Acute mishaps during insulin pump treatment. Lancet 1: 911–912

Leading Article 1985b Diabetes in pregnancy. Lancet 1: 961–962

Levine J H, Michael J R, Guarnieri T 1985a Treatment of multifocal artrial tachycardia with verapamil. New England Journal of Medicine 312: 21–25

Levine J H, Michael J R, Guarnieri T 1985b Multifocal atrial tachycardia: a toxic effect of theophylline. Lancet 1: 12–14

Maddern G J, Horowitz M, Jamieson G G 1985 The effect of domperidone on oesophageal emptying in diabetic autonomic neuropathy. British Journal of Clinical Pharmacology 19: 441–444

Maisel A S, Jotulsky H J, Insel P A 1985 Hypotension after quinidine plus verapamil: possible additive competition at alpha-adrenergic receptors. New England Journal of Medicine 312: 167–170

Manohar M, Parks C M 1984 Porcine brain myocardial perfusion during enflurane anaesthesia without and with nitrous oxide. Journal of Cardiovascular Pharmacology 6: 1092–1101

Marcus F I 1983 Drug interactions with amiodarone. American Heart Journal 106: 924–929

Materson B J 1983 Insights into intrarenal sites and mechanisms of action of diuretic agents. American Heart Journal 106: 188–208

Mazuz M, Friedman H S 1983 Significance of prolonged electrocardiographic pauses in sinoatrial disease: sick sinus syndrome. American Journal of Cardiology 52: 485–489

Metildi L A, Shackford S R, Virgilio R W, Peters R M 1984 Crystalloid versus colloid in fluid resuscitation of patients with severe pulmonary insufficiency. Surgery Gynecology and Obstetrics 158: 207–212

Mondini S, Guilleminault C 1985 Abnormal breathing patterns during sleep in diabetes. Annals of Neurology 17: 391–395

Mueller P S, Vester J W, Fermaglich J 1983 Neuroleptic malignant syndrome. Journal of the American Medical Association 249: 386–388

Murciano D, Aubier M, Lecocguic Y, Pariente R 1984 Effects of theophylline on diaphragmatic strength and fatigue in patients with chronic obstructive pulmonary disease. New England Journal of Medicine 311: 349–353

Mustard J F, Packham M A 1984 Platelets and diabetes mellitus. New England Journal of Medicine 311: 665–667

Naito M, David D, Michelson E L, Schaffenburg M, Dreifus L S 1983 The hemodynamic consequences of cardiac arrhythmias: evaluation of the relative roles of abnormal atrioventricular sequencing, irregularity of ventricular rhythm and atrial fibrillation in a canine model. American Heart Journal 106: 284–291

Nelson T E, Flewellen E H 1983 Current concepts: the malignant hyperthermia syndrome. New England Journal of Medicine 309: 416–418

Nomura K, Kurimoto F, Demura H et al 1984 Effect of metoclopramide on plasma vasopressin in man. Clinical Endocrinology 21: 117–121

Perloff J K, Stevenson W G, Roberts N K, Cabeen W, Weiss J 1984 Cardiac involvement in myotonic muscular dystropy (Steinert's Disease): a prospective study of 25 patients. American Journal of Cardiology 54: 1074–1081

Perrin D G, Becker L E, Madapallimatum A, Cutz E, Bryan A C, Sole M J 1984 Sudden infant death syndrome: increased carotidbody dopamine and noradrenaline content. Lancet 2: 535–537

Phibbs B, Marriott H J L 1985 Complications of permanent transvenous pacing. New England Journal of Medicine 312: 1428–1432

Prorok J J, Trostle D 1984 Operative risk of general surgical procedures in patients with previous myocardial revascularization. Surgery, Gynecology and Obstetrics 159: 214–216

Reynolds A K 1984 On the mechanism of myocardial sensitization to catecholamines by hydrocarbon anesthetics. Canadian Journal of Physiology and Pharmacology 62: 183–198

Ritchie P 1983 Neuroleptic malignant syndrome. British Medical Journal 287: 560

Robertson C E, Sinclair C J, Steedman D, Brown D, Malcom-Smith N 1985 Use of ether in life-threatening acute severe asthma. Lancet 1: 187–188

Salzman R, Manson J E, Griffing G T et al 1985 Intranasal aerosolized insulin: mixed-meal studies and long-term use in type-1 diabetes. New England Journal of Medicine 312: 1078–1084

Schade S D, Eaton R P 1985 Insulin delivery: how, when and where. New England Journal of Medicine 312: 1120–1121

Schuster C J, Weil M H, Bsso J, Carpio M, Henning R J 1984 Blood volume following diuresis induced by furosemide. American Journal of Medicine 76: 585–592

Simonneau G, Vivien A, Sartene R et al 1983 Diaphragm dysfunction induced by upper abdominal surgery. American Review of Respiratory Diseases 128: 899–903

Szabadi E 1984 Neuroleptic malignant syndrome. British Medical Journal 288: 1399–1400

Talbot E M 1984 Dangers of adding insulin to intravenous infusion bags with fixed needle syringes. British Medical Journal 289: 678–680

Talbert D G, Southall D P 1985 A bimodal form of alveolar behaviour induced by a defect in lung surfactant – a possible mechanism for sudden infant death syndrome. Lancet 1: 727–728

Tarnow-Mordi W O, Berril A M, Darby C W, Davis P, Pook J 1985 Precipitation of laryngeal obstruction in acute epiglottitis. British Medical Journal 290: 629

Trigt P Van, Christian C C, Fagraeus L et al 1984a Myocardial depression by anesthetic agents (halothane, enflurane and nitrous oxide): quantitation based on end-systolic pressure-dimension relations. American Journal of Cardiology 53: 243–247

Trigt P Van, Spray T L, Pasque M K, Peyton R B, Wechsler A S 1984b Anesthesia-induced myocardial depression: quantitation of effects on systolic and diastolic function. Surgery 96: 368–374

Verburg M P, Oerlemans F T J J, Van Bennekom C A, Gielen M J M, De Bruyn C H M M, Crul J F 1984 In vivo induced malignant hyperthermia in pigs. 1. Physiological and biochemical changes and the influence of dantrolene sodium. Acta Anaesthesiologica Scandinavica 28: 1–8

Vicken W G, Gauthier S G, Dollfus R E, Hanson R E, Darauay C M, Cosio M G 1984 Involvement of upper-airway muscles in extrapyramidal disorders. A cause of airflow limitation. New England Journal of Medicine 311: 438

Wardle E N 1984 Shock lungs: the post-traumatic respiratory distress syndrome. Quarterly Journal of Medicine, New Series LIII 211: 317–329

Watkins P J 1985 Pros and cons of continuous subcutaneous insulin infusion. British Medical Journal 290: 655–656

Watt A H 1985 Sick sinus syndrome: an adenosine-mediated disease Lancet 1: 786–788

Wilcox C S, Mitch W E, Kelly R A et al 1984 Factors affecting potassium balance during frusemide administration. Clinical Science 67: 195–203

Williams J G, Morris A I, Hayter R C, Ogilvie C M 1984 Respiratory responses of diabetics to hypoxia, hypercapnia, and exercise. Thorax 39: 529–530

Wilson A J, Stevens V, Franks C I, Alexander J, Southall D P 1985 Respiratory and heart rate patterns in infants destined to be victims of sudden infant death syndrome: average rates and their variability measured over 24 hours. British Medical Journal 290: 499–501

Wiseman M J, Saunders A J 1985 Effect of blood glucose control on increased glomerular filtration rate and kidney size in insulin-dependent diabetes. New England Journal of Medicine 312: 617–621

Wolfson J S, Swartz M N 1985 Drug therapy: serum bactericidal activity as a monitor of antibiotic therapy. New England Journal of Medicine 312: 968–975

Anaesthesia and asthma

INTRODUCTION

Asthma is a common disease thought to be present in approximately 5–10% of the population with a higher prevalence in children (Cropp 1985). Dundee et al (1978) reported that 3.5% of 10 000 patients presenting for surgery had bronchial asthma. The asthmatic individual faced with the prospect of surgery presents an increased risk of cardiorespiratory complications to the surgeon and anaesthetist compared to the non asthmatic patient. Approximately 75% of these complications are pulmonary, including some previously asymptomatic asthmatics who develop intra-operative bronchospasm (Shnider & Papper 1961). Stein et al (1962) found that the frequency of post-operative pulmonary complications in patients who present with broncho-pulmonary disease to be 70% as opposed to 3% in patients with normal pre-operative pulmonary status. Thus, careful pre- and post-operative care can reduce complications in these patients.

PATHOPHYSIOLOGY OF ASTHMA

In order to understand more clearly the asthmatic state, the response to anaesthesia and therapeutic strategies, the current concepts of the pathophysiology of asthma are described. Bronchial asthma is a disease of hyperirritable airways manifested by reversible airway obstruction associated with the following pathologic changes: bronchospasm, mucosal oedema, cellular infiltrates, mucus secretion, desquamation of surface epithelial cells, basement membrane thickening and globlet cell hyperplasia (Kaliner 1980). These changes have been attributed to the action of the chemical mediators of anaphylaxis released by lung mast cells.

Histamine and the products of arachidonic acid metabolism, now known as leukotrienes (LT), are thought to be the most important mediators contributing to bronchial airway obstruction (Lewis et al 1982). The pharmacology of LT and histamine is briefly described. (See also Casale & Kaliner 1984.)

Histamine

Histamine causes its pharmacological response by combining with surface cellular membrane receptors designated H_1 and H_2. Histamine causes bronchoconstriction by directly reacting with H_1 receptors on smooth muscle cells and may cause reflex bronchoconstriction by stimulating H_1 vagal afferent irritant receptors. In addition, H_1 receptors mediate increased vascular permeability, PGE_1 and $PGF_{2\alpha}$ generation from human lung and increases in cyclic guanosine monophosphate (CGMP) (Wasserman 1980). H_2 receptor stimulation results in gastric acid secretion, increases in cyclic adenosine monophosphate (CAMP), inhibition of mast cell histamine release, and eosinophilic migration. Vasodilation (hypotension), flush and headache are mediated by both H_1 and H_2 receptors.

Leukotrienes (LT)

The other major mediator in asthma is the biologically active moiety previously referred to as slow reacting substance of anaphylaxis (SRS-A) and now knowm as LT C, D and E metabolites of the lipoxygenase pathway. Studies in human bronchial muscle (Dahlen et al 1980) indicate that LTC and LTD are 1000 times more potent bronchoconstrictors than histamine on a molar basis.

Autonomic Activity

Finally, all normal bronchial tone is mediated by vagal constrictor influence. In asthmatics, the inhalation of muscarinic blocking agents (e.g. atropine) causes bronchodilation which suggests that acetylcholine release might be contributing to the bronchostriction (Storms et al 1975). The potential mechanisms by which cholinergic efferent discharge may be induced in asthma include both mediator stimulated reflex discharges and increased cholinergic hyperreactivity as a reflection of disturbed autonomic balance in asthma (Kaliner et al 1982). This imbalance consists of hypersensitivity to cholinergic and alpha adrenergic stimuli, both of which predispose to increased airway muscle tone, and hyporesponsiveness to beta adrenergic stimuli, which might make the asthmatic less responsive to normal bronchodilator influence. These influences taken as a whole facilitate hyperreactivity of the asthmatic tracheobronchial tree to a variety of immunological and non-immunological stimuli which could result in airway obstruction.

PRE-OPERATIVE ASSESSMENT

Prior to surgery the patient's asthma must be evaluated as to the severity and activity. One must inquire into the duration of the disease, the frequency of coughing and wheezing, both at rest and with exercise, the necessity for medication and the frequency of emergency visits and hospitalization. Sputum, if present, should be characterized as to quantity and quality. A careful smoking history might indicate the degree of bronchitic component. A medication history includes the dosage and frequency of administered theophylline, β adrenergic drugs, sodium cromoglycate (cromolyn sodium) and corticosteroids in addition to a history of drug allergy. Particular attention should be noted to abuse of adrenergic inhalers, which could sensitize the myocardium to arrhythmias during anaesthesia. Steroid coverage will be necessary during the peri-operative period in order to prevent potential adrenal insufficiency if the patient had received corticosteroids in the year prior to surgery (Spark 1975).

Table 3.1 Considerations in use of asthma medication

Drug	Comment
Sympathomimetics Epinephrine and β adrenergic agents	Severe arrhythmias are possible particularly when hypoxaemia and acidosis are present. Halothane increases arrhythmia potential.
Anticholinergics Atropine and glycopyrrolate	Not dangerous to asthmatic if patient adequately hydrated.
Theophylline Aminophylline	Halothane, sympathomimetics and pancuronium increase arrhythmia potential. Monitor theophylline levels since surgery can alter clearance. High dose can antagonize neuromuscular effect of pancuronium.
Corticosteroids	Corticosteroid supplementation recommended in patients receiving suppressive doses within 1 year of surgery. Give hydrocortisol (Solucortef) 125 mg/m² 1 h before induction, then.as continuous infusion over first 24 h. Rapidly taper over next 24–48 h to pre-operative dosage or discontinue.

The physical examination provides an opportunity to rule out infection and assess vital signs, heart and lungs. The patient should be examined in both upright and supine position and asked to breathe out forcibly through the mouth to rule out bronchial obstruction.

In addition to the usual pre-operative laboratory tests, one should examine the sputum, chest radiograph and pulmonary function (PFT) by simple spirometry. The forced expiration volume in one second (FEV_1) and minute ventilatory volume (MVV) give good predictive information regarding the ability to sustain respiratory stress (Geiger & Hedley-Whyte 1985). Response to bronchodilators and arterial blood gases may also be useful indicators particularly if the PFTs are abnormal initially.

PRE-OPERATIVE MANAGEMENT

Certain factors increase the risk of anaesthesia. Smoking or smoke exposure should be discontinued at least 1 week prior to surgery. In addition, active respiratory infection should be treated with physiotherapy (USA = chest therapy) and antibiotics where appropriate. No elective surgery should be scheduled if the asthma is unstable. Appropriate bronchodilators, steroids, if necessary, and physiotherapy may be required to reduce intra-operative and post-operative pulmonary complications. An attempt should be made to correct airway obstruction in the asthmatic patient even prior to emergency surgery. If the FEV_1/FVC ratio is less than 40% ventilatory failure should be anticipated in the post-operative period (Weiss & Faling 1968).

Pre-operative medication

The asthmatic patient may be particularly anxious before operation so that a patient and sympathetic pre-operative visit is advisable to allay the patient's fears. Sedation with diazepam may be prescribed. Antihistamines have also been recommended for sedation and drying secretions in the upper airways. [Promethazine (phenergan) in

(responsible for 60% of normal ventilation) is equally important. In addition, the asthmatic has a decreased ventilatory response to isocapnic hypoxaemia and hypercapnea (Hudgel & Weil 1975). During this period narcotics and sedatives must be used very cautiously or respiratory failure may develop.

Atelectasis is a common complication in the asthmatic because of decreased cough, respiratory depression and the tendency to develop viscous mucus. Airway obstruction can be minimized by administering warmed and humidified air while the patient should be adequately hydrated. The inhalation of a bronchodilator may relieve symptoms (UK = salbutamol, USA = metaproterenol). The physiotherapist should encourage the patient to take slow, deep breaths to ensure uniform distribution of gas throughout the lung.

Asthmatic patients require careful observation at night especially during the early morning hours when sudden death from ventilatory arrest has been reported (Hetzel et al 1977). Deterioration in pulmonary status can be detected by changes in vital signs including pulsus paradoxus, arterial blood gases and spirometry. Thus, tachycardia greater than 110/min, pulsus paradoxus > 30 mm Hg, $FEV_1 < 1L$, FVC $< 1.5L$ and FEV_1/PVC ratio $< 30\%$, MVV $< 50\%$ of predicted and a rising Pa_{CO_2} suggest exhaustion and impending respiratory failure and the need for ventilatory support.

REFERENCES

Barnes P, FitzGerald G, Brown M, Dollery C 1980 Nocturnal asthma and changes in circulating epinephrine, histamine and cortisol. New England Journal of Medicine 303: 263–267
Belani K G, Anderson W W, Buckley J J 1982 Adverse drug interaction involving pancuronium and aminophylline. Anesthesia and Analgesia 61: 473–474
Casale T B, Kaliner M 1984 Allergic reactions in the respiratory tract. In: Bienenstock J (ed) Immunology of the Lung and Upper Respiratory Tract. McGraw Hill, New York, ch 16, p 326–344
Chen W Y, Horton D J 1978 Airways obstruction in asthmatics induced by body cooling. Scandinavian Journal of Respiratory Disease 59: 13–20
Crago R R, Bryan A C, Laws A K, Winestock A E 1972 Respiratory flow resistance after curare and pancuronium measured by forced oscillations. Canadian Anaesthesiology Society Journal 19: 607–614
Cropp G J A 1985 Special features of asthma in children. Chest 87: 55S–62S
Dahlen S E, Hedqvist P, Hammarström S, Samuelsson B 1980 Leukotrienes are potent constrictors of human bronchi. Nature 288: 484–486
Dundee J W, Fee J P, McDonald J R, Clarke R S 1978 Frequency of atopy and allergy in an anesthetic patient population. British Journal of Anaesthesiology 50: 793–798
Geiger K, Hedley-Whyte J 1985 Preoperative and postoperative considerations. In: Weiss E B, Segal M J, Stein M (ed) Bronchial Asthma, 2nd edn. Little Brown, Boston, ch 75, p 892–907
Hetzel M R, Clark T J H, Branthwaite M A 1977 Asthma analysis of sudden deaths and ventilatory arrests in hospital. British Medical Journal 1: 808–811
Hirshman C A 1983 Airway reactivity in humans, anesthetic implications. Anesthesiology 58: 170–177
Hirshman C A, Dunes H, Farbood A, Bergman N A 1979 Ketamine block of bronchospasm in experimental canine asthma. British Journal of Anaesthesia 51: 713–718
Hirshman C A and Bergman N A 1978 Halothane and enflurane protect against bronchospasm in asthma dog model. Anesthesia and Analgesia 57: 629–633
Hudgel D W, Weil J V 1975 Depression of hypoxic and hypercapnic ventilatory drive in severe asthma. Chest 68: 493–497
Kaliner M A 1980 Mast cell-derived mediators and bronchial asthma. In: Hargreave F E (ed) Airway Reactivity. Astra Pharmaceuticals Canada Ltd, Mississauga, Ontario, p 175
Kaliner M, Shelhamer J H, Davis P B, Smith L J, Venter J C 1982 Autonomic nervous system abnormalities and allergy. Annals of Internal Medicine 96: 349–357
Lewis R D, Levine L, Austen K F, Corey E J 1982 Radioimmunoassays (RIA) of the slow reacting substance of anaphylaxis (SRS-A) leukotrienes. Journal of Allergy and Clinical Immunology (supplement) 69: 93

Moss J 1985 Histamine release in anesthesia and surgery. New England and Regional Allergy
    Proceedings 6: 28-36
Moss J, Rosow C E, Savarese J J, Philbin D M, Kniffen K J 1981 Role of histamine in the
    hypotensive action of d-tubocurarine in humans. Anesthesiology 55: 19-25
Philbin D M. Moss J, Atkins C W et al 1981 The use of H_1 and H_2 histamine antagonists with
    morphine anesthesia: A double blind study. Anesthesiology 55: 292-296
Shnider S M, Papper E M 1961 Anesthesia for the asthmatic patient. Anesthesiology 22: 886-892
Spark R F 1975 Hypothalemic-pituitary-adrenal axis in surgery. In: Skillman J J (ed) Intensive Care
    Little Brown, Boston, p 311
Stein M, Koota G M, Simon M, Frank H A 1962 Pulmonary evaluation of surgical patients. Journal
    of American Medical Association 181: 765-770
Storms W W, Do Pico G A, Reed C E 1975 Aerosol SCH 1000 — an anticholinergic bronchodilator.
    American Review of Respiratory Disease 111: 419-422
Wasserman S I 1980 The lung mast cell: its physiology and potential relevance to defense of lung.
    Environmental Health Perspectives 35: 153-164
Weiss E B, Faling L J 1968 Clinical significance of Pa_{CO2} during status asthmaticus: the cross-over
    point. Annals of Allergy 26: 545-551

Treatment of acute severe asthma

Asthma is a common and potentially lethal disease which affects 2–4% of the British population (Gregg 1983). Although individual doctors will rarely see death as the outcome of an acute asthmatic attack, over 1200 asthmatics die of their disease each year (about 2 in every 1000 British asthmatics). It is a sobering thought that despite the many recent advances made in the treatment of this disorder, there has been no improvement in the standardized mortality rates associated with asthma since the introduction of adrenaline in the late 1890s (Clark 1983). In fact this mortality rate increased dramatically in the 1960s due to overreliance on the newly introduced isoprenaline pressurized aerosols (Inman & Adelstein 1969), and more recently there has been a further increase in deaths in New Zealand (Jackson et al 1982) and a lesser trend in Great Britain (Khot & Burn 1984). This latest epidemic is thought to be related to a combination of high dose nebulized β_2-sympathomimetic agents and sustained-release theophylline preparations (Grant 1985). A survey of home nebulizer use in the UK showed that 25% of patients had never received any formal medical instructions on the use of the nebulizer, and some patients were using 50 mg salbutamol a day which is 250 times the manufacturer's recommended dose for this drug (Laroche et al 1985).

ACUTE SEVERE ASTHMA

Retrospective studies of the circumstances surrounding asthma deaths both inside and outside hospital have suggested a number of possible causes (Table 4.1) (Cochrane & Clark 1975, British Thoracic Association 1982). The rate of evolution of a severe attack is characteristically over 24 or more hours, and at post-mortem such patients' lungs have mucosal oedema and airways plugged with tenacious mucus. However some patients die within 30 minutes of an acute attack, and at post-mortem the lungs appear macroscopically and microscopically normal, suggesting that death was due to overwhelming bronchoconstriction. This is one reason why the rather vague term 'status asthmaticus' is no longer used to describe this medical emergency (Clark 1983). The term implied an asthmatic attack which was 'prolonged' (though clearly death can occur early) and unresponsive to treatment (though this was not clearly defined). As a result more clinicians now use the more straightforward term 'acute severe asthma' to describe an episode which is increasing in severity over a short period of time and which does not respond to the patient's usual treatment, or even to an increase in treatment.

Table 4.1 Factors associated with a fatal outcome from acute severe asthma

1. Lack of patient or doctor awareness of the severity of the attack

2. Inadequate or inappropriate treatment:
   e.g. underuse of corticosteroids
       use of sedatives

3. Rapid speed of onset of the terminal attack

4. Failure of diagnosis during life
   e.g. labelled as a chronic bronchitic

Assessment of severity

Analysis of the records of those asthmatics who died in hospital has pointed to inadequate assessment of the asthmatic attack by the doctor which in turn led to undertreatment and, perhaps, inappropriate placement of the patient within the hospital (Ormerod & Stableforth 1980, Cochrane & Clark 1985). No single clinical feature has been shown to be a reliable gauge of severity. The most effective means of assessment involves the use of as many objective measures as possible, which should always be clearly documented. These are summarized below (Clark 1983, Cochrane 1984, Tatham & Gellert 1985).

History

The cardinal features of severe asthma are a history of decreasing exercise tolerance, sleep disturbance due to early morning attacks of wheezing, and failure to obtain relief from the usual (or an increase in) bronchodilator treatment. Patients who have an excessive diurnal variation in airway calibre are known as 'morning dippers' as they tend to have increased bronchospasm between 2–4 a.m. They and patients with brittle disease, i.e. intractable persisting asthma which is resistant to conventional therapy, are thought to be particularly at risk of catastrophically rapid attacks, with the possibility of death within minutes of the onset of the attack (MacDonald et al 1976, Hetzel et al 1977).

Examination

The signs which indicate a severe attack and which must be documented are:

1. Difficulty in speaking a sentence without a pause for breath. In severe cases the patient may be too breathless to be able to say even a single word.

2. Tachypnoea, with the use of accessory muscles.

3. Pulse rate over 110/min. Over 140/min implies severe hypoxia.

4. Pulsus paradoxus more than 10 mm Hg (i.e. an exaggeration of the normal 5–10 mm Hg difference in systolic arterial pressure between inspiration and expiration). The mechanism of this remains uncertain. One theory is that pooling of blood within the lungs, due to dilatation of the pulmonary veins during expiration, leads to a fall in left atrial and thus left ventricular stroke volume. Even this sign may be absent, even in severe asthma, and its disappearance indicates decreasing inspiratory effect caused by exhaustion or improvement.

5. Quiet breath sounds and absent wheezes indicate that the rate of expiratory flow is very slow which is a dangerous sign. It may also mean a pneumothorax is present.

6. Cyanosis and signs of carbon dioxide retention are very late signs, and indicate an impending cardiorespiratory arrest.

Investigations

1. Measure airflow obstruction by recording the peak expiratory flow rate (PEFR) or forced expiratory volume in 1 s (FEV_1). This is the most important single observation to make, as it provides the simplest and best guide to severity. The FEV_1 and FVC are measured on a spirometer, either the conventional water-sealed type or else a 'dry' machine (e.g. Vitalograph). The PEFR is a measure of the maximum speed averaged over 10 ms, as recorded using the Wright peak flow meter. The best of three readings is taken for all of these tests of forced expiration. In a severe attack the PEFR is less than 100 l/min and FEV_1 less than 1 litre (or 30% of the predicted normal) in adults. Some patients with acute severe asthma are too breathless to perform these tests. Serial measurement (i.e. 6 hourly, before and after nebulized brochodilator treatment) is the most useful way of monitoring subsequent progress of the disease.

2. Arterial blood gas analysis should be performed in all patients admitted to hospital. The Pa_{CO_2} correlates well with severity. Any elevation of Pa_{CO_2} (i.e. above 35 mm Hg) indicates that ventilation is proving inadequate and more vigorous treatment is required.

3. A chest X-ray should be performed daily to look for evidence of collapse or infection or the presence of a pneumothorax or pneumomediastinum.

4. An electrocardiogram may reveal changes of acute pulmonary hypertension which should return to normal when the attack improves.

5. Hypokalaemia may occur during acute severe asthma , due in part to the high levels of circulating catecholamines and the renal response to stress (Webb et al 1979). This may be further aggravated by the use of nebulized β-sympathomimetic agents and oral/intravenous corticosteroids.

Treatment of Acute Severe Asthma

There is no simple formula. Every asthmatic demands careful appraisal and choice of treatment, and the clinician must always assess the response to treatment by serial PEFR (or FEV_1) measurements and blood gas analyses, and be on the lookout for associated dangers (e.g. hypokalaemia, pneumothorax). The therapy includes specific treatment to reverse airflow obstruction (Table 4.2) and general measures (Table 4.3).

Specific treatment of reverse airflow obstruction

Step 1. The initial treatment of an acute severe attack should consist of a *bronchodilator* inhaled by a nebulizer.

1. Nebulized selective β_2-adrenergic stimulant: e.g. salbutamol 5 mg or terbutaline 10 mg in a simple nebulizer (e.g. Inspiron 'mini-neb'). Use half these doses if there is a past history of ischaemic heart disease. Peak improvement is reached after 15 minutes and effects last up to 6 hours. This treatment may be repeated 2–4 hourly, as necessary.

Table 4.2   Acute adult severe asthma: specific treatment to reverse airflow obstruction

---

Step 1

Nebulized B$_2$-agonist   (e.g. salbutamol 5 mg, terbutaline 10 mg)
AND
Nebulized ipratropium bromide (500 mg)
	If no response in 30–60 min repeat once
	If still no response — Step 2

Step 2

Corticosteroids:	hydrocortisone 250 mg i.v. 6 hourly AND
	prednisolone 30–60 mg each morning
AND	
i.v. aminophylline:	bolus
	infusion 0.5 mg/kg/h ·
	monitor plasma levels 8–12 hourly

There are a number of practical points to remember. This simple nebulizer has a dead space of 0.6 ml, so the aerosol should be diluted in normal saline or water to make at least 5 ml of solution. The nebulizer must be driven by oxygen at a flow rate of at least 8 1/min, which will deliver the drug in 10–15 minutes. There is no apparent advantage to delivering the drug through an intermittent positive pressure breathing device (e.g. Bird or Bennett ventilator) (Campbell et al 1978). Intravenous administration (5 mg in 500 ml normal saline, 3–20 mg/min) has no advantages even in unconscious patients (nebulizers can be given via a ventilator), and causes more unwanted effects (e.g. tachycardia, hypokalaemia) (Lawford et al 1978, Bloomfield et al 1979).

2. Nebulized *ipratropium bromide* (500 mg–1 mg) gives additional benefit to that provided by nebulized β_2-sympathomimetic drugs (Ward et al 1981, Leahy et al 1983). Ipratropium is an atropine derivative that is a powerful anticholinergic inhibitor of vagally mediated bronchomotor tone. Peak effects are reached after 30–45 minutes, and last 3–5 hours. There are no systemic atropine-like effects (e.g. glaucoma, urinary retention) and no inhibition of mucociliary clearance (Paterson et al 1979). It has been suggested that ipratropium and the β_2-agonists may act preferentially on different parts of the bronchial tree (e.g. large conducting airways — ipratropium; small airways — β_2-agonists) but the evidence is conflicting (Ingram et al 1977, Partridge & Saunders 1981). It is important to remember to dilute the drug in 5 ml N saline and not water, as hypotonic solutions can cause bronchoconstriction.

Table 4.3   Acute adult severe asthma: general measures

Routinely used	Continuous oxygen
	Intravenous rehydration
	Monitor serum potassium
Rarely used	Antibiotics
	Chest physiotherapy
Never used	Sedation
	Bronchial lavage, unless on a ventilator (rarely)

Step 2. This is reserved for those patients who show no improvement in PEFR within 1–2 hours after 10 mg salbutamol (20 mg terbutaline) and 500 mg ipratropium by nebulizer, or for those whose initial assessment indicates that no delay in starting maximal therapy can be contemplated.

1. *Corticosteroids.* An intravenous loading dose of 200–250 mg hydrocortisone (3–4 mg/kg body weight) is given and repeated 6 hourly for 24 hours, and at the same time start oral prednisolone or increase present dose to 30–60 mg each morning. The intravenous steroid will take 4–6 hours to act, and the oral, 12–24 hours. Regular urinalysis to detect diabetes is important. The action of steroids in acute severe asthma is not understood. Certainly in vitro glucocorticoids can effect almost every stage of inflammatory and immunological activity, but only by using often supra-pharmacological doses (Kay 1983). Similarly studies of their use in acute severe asthma have produced contradictory results (Pierson et al 1974, Luksza 1982). At present, however, no firm evidence is available to show that steroids should be withheld and most clinicians feel they are safe and may be life-saving (Grant 1982).

2. *Intravenous aminophylline.* A loading dose of 5.6 mg/kg body weight is given over 15 minutes (i.e. 250–500 mg) followed by a continuous infusion of 0.5 mg/kg body weight/hour with 8–12 hourly estimations of plasma theophylline concentrations to guide dosage. A loading dose must *not* be given to patients already receiving oral sustained-release theophylline preparations (e.g. phyllocontin, uniphylline) unless the result of a plasma theophylline measurement is available. Techniques are now available for on the spot measurements (Cuss et al 1985). These sustained released theophylline preparations are reviewed in Drugs and Theraputics Bulletin 1985.

This drug has an extremely narrow therapeutic range 10–20 mg/l and potentially lethal effects (e.g. dysrythmias, general seizures, cardiac arrests) have been seen with levels as low as 25 mg/l. Such effects are not necessarily preceded by nausea, headache and vomiting (Woodcock et al 1983).

It is not possible to predict with confidence the ideal dose for a particular patient. As a result, estimation of plasma concentration is mandatory if intravenous treatment is contemplated for more than 24 hours (Conway et al 1984).

General Measures (Table 4.3)

Continuous hummidified oxygen (40%) should be given to keep the Pa_{O_2} at least above 8 KPa (when 90% of the haemoglobin is saturated), whilst monitoring the Pa_{CO_2}. If there is concern about possible CO_2 retention, start at 28% oxygen.

Intravenous hydration is necessary both to replace the excessive water losses through the lungs and because the patient may be too breathless to drink (e.g. 1–1.5 l/24h). It will also help preserve an adequate circulating volume and may help mobilize secretions. Routine antibiotic treatment is no longer given, unless there are strong indications of bacterial infection (e.g. lobar pulmonary infiltrate) (Graham et al 1982). Fever may occur in a viral or bacterial infection, and sputum which looks mucopurulent in most cases contains predominantly eosinophils rather than neutrophils. Similarly chest physiotherapy is not indicated unless the chest X-ray indicates lung collapse associated with mucus impaction, or the patient needs help with the use of nebulizers. However severely affected and exhausted patients will not be able to co-operate with the physiotherapist.

Artificial Ventilation

A small number of patients do not improve in spite of maximal therapy and require ventilation. In a 9 year period 1% of all hospital admissions for asthma in one series required ventilation. 40% of these died (Scoggin et al 1977). Other reports have an approximate 10% death rate (Westerman et al; Sheehy et al 1972). The indications for ventilation were surprisingly uniform and were:
1. Drowsiness and confusion
2. CO_2 retention
3. Morbidity in casualty or
4. A cardiac arrest

Table 4.4 outlines a suggested criteria for ventilation. Acute asthmatics have hyperinflated lungs and have to exert a lot of energy in expiration. During an asthmatic attack the mechanical work of ventilation is two to nine times greater than normal and peak airway pressures up to 70 cm of water may be generated (Dacosta & Hedstrand 1970). Expiration is prolonged in an effort to expel the total tidal volume. With artificial ventilation it is important to lengthen the expiratory cycle so as to give a long enough period for inspiration as well as a long expiratory time otherwise inflation pressures will be unacceptably high for the necessary tidal volume. If all the tidal volume is not expired then air trapping will occur giving further lung distension causing lung damage with stretching and distortion of lung capillaries. This causes an increase in pulmonary resistance which will result in a low cardiac output and lead to right heart failure. Thus in the acute asthmatic patient it is important to choose a low minute volume with a slow rate (Branthwaite 1978). The Pa_{CO_2} may well rise from underventilation but the patient will not come to harm unless there is a gross acidosis which can be corrected by sodium bicarbonate (Menitove et al 1983). The hypercarbia will make the patient fight the ventilator so it is important to keep the patient well sedated and if necessary paralysed. Positive end expiratory pressure has no place in asthma as it will cause overdistension of the lungs as the air has greater difficulty in escaping during the expiratory phase time.

Induction of these distressed and hypoxic patients can be difficult and dangerous. A semi-elective procedure should be anticipated before the patient becomes

Table 4.4   Suggested indications for ventilation based on sequential observations

1. Exhaustion, intolerable distress

2. Evidence of deterioration in pulmonary function
        $Pa_{O_2} < 6.7$ KPa (50 mmHg) and falling
        $Pa_{CO_2} > 6.7$ KPa (50 mmHg) and rising

3. Evidence of deterioration in cardiac function
        hypotension, cardiac dysrhythmia

4. Emergency indications
        cardiac or respiratory arrest
        irreversible respiratory depression due to drugs (e.g. sedatives)
        lung collapse, acute pneumothorax, acute pneumomediastinum

moribund. Webb et al (1979) suggested a gas induction in the semi-recumbent position with the bronchodilating agent halothane (Schnider & Papper 1961, Hirshman et al 1978a). However careful monitoring of the heart must be carried out as halothane sensitizes the myocardium to the dysrhythmic effects of catecholamines (Wood et al 1968). Ether also relieves bronchospasm and is useful in experienced hands but it has been generally replaced by newer agents with fewer side effects. Enflurane and isoflurane are known to be bronchodilators in the non-asthmatic (Wood et al 1968) and reported clinical experience with these agents is awaited. They are less likely to give arrythmias in the presence of adrenaline or aminophylline (Stirt et al 1981, 1983). It is important not to intubate too early as reflex bronchospasm may occur with instrumentation or as the endotracheal tube enters the trachea. Webb et al (1979) suggested using a lignocaine spray to prevent reflex (irritant) bronchospasm. Downes & Hirshman (1981) suggest it may not prevent bronchoconstriction. Intravenous lignocaine has been found useful in treating bronchospasm (Brandus et al 1970) but its effect on the myocardium and brain should temper its usage.

Some anaesthetists prefer an intravenous induction with a relaxant. Ketamine has been shown by several workers to improve bronchospasm and be a useful induction agent (Betts & Parkin 1971, Corssen et al 1972). Convulsive-like movements have been noted when an aminophylline infusion has been used at the same time (Hirshman et al 1982b). Suxamethonium is the most frequently used rapidly acting muscle relaxant. There are very few reports of bronchospasm following suxamethonium (Fellini et al 1963, Eustace 1967) so it would appear to be a safe agent.

Once intubation has occurred, total paralysis by a long acting muscle relaxant must be achieved and control of respiration obtained by manual ventilation. D-Tubocurarine is associated with histamine release (Salem et al 1968, Moss & Roscow 1983) and for this reason it is considered to be contra-indicated in asthma. Fung (1985) points out there is no real evidence of D-tubocurarine being associated with worsening of bronchospasm. Pancuronium has been the choice of relaxant for many years because of its low potential for the release of histamine and appears to be a safe agent (Nana et al 1972) but there has been a report of bronchospasm in an asthmatic (Heath 1973). Because of expense and shortness of action there would seem to be no place for vecuronium and atracurium in the long term ventilation of an asthmatic.

Ventilator The ventilator used should be capable of giving a preset volume in the face of increased airways resistance and the facility to increase the expiratory time. Minute volumes should be low, i.e. 6–8 litres in an average adult with a rate of 10–14 breaths per minute. The expiratory phase should be 50% of the respiratory cycle (Branthwaite 1978). The required inspired O_2 level should be determined from several blood gas estimations to keep the PO_2 above 8 kPa. Although oxygen toxicity may be a clinical entity it is vital the patient does not become brain damaged from hypoxia. By accurately measuring the chest wall at regular intervals with a tape measure an increase in diameter by one centimetre or more will suggest air trapping and the minute volume will have to be decreased. On listening to the expiratory ventilator tubing detached from the ventilator it should be possible to hear the end of expiration before the next cycle starts. If this does not happen air trapping will occur.

After the start of artificial ventilation it is not uncommon for the blood pressure to drop to low levels. This may be due to right-sided heart failure but it is more likely to be due to a relative hypovolaemia caused by dehydration giving a low filling pressure. The patient may have a normal or high central venous pressure because of the increased pulmonary vascular pressure from distortion and stretching of the capillaries. The giving of aliquots of colloid such as 4% human albumin solution (plasma protein fraction) or gelatine solutions in 200 ml boluses every 10 minutes should bring the blood pressure to acceptable levels, i.e. high enough to produce a urinary output (Webb et al 1979). Cardiac failure will ensue if too much fluid is given. If possible the patient's low circulating volume should be corrected before intubation.

Opiates are commonly used to maintain sedation and respiratory depression in ventilated patients. Morphine is capable of releasing histamine and has been shown to have a bronchoconstrictive effect in asthma (Blumberg 1973). However morphine (Pontoppidan et al 1972) and phenoperidine (Webb et al 1979) have been used successfully. Fentanyl does not release histamine (Moss & Roscow 1983) and is relatively cardiac stable. Pethidine in spite of its bronchial spasmolytic effect has produced bronchospasm in asthmatics. Diazepam has been shown not to aggravate bronchospasm in asthmatics (Heinonen & Muittari 1972) but it must be remembered it has a long duration of action and active metabolites. Reports are awaited on the usefulness of midazolam. Most clinicians find the combination of a muscle relaxant such as pancuronium and small doses of opiate satisfactory. It is necessary to give the drugs in adequate doses and regularly to prevent 'fighting' and accidents on the ventilator.

Care must be taken to humidify the inspired gases. Mucus plugging associated with atelectasis and poor oxygenation is often a problem before and during ventilation. Bronchial lavage in the hypoxic patient is dangerous as it can give further hypoxia leading to life threatening arrythmias and brain damage. In adults the instillation of 2 ml of normal saline via the endotracheal tube every half hour will, in less than 24 hours, and with the aid of skilled physiotherapy, bring up large amounts of secretions. Bromhexine was used routinely by Webb et al (1979).

Artificial ventilation is only supportive therapy and it is important to continue bronchodilator and steroid treatment at maximal doses. Bronchodilating agents can be given by nebulizers inserted in the ventilator tubing circuit. Antibiotics should not be given unless there is very good evidence of a chest infection or septicaemia. There must be careful attention to cumulative fluid balance as chronic fluid overload will make it difficult to wean the patient off the ventilator. If heart failure develops inotrophic agents will be required.

The complications for artificial ventilation of asthmatics include pneumothorax pneumomediastinum cardiac arrythmias, right-sided heart failure, atelectasis and pneumonia, and accidents with the apparatus. These complications were common to most of the series reviewed including children (Wood et al 1968).

Patients can be classified according to the time they spend on a ventilator into two main groups:

1. Those patients who are ventilated for exhaustion, who have little or no lung damage and respond quickly to therapy usually only require 24–72 hours. These patients have powerful muscles of respiration and weaning is not necessary. Unfortunately extubation is not always easy due to reflex bronchospasm from the

endotracheal tube. Small doses of diazepam and extubation with experienced supervision may be successful. Extubation can be tried with the patient anaesthetized with halothane.

2. The second group are often older, have some degree of pulmonary or cardiac failure, have a poor response to bronchodilator drugs or considerable atelectasis. These patients usually require to be ventilated for 2 weeks or more. Tracheostomy should be carried out if ventilation is predicted to last more than 3 weeks as it makes nursing and weaning (which can be protracted) easier. It must be remembered that these patients may require artificial ventilation again at a later date so the decision to do a tracheostomy should be made with considerable thought.

Lenggenhager (1985) suggests the use of a low pressure chamber with 100% oxygen in the treatment of acute bronchospasm. Clinical observation that the resistance of airways is reduced at high altitudes is supported by model experiments, by theoretical considerations and by measurement of the maximal breathing capacity at low barometric pressures. He has tried the low pressure chamber in acute asthma with some success.

Acute severe asthma is a frightening, complex and difficult condition to treat and the adverse effects of the drugs used are as dangerous as the disease itself. It is important to recognize the danger signs of a severe attack (Table 4.4) and institute artificial ventilation before mucus plugging and atelectasis is widespread or a cardiac arrest has occurred. The significant mortality of the past should now improve with the greater understanding of the disease and its therapy.

Anaesthesia for the asthmatic patient has not been reviewed in this article but is considered by Galant in Chapter 3. See also review articles by Fury (1985) and Kingston & Hirshman (1985).

REFERENCES

Betts E K, Parkin C E 1971 Use of ketamine in an asthmatic child. A case report. Anesthesia and Analgesia 50: 420

Bloomfield P, Charmichael J, Petrie G R, Jewell N P, Crompton G K 1979 Comparison of salbutamol given intravenously and by intermittent positive-pressure breathing in life threatening asthma. British Medical Journal 1: 848–850

Blumberg M Z 1973 Morphine for severe asthma? New England Journal of Medicine 288: 50

Brandus V, Jotte S, Benoit C V, Wolff W I 1970 Bronchial spasm during general anaesthesia. Canadian Anaesthetists Society Journal 17: 269–274

Branthwaite M A 1978 Artificial ventilation for pulmonary disease. Private printing for Brompton Hospital, p 9–11 (Available £2)

British Thoracic Association 1982 Death from asthma in two regions of England. British Medical Journal 285: 1251–1255

Campbell I A, Hill A, Middleton H, Momen M, Prescott R J 1978 Intermittent positive pressure breathing. British Medical Journal 1: 1186–1189

Clark T J H 1983 Acute severe asthma. In: Clark T J H, Godfrey S (eds) Asthma, 2nd edn. Chapman and Hall, London, ch 17, p 393–414

Cochrane G M 1984 The role of bronchodilators in severe acute asthma. In: Clark T J H (ed) Bronchodilator therapy. Adis, New Zealand, Ch 10, p 167–187

Cochrane G H, Clark T J H 1975 A survey of asthma mortality in patients between ages 35 and 64 in the Greater London Hospitals in 1971 Thorax 30: 300–305

Conway S P, Gillies D R N, Littlewood J N 1984 Intravenous aminophylline in acutely ill patients receiving theophylline. British Medical Journal 288: 715–716

Corssen G, Gutierrez J, Reves J G, Huber F C 1972 Ketamine in the anaesthetic management of asthmatic patients. Anesthesia and Analgesia 51: 588–596

Cuss F M, Palmer J B, Barnes P J 1985 Rapid estimation of plasma theophylline concentrations. British Medical Journal 291: 384

Dacosta J, Hedstrand U 1970 The mechanical work of breathing during controlled ventilation in normal subjects and patients with bronchial asthma. Acta Societatis Medicorum Upsaliensis 75: 110–118

Downes H, Hirshman C A 1981 Lidocaine aerosols do not prevent allergic bronchoconstriction. Anesthesia and Analgesia 60: 28–32

Drugs and Therapeutics Bulletin 1985 Sustained release theophylline preparations. 23: 33–36

Eustace B R 1967 Case Report. Suxamethonium induced bronchospasm. Anaesthesia 22: 638–641

Fellini A A, Bernstein R L, Zauder H L 1963 Bronchospasm due to suxamethonium. Report of a case. British Journal of Anaesthesia 35: 657–659

Fung D L 1985 Emergency anaesthesia for asthma patients. Clinical Reviews in Allergy 3: 127–141

Graham V A L, Milton A F, Knowles G K, Davies R J 1982 Routine antibiotics in hospital management of acute asthma. Lancet 1: 418–421

Grant I W B 1985 Asthma in New Zealand. British Medical Journal 286: 374–377

Grant I W B 1982 Are corticosteroids necessary in the treatment of severe acute asthma. British Journal of Diseases of the Chest 76: 125–129

Greenberger P A, Patterson R 1985 Management of asthma during pregnancy. New England Journal of Medicine 312: 897–902

Gregg I 1983 Epidemiology. In: Clark T J H, Godfrey S (eds) Asthma. Chapman and Hall, London, ch 11: p 242–284

Heath M L 1973 Case Report. Bronchospasm in an asthmatic patient following pancuronium. Anaesthesia 28: 437–440

Heinonen J, Muittari A 1972 The effect of diazepam on airways resistance in asthmatics. Anaesthesia 27: 37–40

Hetzel M R, Clark T J H 1983 Adult asthma. In: Clark T J H, Godfrey S (eds) Asthma, 2nd edn. Chapman and Hall, London, ch 19, p 457–489

Hetzel M R, Clark T J H, Branthwaite M A 1977 Asthma: analysis of sudden deaths and ventilatory arrest in hospital. British Medical Journal 1: 808–811

Hirshman C A, Edelstein G, Peetz A, Wayne R, Downes H 1982a Mechanisms of action of inhalation anaesthesia on airways. Anesthesiology 56: 107–111

Hirshman C A, Kreiger W, Littlejohn G, Lee R, Julien R 1982b Ketamine aminophilline induced decrease in seizure threshold. Anesthesiology 56: 464–467

Ingram R H, Wellman J J, McFadden E R, Mead J 1977 Relative contributions of large and small airways to flow limitation in normal subjects and after atropine and isoprenaline. Journal of Clinical Investigation 59: 696–703

Inman W H W, Adelstein A M 1969 Rise and fall of asthma mortality in England and Wales in relation to use of pressurised aerosols. Lancet ii: 279–285

Isles A F, MacLeod S M, Levison H 1982 Theophylline — new thoughts about an old drug. Chest 82: 495–511

Jackson R T, Beaglehole R, Rea H H 1982 Mortality from asthma: a new epidemic in New Zealand. British Medical Journal 285: 771–774

Kay A B 1983 The immunological basis of asthma. In: Clark T J H (ed) Steroids in Asthma. Adis, London, ch 3, p 46–48

Khot A, Burn R 1984 Deaths from asthma. British Medical Journal 289: 557

Kingston R, Hirshman C A 1984 Perioperatic management of asthma. Anesthesia and Analgesia 63: 844–855

Laroche C M, Harries A Y K, Weston R C F, Britton M G 1985 Domicilary nebulisers in asthma: a district survey. British Medical Journal 290: 1611–1613

Lawford P, Jones B J M, Milledge J S 1978 Comparison of intravenous and nebulised salbutamol in initial treatment of severe asthma. British Medical Journal 1: 84

Leahy B C, Gomm S A, Allen S C 1983 Comparison of nebulised salbutamol with nebulised ipratropium bromide in acute asthma. British Journal of Diseases of the Chest 77: 159–163

Lenggenhager K 1985 Treatment of severe bronchial asthma with a low pressure chamber and 100% O$_2$. Anesthesia and Analgesia 64: 551–553

Luksza A R 1982 Acute severe asthma treated without steroids. British Journal of Diseases of the Chest 76: 15–19

MacDonald J B, Seaton A, Williams D A 1976 Asthma deaths in Cardiff 1963–1974: 90 deaths outside hospital. British Medical Journal 1: 1493–1495

Menitove S M, Goldring R M 1983 Combined ventilator and bicarbonate strategy in the management of status asthmaticus. American Journal of Medicine. 74: 398–901

Moss J, Roscow C E 1983 Histamine release by narcotics and muscle relaxants in humans. Anesthesiology 59: 330–339

Nana A, Cardan E, Leittersdorfer T 1972 Pancuronium bromide: its use in asthmatics and patients with liver disease. Anaesthesia 27: 154–158

Ormerod L P, Stableforth D E 1980 Asthma mentality in Birmingham. British Medical Journal 1: 687–690

Partridge M R, Saunders K B 1981 Site of action of ipratropium bromide and clinical and physiological determinants of response in patients with asthma. Thorax 36: 530–533

Paterson J W, Woolcock A J, Shenfield G M 1979 Bronchodilator drugs. American Review of Respiratory Disease 120: 1149–1188

Pierson W E, Bierman C W, Kelley V C 1974 A double-blind trial of corticosteroid therapy in status asthmaticus. Paediatrics 54: 282–287

Pontoppidan H, Geffin B, Lowenstein E 1972 Acute respiratory failure in the adult. New England Journal of Medicine 287: 743–752

Salem M R, El Kim Y, Etr A A 1968 Histamine release following intravenous injection of d-tubocurarine. Anesthesiology 29: 380–382

Scoggin C H, Shalin S A, Petty T L 1977 Status asthmaticus: a nine year experience. Journal of the American Medical Association 283: 1158–1162

Sheehy A F, Di Benedetto R, Lefrak S, Lyons H A 1972 Treatment of status asthmatacus. A report of 70 episodes. Archives of Internal Medicine 130: 37–42

Shnider S M, Papper E M 1961 Anaesthesia for the asthmatic patient. Anesthesiology 22: 886–892

Stirt J A, Berger J M, Roe S D, Ricker S M, Sullivan S F 1981 Safety of enflurane following the administration of aminophyline in experimental animals. Anesthesia and Analgesia 60: 871–873

Stirt J A, Berger J M, Sullivan S F 1983 Lack of arrythmogenicity of isoflurane following the administration of aminophyline in dogs. Anesthesia and Analgesia 62: 586–671

Tatham M E, Gellert A R 1985 The management of acute asthma. Postgraduate Medical Journal 61: 599–606

Ward M J, Fentem P H, Smith R W H, Davies D 1981 Ipratropium bromide in acute asthma. British Medical Journal 282: 598–600

Ward M J, MacFarlane J T, Davies P 1982 Treatment of acute severe asthma with intravenous aminophylline and nebulised ipratropium after salbutamol. Thorax 37: 785

Webb A K, Bilton A H, Hanson G C 1979 Severe bronchial asthma requiring ventilation. A review of 20 cases and advice on management. Postgraduate Medical Journal 55: 161–170

Westerman D E, Benatar S R, Potgieter P D, Ferguson A D 1979 Identification of the high risk asthmatic patient, experience of 39 patients undergoing ventilation for status asthmaticus. American Journal of Medicine 66: 565–572

Wood D W, Downes J J, Lecks H I 1968 The management of respiratory failure in childhood status asthmaticus. Experience with 30 episodes and evaluation of a technique. Journal of Allergy 42: 261–267

Woodcock A A, Johnson M A, Geddes D M 1983 Theophylline prescribing, serum concentrations and toxicity. Lancet ii: 610–612

High frequency ventilation

INTRODUCTION

Effective gas exchange in the lung requires the presention of fresh gas to the alveolus and the removal of used gas from the alveolus. The size and metabolic state of the patient determines the amount of gas required per minute. Conventional artificial ventilation uses rates and tidal volumes within the physiological range. However, recently the idea of using small tidal volumes appeared attractive as this was seen to be associated with a lower peak inspiratory airway pressure and less fluctuation in intrathoracic pressure. This inevitably meant the use of higher ventilatory frequencies.

Ventilation with tidal volumes of less than anatomical dead space volume can provide adequate gas exchange. The giraffe and swan (with their long necks and long tracheas) manage this routinely and it has been known for some time that carbon dioxide removal can be achieved in man with tidal volumes of considerably less than anatomic dead space (Briscoe et al 1954). However, the mechanisms of achieving this are not immediately obvious and have been recently reviewed by Drazen et al (1984).

SMALL TIDAL VOLUME

Traditional

The conventional model of ventilation has fresh gas being presented at pressure at the airway opening; this gas then flows turbulently through the anatomical dead space and 'flushes it out', before appearing at the level of the alveolus. This bulk convection thus takes gas from the airway opening to the alveolar zone. At the alveolus the fresh gas forms an interface with the alveolar gas, further onward movement is then by molecular diffusion, and this takes the gas across the alveolar zone, across the alveolar-capillary membrane and into the pulmonary capillaries. In man molecular diffusion probably takes less than 10 ms. This model works well for larger tidal volumes producing turbulent flow, as can be seen for tidal volumes of above 400 ml in Fig. 5.1. However, at lower tidal volumes more gas gets through to take part in gas exchange than this conventional model predicts.

Nearby alveoli

The transition from the conducting airways, with no gas exchanging function, to the alveolar sacs where only gas exchange takes place, is not sharply demarcated

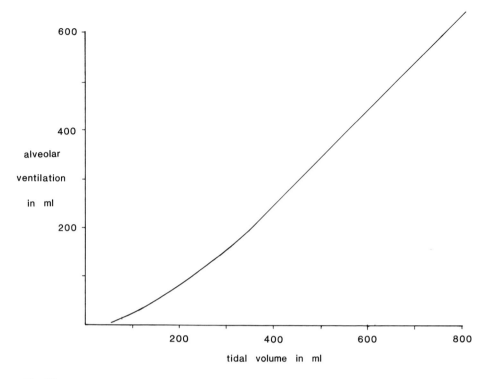

Fig. 5.1

anatomically. The respiratory bronchioles (17th–19th generation) are predominantly conducting airways but do have alveolar sacs opening off them. The alveolar ducts (20th–22nd generation) have gas exchanging epithelium throughout and in addition conduct gas to the alveolar sacs (23rd generation).

Thus any gas that penetrates to the 17th generation air passage by bulk convection will take part in gas exchange and, whereas a small tidal volume will only do this, a normal tidal volume will penetrate right through to the 23rd generation air passages.

Asymmetric velocity profiles

If the flow profiles during inspiration and expiration are different, bulk convection occurs (Haselton & Scherer 1980). If inspiratory flow is laminar, a given block of gas takes up a parobilic profile, and if this is then followed by a turbulent expiratory flow, a square wave pushes the parabolic profile back en bloc. Thus, although the overall result is no net gas movement, in fact there has been axial convection of gas towards the lung periphery and the peripheral gas has moved towards the airway opening. Measurement of velocity profiles in models of the human central airways have shown that inspiratory and expiratory velocity profiles are significantly different from each other, and so this mechanism contributes to bulk convection at small tidal volumes.

Pendelluft

Within the lung differing areas of lung have different time constants. Thus some areas of the lung are fast fillers with short time constants, whereas other areas are slow fillers with long time constants. During early inspiration the fast filling areas become full and during late inspiration are actually emptying into the slow fillers. This is Pendelluft, where the sum of gas movement within the lung is greater than the gas flow down the trachea, i.e. there is gas movement between areas of the lung without gas movement in the trachea. It is apparent that with high frequency ventilation, a slow filler could still be filling from a fast filler which at the same time is providing gas for expiration up the trachea (Lehr 1980).

Taylor dispersion

This is probably the major mechanism of gas movement at small tidal volume and involves augmented axial convection. If the volume of gas entering the airway is small and the flow is laminar the velocity profile in the airway will be parabolic and therefore the gas in the axis of the airway will penetrate much more deeply. The longer the duration of inspiration the more accentuated this spike of augmented axial convection becomes: e.g. over 3 units of time, the gas at the peripheral ring remains stationary whereas the axial gas has travelled three times the distance. As this axial spike of gas penetrates into the gas exchanging units of the lung periphery, the spike of fresh gas exposes a large area of interface with alveolar gas (vide infra).

Molecular diffusion

As we have seen earlier with conventional ventilation, fresh gas is presented at the alveolus and molecular diffusion across the alveolar zone is complete within less than 10 ms. Even with high frequency ventilation, inspiratory times are considerably in excess of 10 ms (see Table 5.2). Thus there is the potential for molecular diffusion to take on a considerably greater role in the movement of gas across the alveoli. It can be appreciated that with Taylor dispersion the spike of augmented axial convection produces a dramatic increase in the interface between fresh gas and alveolar gas thus allowing molecular diffusion an enhanced role in the movement of gas. Slutsky et al (1980) have produced a thereoretical model on these assumptions and compared it convincingly with an experimental preparation.

Summary

These mechanisms can explain adequately how, with a small tidal volume, there is bulk convection of gas from the airway opening to the region of the alveolar zone, where an enhanced role for molecular diffusion permits adequate alveolar gas exchange. Now we must consider the influence of high frequency ventilation on gas exchange.

HIGH VENTILATORY FREQUENCY

Time constant

Our conventional model of ventilation has a mass of gas presented at the airway opening under pressure. This potential energy is converted to kinetic energy to allow the gas to flow along the airway. However, by the end of inspiration this gas is static, the kinetic energy having become potential energy again, stored by distension of the alveoli in the lung units. The time course for this change is determined by the 'time constant' for the lung (the product of total lung compliance and airways resistance). Representative values for time constants can be seen in Table 5.1. It is known that

Table 5.1   Representative values of respiratory parameters for adults in various situations

	Total lung compliance (ml cm $H_2 0^{-1}$)	Airway resistance (cm $H_2 0.1^{-1}$ s)	Time constant (ms)
Normal adult	100	1	100
Anaesthetized adult	50	5	250
Adult respiratory distress syndrome	20	20	400

95% of change has occured within three time constants, and the inspiratory time available with increasing respiratory frequency can be seen from Table 5.2. It can be appreciated rapidly that with higher frequencies, especially with lung pathology (and hence longer time constants) there will be insufficient time for inspiration to go to completion, that is with static gas distending alveoli. It therefore follows that with higher ventilatory frequencies either a reduced tidal volume is delivered to the lung periphery or a higher peak inspiratory pressure is required in order to force the gas to the lung periphery within the available time (two-thirds of the gas will enter within one time constant).

Table 5.2   Duration of inspiration at increasing ventilatory frequency assuming that inspiration occurs over 25% of the ventilatory cycle

Ventilatory min^{-1}	Frequency Hz	Duration of inspiration ms
15	0.25	1000
30	0.5	500
60	1	250
120	2	125
240	4	62
480	8	31
960	16	16
1920	32	8

Airway resistance

This is unfortunately not a constant value and alters as flow changes from laminar to turbulent. Fortunately within the flow range used with conventional ventilation the magnitude of change is small: the assumption of a constant value for airways resistance introduces only minimal error. However, without this flow range a significant error is introduced.

Gas inertia

Of increasing relevance is the kinetic energy of the molecules of gas undergoing bulk convection — as ventilation cycles from inspiration to expiration they will still retain their foreward kinetic energy and continue to progress peripherally into the lung until this kinetic energy has become dissipated. This inertia of the gas molecules is not present with conventional ventilation (where end inpiration is a static state) but obviously will become increasingly important with higher ventilatory frequencies with their shorter inspiratory times, higher inspiratory flow rates and hence higher kinetic energy of gas molecules.

Impedance

The pressure–flow relationships at conventional ventilatory frequencies are adequately expressed by airways resistance. With higher ventilator frequencies airways resistance can no longer be assumed to be constant and inertia makes an increasing contribution — impedance is the parameter used to express the pressure–flow relationships at high ventilatory frequency and it is dependant on both frequency and flow rate.

It has been shown (Gavriely et al 1985) that the largest component of impedance to a high frequency generator lies in the endotracheal tube. This impedance was greatest with a wide bore endotracheal tube and high gas flow rates (large tidal volumes at high frequency). These are the conditions under which the influence of gas inertance would be maximal thus supporting the position of Taylor dispersion with augmented axial convection as the major mechanism of bulk gas movement during high frequency ventilation.

Distribution

Effective respiration, i.e. gas exchange, requires that regional ventilation and perfusion be matched. With conventional respiratory rates, regional ventilation is determined by regional compliance. However, with higher respiratory rates airways resistance and gas inertance have an increasing influence over the distribution of ventilation. Potentially this will result in a change in the distribution of ventilation from areas of high compliance to areas of low impedance. One assumes there is no corresponding change in regional lung perfusion and to date there is no evidence to the contrary, i.e. perfusion remains gravity dependant.

Path length

Another aspect of the higher ventilatory frequencies used is the time available for the bulk convection of gas during inspiration and expiration and the path length it must travel. This is some 20–25 cm from the laryngeal inlet in adult man, but much less in the neonate and most experimental animals. Thus the tidal volume presented at the airway opening will progressively distend the major conducting airways, then the nearby fast alveoli and finally the peripheral slow alveoli. With progressive shortening of inspiration a situation will arise when there is not enough time for the tidal volume to get past the major conducting airways (non gas exchanging units) which therefore act in a way analogous to a capacitor. By implication this also means that the pressure at the airway opening is not transmitted through to the lung periphery and hence the pleural space. However the reverse will also occur during expiration, namely there will not be enough time for intra-alveolar pressure to empty the alveoli. Thus there will be continuous positive airway pressure in the peripheral lung units — the very units where small airway closure is most likely to occur.

EXPERIMENTAL

Delivery system

One cannot look at the experimental results without being aware of the mechanism used to deliver the low tidal volume high frequency ventilation. This area is largely technical: however an appreciation of the delivery systems and their inherent limitations allows a fuller insight into high frequency ventilation.

For adult man it is generally accepted that high frequency ventilation is any rate greater than 1 Hz. It should be noted that the range of ventilatory frequencies given below is arbitary and mostly determined by the technical considerations of the ventilator.

High frequency positive pressure ventilation (HFPPV) (1–2 Hz)

Conventional ventilators deliver a volume of gas during inspiration by occlusion of a relatively wide bore expiratory limb. The mechanical device used to occlude the expiratory limb must function rapidly. It will have a certain mass and hence inertia — the more rapidly it is moved the shorter the response time and the more likely valve bounce is to occur. This is a fundamental problem that means that conventional ventilators are inappropriate for rates above 2 Hz; however they have the advantage of delivering a known volume of gas of known composition.

High frequency jet ventilation (HFJV) (2–6 Hz)

These ventilators allow a high pressure gas source to flow into the airway during part of the respiratory cycle — empirically inspiratory times of 20–35% of the respiratory cycle time have been found most efficient. With a jet ventilator the inspiratory flow is delivered through a narrow (1–2 mm diameter) tube usually at tracheal level. A variant is the flow interrupter type of ventilator when the entire cross section of the airway is utilized for inspiratory flow. These systems require no pneumatic seal of the expiratory limb and hence this is open to atmosphere throughout the cycle. This potentially results in the entrainment of an unknown volume of air (ambient gas): thus the volume and composition of the tidal volume are unknown. However, this open system has the inherent advantage of allowing the patient to take a spontaneous breath at any time during the respiratory cycle which is a major safety feature and of great advantage when weaning a patient.

High frequency oscillation (HFO) (6–40 Hz)

Oscillator type ventilators tend to be used for the higher frequencies. They consist of either a piston driven by a motor or a diaphragm driven electronically at the airway opening generating a to and fro motion of the gas within the airways. The tidal volume is determined by the volume displaced by the piston or diaphragm. This type of ventilator generates a negative pressure during expiration which is thus an active phase of ventilation. Fresh gas is fed in at the airway opening and a low pass filter exhaust port allows excess gas to exit the system (i.e. the low pass filter offers high impedance to high frequencies; thus the tidal volume generated tends to pass down the lower impedance offered by the airway opening rather than be lost through the exhaust port).

External

Certain experimental models have applied high frequency oscillations on the pleural surface of the lung or the chest wall. This latter method can cause damage to the underlying lung. The future of this approach is uncertain.

Carbon dioxide elimination

Low tidal volume

Ventilation with conventional ventilators at low tidal volumes has revealed carbon dioxide in the expired gases with tidal volumes as low as 50–75 ml. However, there does appear to be a cut off point below which no carbon dioxide is detectable in the expiratory gas and conversely all studies have shown increasing carbon dioxide elimination with increasing tidal volume.

High frequency

On theoretical grounds one could anticipate that ventilation with a fixed tidal volume would show a linear relationship between carbon dioxide removal and ventilatory frequency until a time is reached when the duration of inspiration is not long enough to permit the gas to penetrate the conducting airways and reach the gas exchanging areas of the lung. It has been shown in both human adult volunteers (Goldstein et al 1981) and adult patients (Rossing et al 1981) that there was a critical ventilatory frequency of between 3–6 Hz at which carbon dioxide elimination reached a plateau. Above this frequency carbon dioxide elimination was a function of the applied tidal volume and independant of the ventilatory frequency. (NB This critical frequency is undoubtedly dependent on the anatomy of the lung and is certainly higher in dogs and neonates.)

Airway resistance

Goldstein et al (1981) showed that carbon dioxide elimination was dependent on lung volume, being increased at higher lung volume. They also showed increased carbon dioxide elimination after the inhalation of isoprenaline. The presumption is that a reduction in airways' resistance associated with both of these manoeuvres has allowed more effective penetration of the delivered tidal volume to the alveolar zone.

Oxygen delivery

One of the original hopes for high frequency ventilation was that a change from compliance/airway resistance distribution of ventilation to one determined by airway resistance/gas inertia would result in a redistribution and improvement in ventilation. However, it is now accepted that, at equivalent lung volumes, there is no improvement in arterial oxygen tension.

CLINICAL

Introduction

The indiscriminate use of high frequency ventilation offers no theoretical advantage over conventional mechanical ventilation and is not currently recommended (Crawford & Rehder 1985). However, there are a number of special situations where high frequency ventilation may have a role to play.

Endoscopy of the airway

The use of a jet ventilator for airway endoscopy permits near unobstructed access to the surgical field, and the use of high frequency reduces respiratory movement of the surgical field and is enthusiastically used in some centres. (Borg et al 1980). However, ventilation with such an open system denies the anaesthetist accurate information about the tidal volume and the inspiratory gases (vide infra).

Bronchopleural fistula

There are now a number of cases reporting effective ventilation by high frequency jet ventilation after failure of conventional ventilatory techniques. The assumption must be that the distribution of ventilation shifts from compliance determined to airway resistance/gas inertance determined. These patients represent a group where high frequency ventilation offers a lifeline in what was previously often a hopeless situation.

Weaning from ventilator

The majority of patients wean rapidly and readily from ventilatory support on a T-piece circuit. Failure of conventional T-piece weaning can present a clinical problem. The open circuit of the high frequency jet ventilator offers many advantages — it is in essence a circuit which allows the patient to take a breath of fresh gas whenever he will, whilst in addition the jet ventilator can cycle away offering respiratory support and carbon dioxide elimination. Progressive reduction of the high frequency contribution to ventilation will permit progressive weaning. The early literature suggested that high frequency ventilation produced reflexogenic central apnoea per se: however this has now been shown to be secondary to lung distension by the PEEP effect.

Adult respiratory distress syndrome (ARDS)

The adult respiratory distress syndrome patient can be notoriously difficult to achieve good respiratory gas exchange at acceptable airway pressures. The theoretical advantages of high frequency ventilation are: (1) reduced pulmonary barotrauma as small tidal volume ventilation achieves effective minute ventilation at a lower peak and mean airway pressure; (2) less cardiovascular impairment, also because of the lower airway pressure; and (3) a distribution of ventilation less dependent on regional compliance. All three factors should be to the patients advantage in this syndrome: however the hope that high frequency ventilation would revolutionize ventilatory support for these patients has yet to be realized.

Humidification

Several of the early reports of high frequency ventilation reported the problem of profuse secretions shortly after the initiation of high frequency ventilation. This was probably a consequence of the use of dry gases, and more recent reports do not mention this problem. The desirability of humidifying the inspired gases is obvious, but it is often a difficult technical problem with high frequency ventilators.

Monitoring

Conventional ventilators, with their controlled and isolated inspiratory phase, offer many parameters to monitor such as inspiratory tidal volume, FI_{O_2}, airway pressure, or expiratory volume, FE_{CO_2}. With the open system of the high frequency jet ventilator the tidal volume, FI_{O_2}, FE_{CO_2} and airway pressure are not measureable. In addition, with a tidal volume of 100 ml the chest expansion is only of the order of 2–3 mm and barely detectable.

Thus the monitoring of high frequency ventilation on a breath to breath basis is not possible, and one must rely on intermittent measurements, such as arterial blood gases or occasional large tidal volumes to generate an end tidal CO_2. It is possible that future developments with transcutaneous or intravascular electrodes may solve this problem.

CONCLUSION

There are sound theoretical reasons to explain why low volume high frequency ventilation should achieve adequate gas exchange. In adult man tidal volumes as low as 50–75 ml can be used but ventilatory frequencies above 3–6 Hz do not appear to offer any advantage (higher frequencies may be appropriate for a neonate). Consequently the place of high frequency oscillation in adults is unclear.

Clinically the required frequencies can be supplied by a jet ventilator which has the safety advantage of utilizing an open system. The technique has become established for airway endoscopy in the operating theatre. In intensive care it is of accepted value in the ventilation of patients with bronchopleural fistula and possibly in patients proving difficult to wean from a ventilator.

It is, as yet, of no proven advantage in the adult respiratory distress syndrome, despite having a number of theoretical advantages.

The main clinical problem is the monitoring of ventilation and by implication effective gas exchange. As yet this can only be determined intermittently, such as with arterial blood gas analysis, and remains a major problem with a sick unstable patient.

REFERENCES

Borg U, Eriksson I, Sjostrand U 1980 High frequency positive pressure ventilation (HFPPV): a review based on its use during bronchoscopy and for laryngoscopy and microlaryngeal surgery under general anaesthesia. Anesthesia and Analgesia 59: 594–603

Briscoe W A, Forster R E, Comroe J H 1954 Alveolar ventilation at very low tidal volumes. Journal of Applied Physiology 7: 27–30

Crawford M, Rehder K 1985 High frequency small volume ventilation in anaesthetised humans. Anesthesiology 69: 298–304

Drazen J M, Kamm R D, Slutsky A S 1984 High frequency ventilation. Physiological Review 64: 505–543

Gavriely N, Solway S, Loring S H, Butler J P, Slutsky A S, Drazen J M 1985 Pressure flow relationships of endotracheal tubes during high frequency ventilation. Journal of Applied Physiology 59: 3–11

Goldstein D H, Slutsky A S, Ingram R H, Westerman P, Venergas J, Drazen J M 1981 CO_2 elimination high frequency ventilation (4 to 10 Hz) in normal subjects. American Review of R   _.seases 123: 251–255

Haselton F R, Scherer P W 1980 Bronchial bifurcations and respiratory mass transport. Science 208: 69

Lehr J 1980 Circulating currents during high frequency ventilation. Federation Proceedings 39: 576

Rossing T H, Slutsky A S, Lehr J, Drinker P, Kamm R D 1981 Tidal volume and frequency dependance of carbon dioxide elimination by high frequency ventilaion. New England Journal of Medicine 305: 1375–1397

Slutsky A S, Lehr J, Shapiro A H, Ingram R H, Drazen J M 1980 Effective pulmonary ventilation with small volume oscillations at high frequency. Science 209: 609–611

Drugs used in the treatment of hypertension

INTRODUCTION

Data from life insurance statistics clearly demonstrate that patients with high casual blood pressure readings have a reduced life expectancy because of their increased susceptibility to cardiac and cerebrovascular disease. There is ample evidence in moderate and severe hypertension, and increasing evidence in mild hypertension, that reducing the blood pressure will minimize the risk of developing some of these complications. Physicians therefore find themselves obliged to reduce high levels of blood pressure by the chronic prescription of antihypertensive drugs. However, many important questions about the treatment of hypertension remain unanswered. What precise level of blood pressure requires treatment? Is it the systolic, diastolic or mean pressure that is most important? What are the long term metabolic effects of antihypertensive drugs? What is the aetiology of essential hypertension and are the complications caused by, or just associated with the level of pressure? A full discussion of these problems is outside the scope of this review, but the potential prescriber must bear them in mind when he initiates a lifelong course of treatment for an asymptomatic patient.

Many groups of drugs have antihypertensive activity. In this chapter we will confine ourselves to oral preparations that are prescribed specifically for long term treatment of hypertension. The list of drugs discussed is not intended to be comprehensive. We will provide an overview of the different classes of drugs which are currently available and say something about their use both as monotherapy and in combination.

Currently available antihypertensive drugs can be classified according to their mode of action into seven main categories, as follows:

Diuretics
Beta adrenoceptor antagonists
Alpha adrenoceptor antagonists
Calcium channel blocking agents
Angiotensin converting enzyme inhibitors
Direct acting vasodilators
Centrally acting agents

Diuretics

Diuretics have been used in the treatment of hypertension for almost 30 years. The first agent convincingly shown to lower blood pressure was the benzothiadiazine (thiazide), chlorthiazide, but it appears that all classes including loop and potassium sparing diuretics have some antihypertensive activity. Their mechanism of action, however, remains controversial and may differ for different types. It has been known for many years that salt restriction can be an effective method of lowering blood pressure in some individuals, and this group of patients is likely to respond well to diuretic therapy. The fall in pressure is related to a fall in body weight (due to loss of salt and water) and increasing the dose above that which causes maximum weight loss does not increase the hypotensive effect. However, natriuresis alone is not the sole mechanism.

Diuretics, particularly the thiazides, also act on peripheral blood vessels reducing vascular resistance and catecholamine responsiveness. There is little temporal relationship between the changes seen in plasma volume and changes in blood pressure; diuretic potency does not correlate well with antihypertensive effectiveness.

Thiazides are the most effective antihypertensive diuretics but even so their potency is not great. They can be used alone only in mild hypertension though they may be useful in combination with other drugs for more severe forms. Their action is on the distal part of the ascending limb of the loop of Henle promoting loss of potassium as well as sodium and water. They are ineffective in advanced renal failure.

The more powerful loop diuretics such as frusemide and bumetanide paradoxically are less potent antihypertensives. They act on the proximal part of the ascending limb of the loop of Henle increasing urine volume and promoting excretion of sodium, potassium and calcium (as opposed to thiazides which reduce urinary calcium). They have virtually no antihypertensive activity in many patients but are effective in conditions such as chronic renal failure and malignant hypertension when sodium retention contributes towards the raised blood pressure. They also potentiate the antihypertensive activity of angiotensin converting enzyme inhibitors (see below).

Potassium retaining diuretics include amiloride and triamterene which act on the distal convoluted tubule and spironolactone which is an aldosterone antagonist. They are relatively weak antihypertensives and are generally only used in combination with loop or thiazide diuretics to offset increased potassium losses. The importance of diuretic induced potassium depletion is controversial but in general it is more effective to add a potassium sparing diuretic than to give potassium supplementation. As a general rule monotherapy with these agents should be avoided, except in proven hyperaldosteronism when spironolactone is specifically indicated.

Diuretics are commonly prescribed in hypertension because they are cheap, experience with them is great and they are unlikely to cause side effects in the short term. However, longer term complications are common and potentially serious (Table 6.1). In particular, electrolyte imbalance may cause fatal arrhythmias and reduction of the high density to low density lipoprotien ratio may accelerate atherosclerosis. There has been some debate about whether these risks could outweigh the benefits in mild hypertension. Many of the adverse effects can be minimized by using low doses, and this policy makes sense as they have a flat dose response curve. MacGregor et al (1983) have shown that 12.5 mg per day of hydrochlorthiazide has the same antihypertensive potency as 50 mg per day.

The use of diuretics in hypertension is likely to decline as newer antihypertensives of other classes become available.

Table 6.1   Side effects of diuretic therapy

	Thiazide diuretics	Loop diuretics	Potassium sparing diuretics
Electrolytes	Hypokalaemia Hyponatraemia Hyperchloraemic alkalosis	Hyponatraemia Hypokalaemia Hypercaluria	Hyperkalaemia Hyponatraemia
Metabolic	Hyperuricaemia Glucose intolerance Hyperlipidaemia Pre-renal azotaemia	Hyperuricaemia Glucose intolerance Pre-renal azotaemia	
Others	Impotence Skin rashes Thrombocytopaenia	Ototoxicity Nephrotoxicity Myalgia (bumetanide)	Renal stones (triamterene) ·Gynaecomastia (spironolactone) Impotence Gastrointestinal disturbance

β Adrenergic antagonists (β blockers)

The sympathetic nervous system is important in short and long term regulation of blood pressure. End organ receptors for the system are broadly divided into two groups, α and β by their function and relative sensitivity to pressor amines. Drugs which block β receptors reduce blood pressure by a variety of mechanisms. They lower cardiac output by reducing myocardial contractility and heart rate, inhibit renin secretion, reset baroreceptors and reduce noradrenaline release from sympathetic neurones. They may also have a central effect on blood pressure regulation.

Since Pritchard & Gillam (1964) first demonstrated the antihypertensive effect of propranolol over 20 years ago more than 10 different β blockers have appeared on the market in most western countries, although only three, atenolol, propranolol and metoprolol are available in North America.

This proliferation of agents with a similar mode of action has led — often for commercial reasons — to extensive debate about the relative merits of other pharmacological properties which differ from drug to drug (Table 6.2). These properties include:
Cardioselectivity
Intrinsic sympathomimetic activity
Membrane stabilising activity and
Lipid solubility

Cardioselectivity. Cardioselective β blockers antagonize the mainly cardiac β_1 receptors more than pulmonary β_2 receptors. In theory this should reduce their adverse effects on peripheral vascular tone, glucose homeostasis and bronchial smooth muscle. However in practice these advantages are difficult to detect. Certainly even the most cardioselective β blockers can produce dangerous deterioration of pulmonary function in asthmatics.

Intrinsic sympathomimetic activity (ISA). The term intrinsic sympathomimetic activity (ISA) refers to the partial agonist effect possessed by some of these drugs. It has been claimed that this may offset side effects due to the lowered cardiac output and heart rate, and the increased peripheral resistance. It has not been convincingly

Table 6.2   Pharmacological properties of some β blockers

	Cardioselectivity	ISA	Membrane stabilizing activity	Lipid solubility
Acebutolol	–	+	+	+
Atenolol	+	–	–	–
Metoprolol	+	–	–	+
Nadolol	–	–	–	±
Oxprenolol	–	+	+	+
Pindolol	–	++	+	+
Propranolol	–	–	+	++
Sotalol	–	–	–	–
Timolol	–	–	+	+

shown that ISA is of clinical importance and in practice excessive ISA may reduce the antihypertensive effectiveness.

Membrane stabilising activity. Some β blockers possess a membrane stabilizing or 'quinidine like' activity which should in theory enhance their antiarrhythmic properties. However, the doses required to achieve this action are many times greater than those which give full β blockade. This pharmacological characteristic of some β blockers can be considered clinically irrelevant.

Lipid solubility. The importance of lipid solubility is better established. Highly lipid soluble drugs such as propranolol pass easily into the central nervous system and are rapidly eliminated by biotransformation in the liver. They are more likely to produce central adverse effects such as vivid dreams and should be avoided in liver failure. Lipid insoluble drugs such as atenolol are excreted unchanged by the kidneys and should be used in reduced dosage when renal failure is present.

In general there is little to choose between one β blocker and the next.

Side effects are common and may occasionally be catastrophic. Many are predictable and related to β blocking activity. These include excessive hypotension and bradycardia, decompensation of cardiac failure, bronchospasm and cold hands and feet. Other less predictable problems include vivid dreams, fatigue, sleep disturbance and impotence. In diabetes they mask the tachycardia and sweating of acute hypoglycaemia and inhibit glucose mobilization, thus increasing the risk of severe neuroglycopaenia.

Several large trials have shown that long term β blockade can improve survival following acute myocardial infarction (Singh & Vencatesh 1984). It is tempting to speculate that this cardioprotective action may also apply in hypertensive patients with ischaemic heart disease, but this hypothesis has not yet been confirmed in clinical trials.

Sudden withdrawal of β blockers in patients with ischaemic heart disease may be associated with an increase in angina, arrhythmias and sudden death (Nattel et al 1979). These drugs, like thiazide diuretics, will reduce the high density to low density lipoprotein ratio. This could theoretically increase the rate of atheroma formation, but the clinical significance of this effect is uncertain.

β-Blocking agents are potent drugs and remain in the first line of treatment for all grades of hypertension despite their many side effects.

α Adrenoceptor antagonists (α blockers)

In the context of blood pressure control α adrenergic receptors increase peripheral vascular tone and inhibit renin release. Over the last 10 years it has become apparent that there are two types of receptors termed α_1 and α_2 with different anatomical distribution and functions. The α_1 receptors are post-synaptic mediators of vasoconstriction while the pre-synaptic α_2 receptors inhibit noradrenaline release from nerve endings forming the basis of a negative feedback loop. Drugs which block α_1 receptors reduce peripheral resistance promoting a fall in blood pressure. They are particularly useful in combination with β blockers whose central cardiac β_1 hypotensive action tends to be opposed by peripheral vasoconstriction due to β_2 blockade. The most important drugs of this class are prazosin, indoramin and the combined α and β blocker labetalol. Phenoxybenzamine, a much older drug, is now rarely used because of its tendency to produce tachycardia, drowsiness and ejaculatory failure.

Prazosin is a quinazoline derivative that specifically antagonises α_1 receptors. It has a half life of about 4 hours and has to be given in divided doses. The tachycardia normally seen with peripheral vasodilation does not occur, though the reasons for this are not entirely clear (Graham & Pettinger 1979). The drug causes venous as well as arterial dilatation. The resulting decrease in venous return inhibits the expected reflex rise in cardiac output and further potentiates its hypotensive effect.

A proportion of patients get profound postural hypotension with the first dose but this becomes less obvious with continued treatment. In practice, this problem can be circumvented by starting with a small dose (0.5 mg) given immediately before the patient goes to bed. The dose can then gradually be increased over several weeks until therapeutic levels are reached. Fluid retention may occur at higher doses but this generally responds to diuretic therapy. Other problems include drowsiness, nasal congestion, blurred vision and impotence, though these are uncommon. Although not a very potent antihypertensive in its own right, prazosin remains a valuable adjunct to conventional therapy with β blockers and diuretics.

Indoramin is very similar to prazosin in action and side effects although they are not structurally related. It has potent selective α_1 blocking actions and is effective when given twice daily. It has a local anaesthetic type membrane stabilizing action which gives it class 1 antiarrhythmic properties.

Labetalol has combined α and β blocking properties, though the β effect is between three and seven times more potent than the α effect. Its action and side effects are identical to those that would be produced by a combination of prazosin and propranolol. The potential advantage of having both actions in one drug is to a certain extent minimized by the fact that it needs to be given in divided doses, often of more than one tablet. For practical purposes it can be regarded as a nonselective β blocking agent and it has typical β blocker side effects in asthma, heart failure and peripheral vascular disease. The usual α side effect of postural hypotension only occurs if high doses (in excess of 1500 mg/d) are used.

The α adrenergic blockers, though theoretically attractive, have a limited application in hypertension. The main problems are their side effects and lack of potency. Nevertheless, there is a place, particularly for prazosin, in selected resistant cases.

Calcium slow channel blocking agents (calcium antagonists)

Calcium antagonists have only recently been widely recognized as useful antihypertensive drugs. This heterogenous group of drugs act by blocking uptake of calcium into vascular and cardiac smooth muscle cells during the slow phase of depolarization. This lowers intracellular calcium concentration reducing contractility and automaticity. Their mechanism of action is not fully understood but it is certainly more complicated than simple blockade of slow calcium channels as was once thought (Braunwald 1982) and it appears to vary from drug to drug. Nevertheless, the final result is reduction of vascular smooth muscle tone and thus peripheral vascular resistance. The two most important calcium antagonists currently available for treating hypertension are *nifedipine* and *verapamil*, although two newer agents, felodipine and nicardipine will probably soon appear.

These drugs vary in their effect on vascular, myocardial and conducting tissue, though all exhibit a hypotensive effect which is mainly related to their vasodilating potency (Table 6.3).

Nifedipine is the calcium antagonist most widely used for hypertension. The standard preparation is rapidly and completely absorbed but has a half life of only 5–6 hours, necessitating thrice daily doses. A slow release preparation (Adalat retard) which can be given twice per 24 h is now available. It affects peripheral blood vessels much more than myocardial and conducting tissue and it can quite safely and effectively be combined with β blockers without the risk of bradycardia or excessive depression of cardiac output.

The main side effects, which comprise headache, flushing, palpitations and peripheral oedema, are related to the vasodilation and so are dose dependent (and relatively common). Occasionally, idiosyncratic hepatotoxicity can occur, and fatigue, depression and nausea, though unusual, can be troublesome.

Nifedipine is also a potent antianginal and is particularly useful for treating hypertensive patients with ischaemic heart disease. It has been shown to be effective when used alone (Bayley et al 1982) and in combination with β blockers, diuretics and methyl dopa (Guazzi et al 1980), but it may cause dangerous hypotension when combined with prazosin (Jee & Opie 1983).

The use of this drug is increasing and it may soon be competing with β blockers and diuretics for a place amongst the first line agents.

Verapamil is much less firmly established as an antihypertensive agent but still has a place when other agents have failed or are contraindicated. The oral preparation is rapidly absorbed but an extensive first pass metabolism allows only 10%–20% of the

Table 6.3 Calcium antagonists used for treating hypertension

Effect	Nifedipine	Verapamil	Felodipine	Nicardipine
Peripheral vasodilatation	++	+	+++	++
Negative inotropism	+	++	+	±
Conduction disturbance	0	+++	0	0
Reflex tachycardia	±	0	++	0
Hypotension	++	+	+++	++
Compatible with blocker	Yes	No	Yes	Yes

drug to reach the circulation leading to considerable variability in blood levels. Its chief metabolite, norverapamil, is pharmacologically active with about one tenth the potency of its parent. This may accumulate with time and cause further unpredictability of effect. The half life is short (2–4 h) and divided dosing is required, though a sustained release preparation is currently undergoing clinical trials and should soon be available (Schutz et al 1982).

The major problems with verapamil in hypertension are its negative inotropism which can precipitate cardiac failure, and its depressant effect on the AV node, causing heart block. It should not be used in patients with myocardial deficiency or conduction disturbances and only used with caution in patients concurrently receiving β blockers or digoxin. Constipation due to inhibition of gut motility occurs in 30% of patients and rashes, joint pains and hepatotoxicity have been described. Its action on conducting tissues however may make it a particularly useful drug for treating hypertension in patients with episodic supraventricular arrhythmias.

Felodipine and nicardipine are not yet commercially available but show considerable promise in clinical trials. They both act predominantly on vascular smooth muscle with minimal myocardial and conduction depressant effects.

Buhler (1983) maintains that calcium antagonists are more effective in elderly patients, particularly those with low plasma renin activity, but this view is by no means universally accepted. There is no doubt that interest in these drugs is increasing and they may well assume as much importance in the 1980s as thiazides and β blockers did in the 1970s.

Angiotensin converting enzyme inhibitors

The renin angiotensin system plays an important role in blood pressure homeostasis and salt and water balance (Fig. 6.1). Renin is produced by the kidney response, amongst other things, to a decrease in renal arterial pressure, and it acts upon circulating globulin, angiotensinogen, to produce angiotensin I. This is then converted to the octapeptide angiotensin II (AII) by angiotensin converting enzyme (ACE). AII is a powerful vasoconstrictor and also stimulates aldosterone release from the adrenal cortex. In normal individuals AII levels are sufficient to exert a significant pressor effect. In hypertension the significance of AII will depend upon the plasma renin activity (PRA) which is determined by renal perfusion and salt water balance. Very high renin levels are found in patients with compromised renal perfusion due to renal artery stenosis, and some renal parenchymal disease (renovascular hypertension). ACE is also responsible for breaking down the naturally occurring vasodilator bradykinin, though the significance of this in hypertension is not clear. Currently, only two oral ACE inhibitors, *captopril* and *enalapril* are available.

Captopril has been shown to be a potent antihypertensive agent in man. It has a short half life and needs to be given 3 times/24 h. The antihypertensive effect is mediated by vasodilatation due to the virtual abolition of circulating AII. The magnitude of the hypotensive effect is related to pre-treatment PRA and thus AII levels. Diuretic drugs which tend to increase PRA will potentiate its action whereas β blockers which inhibit PRA do not work well in combination (MacGregor 1982).

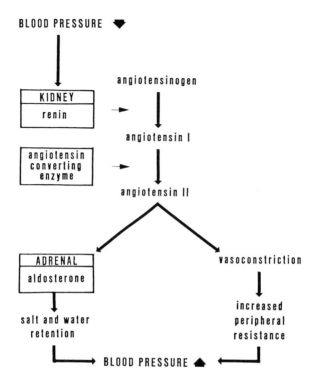

Fig. 6.1 Simplified diagram of the renin/angiotensin system.

Experience with captopril and calcium antagonists together is limited, but our initial impression of this combination is favourable.

Patients with renovascular hypertension are particularly suitable for ACE inhibitors but caution must be applied as captopril can impair renal function by deranging autoregulation of intrarenal blood flow. Acute renal failure has been caused in patients with bilateral disease or renal artery stenosis in a solitary kidney (Hricik et al 1983). Rather surprisingly, captopril has been shown to be of benefit in some cases of low renin hypertension, and even in anephric patients with limited endogenous sources of renin (Man in't Veld et al 1980).

Side effects are common and similar to those of penicillamine. They include rashes, arthralgia, fevers, leucopoenia, disturbance of taste and proteinuria. Their incidence can be greatly reduced by restricting the maximum daily dose to 150 mg. Many authors now regard the dose of 450 mg originally recommended by the manufacturers as being too high and well into the flat part of the dose response curve (Domby 1983).

A dramatic fall in blood pressure may occur on initial dosing. This is more likely to occur in patients rendered sodium deficient by previous diuretic therapy, but there is no good way of predicting exactly who will develop this problem (Hodsman et al 1983). Consequently, it is sensible to ensure that the first dose is given under medical supervision in a setting where resuscitation facilities (i.v. saline) are available. Potassium retention can occasionally occur and the serum electrolytes should be regularly monitored as should the white cell count and urine protein excretion. In view of these problems it is unlikely that captopril will find a large place in the treatment of hypertension outside hospital.

Enalapril, a newer drug, may have some advantages over captopril. It has a longer half life and so can be given only once or twice per 24 h, and it does not possess the sulphydryl group which is responsible for many of the side effects which have limited the usefulness of captopril. It does however have the same adverse effect on renal haemodynamics in renal artery stenosis and causes first dose hypotension which may be more severe and prolonged than with captopril.

ACE inhibition provides a new and pathophysiologically satisfying way of treating hypertension. Considerable research is going on in this branch of pharmacology and new drugs with fewer side effects will become available. ACE inhibitors will certainly play an increasing part in future antihypertensive prescribing.

Direct acting vasodilators

These structurally unrelated drugs are arbitrarily grouped together here because they all cause vasodilatation by direct action on vascular smooth muscle, reducing peripheral resistance and hence blood pressure. They achieve this by a variety of mechanisms which are often poorly understood. Their haemodynamic effects however are similar. The main contenders in this class are hydralazine, minoxidil and diazoxide, though the latter has lost popularity over the last 10 years because of side effects. The newer drug endralazine has been extensively studied recently but is not yet commercially available.

These drugs act predominantly on the arterial system with minimal effect on veins. If used alone they cause a reflex tachycardia and an increase in cardiac output which antagonizes their antihypertensive effect. They also cause salt and water retention by stimulating the renin angiotensin system and increasing intracapillary pressure.

Hydralazine has been used for over 30 years. It is rapidly absorbed from the gut, but has a half life of only 3 hours and divided dosing is necessary. Metabolism occurs in the liver by N-acetylation. This is of clinical significance as about 50% of the population are 'slow' acetylators and tend to accumulate the drug. A patient's acetylator status is determined genetically and can be measured by laboratory tests.

In some cases hydralazine produces a hypersensitivity syndrome which resembles systemic lupus erythematosis with skin rashes, arthralgia, fever, leucopaenia and lymphadenopathy. Antinuclear antibodies are usually found but the DNA binding test is characteristically negative. The syndrome will regress slowly after stopping the drug. It was once thought to be avoidable if the patient's acetylator status is checked and the total daily dose kept below 200 mg but Cameron & Ramsay (1984) have disputed this. Other side effects are related to vasodilatation and include flushing, tachycardia, headache, nasal congestion and postural hypotension.

Minoxidil is the most potent orally active vasodilator currently available and is often found to be effective when other treatment has failed. It is given 2 or 3 times per 24 h up to a maximum of 50 mg/24 h, although often as little as 2.5 mg bd will be effective. Unfortunately side effects are common. Reflex tachycardia and fluid retention are usually severe, making concomitant therapy with a β blocker and a diuretic mandatory, Glomerular filtration fraction is reduced, so considerable

C

caution must be employed when using this drug in the presence of renal impairment. Pericardial and pleural effusions can occur, particularly in uraemic patients, and occasionally cardiac tamponade is produced.

A major problem with minoxidil is its ability to stimulate growth of body hair. Hypertrichosis is almost always found and can be severe. It may not be such a great problem in men who occasionally welcome the rugged appearance and reversal of male pattern baldness, though some baulk at the prospect of having to shave their foreheads! It is difficult to use this drug in women, even with the use of depilatory creams.

Diazoxide is structurally related to thiazide diuretics and shares some of their side effects. It has a greater adverse effect on glucose tolerance and most patients will require an oral hypoglycaemic agent to control blood sugar levels. It also produces hyperuricaemia and has theoretical adverse effects on lipid metabolism. Unfortunately it does not have the thiazide diuretic action and in fact will cause fluid retention secondary to vasodilatation. Largely because of these problems, diazoxide does not have a major part to play in the long-term management of hypertension, but it remains a very potent agent which can be employed in difficult hypertension.

Centrally acting agents

These drugs have been used for many years and are declining in popularity. They will not be discussed in detail here. The two most important are *methyl dopa*, which inhibits noradrenaline synthesis, and *clonidine*, which is a central α_2 stimulant that reduces sympathetic outflow. Methyl dopa was one of the drugs used in a large recently published trial of treatment of high blood pressure in the elderly (Amery et al 1985). The results suggest a significant reduction in cardiovascular mortality with active treatment. This may be a specific property of the drug or the effect of pressure reduction. We do not believe that the evidence from this trial alone is sufficient to justify a major revival of this drug.

These drugs have many side effects which include sedation, haemolytic anaemia, jaundice and impotence for methyl dopa, and sedation, dry mouth and rebound hypertension on withdrawal for clonidine.

GENERAL PRINCIPLES

Long term treatment of sustained hypertension

Current practice has evolved from a stepwise approach to therapy, starting with a β adrenergic blocking agent and/or a diuretic, and progressing to so-called 'third-line' drugs in resistant cases, or where first line drugs are contraindicated. This method may need to be modified in future as new calcium antagonists and angiotensin converting enzyme inhibitors become available. However, for the moment, the stepped therapy method remains a widely accepted and useful approach to planning antihypertensive treatment.

Most of the third line agents work, to a greater or lesser degree, by reducing peripheral vascular resistance, though often together with a variety of other actions. Some, such as minoxidil and hydralazine, will cause a reflex tachycardia and are best used only in combination with a β adrenergic blocking agent. Others, such as nifedipine and captopril, do not cause a tachycardia and can be used safely in patients when β blockers are contraindicated. The use of verapamil with β blockade can produce conduction disturbances and myocardial depression (Hutchinson et al 1984).

Unfortunately, choosing a third line drug to complement an antihypertensive regime remains as much an art as a science. The theoretically appealing approach of trying to correct underlying haemodynamic and endocrine abnormalities has proved difficult to achieve, and has not yet been demonstrated to be of benefit. Simply reducing the blood pressure is far more important than the manner in which the reduction is achieved.

When selecting a drug it is important to consider the patient as a whole. His age, ethnic origin, intelligence, occupation and the presence of other pathology could all influence the choice. Older patients often do not tolerate β blockers and potent vasodilators, but respond well to calcium antagonists and diuretics. Afrocaribbeans often have low renin hypertension which responds better to diuretics (and perhaps calcium antagonists) than to β blockers and captopril. Drugs which cause drowsiness or mental slowing should be avoided in those with occupations demanding high levels of mental or physical performance. The opportunity to use drugs which will also treat concurrent pathology should not be missed. Calcium antagonists are appropriate for patients with angina or tachyarrhythmias and diuretics for those with heart failure.

In general it is wise to adhere to a combination of drugs which has proved effective and to avoid those likely to be antagonistic. A summary of desirable and undesirable combinations is given in Fig. 6.2.

Once antihypertensive drugs are commenced, they will usually need to be

	Diuretics	Beta Blockers	Nifedipine	Verapamil	Captopril	Prazosin	Indoramin	Labetolol	Hydralazine	Minoxidil	Diazoxide	Methyl Dopa
Nifedipine	P	P										
Verapamil	P	D	N									
Captopril	R	P	P	P								
Prazosin	P	P	P	P	P							
Indoramin	P	P	P	P	P	N						
Labetolol	P	N	P	D	P	N	N					
Hydralazine	P	R	N	R	P	P	P	R				
Minoxidil	M	M	P	R	P	?	?	R	N			
Diazoxide	P	R	P	R	P	?	?	R	N	N		
Methyl Dopa	P	P	P	P	P	P	P	P	P	P	P	
Clonidine	P	N	P	P	P	P	P	P	P	P	P	N

Fig. 6.2 Possible combinations of antihypertensive drugs. M = mandatory, R = recommended, P = possible, N = not recommended D = dangerous, ? = unknown.

continued for life. In this context, even apparently very minor side effects will, with the passage of time, increase in their importance to the patient and the impact that they have on his quality of life. Adequate therapy should not only control his blood pressure, but leave him feeling as healthy as possible. Clinicians should be receptive to patients' complaints and be prepared to change offending drugs even if it seems that control may not be quite as good. The 'you will have to grin and bear it' approach eventually leads to poor compliance and increases the number of clinic defaulters. Thus, it is an advantage for the doctor to have a wide range of drugs in his therapeutic armamentarium. However, familiarity with a drug, its side effects, its interactions and its metabolism is also desirable, and we would encourage clinicians to familiarize themselves with a small range of agents which they use routinely, venturing outside this range only when circumstances dictate.

Patients with severe hypertension often require a considerable number of tablets. The chance of getting good compliance will diminish as the number of tablets and daily dosing times increase. In these patients particularly, the use of combination and sustained action preparations is desirable, despite their relative lack of flexibility and increased cost.

Short term treatment of hypertensive emergencies

As a rule high blood pressure should be gradually reduced over a period of weeks or months by careful titration of appropriate oral antihypertensive agents. This is important as cerebral blood flow can only be kept constant by autoregulation within a range of about 50 mmHg around the prevailing mean pressure. Rapid reduction of pressure will not allow the autoregulatory mechanisms to reset and may cause acute cerebral ischaemia or infarction (Graham 1975). Very rarely a sudden large increase in pressure may break through autoregulation at the upper end of this range and cause symptoms such as headache, confusion, seizures and coma (hypertensive encephalopathy). In this situation rapid reduction of blood pressure is essential, although it should only be brought down to the middle of the autoregulatory range for that patient, rather than to normotensive levels. Even more rarely, acute reduction of pressure is required to prevent progression of life threatening complications. A list of conditions which may require emergency antihypertensive therapy is given in Table 6.4.

When blood pressure reduction is required within minutes, parenteral therapy is necessary. Many drugs have been used for this purpose over the years (Table 6.5) and all have their advocates. We favour labetalol, sodium nitroprusside, diazoxide and hydralazine.

Labetalol is very effective but must be given cautiously as its onset of action may be delayed by as much as 20 minutes following an intravenous bolus. It has been (incorrectly) advocated for use in patients with a phaeochromocytoma because of its combined α and β adrenergic blocking effects. In these patients β blockade alone can paradoxically increase blood pressure by blocking the vasodilating peripheral vascular β_2 receptors while allowing unopposed a mediated vasoconstriction. Labetalol in fact has much more potent β blocking than α blocking activity and has the same problems as the pure β blockers. The α blocker phentolamine is much safer choice in this condition.

Table 6.4 Conditions in which emergency antihypertensive therapy may be required

Reason for emergency antihypertensive therapy	Associated disease process	Timescale for pressure reduction
Hypertensive encephalopathy	Malignant hypertension Acute glomerulonephritis Eclampsia Phaeochromocytoma Clonidine withdrawal Food interaction with MAO inhibitors	Minutes to hours
Prevention of life threatening complications	Acute aortic dissection	Minutes
	Intracerebral or subarachnoid haemorrhage	Minutes to hours
	Bleeding post cardiovascular surgery	
	Eclampsia	
	Malignant hypertension	Hours
	Acute left ventricular failure	

Nitroprusside and trimetaphan have the advantage of a rapid onset and short duration of action. They are given by continuous infusion and the desired level of blood pressure can be easily achieved by adjusting the infusion rate. Unwanted hypotension is treated simply by temporarily discontinuing the infusion. Nitroprusside solution reacts with light and it is necessary to cover the administration set with aluminium foil. Trimetaphan is a ganglion blocker and its hypotensive action may be augmented by head up tilting.

Diazoxide and hydralazine reduce blood pressure by reducing peripheral vascular resistance. They cause a reflex increase in heart rate and cardiac output which increases cardiac oxygen consumption and so are less desirable in patients with reduced coronary reserve or acute left ventricular failure.

Table 6.5 Drugs which may be used parenterally in a hypertension emergency

Drug	Route of administration	Dose (mg)	Onset of action	Duration	Comments
Diazoxide	i.v.	100–600	3–5 min	4–18 h	Increase cardiac oxygen consumption
Hydralazine	i.v., i.m.	10–30	3–6 min	3–8 h	Tachycardia
Labetalol	i.v., i.m., infusion	10–50	5–20 min	3–4 h	Not for phaeochromocytoma
Sodium nitroprusside	Infusion	0.03–0.5/mm	1–5 min	5–10 min	Unstable solution
Trimetaphan	Infusion	1–15	1–5 min	5–10 min	Helped by head up tilt; easy to control
Clonidine	i.v.	0.15–0.3	15–30 min	3–4 h	May cause an initial increase in pressure; sedative
Phentolamine	i.m.	10–20	15–30 min	5–6 h	Good for phaeochromocytoma; may cause severe hypotension

Parenteral therapy should only be used on the rare occasions that rapid reduction of blood pressure is imperative. In less urgent situations a nifedipine capsule crushed between the teeth is effective in about 30 minutes. The capsules contain a solution of nifedipine which is rapidly absorbed through the buccal mucosa. Treatment can be continued with slow release nifedipine tablets.

FUTURE DEVELOPMENTS

Many new compounds are currently undergoing clinical evaluation although only a small proportion will ever become available for general use. A few drugs with entirely novel modes of action are also being developed. One example of this is the serotonin antagonist ketanserin which has been found to lower blood pressure (Wenting et al 1982).

The range of antihypertensive agents currently available is large and this is bound to generate some confusion. As research into the pathophysiology of blood pressure and new drugs continues, the day is approaching when scientific and side effect free therapy will be possible. In the meantime we should try to remember that a patient's quality of life may be more important to him than the statistical risks of less than perfect control, and we should never forget that the people who sell us these drugs may have more than just our patient's health in mind.

REFERENCES

Amery A, Birkenhager W et al 1985 Mortality and morbidity results from the European Working Party on high blood pressure in the elderly trial. Lancet i: 1349–1354
Bayley S, Dobbs R J, Robinson B F 1982 Nifedipine in the treatment of hypertension: report of a double blind controlled trial. British Journal of Clinical Pharmacology 14: 509–512
Braunwald E 1982 Mechanism of action of calcium channel blocking agents. New England Journal of Medicine 307: 1618–1627
Buhler F R 1983 Age and cardiovascular response adaptation. Determinants of an antihypertensive treatment concept primarily based on beta-blockers and calcium entry blockers. Hypertension 5: 94–100
Cameron H A, Ramsey L E 1984 The lupus syndrome induced by hydralazine: a common complication with low dose treatment. British Medical Journal 289: 410–412
Dombey S 1983 Optimal dose of captopril. Lancet i: 355
Graham D I 1975 Ischeamic brain damage of cerebral perfusion type after treatment of severe hypertension. British Medical Journal iv: 739
Graham R M, Pettinger W A 1979 Prazosin, New England Journal of Medicine 300: 232–236
Guazzi M D, Fiorentine C, Olivari M T, Bastorelli A, Necchi G, Polese A 1980 Short- and long-term efficacy of a Ca-antagonistic agent (nifedipine) combined with methyldopa in the treatment of severe hypertension. Circulation 61: 913–919
Hodsman G P, Isles C G, Murray G D, Usherwood T P, Webb D J, Robertson J I S 1983 Factors related to the first dose hypotensive effect of captopril: prediction and treatment. British Medical Journal 286: 832–835
Hricik D G, Browning P J, Kopelman R, Goorno W E, Madias M E, Dzau V J 1983 Captopril induced functional renal insufficiency in patients with bilateral renal artery stenosis or renal artery stenosis in a solitary kidney. New England Journal of Medicine 308: 373–376
Hutchinson S J, Lorimer A R, Lakhdar A, McAlpin S G 1984 Beta blockers and verapamil: a cautionary tale. British Medical Journal 289: 659–660
Jee L D, Opie L H 1983 Acute hypotensive response to nifedipine added to prazosin in treatment of hypertension. British Medical Journal 287: 1514
MacGregor G A, Markandu N D, Banks R A, Roulston J E, Jones J C 1982 Captopril in essential hypertension; contrasting effects of adding hydrochlorthiazide or propranolol. British Medical Journal 284: 693–696

MacGregor G A, Banks R A, Markandu N D, Bayliss J, Roulston J 1983 Lack of effect of beta-blocker on flat dose response to thiazide in hypertension: efficacy of low dose thiazide combined with beta-blocker. British Medical Journal 286: 1535–1538

Man in't Veld A J, Schicht I M, Derkx F H M, de Bruyn J H B, Schalekamp M A D H 1980 Effects of an angiotensin-converting enzyme inhibitor (captopril) on blood pressure in anephric subjects. British Medical Journal 280: 288–290

Nattel S, Rangno R E, Van Loon G 1979 Mechanism of propranolol withdrawal phenomena. Circulation 59: 1158–1164

Pritchard B N C, Gillam P M S 1963 Use of propranolol (inderal) in treatment of hypertension. British Medical Journal 2: 725–727

Schutz et al 1982 Serum concentration and antihypertensive effect of slow release verapamil. Journal of Cardiovascular Pharmacology 3: 346–349

Singh B N, Venkatesh N 1984 Prevention of myocardial reinfarction and sudden death in survivors of acute myocardial infarction: role of prophylactic beta-adrenoceptor blockade. American Heart Journal 107: 189–200

Wenting G J, Man in't Veld A J, Woittiez A J 1982 Haemodynamic effects of ketanserin, a 5-hydroxytryptamine (serotonin) receptor antagonist, in essential hypertension. Clinical Science 63: 435s–438s.

C. W. Howell

Hypotensive anaesthesia

INTRODUCTION

The use of deliberate hypotension in anaesthesia is still a contentious issue. The advocates of the technique consider it to be without special problems, whereas others feel that its use should be restricted to making the impossible operation possible.

Awareness of blood pressure change during anaesthesia is surprisingly recent and its recording has become established practice only in the professional lifetime of practising anaesthetists. The concept of flow and perfusion is even more recent. Although initial reports on the use of deliberate hypotension appeared in the early 1950s, these were mainly clinical observations. It was almost 15 years later that significant scientific observations were made into the physiological changes produced by this technique.

PHYSIOLOGY

Measured blood pressure is a function of cardiac output and peripheral vascular resistance. Other factors —particularly viscosity — affect flow. Frequent references are made to Poiseuille's law and Reynolds number but they are of little relevance to clinical practice. Poiseuille stated that the pressure in a conduit was in inverse ratio to the fourth power of its radius. It relates to a homogenous fluid, with laminar flow in a non-distensible conduit. Reynolds number relates the velocity, viscosity and density of a fluid to the diameter of the conduit. A figure below 2000 indicates laminar (efficient) flow, and greater than 2000 turbulent (poor) flow.

The viscosity of blood is stated as four times that of water. However, this varies at different sites within the circulation (it falls in small vessels) and with different packed cell volumes (PCV). A rise in the PCV of 50% doubles the viscosity, while a fall of 25% halves it. Therefore, haemodilution by removing 1 litre of blood and replacing with a balanced salt solution materially improves perfusion.

Peripheral vascular resistance (PVR)

The major component of PVR occurs in the arterioles. Arterioles are classified as first to fourth order depending on their branching and size (200 μm down to 10 μm of the precapillary arteriole). Control tends to be neural in the larger arterioles and humoral in the smaller. The neural effect varies at different sites, but local metabolites act at the humoral sites. Catecholamines, vasopressin and angiotensin II are amongst the humoral pressors, while histamine, 5HT, bradykinin and prostaglandins have the reverse effect. Although cholinergic, vasodilator fibres have been described in skeletal muscle, the vast majority of neural control is via the sympathetic system. At rest, sympathetic impulses are about one per second, but under stress this multiplies 10-fold. Intense sympathetic activity increases the critical closing pressure (the pressure at which a vessel closes to redirect flow to essential circulation) to about 60 mmHg. Any agent which inhibits sympathetic discharge or its effects will, necessarily, reduce the closing pressure, usually to about 15 mmHg. Thus deliberate hypotension will maintain perfusion. Baroreceptors in the aortic arch and carotid sinus detect pressure changes and, via the vasomotor centre, affect sympathetic and vagal tone. The effects are seen mainly in the PVR and capacitance vessels and, to a lesser extent, in the cardiac output. Chemoreceptors in similar sites relate mainly to respiratory control, though hypoxia does cause a reflex increase in cardiac output and PVR. The role of peripheral chemoreceptors is uncertain, but hypoxia or hypotension may activate them and may be the mechanism by which respiratory effort — seen with very low blood pressure — is mediated. Anaesthesia itself inhibits both baroreceptor and chemoreceptor response.

Capacitance

75% of the blood volume is contained within the veins, the remainder in arteries and capillaries. The veins are rich in smooth muscle and are controlled in a manner similar to the arteries. They are able to contract around a diminishing blood volume to maintain cardiac return and output. Indeed, a blood loss of 500 ml reduces the central venous pressure by only 1 mmHg.

Cardiac output

The output of the heart produces the pressure head in the vasculature, and with a constant PVR the blood pressure will change accordingly. Venous return is the major factor determining cardiac output, with chronotropic and inotropic states playing a lesser role. Within specific limits, increased sympathetic tone will increase cardiac output. However, any abnormally high or low rate will result in a reduced cardiac output as a result of opposite changes in stroke volume.

Posture

Posture significantly affects pressures in both arteries and veins. In the arteries, for every 1 cm elevation above the site of measurement, the blood pressure will fall approximately 1 mmHg. Below the level of measurement, the converse is true. Similar changes are seen in venous pressure, but of a lesser magnitude in the elevated parts of the body. This orthostatic effect is magnified in association with general anaesthesia. In the head up position, the reduced muscle tone in the legs deprives the veins of external compression, and combined with reduced sympathetic tone, the veins dilate allowing more than 250 ml of blood to pool in the legs.

Ventilation

Intermittent positive pressure ventilation (IPPV) produces an initial rise in blood pressure as the increased intrathoracic pressure is transmitted directly to the arteries. However, the blood pressure then falls as venous return to the heart is reduced. Positive end expiratory pressure (PEEP) enhances the hypotension.

Auto regulatory blood flow

Blood flow varies in different tissue for different reasons, e.g. in skeletal muscle it is almost independent of sympathetic activity, but very responsive to vasodilator metabolites, whilst cutaneous blood is under mainly sympathetic control.

Coronary blood flow

Like systemic flow, coronary blood flow is a function of perfusion pressure and coronary vascular resistance. Coronary perfusion pressure is relatively high as both coronary arteries arise from the ascending aorta. At rest, 75% of flow in the coronary arteries occurs during diastole. Consequently a raised ventricular end diastolic pressure will effectively reduce the coronary perfusion pressure. During stress, the flow pattern is reversed, with most of the flow occurring during systole.

Coronary vascular resistance is determined by the lumen of the vessel and the muscle tone of the myocardium through which the vessel passes. Dynamic changes in the lumen occur mainly by changes in sympathetic tone, but metabolites — particularly carbon dioxide — play a major role. Static changes (e.g. atheroma) are more predictable in their effect. Stenosis, less than 70% of a vessel, produces little fall in flow unless under stress, but in excess of 90%, collateral vessels are required to maintain flow. The major coronary arteries are on the external surface of the heart, and must pass through the myocardium to supply the endocardium. It is the endocardium which is the first to suffer the effects of hypoperfusion. Myocardial hypertrophy per se reduces endocardial perfusion as does tachycardia — often seen as a baroreceptor response to hypotension.

The rate pressure product (heart rate × systolic blood pressure) is regarded as a useful indicator of ischaemia and, during anaesthesia, it should be kept below 12 000. It is not of great value during hypotensive anaesthesia.

Cerebral blood flow

Autoregulation of cerebral blood flow (CBF) is marked. There is relatively poor sympathetic innervation, but the vasodilator response to metabolites — especially carbon dioxide — is impressive. CBF is constant, with a mean arterial pressure above 60 mmHg. Below 50 mmHg, the CBF falls in direct proportion to the blood pressure, as autoregulation has failed and carbon dioxide now has no effect. In the untreated hypertensive, the pressure limits for maintaining CBF are higher, but with treatment they return to normal. In the presence of raised intracranial or venous pressure, a higher blood pressure is needed to maintain perfusion pressure and CBF. IPPV and PEEP have little effect on the cerebral venous pressure in the upright person, but in the supine position, the raised intrathoracic pressure is transmitted directly to the cerebral veins.

Blood flow to the parts of the brain furthest from the blood supply is relatively poor and more likely to be affected by a rise in intracranial pressure. These 'boundary areas' include the sulci of the cortex, hippocampus and basal ganglia. Consequently, survival following hypoperfusion states results in personality change and memory loss.

A sudden fall in the blood pressure has a greater effect on CBF as there is a lag in autoregulation. However, if the hypotension is drug induced, as opposed to haemorrhagic, the vasodilator effect of the drug (especially sodium nitroprusside) to some extent preserves the CBF and cortical oxygenation.

If hypotension is profound and prolonged, cerebral oedema occurs. Although this may not produce permanent damage, when perfusion is restored, recovery may occur in an erratic manner. The protective role of barbiturates in reducing cerebral oxidative metabolism has few advocates when induced hypotensive anaesthesia is used.

Renal blood flow

Like the brain, the kidneys have a marked metabolic autoregulatory system. Unlike the brain, it can be overridden by powerful sympathetic control. There is little change in perfusion and filtration above a mean arterial pressure of 70 mmHg, but below 60 mmHg, filtration stops and perfusion is dramatically reduced. Severe and prolonged hypotension results in acute tubular necrosis. Most volatile anaesthetic agents reduce renal blood flow independent of any hypotension they may produce.

Hypotension activates the renin–angiotensin axis. Renin is released from the juxtaglomerular apparatus and this activates angiotensin I. Although angiotensin I is not very vasoactive, passage through the lungs converts it to angiotensin II — a highly vasopressor substance. This sequence is seen particularly following hypotension with sodium nitroprusside.

Pulmonary function

It is well established that arterial oxygenation falls during general anaesthesia, as a result of reduced alveolar ventilation and an imbalance in the ventilation–perfusion ratio ($\dot{V}/\dot{Q}$) in dependent parts of the lungs. This fall in oxygenation is increased during induced hypotension. The change in $\dot{V}/\dot{Q}$ ratio is particularly marked during spontaneous ventilation and may well be affected by the action of drugs on the bronchial musculature. IPPV and PEEP improve the $\dot{V}/\dot{Q}$ ratio, but at the expense of obstructing pulmonary capillary flow and cardiac output. Drug induced, as opposed to haemorrhagic, hypotension obtunds the hypoxic pulmonary vasoconstrictor response — an homeostatic response to improve the $\dot{V}/\dot{Q}$ ratio. In a spontaneously breathing patient, the P_{CO_2}, rises while the pO_2 falls. The $\dot{V}/\dot{Q}$ ratio improves with IPPV but at the expense of cardiac output.

Patients lying supine do not show as great a change in blood gases as those in a head up tilt. IPPV does not improve the gas picture as much as would be expected, but it does reduce cardiac output, which is said to improve the bloodless field. The raised

$P\text{CO}_2$ seen in spontaneous ventilation does maintain cardiac output and cerebral blood flow but may increase bleeding.

INDICATIONS AND COMPLICATIONS

All indications and contraindications must be relative rather than absolute. By far the most common indication for the use of induced hypotension is to improve the surgical field. Less blood obscuring the field allows for greater surgical accuracy in identifying vital structures and pathological tissue. During microsurgery, a minute quantity of blood may totally obscure the surgical field. A relatively bloodless field can substantially reduce operating times, but this should never be a prime consideration.

Reduced blood loss during surgery leads to a reduced requirement for blood transfusion — an expensive and potential dangerous fluid. In a Jehovah's Witness, it is a major factor to consider.

Reduced blood loss is usually associated with less peri-and post-operative oedema — crucial in skin flaps and pedicles or in other areas of compromised blood supply. Within limits, perfusion is improved despite reduction in systemic blood pressure.

The use of deliberate hypotension during anaesthesia must relate not only to the experience of the anaesthetist, but the familiarity of the surgeon with the technique. The ultimate decision is that of the anaesthetist.

Patients with pre-existing cardiovascular disease must be considered carefully. The hypertensive patient will probably have reset his autoregulation at a higher level than normal. In these patients hypotension must be regarded as a fall in the systolic pressure of 33%. Treated hypertensives should continue their treatment, but their medication will be potentiated by anaesthesia, and per-operative hypotensive agents. Drugs blocking α adrenergic receptors may well have bound permanently to the receptor. Existing myocardial ischaemia is often improved by hypotension — provided any infarct occurred more than 6 months pre-operatively. Intra-operative myocardial ischaemia can be identified by careful monitoring (vide infra) and if the fall in blood pressure is corrected immediately, there is no evidence that permanent damage ensues (Rollason & Hough 1969). Thompson (1978) and her colleagues, in a wider study, found no biochemical changes as a result of hypotension. Cerebral disease, whether vascular or from other pathology, tolerates hypotension well as cerebral blood flow (down to a mean pressure of 50 mmHg) is more dependent on changes in carbon dioxide than blood pressure. However, patients with established infarcts should not be hypotensed for fear of extending the infarct in areas of critical flow. Eckenhoff et al (1964) and Rollason et al (1971) could not detect changes by psychometric testing following hypotensive anaesthesia, even in the elderly.

Consideration must be given to the changes in blood gases in patients with respiratory disease and the effects of hypotensive agents on bronchical structures, particularly in asthmatics. Asthmatics should not be given ganglion blockers and; there is usually some hyperinflation and they do not readily tolerate the valsalva manoeuvre associated with IPPV. Poor renal and hepatic function may be aggravated by hypotension in these organs.

Ganglion blockade enhances the effect of insulin in diabetic patients. Those insulin dependent are further at risk, as they have abnormally responsive cerebral blood flow and may be unable to compensate for hypotension (Dandona et al 1978).

Anaesthesia itself can produce problems in pregnant women. Induced hypotension should be avoided, as the effects of the technique on the blood flow in the uterus and fetus are uncertain. Initial enthusiasm for induced hypotension in the early 1950s rapidly waned as a result of the apparent increase in morbidity and mortality. Hampton & Little (1953) quoted a mortality of approximately 1:500, three times that of normotensive anaesthesia, and with a comparable rise in morbidity. Although figures produced at this time are open to criticism, the use of controlled hypotension would have become historical interest but for the enthusiasm of anaesthetists such as Enderby and later Eckenhoff, to whom must go credit for their scientific application. More recent estimates of mortality and morbidity (Enderby 1980) show that with improved selection and technique, morbidity and mortality compares favourably with normotensive anaesthesia.

MONITORING

The mercury manometer or its aneroid equivalent has been the time honoured method of measuring blood pressure during anaesthesia, but it is slow and cumbersome. The oscillotonometer is a better instrument and, suitably modified, it can be left inflated and can act as a continuous monitor of pulse and trends in the systolic pressure. Most automatic blood pressure recorders work on the oscillotonometric principle, but they have a slow cycle time. Most read low — a safety factor. Intra-arterial cannulation — usually of the radial artery — gives a direct and accurate reading of the blood pressure. Although considered to have complications, the incidence of permanent sequelae is minimal. It is a low risk, high benefit, method of patient monitoring (Slogoff et al 1983). Allens test to predict occlusive problems is probably of little value (Wilkins 1985). A 20G Teflon cannula is preferred. Intra-arterial cannulation also offers the facility for blood gas measurement. Hypoxia produced by hypotension can be accurately compensated for and PCO_2 measurement is particularly useful to monitor cerebral blood flow. In the absence of blood gas estimation, an end tidal CO_2 meter is of value.

The electrocardiogram is a good indicator of cardiac rhythm. During anaesthesia the selection of lead II is often standard. However, this is a poor indicator of myocardial ischaemia, for which lead V5 should be selected. A modified V5 may be produced by placing RA on the right side of the chest, LA in the position of V5 with the reference electrode in any convenient position.

The cerebral function monitor (CFM) is an integrated electroencephalogram with both amplitude and frequency modified. Two electrodes are placed parietally. It indicates global rather than focal hypoperfusion or hypoxia and the record will remain normal while the systolic blood pressure is above 60 mmHg, and the fall in blood pressure is not greater than 10 mmHg/min. The CFM tells of hypoxic events, but some time after they have occurred and consequently it is of little value in clinical practice.

Changes in blood volume in an hypotensed patient have a greater physiological effect than in a normotensive patient. A blood loss as small as 100 ml may have a significant hypotensive effect, while the infusion of a similar amount of fluid have the reverse effect. It is important to estimate blood loss accurately and have a reliable intravenous infusion.

It is easy to concentrate on instruments and their readings. But this should not replace observation of the patient.

TECHNIQUES

Hypotension has been defined as a fall of 33% in the systolic blood pressure. In practice the systolic blood pressure is reduced to 60–80 mmHg. During normal sleep, the systolic pressure may not be much greater than 80 mmHg. There has been debate as to whether a reduced cardiac output, rather than hypotension, is the greater factor in producing a relatively ischaemic surgical field. Hypotension is probably the major factor (Sivarajam et al 1980). The use of high spinal analgesia to create intra-operative hypotension — though hardly controlled — was reported by Koster as early as 1928. Griffiths & Gilles used the technique more widely and reported their results in 1948. Three years later Bromage advocated the use of epidural anaesthesia for hypotension (Bromage 1951). Total sympathetic blockade has now no part in hypotensive anaesthesia, but more limited blockade, using spinal or epidural analgesia, has an important role in producing an ischaemic surgical field.

Haemorrhagic hypotension —arteriotomy following by autotransfusion — as practiced by Gardner in 1946, was associated with many problems and gained little acceptance. However haemodilution, with or without hypotension, is now practised. The viscosity of blood is reduced and tissue perfusion improved at the expense of a slight rise in cardiac output. Fahmy and colleagues (1980) report a marked reduction in the need for donor blood transfusion using this technique. The development of drugs able to block the autonomic ganglia (Paton & Zaimis 1948) heralded the hesitant beginning of hypotensive anaesthesia as we know it today, and became the basis of 'pharmacological blockade' (vide infra). Of the early ganglion blocking drugs, pentolinium and trimetaphan are the only ones which remain in use. Both have a competitive blocking action on the autonomic ganglion, similar to that of D-tubocurarine.

Pentolinium

Pentolinium is presented as 10 mg in 2 ml ampoules. The dose required is variously reported as between 0.05 mg/kg and 0.3 mg/kg. Fit, young patients require higher doses, as do patients breathing spontaneously. Posture and volatile anaesthetic agents have a marked synergism on the hypotensive effect. An accurate initial dose is important, as it is usually the only effective dose, subsequent doses having a lesser effect. When used during cardiopulmonary bypass, only 0.5 mg may be required to achieve hypotension. There is a great individual response to pentolinium, the maximal effect occurring within 30 minutes and lasting between 1 and 4 hours. However, many of its effects may last much longer — e.g. cycloplegia up to 12 hours. The drug is not metabolized and is excreted unchanged in the urine.

Trimetaphan

Trimetaphan was first used clinically in the early 1950s (Magill et al 1953). The hypotensive effect of ganglion blockade is potentiated by a direct vasodilator action and histamine release. An ampoule of 250 mg in 5 ml is normally diluted to 0.1% solution and administered by continuous infusion. Hypotension occurs in approximately 3 minutes and last 15 minutes. Tachyphalaxis is marked, but the total dose should never exceed 1 g. As a sole hypotensive agent, it has been superseded by sodium nitroprusside, but is returning in popularity in a combination of trimetaphan/sodium nitroprusside. Its metabolic pathway has never been established, but one-third appears unchanged in the urine.

β Adrenergic blockade

β Adrenergic blockade with propanolol has been used to control the tachycardia produced by hypotensive anaesthesia for 20 years (Hellewell & Potts 1966). Its use is firmly established, although there are now more cardioselective blockers such as practolol and atenolol. The effects of these drugs on cardiac output is an important factor in achieving a bloodless surgical field.

β Adrenergic stimulation is usually seen during hypotension with the directly acting vasodilators, sodium nitroprusside and trinitroglycerine. This activates the renin–angiotensin axis, making per-operative control of blood pressure difficult and post-operative hypertension frequent. β Blockade inhibits the renin–angiotensin reaction (Khambatta et al 1981), but renin activation is not seen as a result of ganglion blockade.

Labetolol

Labetolol, an α and β adrenergic blocker, is marketed to treat hypertensive patients with associated angina. Its use in anaesthesia was first reported by Scott and colleagues in 1976. Its hypotensive effect occurs mainly by lowering the peripheral resistance and, to a lesser extent, reducing the cardiac output. Given in 5 mg increments, to a maximum of 25 mg in 10 minutes, its effects are rapidly apparent. The hypotension is potentiated by volatile anaesthetic agents and head up tilt. As with most other hypotensive agents, moderation is needed in the elderly, in whom bradycardia, extending well into the post-operative period, may be marked. Although this can be effectively treated with atropine, profound hypotension may be difficult to reverse because of the α blockade.

Direct acting vasodilators

It was a quarter of a century after its initial clinical use that sodium nitroprusside (SNP) was first used in anaesthesia (Moraca et al 1962). It is now probably the most common agent used to reduce the blood pressure during anaesthesia. It is presented as a powder, 50 mg, which must be reconstituted immediately prior to use and is normally diluted in 500 ml, 5% dextrose for use as a continuous infusion. SNP is rapidly decomposed by light to aquapentaferrocyanate, a substance which releases free cyanide. Protected from light, a solution should be discarded 4 hours after preparation.

In the body, toxic cyanide ions are converted into non-toxic cyanmethaemoglobin and thiocyanate. Although excess cyanide ions can be absorbed by the administration of sodium thiocyanate, it is better to reduce the risk of cyanide poisoning by restricting the dose of SNP. The recommended maximum safe dose is accepted as 1.5 mg/kg over a short period and 0.5 mg/kg per hour for prolonged administration. The rate of infusion should not exceed 10 μg/kg per minute. Patients with B_{12} deficiency should not be given SNP as they have defective cyanide metabolism. The effect of SNP is directly on the vascular smooth muscle, the exact mode being of continuing speculation, but effects are more marked in arterial muscle than venous muscle. Onset of hypotension occurs within 1 minute and normal blood pressure is restored within 4 minutes of ceasing the infusion. Moderate tachycardia, a baroceptor response, is usual. The effects on myocardial perfusion and oxygen demand are confusing, but both are probably reduced, making it useful to control ischaemic hypertension during coronary artery surgery.

Cerebral perfusion is increased, causing an initial rise in the intracranial pressure. For this reason it has been suggested that an infusion of SNP does not begin until the dura has been opened. SNP is the drug of choice for hypotension during neurosurgery, as its effects are solely vascular and readily reversible. MacRae (1981) has described the combined infusion of trimetaphan/SNP (10:1). Hypotension is good with greatly reduced doses of individual drugs, thereby reducing their individual toxicities.

Trinitroglycerin (TNG), like SNP, has a direct action on blood vessel musculature, but unlike SNP its effects are mainly venous. It is presented as 1 mg/ml in 10 ml ampoules, which is diluted to 100 ml for infusion. It is usually used to reduce cardiac pre-load and dilate the coronary vasculature and to this end, it is an excellent hypotensive agent during cardiac surgery. Its role as an hypotensive agent in patients without ischaemic heart disease is less certain. It will produce hypotension in normal patients, but the dose required is often very high. In combination with SNP, good hypotension is achieved and this use may be indicated for hypotensive anaesthesia in patients with proven myocardial ischaemia.

Volatile agents

Halothane potentiates most hypotensive drugs, though its use as a sole hypotensive agent is limited. Autonomic depression does occur, together with a reduced cardiac output and reduced peripheral vascular resistance. Alone it produces adequate hypotension, particularly in the elderly, but a concentration of 3% with intermittent positive pressure ventilation may be necessary. Enflurane produces similar effects but it is unlikely that isoflurane will contribute much to hypotensive anaesthesia, as with this drug cardiac output is maintained.

Pharmacological blockade (Enderby 1974)

This is a concept which explains hypotension produced by several factors:
Ganglion blockade (pentolinium, trimetaphan).

Beta blockade (propanolol, practolol, atenolol).

Volatile agents (halothane, enflurane).

Posture (25% head up tilt).

IPPV ± PEEP.

It is a safe technique employed and recorded during many thousands of general anaesthetics by anaesthetists trained and experienced in the technique. However, for the anaesthetist called upon to provide hypotensive anaesthesia less frequently, it is probably not the method of choice. Instead they should consider a technique involving IPPV with a volatile agent and posture which produce satisfactory results, supplemented by SNP or labetolol. There can be no dogma relating to spontaneous respiration or intermittent ventilation, but oxygen enriched mixtures are mandatory.

Deliberate hypotension in anaesthesia is a safe procedure which can materially benefit surgery. With careful selection of patients and administration of the technique, there should be no greater incidence of complications than with normotensive anaesthesia.

REFERENCES

Bromage P R 1951 Anaesthesia 6: 26

Dandona P, James A M, Newbury M L, Beckett A G 1979 Instability of cerebral blood flow in insulin dependent diabetics. Lancet ii: 1203–1205

Eckenhoff J E, Compton J R, Larson A, Davies R M 1964 Assessment of cerebral effects of deliberate hypotension by psychological measurements. Lancet ii: 711–714

Enderby G E H 1974 Pharmacological blockade. Postgraduate Medical Journal 50: 572–575

Fahmy N R, Chandler H P, Patel D G, Lappas D G 1980 Haemodynamics and oxygen availability during acute haemodilution in conscious man. Anesthesiology 53: 584

Gardner W J 1946 The control of bleeding during operation by induced hypotension. Journal of the American Medical Association 132: 572–574

Griffiths H W C, Gilles J 1948 Thoracolumbar splanchnicectomy and sympathectomy: anaesthetic procedure. Anaesthesia 3: 134–136

Hampton L J, Little D M 1953 Results of a questionnaire concerning controlled hypotension in anaesthesia Lancet : 1299–1300

Hellewell J, Potts M 1966 Propanolol during controlled hypotension. British Journal of Anaesthesia 38: 794–801

Khambatta H J, Stone J G, Khan E 1981 Propanolol alters renin release during nitroprusside induced hypotension and prevents hypertension on discontinuation of nitroprusside. Anesthesia and Analgesia (current researches) 60: 569–573

Koster H 1928 Spinal anaesthesia with special reference to its use in surgery of the head, neck and thorax. American Journal of Surgery 5: 554

Magill I W, Scurr C F, Wyman J B 1953 Controlled hypotension by a thiophanium derivative. Lancet i: 219–220

Moraca P P, Elmers M B, Hale D E, Wasmuth C E, Pontasse E F 1962 Clinical evaluation of sodium nitroprusside as an hypotensive agent Anesthesiology. 23: 193

Paton W D M, Ziamis E J 1948 Curare like action of polymethylene bis-quaternary ammonium salts. Nature 161: 718–719

Rollason W N, Hough J M 1969 A re-examination of some electrocardiographic studies during hypotensive anaesthesia. British Journal of Anaesthesia 41: 985–986

Sivarajan M, Amory D W, Everett G B, Buffington C 1980 Blood pressure not cardiac output determines blood loss during induced hypotension. Anesthesia and Analgesia (current researches) 59: 203–206

Slogoff S, Keats A S, Arlund C 1983 On the safety of radial artery cannulation. Anesthesiology 59: 42–47

Thompson G E, Miller R D, Stevens W C, Murray W R 1978 Hypotensive anaesthesia for total hip arthroplasty. A study of blood loss and organ function (brain, heart, liver and kidney). Anesthesiology 48: 91–96

Wilkins R G 1985 Radial artery cannulation and ischaemic damage: a review. Anaesthesia 40: 896–899

Further reading
Enderby G E H (ed) 1985 Hypotensive anaesthesia. Churchill Livingstone, Edinburgh

A classification of antiarrhythmic drugs and common cardiac arrhythmias

INTRODUCTION

A large number of antiarrhythmic drugs are currently available to practising doctors, with even more waiting in the wings at the stage of clinical research. None is a panacea and all may aggravate or provoke rather than relieve arrhythmias (proarrhythmic effect). Careful consideration is therefore necessary before prescription of any of these agents. The most important factor in choosing an antiarrhythmic drug is the arrhythmia to be treated. In this chapter, we will therefore briefly outline the arrhythmias commonly encountered in terms of their presentation, electrocardiographic features, underlying mechanism and their clinical significance before proceeding to discuss the drugs used in their management in Chapter 9. The references for both chapters are listed at the end of Chapter 9.

Many drugs have lapsed into relative disuse because of the advent of a more effective or less toxic agent while other new agents do not offer any advantage over existing, tried and trusted drugs. Such new drugs are unlikely to replace these successful older drugs and so we will concentrate on those drugs commonly used and the new drugs likely to become future favourites. For the sake of convenience, we will employ the *Vaughan-Williams classification* of antiarrhythmic drugs (Table 8.1). This classification is based on the cellular electrophysiological effects of antiarrhythmic drugs whereby they are divided in to four main classes, I–IV. This is not always clinically relevant because, although the classification usually allows prediction of sites of action and inaction, there is often considerable variation in terms of clinical effect even between drugs in the same group.

The class I antiarrhythmic drugs

The class I antiarrhythmic drugs are subclassified as Ia, Ib and Ic drugs according to their different effects upon the action potential (Table 8.1). The predominant effect of all class I drugs is to reduce the rate of maximal depolarization of myocardial cells and Purkinje tissue. The resting transmembrane potential is little affected whereas the threshold for depolarization is raised, conduction velocity depressed and refractory periods prolonged. The spontaneous diastolic deplorization which identifies pacemaker cells is also inhibited, this effect being seen more readily, i.e. at lower drug concentrations, than the effects on threshold and conduction. This combination of properties makes arrhythmias which are due both to foci of increased

Table 8.1 The Vaughan-Williams classification

Class	Effects	Examples
Ia	Membrane stabilization Prolonged action potential duration	Quinidine Procainamide Disopyramide
Ib	Membrane stabilization Shortened action potential duration	Lignocaine Mexiletine Tocainide
Ic	Membrane stabilization Varied action potential duration	Flecainide Propafenone
II	β-Blockade	Propranolol Metoprolol Acebutolol
III	Action potential prolongation	Amiodarone Bretylium
IV	Calcium antagonism	Verapamil Diltiazem Tiapamil

automaticity and to reentry susceptible to class I agents. The subdivision of the group is based upon their differing effects on the subsequent duration of the action potential, i.e. the time taken for repolarization. Drugs of class Ia prolong this, class Ib agents shorten it and flecainide acetate (the first representative of class Ic) prolongs repolarization of atrial and ventricular myocardium while shortening that of the His-Purkinje system. These different properties might explain some of the differences in their antiarrhythmic effects. Those which prolong the action potential duration are effective against atrial flutter and fibrillation (reentry atrial arrhythmias influenced by tissue refractoriness) whereas those of class Ib are not, their activity being restricted to ventricular arrhythmias. Because of their effects upon the His-Purkinje system, all class I drugs are relatively contraindicated in patients with conduction abnormalities. The efficacy of class I drugs depends upon the potassium concentration. Hypokalaemia renders them less effective (and is itself proarrhythmic) whereas hyperkalaemia accentuates their effects. With the possible exceptions of mexiletine, tocainide (Ib) and propafenone (Ic), all are at least moderately negatively inotropic and left ventricular function should be considered before prescription of class I agents.

The class II antiarrhythmic effect

The class II antiarrhythmic effect is β-blockade. Thus arrhythmias influenced by sympathetic tone are affected. These include those related to ischaemia and the long QT syndromes although the most profound electrophysiological effect of β-blockade is seen at the AV node where refractoriness is increased and conduction velocity decreased. The antiarrhythmic uses of the drugs are therefore mainly to reduce conduction of atrial arrhythmias via the AV node (and thereby the ventricular response rate) or more rarely to prevent or terminate junctional reentry arrhythmias which use the AV node as at least one limb of the reentry circuit. The major cellular electrophysiological effect is to depress phase 4 depolarization, although at relatively high concentrations some have independent class I effects. Chronic β-blockade (>3 weeks) results in slight but statistically significant QT prolongation.

The class III antiarrhythmic effect

The class III antiarrhythmic effect is related purely to prolongation of action potential duration. The major example of the group is amiodarone (Singh & Vaughan-Williams 1970) which also exhibits a mild non-competitive antisympathetic effect (Charlier 1970). The effect is widespread throughout the heart, this being mirrored by the diversity of the arrhythmias treatable by amiodarone.

The class IV antiarrhythmic drugs

The class IV antiarrhythmic drugs are a heterogenous group sharing the property of slow-channel (calcium dependent) blockade. Accordingly, they are collectively known as calcium antagonists. The slow channel is particularly important in the formation and propagation of impulses at the sinus and AV nodes. The major antiarrhythmic effects of slow channel blocking drugs are seen at the AV node; they are often used in the same context as β-blockers. Not all possess antiarrhythmic effects. Nifedipine is a member of the group which does not have an antiarrhythmic property at clinical concentrations, exerting its effect predominantly upon vascular smooth muscle.

ARRHYTHMIAS

Only the drug treatment of *tachy*arrhythmias will be considered because, with the exception of acute emergencies, drugs have no place in the modern management of bradyarrhythmias. Cardiac arrhythmias may usefully be described according to their site of origin: atrial, AV junctional and ventricular (Table 8.2).

Atrial arrhythmias

Atrial arrhythmias share a number of features in terms of both presenting symptoms, electrocardiographic appearance and therapy. They usually present as paroxysmal palpitations although all (particularly atrial fibrillation) may become incessant or permanent. As with all arrhythmias however, they are often asymptomatic and non-specific symptoms such as fatigue, dyspnoea or dizziness are common. Symptoms depend to a large extent on the rate of the atrial arrhythmia, its conduction to the ventricles and ventricular function. Most are facilitated by increased sympathetic tone, caffeine and alcohol. Electrocardiographically, they usually appear as narrow QRS complex tachycardias unless there is bundle branch block or ventricular pre-excitation via an accessory AV pathway as in the Wolff-Parkinson-White syndrome. The ventricular rate is dependent on both the rate of the atrial arrhythmia and the degree of AV block. In all cases, therapy may be aimed at either

Table 8.2   Common cardiac arrhythmias

Arrhythmia	Causes	R aim	Drugs	Disadvantages
Atrial				
Tachycardia	—	Prevention ↓Vr	Ia, Ic, III II, IV, dig	Ia may ↑Vr No prevention
Flutter	CAD, MVD	Prevention ↓Vr	Ia, III II, IV, dig	Ia may ↑Vr No prevention
Fibrillation	CAD, MVD, Ca, ↑T4	Prevention ↓Vr	Ia, Ic, III II, IV, dig	Ia may ↑Vr No prevention
Junctional				
AVNRT	DAH	Prevention	All but Ib	Potentially proarrhythmic
AVRT	AP	Prevention	All but Ib	Potentially proarrhythmic
Ventricular				
Ectopics	None, CAD, CM, MVP	Prevention	I, II, III	Potentially proarrhythmic
Tachycardia	CAD, CM, None	Prevention	I, II, III	Potentially proarrhythmic

Abbreviations: AP = accessory pathway; AVNRT = atrioventricular nodal reentry tachycardia; AVRT = atrioventricular reentry tachycardia; Ca = bronchial carcinoma; CAD = coronary artery disease; CM = cardiomyopathy; DAH = dual atrio-His conduction; dig = digoxin; MVP = mitral valve prolapse; R = treatment; ↑T4 = thyrotoxicosis; Vr = ventricular response rate; (↑) = increase; (↓) = decrease.

prevention of paroxysms (with class I, II or III drugs) or reduction of ventricular rate (and thereby the severity of associated symptoms and haemodynamic disturbance) in the event of sustained arrhythmia (with digoxin or class II or IV drugs).

Atrial ectopic beats

Atrial ectopic beats commonly occur in normal people but increase with atrial disease and occur frequently in the sick sinus syndrome. They are mostly asymptomatic and usually require no treatment. The exceptions are those which are symptomatic or initiate other arrhythmias.

Atrial tachycardia

Atrial tachycardia is a relatively uncommon but sometimes distressing clinical problem. As a rule, there is no underlying heart disease to be found but cardiomyopathy is an occasional association. The relationship between the abnormal P waves and the QRS complexes may be fixed or variable (Wenckebach phenomenon) and is determined by AV nodal conduction properties. Both the diagnosis and choice of preventative therapy can often only be made following electrophysiological study.

Atrial flutter and fibrillation

Atrial flutter and fibrillation often coexist in the same patient, atrial flutter sometimes being a transitory stage between atrial fibrillation and sinus rhythm. Atrial flutter represents a single intra-atrial macroreentry circuit whereas atrial fibrillation is the result of multiple, discrete and independent microreentry circuits. Chronically, they both usually reflect atrial disease, most commonly in association with coronary artery disease, mitral valve disease, bronchial carcinoma, the sick sinus ('brady-tachy') syndrome, or more rarely with cardiomyopathies, mitral valve prolapse or atrial tumours. The 'lone fibrillators' do not suffer from identifiable pathology and in acute toxic illnesses such as pneumonia or hyperthyroidism, both arrhythmias may occur transiently without cardiac disorder, only to disappear once the toxicity is treated. Both may complicate pericarditis. Symptomatic paroxysmal palpitations are typically described as irregular and a common occasional presenting complication of atrial fibrillation is thromboembolism especially in the presence of mitral stenosis. The development of atrial fibrillation under any circumstance is associated with an adverse prognosis.

Junctional arrhythmias

Junctional arrhythmias are of three types, two of which are reentrant in nature. The exception is the rare *ectopic junctional or His bundle tachycardia*. This arrhythmia is almost always seen in children either after surgical correction of congenital heart disease or spontaneously within the first 6 months of life. The diagnosis is made from the surface ECG which reveals a narrow complex tachycardia with dissociated P waves. It is virtually unresponsive to drug therapy with the exception of the experience of Bucknall and co-workers (Bucknall et al 1985) who were able to control the arrhythmia with either disopyramide, mexilitine or amiodarone, alone or in combination, in four out of five patients. Apart from their experience, the prognosis is generally regarded as poor.

The reentrant junctional tachycardias

The reentrant junctional tachycardias occur as the result of there being at least two routes for conduction between atria and ventricles. Normally, both pathways conduct anterogradely in response to sinus rhythm. However, the refractory properties of the two pathways differ so that a premature (usually atrial) ectopic depolarization may find one of them still refractory. Conduction will proceed along the other, 'second' pathway and is then able to return to the atria via the now recovered 'first' pathway. A reentry circuit is thereby created for a depolarization wavefront which, in the case of intra-AV nodal reentry tachycardia, is contained within the AV node and in the case of atrioventricular reentry tachycardia consists of the AV node, ventricular myocardium, an extra-AV nodal accessory pathway and atrial myocardium. Typically, symptoms are of paroxysmal regular palpitations of sudden onset and termination. Associated symptoms depend on the rate and duration of tachycardia and include dizziness, chest pain and dyspnoea. Electrocardiographically they appear as regular, usually narrow QRS complex, tachycardia. In the case of atrioventricular reentry tachycardia, the retrograde P wave may sometimes be seen in the ST segment but this can be difficult to detect with certainty especially with very rapid tachycardia. In typical intranodal tachycardia, QRS complexes and P waves occur simultaneously and so the P wave is obscured.

It can be seen that choice of therapy will be considerably helped by knowledge of which type of reentry circuit is operating. For intra AV nodal reentry, digoxin and drugs of classes II or IV might be expected to be particularly useful whereas with atrioventricular reentry, drugs of classes Ia, Ic and III would be preferable because of their depressant effects on the other parts of the circuit, especially the accessory pathway. Rapid anterograde conduction to the ventricles of atrial tachyarrhythmias such as atrial fibrillation is a potentially dangerous complication of such pathways which might therefore be avoided by the use of such agents. Because reentry arrhythmias are likely to be initiated by atrial (or ventricular) ectopic beats, suppression of these by drugs of classes Ia, Ic and III might also reduce the incidence of both types of junctional reentry tachycardia. It must be remembered that the propensity to reentry in an individual patient depends, not only on the number of ectopic beats occurring, but also upon the relative properties of the two pathways. By altering these with a drug, it is possible to paradoxically facilitate the onset and perpetuation of reentry.

Ventricular tachyarrhythmias

The significance of ventricular tachyarrhythmias depends on underlying heart disease. Ventricular tachycardia does not much affect prognosis in individuals with otherwise normal hearts whereas, in the presence of underlying heart disease, lesser grades of arrhythmia such as frequent ventricular ectopic beats are associated with an increased risk of sudden death. Different ventricular arrhythmias carry different prognostic weight, especially in the context of acute myocardial infarction and the Lown classification was designed accordingly so that therapy could be aimed at the higher risk patients identified by the higher grades of their arrhythmias (Table 8.3). However, the applicability of this classification to patients without recent myocardial infarction is limited. Treatment of ventricular arrhythmias can be achieved with drugs of classes I, II and III and, in the rare fascicular ventricular tachycardia, with class IV (Ward et al 1984) drugs. Electrocardiographic diagnosis of ventricular tachycardia traditionally causes problems. These problems appear to be related to difficulty in distinguishing ventricular tachycardia from aberrantly conducted supraventricular tachycardia. Points which are specific for ventricular tachycardia are: dissociated P waves, fusion beats which are narrower than the tachycardia complexes, R wave only in V_1, initial R taller than R' of an RSR' deflection in V_1, Rs deflection in V_1, QS deflection in V_6 and QRS width > 140 ms. A feature which is

Table 3   The Lown classification of ventricular arrhythmias

Grade	Arrhythmia
0	None
1	Occasional, isolated, unimorphological VPB
2	Frequent ($> 1/$min or 30/h), isolated, unimorphological VPB
3	Multimorphological VPB
4	Repetitive VPB (a) couplets (b) salvoes
5	Early VPB (R-on-T)

VPB = ventricular premature beats

highly suggestive of, but not totally specific for, ventricular tachycardia is a frontal plane axis of < -30 (left axis deviation) of a wide complex tachycardia.

Torsade de pointes is an unusual arrhythmia which resembles non-sustained ventricular fibrillation. It is typically seen in situations associated with a long QT interval such as the congenital syndromes of Romano and Ward, and Jervell and Lange-Nielsen, bradycardia or phenothiazine toxicity. For the congenital variety, the treatment varies from atrial pacing with β-blockade to left stellectomy to correct a supposed autonomic imbalance. Treatment for the aquired forms is in many ways opposite to that of the congenital forms since the aim is to shorten the QT interval. The treatment of choice is therefore isoprenaline which is contraindicated for the congenital syndromes.

Acknowledgement

The authors are supported by the British Heart Foundation.

REFERENCES

See end of chapter 9.

Individual antiarrhythmic drugs

QUINIDINE

Quinidine is a cinchona derivative, being the dextroisomer of quinine. It is usually prescribed as the sulphate or gluconate forms and may be given both orally and parenterally although intramuscular injection is painful and intravenous injection is dangerous. The gluconate form is absorbed rather more slowly after oral dosage with peak blood levels occurring at 3–4 hours compared with 1.5–2 hours for the sulphate. The relatively short half life of 6–10 hours, increasing with age, has led to the use of slow release preparations (e.g. Kinidin Durules) which may be taken twice a day. The therapeutic range lies between 2 and 5 $\mu g/ml$ (plasma concentration) and a high incidence of toxic side effects is encountered at levels above 7 $\mu g/ml$. The majority of the drug undergoes hepatic oxidative metabolism with only 10–20% being excreted unchanged in the urine. Because the drug is basic, this may be increased by acidification of the urine. The metabolites are incompletely investigated but do possess antiarrhythmic activity. Quinidine is useful against atrial and ventricular arrhythmias as well as those junctional reentry tachycardias utilizing an accessory AV pathway. The drug also possesses some α-blocking properties.

Quinidine has a low toxic/therapeutic ratio and this has led to its decreased use. Approximately one-third of patients have to discontinue therapy because of unwanted side effects. These include gastrointestinal disturbance, thrombocytopenia, haemolytic anaemia and arrhythmogenic effects, either aggravation or development

Table 9.1 Pharmacokinetics and dynamics of antiarrhythmic drugs

Drug	Administration route	Therapeutic level	Plasma half life	Elimination route
Quinidine	oral	2–5 $\mu g/ml$	6–7 hours	Hepatic
Procainamide	oral/i.v.	4–10 $\mu g/ml$	3–5 hours	Renal/hepatic
Disopyramide	oral/i.v.	2–4 $\mu g/ml$	4–10 hours	Renal/hepatic
Lignocaine	i.v.	1–5 $\mu g/ml$	1–3 hours	Hepatic
Mexiletine	oral/i.v.	1–2 $\mu g/ml$	9–12 hours	Hepatic
Tocainide	oral/i.v.	3–10 $\mu g/ml$	10–15 hours	Renal/hepatic
Flecainide	oral/i.v.	160–980 ng/ml	12–27 hours	Hepatic/renal
Propafenone	oral/i.v.	Variable	3–6 hours	Hepatic
Amiodarone	oral/i.v.	1.5–3 mg/l	7–40 days	Hepatic/renal
Verapamil	oral/i.v.	50–250 ng/ml	5–10 hours	Hepatic/renal
Diltiazem	oral/i.v.	100–200 ng/ml	4–6 hours	Hepatic/renal
Digoxin	oral/i.v.	2–2.5 ng/ml	33–38 hours	Renal

of conduction abnormalities, or provocation of ventricular arrhythmias. It interacts with other drugs, in particular with digoxin, the serum levels of which approximately double when quinidine is added at a therapeutic dose. The dose of digoxin should therefore be halved. Quinidine potentiates the neuromuscular blocking effects and ventilatory depression of succinylcholine, decamethonium and tubocurarine (Miller et al 1967). By depressing clotting factor synthesis, quinidine potentiates the effect of coumarin anticoagulants. As with any drug extensively metabolised in the liver, its levels are significantly influenced by those drugs which induce hepatic enzymes such as phenytoin and rifampicin.

PROCAINAMIDE

Procainamide was developed after the antiarrhythmic effects of the local anaesthetic procaine were noticed 30 years ago. It is available both for oral and parenteral use. As with quinidine it needs to be given frequently because of its short half life of 6–10 hours. This is dependent upon whether the patient is a fast or a slow acetylator. However, its major metabolite, N–acetylprocainamide (NAPA) has significant electrophysiological effects which are similar to the effects of procainamide except for the absence of effect on the His-Purkinje system. A sustained release preparation is available. The therapeutic plasma concentration lies between 4 and 10 μg/ml, toxic side effects being uncommon with levels below 12 μg/ml but common above 16 μg/ml. Procainamide is eliminated by both renal and hepatic routes with approximately 75–95% appearing in the urine, 30–60% as the unchanged drug. Its spectrum of antiarrhythmic activity is identical to quinidine and so are its electrocardiographic side effects. Other common unwanted side effects include gastrointestinal disturbance and a drug-induced lupus syndrome. It is contra-indicated in patients with myasthenia gravis for fear of aggravation. There are relatively few drugs which interact with procainamide although, being weakly basic, its renal elimination will be increased by urinary acidification.

DISOPYRAMIDE

Disopyramide's antiarrhythmic effects were first described in 1962 and are essentially the same as those of quinidine and procainamide. It is active both parenterally and orally, the majority of the drug being excreted unchanged in the urine. At therapeutic levels (2–4 μg/ml), about 40% of the drug is protein bound. It follows first order kinetics for a two compartment model, the unbound drug level being proportional to the dose. Plasma half life is in the order of 7 hours but increases with renal impairment and after myocardial infarction. A sustained release preparation is available allowing twice daily dosage.

Although disopyramide's antiarrhythmic effects are similar to those of quinidine and procainamide, its unwanted side effects, although numerous, are not so serious. It has therefore become the most widely used drug in this group in the UK. The major unwanted effects are anticholinergic, due mostly to its N-monodealkylated metabolite. Dry mouth is the commonest of these but the drug should be avoided in patients with narrow angle glaucoma and used with great care in those with symptoms of prostatism.

LIGNOCAINE

Lignocaine has been used as an antiarrhythmic agent since 1950. Because of an extensive first pass hepatic metabolism to inactive metabolites, it is only active when given parenterally. This limits its use to the treatment of non-recurrent ventricular arrhythmias such as those seen in the acute stages of myocardial infarction. The half life following intravenous bolus is about 1.5 hours so that a continuous infusion is required after the initial bolus, the infusion being titrated against the clinical effect. The usual infusion rate is between 1 and 4 mg per hour. Toxic side effects, especially central nervous effects, occur at whole blood levels in excess of 5–6 $\mu g/ml$, therapeutic levels needing to be in excess of 1–2 $\mu g/ml$. It is certainly active against reentrant ventricular arrhythmias but its ability to suppress automatic foci is controversial. The most important unwanted effects are related to the central nervous system. These range from drowziness to psychotic reactions and grand mal seizures. They are dose related. Predictable electrocardiographic side effects such as the development or aggravation of conduction disorders occur as well as sinus bradycardia, sinus arrest and sinoatrial block. It is a negatively inotropic drug which requires careful monitoring when used in patients with impaired left ventricular function. In terms of drug interactions, there are relatively few which are unexpected. One is the reduction in lignocaine clearance seen in people also receiving propranolol; sympathomimetic agents have the opposite effect. Halothane decreases lignocaine metabolism, perhaps by reducing hepatic blood flow or possibly by affecting hepatic enzymes. Therapeutic serum lignocaine concentrations reduce the anaesthetic requirements for halothane and nitrous oxide (Himes et al 1977).

MEXILETINE

Mexiletine, a relatively new drug, was synthesized and studied in the early 1970s and has similar local anaesthetic and antiarrhythmic effects to lignocaine. It differs from lignocaine in having a high degree of bioavailability after oral administration followed by a long half life of about 9 hours when taken via this route. It is extensively metabolized in the liver both by oxidation and reduction, the metabolites being unlikely to have significant antiarrhythmic properties since these decline with the mexiletine level. Under 10% of the drug appears unchanged in the urine, a figure which could also be increased by urinary acidification since mexiletine is weakly basic. Therapeutic plasma levels lie between 0.75 and 2.0 $\mu g/l$, above which side effects are seen with increasing frequency. The drug is negatively inotropic but this effect is not marked at clinical antiarrhythmic doses. However, the effects in patients with pre-existing heart failure are not known. Like lignocaine it is only active on ventricular arrhythmias with perhaps a particular effect on torsade de pointes (Krikler & Curry 1976). Its unwanted side effects, both intra- and extra-cardiac, are also similar although the commonest side effect is nausea and vomiting. There is little evidence of significant interaction with other drugs other than those which induce its hepatic metabolism.

TOCAINIDE

Tocainide is the result of an attempt to find a drug similar to lignocaine but which would be active when taken orally. Elimination half life after oral dosage is between 10 and 15 hours and is little affected by myocardial infarction. Elimination of the drug is shared between renal excretion of the unchanged drug and glucuronidation in the liver. Therapeutic levels are above 3 μg/l with toxic effects being increasingly encountered above 10 μg/l. As with mexiletine, negative inotropism is a property but the effect at a clinical dosage is minimal. It is active primarily against ventricular arrhythmias but has been shown to terminate junctional reentry tachycardia (Waleffe et al 1979). It has also been used to treat paramyotonia congenita (Dengler & Rudel 1979) and tinnitus although this can paradoxically also be a side effect of tocainide therapy. Not surprisingly, in patients with pre-existing conduction abnormalities, these were aggravated when tocainide was given in combination with the β-blocker metoprolol with three patients developing asystole (Ikram 1980). Hepatic enzyme induction does not appear to significantly affect its metabolism (Elvin et al 1980).

ENCAINIDE

Encainide is an interesting antiarrhythmic drug whose antiarrhythmic effects after oral dosage differ from those after intravenous use because of the significant antiarrhythmic effects of two metabolites produced after oral administration. Its propensity to provoke malignant ventricular arrhythmias will probably severely limit its future use.

FLECAINIDE

First reported 10 years ago, flecainide is a potent antiarrhythmic agent with a wide spectrum of activity. It is active after both intravenous and oral administration with a half life after a single dose (via either route) of about 13 hours, this ranging from 12 to 27 hours after repeated oral doses. It can therefore be taken twice a day with peak plasma levels being seen at 2–4 hours after a single oral dose. Approximately 60% is metabolized by the liver (one of the products having weak antiarrhythmic activity) the rest being excreted unchanged, mostly in the urine but a small amount appearing in the faeces. The clinical effects of flecainide within the heart are widespread: PA (right atrial conduction time), AH (AV nodal conduction time), HV (His-Purkinje conduction time) and QRS duration are all prolonged. Although the QT is also prolonged, this prolongation is composed solely of increased QRS duration, the JT being unaffected. Atrial and ventricular refractory periods are slightly prolonged, but the most marked prolongation of refractoriness in response to flecainide is seen with anomalous AV pathways both intra- and extra-AV nodal with the effect being most marked in the retrograde direction. In terms of arrhythmias, all are susceptible to flecainide therapy, the least sensitive being atrial flutter.

Of its unwanted effects, a strong negative inotropic effect is the most commonly encountered and is therefore an important limiting factor especially in patients with impaired left ventricular function. Proarrhythmic effects include the aggravation/ production of conduction abnormalities, and a nearly 5% incidence of serious ventricular arrhythmias is reported (Nathan & Hellestrand 1984). The risks of these effects are greater in patients with coronary artery disease, myocardial infarction, pre-existing ventricular tachycardia or congestive cardiac failure. Pacing thresholds are elevated by the drug to a far greater extent than observed with other drugs. Extracardiac effects are predominantly central nervous with dizziness and visual disturbance being the commonest. Little is known of drug interactions but there is an increase in both rate of absorption and peak levels of digoxin in patients given flecainide although steady state levels are little affected.

PROPAFENONE

Propafenone, although being predominantly a class Ic agent, also possesses class II, III and weak class IV activities (Dukes & Vaughan Williams 1984). It slows conduction and prolongs the refractory periods of atrial, junctional (including accessory) and ventricular tissue. After a single dose, its half life is about 3.5 hours increasing to 6 hours with multiple doses but appears to have considerable and unpredictable individual variation. Elimination is by hepatic metabolism. The drug is of limited use in the conversion or prophylaxis of atrial flutter or fibrillation although its effects upon accessory AV conduction are sufficient for it to be effective in the termination and prevention of reentry arrhythmias using such pathways (Rudolph et al 1979). Ventricular premature beats and sustained tachycardia are also susceptible to the drug (Salerno et al 1984, Shen et al 1984). The significant side effects are related to its negative inotropism, the production of conduction abnormalities and proarrhythmic effects. The unwanted effects encountered most commonly are gastrointestinal disturbance and dizziness.

THE β-BLOCKERS

The β-blockers: this group contains more drugs than any of the other groups, reflecting the diverse uses of β-blockade. The electrophysiological effects of β-blockers are principally the result of sympathetic blockade. Thereby, they depress excitability and conduction, especially of the sinus and AV nodes, and by reducing ischaemic episodes, perhaps decrease automaticity and inhibit reentry at ventricular level. Although all these drugs appear as racemic mixtures for clinical use, virtually all the β-blocking effect is within the L-isomer. Many has class I 'local anaesthetic' properties but these are not seen at therapeutic levels. They are subdivided according to properties such as cardioselectivity and intrinsic sympathomimetic activity but these do not effect their antiarrhythmic effects. Sotalol, however, is interesting among the β-blockers because of its additional class III effect (predominantly in the D-isomer). All β-blockers prolong the QT interval when used chronically. However, this is a small effect compared with that observed with sotalol, with which the effect is most marked after an intravenous bolus. Specific β-blockers and their special properties are well described elsewhere (Silverman & Frishman 1983) and so they will not be discussed further here.

AMIODARONE

Amiodarone is the most important representative of those drugs possessing class III effect. It is unique amongst antiarrhythmic drugs in many ways. It was introduced as an antianginal drug, the mode of action being vasodilatation, its antiarrhythmic effect being discovered later (Charlier et al 1968, Marcus et al 1981, Ward et al 1980). It has a major advantage in that it is not negatively inotropic and therefore may even be haemodynamically beneficial to patients with impaired left ventricular function through its vasodilatory effect. Amiodarone has an extremely long half life of 7–40 days (Holt et al 1983). This probably reflects the high lipid solubility of the drug, its volume of distribution being estimated at about 5000 litres. The sole metabolite is des-ethyl amiodarone, the antiarrhythmic effects of which are unknown. Its concentration, however, always exceeds that of the parent compound except in fat. Elimination is by biliary and gastrointestinal excretion, little of either the drug or its metabolite appearing in the urine. The extremely long half life is both a disadvantage and an advantage. The problems associated with this are that antiarrhythmic effects may not be seen for several days because of its wide distribution and the electrophysiological confirmation of effect is therefore difficult. In fact, conventional assessment by electrophysiological study may not be possible for this drug. Furthermore, discontinuation of the drug is followed by a long and variable period during which the extent of amiodarone's remaining effects are uncertain and the patient therefore extremely difficult to assess. The advantage of the long half life is seen with chronic therapy, which can be with once daily dosage, and in those patients with life threatening arrhythmias where short term cessation of therapy (inadvertent, diarrhoea) will not have catastrophic results. In order to shorten the delay to onset of antiarrhythmic efficacy, loading by means of high dose intravenous infusion has recently been used. Any advantage that this may have over oral loading is uncertain as is the nature of action of the drug when used intravenously.

The spectrum of oral amiodarone's antiarrhythmic effects is wide. It has been used successfully in the prophylactic treatment of all forms of atrial, junctional and ventricular arrhythmias. However, because of the problems associated with its use, this has tended to be restricted to arrhythmias refractory to more conventional drugs and to those patients with life threatening arrhythmias in the presence of poor left ventricular function. Apart from the difficulties associated with the drug's pharmacokinetics, unwanted effects are another discouragement to its use. Although many are dose related, it is their serious nature and the difficulty associated with their reversal (because of the long drug elimination time) which present the problems. The commonest is photosensitivity of varying severity but which may necessitate withdrawal of the drug in extreme cases. In some, this may be followed by an asymptomatic and permanent bluish discolouration of exposed areas associated with deposition of lipofuschin. Next commonest are side effects on the neurological system varying from insomnia and nightmares to axonal degeneration producing peripheral neuropathy which may also leave permanent sequelae unless the drug is stopped promptly. Corneal microcrystal deposition is seen in every patient but seldom causes a problem. Occasionally, 'gritty' eye occurs or the crystals are deposited in the visual field; short term discontinuation of the drug is usually enough to relieve the latter problem as the crystals are reabsorbed. Hyperthyroidism is increasingly recognized among patients on long term amiodarone although hypothyroidism is also seen. Rarely, severe alveolitis with associated radiological shadowing is encountered, usually with high doses. However, subclinical deterioration in lung function is seen more often, the changes being in diffusing capacity. The problem resolves with steroid therapy and withdrawal of treatment which is only necessary for the clinically overt syndrome. Gastrointestinal effects are uncommon with nausea, vomiting and constipation occasionally encountered, especially during the loading period. A chemical hepatocellular hepatitis is seen in about 15% of patients but resolves spontaneously even with continued therapy at the same dose in the majority of cases.

In terms of drug interactions, because of its protein and tissue binding, other drugs such as digoxin and anticoagulants are displaced although alternative explanations for the increased plasma concentrations of the drugs exist. With digoxin, there may be interference with its excretion and warfarin's metabolism may be impaired by hepatic microsomal enzyme inhibition. Amiodarone has an additive effect with beta blockers, verapamil and diltiazem in depressing sinus and AV nodal function.

BRETYLIUM

Bretylium is an adrenergic neurone blocker with a class III antiarrhythmic effect. Its use is restricted to refractory ventricular arrhythmias, especially fibrillation. It is given intravenously because of poor and unpredictable absorption after oral administration. The adrenergic neurone blocking effects are probably not relevant to its antiarrhythmic effects which are seen in denervated and catecholamine depleted hearts as well as those where the blockade has been reversed by using tricyclic antidepressant drugs. Also, the antifibrillatory effects of bretylium are not shared by guanethidine (Bacaner 1968) despite the shared adrenergic neurone blocking effects. It has an advantage that, being a vasodilator, it can be used readily in patients with poor left ventricular function.

Drug interactions such as the increased efficacy of supportive pressor amines can be predicted from knowledge of the adrenergic neurone blocking properties of bretylium.

SOTALOL

Initially introduced as a non-selective β-blocker without sympathomimetic nor membrane stabilizing activity, Sotalol has subsequently been demonstrated to possess class III antiarrhythmic properties seen especially after intravenous bolus administration although also detectable after chronic oral use to a degree more than might be expected from pure β-blockade. However, the clinical advantage of this additional property is uncertain, there being no demonstrable increase in antiarrhythmic effect compared with β-blockade alone. The drug is a racemic mixture of the L and D-isomers. It appears that most of the β-blocking effect is contributed by L-sotalol and so D-sotalol is currently being investigated as a class III antiarrhythmic drug (although it is also a weak β-blocker).

VERAPAMIL

Verapamil is a synthetic papaverine derivative and is used for its antiarrhythmic, antianginal and antihypertensive properties. Of the calcium antagonist group, it is the most widely used for an antiarrhythmic effect. It reduces the amplitude of transmembrane action potentials in the SA and AV nodes and prolongs AV nodal refractoriness. Although active both orally and parenterally, there is extensive first pass metabolism so that, to overcome this, the oral dose used is about 10 times the intravenous dose. After either oral or intravenous administration there is an initial distribution phase (α) of about 20–30 minutes followed by a more gradual elimination (β) phase with a half life of 3–7 hours. Following an intravenous bolus, the onset of effect is within 2 minutes, peaking at 10 minutes. After oral administration, effects appear at 2 hours and peak at 5 hours. Less than 5% of the drug appears unchanged in the urine. With the solitary exception of fascicular tachycardia (Ward et al 1984), verapamil is inactive on ventricular arrhythmias. The effect upon supraventricular arrhythmias depends on the role of the AV node in the arrhythmia. When it merely transmits an atrial tachyarrhythmia to the ventricles, the effect will be to decrease the ventricular response rate as in atrial tachycardia, flutter and fibrillation whereas, when the AV node is an essential part of the tachycardia substrate as in junctional reentry tachycardia, then the arrhythmia will usually be terminated.

In that it interferes with calcium mediated myocardial excitation–contraction coupling, verapamil has a negatively inotropic effect. This has to be balanced against the beneficial effects of vasodilatation but overall, care is required in patients with impaired left ventricular function. Other unwanted cardiac effects include sinus bradycardia, sinus arrest, AV block, ventricular arrhythmias and ventricular asystole. Extracardiac side effects include headaches, flushing and constipation. Verapamil interacts with other drugs, principally digoxin and β-blockers. The former interaction is synergistic whereas the combination of β-blockade and verapamil aggravates the negatively inotropic properties of both drugs and can lead to asystole (Krikler & Spurrell 1972). The usual interactions with protein bound drugs such as warfarin occur and verapamil when used in conjunction with class I agents may cause profound myocardial depression. Although reentry tachycardias in association with the Wolff-Parkinson-White syndrome may be terminated by verapamil, conduction of atrial tachyarrhythmias via an accessory pathway may be accelerated. A number of verapamil derivatives, such as Tiapamil, have been developed and are structurally and functionally similar to the parent compound.

DILTIAZEM

Apart from verapamil, diltiazem is the only calcium antagonist used frequently for its antiarrhythmic effect. This is similar to verapamil although it differs in being a less effective peripheral vasodilator and is less negatively inotropic. Also, as a result of less hypotensive effect, there is less reflex sympathetic response. However, the pharmacokinetics and pharmacodynamics of the two drugs are very similar. Diltiazem has fewer side effects, thanks mainly to the relative absence of negative inotropism although constipation is also uncommon.

D

DIGOXIN

The antiarrhythmic effects of digoxin in man are restricted to the AV node. There is no overall effect upon sinus cycle length nor atrial refractory period. Ventricular refractoriness is reduced and there is evidence that accessory pathway refractoriness is also reduced (Wellens & Durrer 1973). Both these effects may be proarrhythmic whereas the effects on the AV node are to prolong its conduction time and refractory period. Thereby, digoxin may be useful in controlling the ventricular response rate to an atrial tachyarrhythmia such as atrial fibrillation (its main use) or in terminating or preventing an arrhythmia for which the AV node is essential such as AV nodal reentry tachycardia. Its uses are therefore similar to those of verapamil but its side effect profile is different. It has been used in the past for a positive intropic effect. Whilst the importance of this effect is controversial, it is certainly not negatively inotropic like verapamil and so the same precautions about its use in patients with poor left ventricular function are not necesary. However, its proarrhythmic effects should be remembered in such patients. Any arrhythmia can be provoked by digoxin and the proarrhythmic effects are usually dose related. Heart block is particularly common as are ventricular ectopics. Hypokalaemia potentiates the effects of digoxin and may therefore be responsible for toxic effects being seen at therapeutic levels. Serum levels are of little benefit in making the diagnosis of digoxin toxicity which should therefore be made from clinical criteria such as arrhythmias and symptoms which include nausea, anorexia and central nervous effects. Subjective visual disturbances are common. These unwanted effects are particularly common amongst the elderly, especially those with renal impairment in whom renal excretion of the unchanged drug is reduced.

The drug is given orally once daily in view of its long half life with the usual dosage being between 0.125 and 0.5 mg daily. Loading is required, usually with an increased oral dose for the first 24 to 36 hours but this can be accelerated by intravenous use, care being taken to monitor for potential proarrhythmic effects. Treatment of intoxication consists of correcting electrolyte abnormalities where they exist and withdrawal of the drug. These measures usually suffice but where more urgent treament is required, class Ib agents may be used to treat ventricular arryythmias until the drug level subsides and temporary pacing may also be indicated for a similar period in patients with severe bradycardia as a result of digoxin induced heart block. More aggressive methods such as adsorptive haemoperfusion or the use of digoxin-specific antibodies are rarely required.

CONCLUSIONS

We have attempted to summarize the antiarrhythmic drugs in common use. Although the anticipated antiarrhythmic effects of many of them will overlap, there is considerable individual variation in the observed clinical antiarrhythmic effects, even of drugs with apparently identical spectra of basic and clinical electrophysiological effects. This implies that much remains to be learnt regarding the precise mechanism of the arrhythmias and the drugs' antiarrhythmic effects.

Acknowledgements

The authors are supported by the British Heart Foundation.

REFERENCES

Bacaner M B 1968 Quantitative comparison of bretylium with other antifibrillatory drugs. American Journal of Cardiology 21: 504–512

Bucknall C, Laudusans E, Tynan M J, Curry P V L 1985 Ventricular tachycardia masquerading as supraventricular tachycardia: management of His bundle tachycardia. British Heart Journal 53: 681 (Abstract)

Charlier R, Deltour G, Baudine A, Chaillet F 1968 Pharmacology of amiodarone, an antianginal drug with a new biological profile. Arzneimittel Forschung 18: 1408–1417

Charlier R 1970 Cardiac actions in the dog of a new antagonist of adrenergic excitation which does not produce competitive blockade of adrenoreceptors. British Journal of Pharmacology 39: 668–673

Dengler R, Rudel R 1979 Effects of tocainide on normal and myotonic mammalian skeletal muscle.Drug Research 29: 270–273

Dukes I D, Vaughan Williams E M 1984 The multiple modes of action of propafenone. European Heart Journal 5: 115–125

Elvin A T, Lalka D, Stoeckel K et al 1980 Tocainide kinetics and metabolism: effects of phenobarbital and substrates for glucuronyl transferase. Clinical Pharmacology and Therapeutics 28: 652–658

Himes R S, DiFazio C A, Burney R G 1977 Effects of lidocaine on the anaesthetic requirements for nitrous oxide and halothane. Anesthesiology 47: 437–440

Holt D W, Tucker G T, Jackson P R, Storey G C A 1983 Amiodarone pharmacokinetics. American Heart Journal 106: 840–847

Ikram H 1980 Hemodynamic and electrophysiologic interactions between antiarrhythmic drugs and beta-blockers, with special reference to tocainide. American Heart Journal 100: 1076–1080

Krikler D, Spurrell R 1972 Asystole after verapamil. British Medical Journal 2: 405 (Letter)

Krikler D M, Curry P V L 1976 Torsade de pointes, an atypical ventricular tachycardia. British Heart Journal 38: 117–120

Marcus F, Fontaine G H, Frank R, Grosgogeat Y 1981 Clinical pharmacology and therapeutic applications of the antiarrhythmic agent, amiodarone. American Heart Journal 101: 480–493

Miller R D, Way W L, Katzung B G 1967 The potentiation of neuromuscular blocking agents by quinidine. Anesthesiology 28: 1036–1041

Nathan A W, Hellestrand K J 1984 Flecainide acetate: a review. Clinical Progress in Pacing and Electrophysiology 2: 43–53

Rudolph W, Petri H, Kofk W, Hall D 1979 Effects of Propafenone on the accessory pathway in patients with WPW syndrome. American Journal of Cardiology 43: 430 (Abstract)

Salerno D M, Granrud G, Sharkey P et al 1984 A controlled trail of propafenone for treatment of frequent and repetitive ventricular premature complexes. American Journal of Cardiology 53: 77–83

Shen E N, Sung R J, Morady F et al 1984 Electrophysiologic and hemodynamic effects of intravenous propafenone in patients with recurrent ventricular tachycardia. Journal of the American College of Cardiology 3: 1291–1297

Singh B N, Vaughan-Williams E M 1970 The effect of amiodarone, a new anti-anginal drug on cardiac muscle. British Journal of Pharmacology 39: 657–667

Silverman R, Frishman W H 1983 Drug treament of cardiac arrhythmias: propranolol and other beta-blockers. In: Gould L A (ed) Drug treatment of cardiac arrhythmias. Futura Press, New York, ch 11

Waleffe A, Brunix P, Mary-Rabine L, Kulbertus H E 1979 Effects of tocainide studied with programmed electrical stimulation of the heart in patients with reentrant tachyarrhythmias. American Journal of Cardiology 43: 292–299

Ward D E, Camm A J, Spurrell R A J 1980 Clinical antiarrhythmic effects of amiodarone in patients with resistant paroxysmal tachycardias. British Heart Journal 44: 91–95

Ward D E, Nathan A W, Camm A J 1984 Fascicular tachycardia sensitive to calcium antagonists. European Heart Journal 5: 896–905

Wellens H J J, Durrer D 1973 Effect of digitalis on atrioventricular conduction in circus movement tachycardia in patients with Wolff-Parkinson-White syndrome. Circulation 47: 1229–1233

Young M D, Hadidian Z, Horn H R et al 1980 Treatment of ventricular arrhythmias with oral tocainide. American Heart Journal 100: 1041–1045

Anaesthesia for cardiac surgery and post-operative care

During the last 10 years there has been a marked increase in the number of cardiac surgical operations performed in the UK. The UK cardiac surgical register shows that in 1977 there were 11 602 operations and by 1982 this had risen to 15 890. Adult valvular heart disease and congenital heart disease have remained constant at 4000–5000 and 3000–3500 per year respectively while those for ischaemic heart disease have increased from 3000 to 7000 cases per year (English et al 1984) (Table 10.1). This review is concerned with anaesthesia for adult cardiac surgery and does not include anaesthesia for congenital heart disease in paediatric patients.

The risks of cardiac surgery have fallen with the result that in 1982 in the UK, the mortality within 30 days of operation was 7.1%. Factors associated with increased risk include unstable angina, recent myocardial infarction, poor left ventricular function, age greater than 65 years, obesity, reoperation, heart failure and emergency surgery. A prospective study from the Montreal Heart Institute demonstrated a mortality of 0.4% in the absence of any of these factors rising to 3.1% if one is present and to 12.2% for more than one factor (Paiement et al 1983).

Newer surgical operations include ablation of arrhythmogenic areas in refractory arrhythmias, heart transplants, combined heart and lung transplants and implantation of a total artificial heart.

Table 10.1   Number (percentage mortality[a]) of all cardiac operations 1977–82 (Reproduced from English 1984 with kind permission of the British Medical Journal)

Year	Valvular heart disease	Ishaemic heart disease	Congenital valve disease	Adult miscellaneous acquired heart disease	Total
1977	4832 (8.9)	3040 (9.3)	3344 (10.1)	386 (22.8)	11 602 (9.8)
1978	4873 (9.5)	3345 (7.5)	3385 (11.1)	340 (22.1)	11 943 (9.8)
1979	4791 (9.2)	3688 (8.4)	3275 (10.5)	402 (19.2)	12 156 (9.6)
1980	4814 (7.0)	5011 (6.1)	3472 (11.0)	442 (15.4)	13 739 (7.9)
1981	4762 (6.9)	6123 (5.8)	3501 (8.9)	401 (19.7)	14 787 (7.3)
1982	4652 (6.7)	7403 (5.2)	3392 (9.9)	443 (21.7)	15 890 (7.1)
Total	28 724 (8.1)	28 610 (6.6)	20 369 (10.0)	2414 (19.9)	80 117 (8.5)

[a]Mortality (%) defined as death at or within 30 days after operation (English 1984).

PRE-OPERATIVE PREPARATION

Patients presenting for operation should be prepared with the adjunct of drugs to be in the best possible medical condition.

Valvular disease

Patients with valvular disease may have heart failure which is controlled by diuretics and digoxin. Those with mitral valve disease may be in atrial fibrillation and although digoxin has traditionally been used to slow the ventricular rate by blocking atrioventricular conduction, verapamil, a calcium antagonist may be used instead or in combination with digoxin.

Coronary artery disease

Patients with coronary artery disease are commonly treated with β adrenergic antagonists, calcium antagonists and vasodilators, i.e. triple-drug therapy.

β Adrenergic blocking drugs

Newer β blocking drugs may have a greater selectivity for the β_1 adrenergic receptors in cardiac muscle and have a lesser effect on the β_2 receptors in bronchial muscle and the peripheral circulation. Such cardioselective drugs are metoprolol and atenolol and these may be useful in patients with respiratory disease leaving the β_2 receptor unblocked. However, even the cardioselective β blockers may aggravate bronchospasm in some asthmatic patients. Absorption and metabolism of β blockers determine the onset of their action and the duration of clinical effect. Propranolol is lipid soluble, is almost completely absorbed from the gastrointestinal tract and is metabolized by the liver. Its elimination half life is 3.5–6 h and administration is 3–4 times per 24 h. Atenolol is water soluble, is completely absorbed from the gut and is eliminated unchanged by the kidney. Its elimination half life is 6–9 h and once per 24 h dosage is appropriate (Frishman 1981). All β blockers can be dangerous in patients with compromised ventricular function and can precipitate pulmonary oedema, hypotension and heart block.

Calcium antagonists

The calcium antagonists are a developing group of drugs which inhibit calcium ion flux in smooth muscle. Verapamil, nifedipine and diltiazem are used for their effects on cardiac and vascular muscle. They produce vasodilation, particularly of the arterial bed and the coronary arteries appear to be especially sensitive. In stable, effort induced angina, coronary vasospasm may add to the obstructive coronary artery disease and further reduce blood flow. Other factors involved in amelioration of anginal symptoms are given in Table 10.2. Headache from cerebral arteriolar dilatation and postural hypotension are among the side effects. Since verapamil and β blockers both have negative inotropic activity and depress atrioventricular conduction, the effects of verapamil and β blockers can be additive. If intravenous β

Table 10.2   Mechanisms of beneficial effect of Ca^{2+} channel blocking agents in stable angina (Reproduced from Braunwald 1982 with kind permission of the New England Journal of Medicine)

1   Increased myocardial oxygen supply
   A   Coronary artery dilation
   B   Improved subendocardial perfusion

2   Decreased myocardial oxygen demand
   A   Decreased peripheral resistance
   B   Decreased myocardial contractility (verapamil)
   C   Decreased heart rate (verapamil)

Table 10.3   Side effects of anti-anginal drugs (Reproduced from Braunwald 1982 with kind permission of the New England Journal of Medicine)

Drug	Hypotension; flushing; headache	Left ventricular dysfunction	Decreased heart rate; atrioventricular block	Gastrointestinal symptoms	Broncho-constriction
β Blockers	0	++	+++	+	+++
Diltiazem	+	+	+	0	0
Nifedipine	+++	0	0	0	0
Verapamil	+	+	++	++	0
Nitrates	+++	0	0	0	0

Abbreviations: 0 = absent; + = mild; ++ = moderate; +++ = sometimes severe.

blocking drugs are given to patients taking verapamil, there is an increased risk of bradycardia and hypotension. (Table 10.3).

Nitrates

Nitrates, because of their vasodilating properties, are the third group of drugs used to stabilize patients with coronary artery disease. The vasodilation is mainly venous, producing a lower ventricular filling pressure and a reduction in myocardial wall tension and myocardial work. By lowering ventricular diastolic pressure they also permit an increased blood flow as this depends on the difference between aortic and ventricular diastolic pressures. Sublingual glyceryl trinitrate is particularly useful for acute events and more sustained action can be achieved with absorption through the skin. Other longer acting nitrates have been developed for oral use such as isosorbide dinitrate and pentaerythritol tetranitrate. Glyceryl trinitrate and isosorbide dinitrate can also be administered by intravenous infusion to patients with prolonged or severe rest pain.

Patients may find waiting for surgery to be stressful and it is therefore routine practice to continue all medical therapy up to and including the day of operation.

OPERATIVE MANAGEMENT

Monitoring

One of the major tasks of an anaesthetist is to maintain the patients' physiological functions during surgical events. Being forewarned is to be forearmed and it is

Table 10.4   Monitoring during cardiac anaesthesia

System monitored	Measurement
Cardiovascular	Electrocardiogram Blood pressure via arterial cannulae Central venous pressure Pulmonary artery pressure Left atrial pressure
Respiratory	Airway pressure Blood gases
Central nervous	Pupil size Cerebral function trace
Renal	Urine output Serum potassium Acid–base status

advantageous to have adequate monitoring established before anaesthesia and surgery commence. Parameters often monitored are shown in Table 10.4. In addition central temperature is usually measured by a nasopharyngeal probe. Some of the measurements such as pupil size cause minimal hazard to the patient but invasive cardiovascular monitoring carries the risk of infection, haemorrhage, thrombosis and embolism. The risks associated with monitoring have to be balanced against the protection they afford the patient.

Arterial lines

The radial artery is most used and a number of workers have looked at the incidence of complications, especially that of thrombosis. Bedford & Woolmer (1973) in a prospective study found that 40 of their 105 patients (38%) developed vessel occlusion. In a study in New Zealand the incidence was reduced to 3% when 20 g teflon canulae were used instead of 18 g polypropylene (Davis & Stewart 1980). However, workers in Texas with the largest prospective series yet reported comprizing 1699 patients, had an overall incidence of 25% abnormal flow and also failed to show that catheter size (18 g or 20 g) or material altered this figure (Slogoff et al 1983). These authors also queried the predictive value of an Allen's test for collateral ulnar flow. 411 patients were examined and 16 had an abnormal test (> 15 s). None of the 16 patients had signs of ischaemia during cannulation and none had abnormal flow after decannulation.

Regular examination of the patients hand will identify ischaemic changes. If there is a persistent decrease in temperature or skin mottling, the catheter should be removed,

Central venous lines

As well as providing a measurement of right atrial pressure, these lines are used for administration of vasoactive and irritant drugs. The insertion of more than one line allows infusion of such drugs to proceed without interference from CVP measurements or bolus injections. A trilumen catheter has recently been introduced (Deseret Medical). Luer-lock connections help to reduce the risk of accidental disconnection and air embolism.

Pulmonary artery catheters

These are widely used in patients in North America having heart surgery but are inserted more selectively in the UK. Pulmonary artery wedge pressure is considered equivalent to left atrial pressure providing the catheter is free from clot and its tip is at or below the level of the left atrium. In the upper zones of the lung, alveolar pressure may occlude pulmonary vasculature, produce intermittent blood flow and a wedge pressure may not correspond to that of the left atrium. In conditions where the right and left filling pressures of the heart are different (e.g. mitral valve disease, poor left ventricular function) pulmonary artery catheters may provide additional information to that obtained from central venous pressure measurement. The main indication for use in cardiac surgical patients is an unstable patient with poor left ventricular function. Arrhythmias and premature beats can occur during catheter insertion. In a recent prospective study of 500 patients (Boyd et al 1983) this occured in 100 (20%). Serious complications occured in 23 patients (4.3%) including ventricular tachycardia (8), pulmonary heamorrhage (1), pulmonary infarction (7) and septicaemia(7).

Left atrial lines

These are placed by the surgeon at operation either directly into the left atrium or via the right atrium through the atrial septum. Embolism is a particular danger. The line should be clearly labelled, not used for infusion or injection of drugs and removed as soon as possible.

Choice of anaesthetic agent

Many techniques and anaesthetic drugs have been described for use in cardiac anaethesia. It would appear to be primarily important to maintain cardiovascular stability and the drugs an individual anaesthetist chooses to achieve this are of less importance. It has been shown that peri-operative ischaemia can influence the incidence of post-operative myocardial infarction after coronary artery bypass surgery (Slogoff & Keats 1985). Endotracheal intubation, the surgical stimulation of skin incision and sternal splitting accounted for half of these ischaemic episodes. Nine anaesthesiologists were involved in the study, one being associated with a significantly higher incidence of peri-operative ischaemia and post-operative myocardial infarction.

Anaesthetic drugs (recently introduced)

Isoflurane This inhalational agent causes a drop in blood pressure by a reduction of systemic vascular resistance. Vasodilation and a fall in blood pressure would seem very useful properties in patients with coronary artery disease but a clinical study by

Reiz et al (1983) revealed possible hazards. Isoflurane (1% end-tidal) was administered to patients with coronary artery disease having major vascular surgery of the aorta or iliac vessels. Ten of the 21 patients developed ECG changes with ST depression or T wave inversion. Coronary blood flow did not alter in these 10 patients but lactate extraction decreased markedly. It is postulated that a reduction in perfusion pressure may lead to regional ischaemia in the area supplied by the stenosed coronary arteries. Thus isoflurane should be used with care in patients with critical coronary stenoses.

Vecuronium Pancuronium has been used as a neuromuscular blocking agent in many cardiac procedures as blood pressure is maintained but the tachycardia is a disadvantage. This does not occur with the short acting monoquaternary steroid, vecuronium bromide.

Atracurium This interesting muscle relaxant undergoes degradation with increase in pH and temperature. If given to a cold patient, e.g. one cooled in cardiopulmonary bypass, additional incremental doses are unnecessary.

Management on bypass

One of the important tasks is to ensure adequate heparinization during cardiopulmonary bypass. Too little leads to the risk of emboli while the filters of the bypass system may become blocked; too much leads to difficulty in reversing the anticoagulation at the end of the procedure. The management of the anticoagulation has been greatly assisted by the use of the activated clotting time (ACT). 2–3 ml of blood are added to 12 mg of celite which reduces the whole blood clotting time to 90–130 s. Adequate heparinization is achieved by prolonging the ACT to at least three times baseline levels. Two groups of patients, those with recent subacute bacterial endocarditis and those on intra-aortic balloon pumps may require more than 3 mg/kg to do this (Kamath & Fozard 1980).

Perfusion of vital organs such as the brain, heart and kidneys may alter when the patient is no longer dependent on his own cardiac output but that produced by a machine. Patients with cerebrovascular disease are more susceptible to falls in blood pressure and reduced blood flow. In a retrospective survey of 3206 patients Bojar et al (1983) showed that of the 32 patients developing a major neurological deficit, six had carotid or vertebral stenoses. Monitoring pupil size is a simple way of assessing gross cerebral perfusion and more sophisticated methods involve the electroencephalogram (EEG). The EEG during ischaemia shows a decline in electrical activity. There is a decrease in faster waves with an increase in slower waves leading to electrical silence. The cerebral function monitor was designed to follow the electrical activity from a single pair of parietal electrodes. The position of the trace indicates the overall cerebral energy and the width of the trace shows the variation in wave form. The cerebral function analysing monitor is a further refinement giving a more detailed plot of the β, α, θ and δ waves (Maynard & Jenkinson 1984) (Fig. 10.1).

For some procedures involving the correction of complex cardiac defects in small infants or for aortic arch surgery in adults, complete cessation of the cerebral circulation may be required. What is considered to be a safe time of circulatory arrest at a given temperature? Treasure (1984) studied hypothermic circulatory arrest in a small mammal, the gerbil. There is no circle of Willis in these animals and carotid

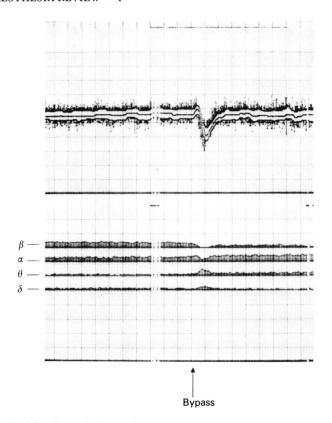

β —
α —
θ —
δ —

↑
Bypass

Fig. 10.1 Cerebral function analysing monitor trace recorded as patient went on to cardiopulmonary bypass. A fall in mean voltage is seen which recovers within 2 minutes. There is also a decrease in β and α wave activity with an increase in θ and δ wave activity.

occlusion leads to a cessation of cerebral blood flow. With brain temperatures of 20°C the gerbils had periods of cerebral ischaemia for times ranging from 0 to 180 minutes and were then examined metabolically, functionally and morphologically. Although one hour of circulatory arrest at temperatures below 20°C has been considered acceptable, Treasure convincingly suggested that this time should be revised to 20 minutes if the risk of irreversible brain damage is to be minimized.

Technical developments

Oxygenators

The most common type of oxygenator is a bubble oxygenator and damage to blood cell membranes can occur at the blood-gas interface. Membrane oxygenators are similar to the physiological arrangements in the lungs as the perfusion fluid and gases are separated by a polypropylene or silicone membrane. Membrane oxygenators cause less haemolysis, and higher platelet and white cell counts are obtained than when bubble oxygenators are used (Pierce 1980). Although they are more expensive and may take longer to set up, membrane oxygenators offer advantages for lengthy procedures. They can also be used for patients requiring long term support on partial cardiopulmonary bypass.

Biochemical measurements

Advancing technology allows for the rapid determination of pH, serum potassium and blood gases. Intravascular probes have been used to give continuous measurement of PO_2 but because probes are 5 French gauge, they are more suited to insertion into a central line than into a peripheral artery. An on-line oxygen electrode has been developed for use in the arterial or venous lines of the cardiopulmonary bypass to give a continuous reading of PO_2 (Claremont et al 1984).

POST-OPERATIVE CARE

Cardiovascular system

Following cardiac surgery the patient may develop life threatening complications and is therefore transferred to an intensive care ward for intensive monitoring. Monitoring is continued during the transfer using portable ECG and blood pressure equipment. Cardiovascular stability may require the use of ionotropes, vasodilators, dysrhythmic drugs, pacemakers and intra-aortic balloon pumps. Replacement of blood loss is important to maintain adequate filling pressures for the right and left atria. Autologous blood transfusion using a closed collecting system (commercially available from Sorensen Research Corporation) can reduce the requirement for banked blood (Johnson et al 1983).

Respiration

In the early post-operative period respiration is usually maintained by mechanical ventilators. The reasons for this are as follows.
1. The patient may have been in heart failure pre-operatively and still have residual pulmonary oedema.
2. High doses of opiates can markedly depress respiration.
3. Patients may be cool both centrally and peripherally and have greater oxygen demands when warming.
4. An increase in alveolar-arterial PO_2 difference is seen for at least 3 hours after operation (Sanchez de Leon et al 1982). A fall in colloid osmotic pressure occurs during bypass due to crystalloid fluid in the bypass priming fluid. This may cause interstitial pulmonary oedema and an increasing alveolar-arterial oxygen gradient. The time course for the restoration of colloid osmotic pressure and haematocrit parallels that of the restoration of the alveolar-arterial gradient.

More recently a number of workers have questioned the need to ventilate post-operatively patients with stable cardiovascular systems. Quasha et al (1980) studied patients following elective coronary artery bypass grafts. They compared a group of 18 patients who had spontaenous respiration and early extubation at a mean of 2 hours post-operatively with 20 patients who were ventilated overnight and extubated late at a mean of 18 hours. There was a significant reduction in the amount of analgesics and sedatives given to the early extubation group. There was a reduction in the amount of vasodilator drugs required to keep the systolic and diastolic pressures under 160 mmHg and 90 mmHg respectivelyalthough this did not reach statistical significance. One patient was reintubated when his PCO_2 reached 56 mmHg (kPa 7.46). Schuller et al (1984) have also reported on their practice of early extubation in children and had a low reintubation rate of 2.4%.

Humidification

For those who are ventilated, heating and humidification of the inspired gases has been provided by heated water reservoirs. Simpler hygroscopic condenser humidifiers are now available. Fully humidified expired gases at 37°C contain 44 mg water vapour per litre, which falls to 19.5 mg/l at 22°C in a heat and moisture exchanger. If a hygroscopic surface is added expired gases leave with only 9 mg water vapour per litre leaving 35 mg for humidification of inspired gas. A device such as the Servo-Humidifier (Siemans-Elema) is thus capable of providing inspired gases at a temperature of 5°C below body temperature and a relative humidity of 75% (Gedeon & Mebius 1979). The Pall Ultipor hydrophobic filter, designed as a bacterial filter, can also function as a humidifier. It compares well with other devices and has the added advantage of bacterial filtration (Bethune 1985).

Sedation

Regimes for sedation of patients on ventilators have come under scrutiny following the increased mortality in patients receiving etomidate infusions (Ledingham & Watt 1983). Etomidate infusions can no longer be recommended while althesin has been withdrawn from use (see also p. 219).

Combinations of analgesics and sedatives are often employed. A water soluble benzodiazepine, midazolam, has the advantage of being non-irritant to veins and has a short half life of 60–90 minutes (Dundee et al 1980) (see also p. 218).

REFERENCES

Bedford R F, Woolmer H 1973 Complications of percutaneous radial artery cannulation: an objective prospective study in man. Anesthesiology 38: 228–236
Bethune D W 1985 Hydrophobic versus hygroscopic heat and moisture exchangers. Anaesthesia 40: 210–211
Bojar R M, Hassan N, DeLaria G A, Sherry C, Goldin M D 1983 Neurological complications of coronary revascularisation. Annals of Thoracic Surgery 36: 427–432
Boyd K D, Thomas S J, Gold J, Boyd A D 1983 A prospective study of complications of pulmonary artery catheterisations in 500 consecutive patients. Chest 84: 245–249
Braunwald E 1982 Mechanism of action of calcium-channel blocking agents. New England Journal of Medicine 307: 1618–1627
Claremont D J, Pagdin T M, Walton N 1984 Continuous monitoring of blood Po_2 in extracorporeal systems. An in vitro evaluation of a re-usable oxygen electrode. Anaesthesia 39: 262–269
Davis F M, Stewart J M 1980 Radial artery canulation: a prospective study in patients undergoing cardiothoracic surgery. British Journal of Anaesthesia 52: 41–46
Dundee J W, Samuel I O, Toner W, Howard P J 1980 Midazolam: a water-soluble benzodiazepine. Studies in volunteers. Anaesthesia 35: 454–458

English T A H, Bailey A R, Dark J F, Williams W G 1984 The UK cardiac surgical register 1977–82. British Medical Journal 289: 1205–1208

Frishman W H 1981 β-Adrenoceptor antagonists: new drugs and new indications. New England Journal of Medicine 305: 500–506

Gedeon A, Mebius C 1979 The hygroscopic condenser humidifier. A new device for general use in anaesthesia and intensive care. Anaesthesia 34: 1043–1047

Johnson R G, Rosenkrantz B A, Preston R A, Hopkins C, Daggett W M 1983 The efficacy of postoperative autotransfusion in patients undergoing cardiac operations. Annals of Thoracic Surgery 36: 173–179

Kamath B S K, Fozard J R 1980 Control of heparinisation during cardiopulmonary bypass. Experience with the activated clotting time method. Anaesthesia 35: 250–256

Ledingham I M, Watt I 1983 Influence of sedation on mortality in critically ill multiple trauma patients. Lancet i: 1270

Maynard D E, Jenkinson J L 1984 The cerebral function analysing monitor. Initial clinical experience, application and further development. Anaesthesia 39: 678–690

Paiement B et al 1983 A simple classification of the risk in cardiac surgery. Canadian Anaesthetists Society Journal 30: 61–68

Pierce E C 1980 The membrane versus bubble oxygenator controversy. Annals of Thoracic Surgery 29: 497–499

Quasha A L, Loeber N, Feeley T W, Ullyot D J, Roizen M F 1980 Postoperative respiratory care: A controlled trial of early and late extubation following coronary-artery bypass grafting. Anesthesiology 52: 135–141

Reiz S, Balfors E, Sorenson M B, Ariola S, Friedman A, Truedsson H 1983 Isoflurane — a powerful coronary vasodilator in patients with coronary artery disease. Anesthesiology 59: 91–97

Sanchez De Leon R, Paterson J L, Sykes M K 1982 Changes in colloid osmotic pressure and plasma albumin concentration associated with extracorporeal circulation. British Journal of Anaesthesia 54: 465–473

Schuller J L, Bovill J G, Nijveld A, Patrick M R, Macelletti C 1984 Early extubation of the trachea after open heart surgery for congenital heart disease. A review of 3 years experience. British Journal of Anaesthesia 56: 1101

Slogoff S, Keats A S, Arlund C 1983 On the safety of radial artery cannulation. Anesthesiology 59: 42–47

Slogoff S, Keats A S 1985 Does perioperative myocardial ischemia lead to postoperative myocardial infarction? Anesthesiology 62: 107–114

Treasure T 1984 The safe duration of total circulatory arrest with profound hypothermia. Annals of the Royal College of Surgeons of England 66: 235–240

Endocrine response to surgery

INTRODUCTION

Trauma, whether accidental or surgical, induces a series of circulatory, endocrine, metabolic and immunological changes which are known as the stress response.

In 1942 Cuthbertson divided the stress response into an ebb phase and a flow phase (Cuthbertson 1942) and in 1953 Moore described four separate metabolic phases associated with surgery — an adrenergic/corticoid phase, a corticoid withdrawal phase, an anabolic phase and a fat gain phase (Moore 1953). The ebb or adrenergic/corticoid phase lasts between 1 and 5 days and consists of a series of emergency responses designed to ensure the immediate survival of the individual. Four catabolic hormones, catecholamines, cortisol, growth hormone and glucagon are involved while the actions of insulin are suppressed. This results in an increase in the production and delivery of readily utilizable fuel sources (glucose and free fatty acids) to organs vital for fight and flight, namely the heart, the brain and muscle. The ebb phase gives way to the flow phase (the corticoid withdrawal phase, Moore 1983) when catabolic activity decreases and the metabolic effects of insulin predominate.

The flow or anabolic phase usually starts 2–10 days after the initial trauma and during this time there is a positive nitrogen balance as the protein loss which occurred in the initial phases is replaced. Fat stores are also replaced during the fat gain phase. However the whole anabolic process may be protracted and last several months.

Metabolic changes may occur in the pre-operative phase in the patient who is unduly anxious or who has severe pain, but the main causative factor is surgical trauma resulting in afferent nerve stimulation and increased production of hormones. Anaesthesia per se has little overall effect in initiating the response although different techniques have been developed to limit the consequences thereof.

It has been suggested that attempts to suppress the metabolic response to surgery are unnecessary or even harmful. The view has been propounded that the traumatized animal survives because of the response but it is illogical to suggest that this situation is analogous to that seen following surgical trauma. By suppressing the stress response the loss of energy stores and tissues can be minimized. During the stress response extensive amounts of hormones are released and the management of this at operation appears to be turning towards preventing their release and instituting appropriate hormone therapy as necessary.

In addition Akiyoshi et al (1984) have shown that immunodepression occurs following major abdominal surgery while Mani et al (1984) found a similar effect occurred following cardiac surgery. Animal studies by Fujiwara (1984) referred to the role of surgical stress in suppressing the graft versus host reaction following trauma. All these findings suggest that modifying the endocrine response to trauma may be an important factor in improving immunocompetence following major surgical procedures, with the result that post-operative complications may be minimized so that an earlier return to normal health is possible.

This chapter reviews the hormonal and metabolic changes that occur in response to surgery and trauma, the mechanisms by which they are initiated and the methods available to clinical anaesthetists to modify the response.

HORMONAL RESPONSE TO SURGERY

Catecholamines

The adrenergic response to surgery provides the anaesthetist with an important guide to depth of anaesthesia and adequacy of analgesia. The response is initiated by emotion, pain, hypotension, hypoxia and exposure to cold, and it manifests itself as an increase in adrenergic nerve activity intra-operatively with an increase in the circulating levels of catecholamines post-operatively (Halter et al 1977, Nistrup Marsden et al 1978).

The effects of a large increase in adrenergic activity are widespread. There is an increase in peripheral vascular resistance, heart rate and cardiac contractility, resulting in a rise in blood pressure, cardiac output and myocardial oxygen consumption. There are a series of metabolic responses which are mainly attributed to the effects of adrenaline, which has eight times the metabolic effects of noradrenaline and which is secreted in larger quantities, especially in the immediate post-operative period. Glycogenolysis is stimulated in muscle by a β adrenergic effect, which is cyclic AMP dependent and results in a rise in plasma lactate. In the liver glycogenolysis is an α effect and is dependent on calcium ions. There are several factors which reduce glucose utilization: these include a reduction in glycogen formation by an inhibition of glycogen synthase activity, a reduction in insulin activity and an increase in the formation of alternative fuel sources, e.g. fat. In addition, α adrenergic stimulation suppresses the release of insulin from the β pancreatic cells (Allison 1971).

The lipolytic effects of catecholamines produce free fatty acids, glycerol and 3-hydroxybutyrate. This β adrenergic, cyclic AMP effect of noradrenaline is brought about by an increase in activity of the enzyme triglyceride lipase. Noradrenaline is also responsible for an increase in heat production in brown adipose tissue (Hall & Lucke 1982). Although this may be important to the neonates the significance to the adults has yet to be determined.

The release of other catabolic hormones, such as growth hormone and cortisol, may be affected by an increase in adrenergic activity. α Adrenergic stimulation promotes the release of growth hormone from the pituitary while α adrenergic blockade has the opposite effect (Ishibashi & Yamaji 1984). These effects are probably mediated by dopamine receptors. Release of corticotrophin releasing factor (CRF), and hence adrenocorticotropic hormone and cortisol, may also be affected by adrenergic activity (Rees et al 1970, Gignere et al 1981, Grossman & Besser 1982).

The methods of assaying hormones and technical difficulties may make studies on plasma catecholamine levels appear contradictory. Thus the precise role of catecholamines in the metabolic response to surgery is still unclear.

Adrenocorticotropic hormone and cortisol

Early investigations into cortisol and adrenocorticotropic hormone (ACTH) secretion relied on the presence of indirect indicators such as eosinophilia. It was not until 1953 that plasma cortisol could be measured directly in the plasma (Weichselbaum et al 1953), while it was not until 1962 that ACTH could be assayed (Cooper & Nelson 1962). However, inspite of these difficulties, Long (1946) recognized that afferent nerve stimulation produced an endocrine response, including the release of ACTH and cortisol. In 1953 Hume described the nervous pathways in an elegant series of experiments on dogs. In the same year Moore (1953) measured urinary excretion of 17 ketosteroids during the 'four phases of convalescence'. The steroid excretion rose intra-operatively and in the immediate post-operative period, and then fell 2–5 days post-operatively to subnormal levels, for a period of several weeks.

It has since been shown that the plasma cortisol levels may rise to five times normal or more during major surgery (Thomasson 1959) or severe trauma. Levels usually remain high for 1–3 days. Factors such as severity of trauma, pain, hypovolaemia, hypoxia, sepsis and emotional upset increase cortisol production. Afferent neuronal blockade, central depression of ACTH and pharmacological inhibition of cortisol manufacture by agents such as etomidate (Ledingham & Watt 1983, Fry & Griffiths 1984) will reduce the adrenocortical response. There is a loss of the normal negative feedback control of ACTH by cortisol.

High circulating levels of corticosteroids will have effects on the circulation, on metabolism and on immunological systems. Some of their effects are due to a selective increase in enzyme production brought about when the steroids bind to nuclear chromatin and increase RNA production.

Cortisol is 95% bound to transcortin and albumin and only the free, unbound form is active. Its half life is 80 minutes and it is excreted by the kidneys after conjugation in the liver. It is possible that part of the increase in circulating cortisol levels during surgery is due to a decrease in liver blood flow resulting in a decrease in conjugation and excretion with also a reduction in protein binding.

The circulatory effects of the corticosteroids result in a decrease in peripheral resistance, an increase in cardiac output and an improvement in tissue perfusion (Samblin et al 1965). Hepatic blood flow and glomerular filtration rate are also improved. Cortisol has a mild sodium and water retaining effect and is important in maintaining the correct cell membrane permeability to electrolytes.

Cortisol inhibits the actions of insulin. This results in hyperglycaemia as glucose utilization is reduced while gluconeogenesis in the liver results in increased production of glucose. Cortisol also increases lipolysis. The effects of cortisol on protein metabolism are controversial and it is unclear whether it increases catabolism (Simmons et al 1984), decreases anabolism, or both (Hoover-Plow & Clifford 1978). The increase in protein catabolism may only occur when circulating levels are very high (Munro 1979) and in the absence of ketosis (Williamson et al 1977). That is, it may not be of significance in starved patients. It has been suggested that cortisol may not be an important wound hormone leading to post-traumatic muscle wasting and nitrogen loss (Baracos et al 1983, Clowes 1983, Bessey 1984, Gelfand et al 1984).

High circulating levels of cortisol will also have immunological effects, depressing allergic and inflammatory responses by decreasing antibody formation, decreasing numbers of circulating lymphocytes and eosinophils, decreasing the mass of lymphatic tissue, and inhibiting the activity of granulocytes and monocytes. There is a selective depression of T lymphocytes relative to B lymphocytes. If the high level of cortisol persists there may be an increased susceptibility to infection, inhibition of fibroblasts with an increased risk of bruising and poor wound healing, increased erythropoeisis, osteoporosis, and mood and behaviour changes.

Growth hormone

The factors effecting the release of growth hormone from the pituitary include pain, trauma, emotional stress, α adrenergic stimulation, antidiuretic hormone, deep sleep, fasting, protein depletion, glucagon, arginine and insulin hypoglycaemia (Newsome 1975). Growth hormone (GH) is inhibited by somatostatin, produced by 'D' cells of the pancreas (it also inhibits release of insulin and glucagon), and hyperglycaemia. However, during surgical stimulation GH is released in large quantities despite the presence of hyperglycaemia. Its effects are to produce hyperglycaemia, lipolysis and protein synthesis. Its effects on protein metabolism are probably mediated by somatomedin, a peptide produced by the liver, which stimulates amino acid uptake and protein synthesis and hence promotes nitrogen retention. It is likely that GH reduces the loss of protein that would otherwise occur during surgery, for it has been found that infusions of GH and calories in patients with burns will result in a positive nitrogen balance (Wilmore et al 1974). GH produces a decrease in glucose utilization but this tendency is partially offset as it sensitizes the β islet cells to produce more insulin. As with ACTH, the normal control of growth hormone release is lost during stress. There may be an adrenergic mechanism in GH release during stress as α blockade can inhibit its release. (Imura et al 1971, Ishibashi & Yamaji 1984).

Glucagon

Glucagon is a polypeptide secreted from the α islet cells of the pancreas (Unger & Orci 1976). The difficulty in assaying glucagon has meant that little information about its role in the stress response is available. Its release is initiated by hypoglycaemia, starvation, exercise, stress, pancreozymin and gastroinhibitory peptide: glucagon release is inhibited by somatostatin, α stimulation and β blockade.

Its effects are to produce hyperglycaemia by stimulation of hepatic glycogenolysis and gluconeogenesis, to produce a rise in circulating free fatty acids by increasing lipolysis, and to increase protein breakdown. It also stimulates insulin release.

Insulin

Insulin is the main anabolic hormone involved in the stress response. Under non-stressful circumstances it is released from the β islet cells of the pancreas by an increase in blood glucose levels, by amino acids such as phenylalanine, anginine, leucine and histidine, by other hormones such as ACTH, growth hormone and glucagon, by β adrenergic stimulation, vagal stimulation, and by secretin and cholecystokinin. Its release is inhibited by somatostatin, α adrenergic stimulation, β adrenergic blockade, thiazide diuretics and phenytoin.

During surgery the circulating levels of insulin fall probably as a result of α adrenergic stimulation on insulin release (Allison 1971, Walsh et al 1982). This fall in circulating levels occurs despite the high blood glucose levels, suggesting there is resistance of the normal response to hyperglycamia. This is most marked during major surgery and severe trauma and results in a decrease in the body's glucose utilization. The cells are unresponsive to the effects of endogenous insulin in the non-diabetic patient (Buckingham et al 1976, Meyer et al 1979) and to endogenous insulin in diabetics (Meyer et al 1979). Those cells not solely dependent on glucose metabolism will therefore turn to alternative fuel sources such as free fatty acids.

In the hours following surgery the circulating levels of insulin begin to increase but its effects are still inhibited by cortisol, growth hormone, glucagon and adrenaline with hyperglycaemia persisting for 24–48 hours.

The actions of insulin are wide ranging. Its main effects are on carbohydrate metabolism increasing membrane transport, and hence the utilization of glucose. It also increases the storage of glucose in the form of fat or glycogen. It stimulates RNA and DNA synthesis and inhibits lipolysis.

Many of the metabolic effects of surgery are brought about by the actions of catabolic hormones during a time of inhibition of the main anabolic hormone, insulin. Indeed, many of the metabolic effects of surgery can be prevented by a per-operative infusion of insulin (Woolfson et al 1979).

There are certain problems with the interpretation of the measurements of peripheral levels of insulin because of the very rapid and variable liver metabolism of the hormone and it has been suggested that measurement of C-peptide, which is secreted molecule for molecule with insulin, may provide a better guide to the insulin secretion rate (Jaspan et al 1984).

Antidiuretic hormone

Antidiuretic hormone, or arginine vasopressin (ADH) is an octopeptide released from the neurosecretory cells of the supraoptic and paraventricular nuclei of the hypothalamus (Moran 1971). After combining with a polypeptide carrier, neurophysin, it migrates down the axons and is stored in the posterior pituitary. Its release from the posterior pituitary is controlled by osmoreceptors in the hypothalamic region, volume receptors in the left atrium, and baroreceptors in the carotid sinus and aortic arch. In low doses ADH increases the distal tubular reabsorbtion of water without solute. In intermediate doses it increases solute reabsorbtion, possibly through a reduction in glomerular filtration rate, while in high doses it produces vasoconstriction. It is one of the most potent endogenous vasoconstrictors in man. The half life of ADH is only 3 minutes and it undergoes rapid hepatic clearance. 10% is excreted unchanged in the urine. Normal circulating levels are 1–2 pg/ml. During surgical stress levels of up to 120 pg/ml have been reported. Anaesthesia per se has little effect on ADH secretion in comparison to surgery (Philbin & Coggins 1978).

The effects of artificial ventilation on ADH levels is still uncertain but IPPV and PEEP probably do not directly increase secretion (Kumar et al 1974), though there is a secondary increase due to a fall in cardiac index. Hypercarbia is a stimulus to ADH secretion (Philbin et al 1970).

Renin angiotensin aldosterone

Renin is a proteolytic enzyme released from the macula densa of the juxtaglomerullar apparatus (Miller 1981). It breaks an α 2 globulin in the plasma to produce angiotensin I, a decapeptide which releases catecholamines from the adrenal gland resulting in an increased blood pressure and also producing a sensation of thirst. Most angiotensin I is transformed into angiotensin II in the lungs by a converting enzyme. Angiotensin II is a vasoconstrictor with a positive inotropic effect on the heart; it also stimulates catecholamine release. It has a short half life, measured in seconds. It is metabolized in the plasma and the liver by angiotensinase to angiotensin III. Both angiotensin II and angiotensin III are potent releasers of aldosterone while angiotensin II will also stimulate ACTH release. The ACTH releases cortisol which also has a mild mineralocorticoid effect. Aldosterone will increase water and sodium retention in the kidney resulting in an increased potassium loss.

During major surgery there may be a three- to four-fold increase in plasma renin activity (PRA). The actual increase will depend on the blood loss and fluid administration during surgery (Bailey et al 1975). Renin is released by a variety of mechanisms. A fall in renal perfusion will be detected by renal baroreceptors. Hypovolaemia will result in increased renal nerve and adrenergic receptor activity with an increased level of circulating catecholamines. Decreased glomular filtration rate results in a rise in urinary sodium concentration. All these mechanisms result in increased renin secretion, conserving sodium and water in an attempt to maintain cardiac output. In addition, hormones such as ACTH, adrenal corticosteroids and oestrogens, and changes in posture as well as hypothermia will all increase PRA. Angiotensin II and ADH reduce PRA.

Other hormones

Plasma thyroxine remains unchanged during surgery whilst plasma triiodothyronine is decreased during and after surgery. Testosterone and prolactin are increased. The significance of these findings is not fully understood.

Wound hormones

Phagocytic cells can produce hormone like substances which have been given a variety of names such as leucocytic pyrogen, endogenous pyrogen, leukocytic andogenous mediator, neutrophil releasing factor, lymphocyte activating factor and, most recently, Interleukin I (ILI) (Dinarello 1984).

These substances are involved in the production of post-traumatic fever, the production and release of neutrophils from the bone marrow, hepatic production of acute phase reactants such as C-reactive protein, fibrinogen, ceruloplasmin and antitrypsin, and also in the hepatic uptake of amino acids. They are also capable of proteolysis (Clowes 1983), possibly by increasing prostaglandin E_2 synthesis in muscle (Baracos et al 1983).

The extent to which ILI is involved in the metabolic response to surgery has yet to be decided. If wound hormones do play a large part then it is obvious that afferent blockade alone will not completely obtund the stress response.

METABOLIC RESPONSE TO SURGERY

Carbohydrate metabolism

Following surgery the blood glucose levels rise from their normal level of 3.5–5 mmol/l to as much as 8–10 mmol/l. The blood sugar may remain elevated for more than 24 hours if trauma is severe. Under resting conditions in the adult, the liver can produce about 10 g of glucose per hour, of which 6 g/h will be utilized by the brain while both white cells and red blood cells will each use 2 g/h. The glucose is manufactured from lactate and pyruvate, which is released from muscle (following the breakdown of muscle glycogen), from amino acids (from muscle protein breakdown), from liver glycogen stores, and from glycerol and free fatty acids (from adipose tissue breakdown).

During surgery the break down of liver glycogen increases, though liver glycogen stores may be low following pre-operative fasting. This is an α adrenergic, calcium ion dependent effect which is enhanced by glucagon and cortisol. In muscle β adrenergic cyclic AMP dependent mechanisms cause glycogen breakdown with the production of lactate and pyruvate, which are metabolized to glucose in the liver. Protein breakdown is increased with the production of more amino acids for gluconeogenesis and lipolysis while glyceroland free fatty acid production is also increased. Gluconeogenesis is encouraged by cortisol, growth hormone and thyroxine. Glucose utilization is reduced due to the low level of insulin and the inhibition of the effects of insulin by catecholamines, glucagon, growth hormone and cortisol (Stanley 1981a–c).

Protein metabolism

Following surgery there is a period of negative nitrogen balance which lasts for 2–10 days, depending on the severity of the surgery and on the presence of complications such as sepsis. The loss of nitrogen may vary from 5–20 g/d, following uncomplicated surgery in patients on no nutritional support, to 50 g/d in hypercatabolic states after burns or multiple injuries. Each 50 g of nitrogen represents a loss of 1.25 kg of muscle mass.

There are many causes of protein loss including hypothermia. Prevention of heat loss in the elderly during surgery prevents most of the peri-operative nitrogen losses (Carli et al 1982). Cortisol levels which are elevated two to three times that of normal during surgical stress may effect the breakdown of myofibrillar protein (Thomas et al 1979). The increased secretion of glucagon, growth hormone and catecholamines and reduced insulin activity all promote catabolism.

However, infusions of the four catabolic hormones, adrenaline, growth hormone, cortisol and glucagon, resulted in blood glucose changes similar to those seen during surgery, but the effects on protein metabolism were less impressive (Bessey et al 1984, Gelfand et al 1984). It was concluded that the catabolic hormones contributed to, but were not the sole mediators of, the muscle wasting seen after surgery.

There has been a recent resurgence of interest in wound hormone. Interleukin I is a polypeptide which is released from activated monocytes and macrophages after tissue damage (Dinarello 1984). It is known to increase protein breakdown in muscle without affecting synthesis and it is possible that it plays a role in post-operative nitrogen loss (Baracos et al 1983, Clowes et al 1983).

The breakdown of muscle protein in the peri-operative period is of value because this source of amino acids can be used to provide glucose by gluconeogenesis in the liver and to provide new proteins for antibody production, wound healing and acute phase reactant synthesis. In addition, the branched chain amino acids can be used within the muscle as a direct source of energy.

Fat metabolism

Lipid metabolism shows great individual variation and is affected by fasting and by surgery (Stanley 1981a,b). Lipolysis is caused by all the catabolic hormones — cortisol, glucagon, growth hormone and especially the catecholamines. The increase in lipolysis is due to increased activity of triglyceride lipase caused by a cyclic AMP dependent mechanism. Free fatty acids are released into the circulation and are converted to ketone bodies in the liver. Both free fatty acids and ketone bodies are used by muscle as fuel sources and ketone bodies are used by the brain and renal cortex as a fuel source during prolonged fasting. The brain's glucose requirements during starvation may fall from 120 g/d to 35 g/d as the brain begins to utilize ketone bodies. Insulin is released in response to high circulating levels of ketone bodies and it will also inhibit further lipolysis.

The changes in lipid metabolism during the surgical period will depend on the individual's fat content and individual metabolic profile, the severity of trauma, the level of anxiety and the duration of fasting. Fasting alone may raise free fatty acid and ketone body levels in the plasma more than surgery (Foster et al 1979).

CAUSES OF THE STRESS RESPONSE

The most important initiator of the stress response to surgery or trauma is stimulation of afferent somatic and autonomic nerves at the wound site. The relative importance of somatic, parasympathetic and sympathetic afferent pathways has still to be determined although complete blockade of all afferent pathways is necessary for obtunding the stress response (Engquvist et al 1977).

Physiological factors such as hypothermia (Carli et al 1982), hypoxia, hypercarbia, infection and hypotension all play a part in initiating the stress response, as will psychological factors such as fear and anxiety. Clowes et al (1983) and Baracos (1983) have recently scrutinized the part played by wound hormones and bacterial forms.

Advantages and disadvantages of the stress response

The immediate benefits of the stress response are to maintain blood flow to vital organs and to increase the delivery of readily utilizable fuel sourses to those organs. The supply of fuel is then maintained by the continuing process of catabolism until the patient has recovered sufficiently to resume normal food intake. However, afferent nerve stimulation activates the response in an indiscriminant manner producing an excess of hormones which may be life saving in some patients, but can be disadvantageous in others. Oxygen consumption is increased and there is a relative hypoxia in non vital organs such as the lung, kidney and intestines. An increase in blood pressure, heart rate and hence cardiac work may be detrimental to patients with coronary artery disease while the high level of adrenergic activity may increase the risk of life threatening arrhythmias. Acidosis may inhibit myocardial function. Muscle protein breakdown produces negative nitrogen balance which, if severe, may delay wound healing. There may be an increased risk of infection due to depression of the immune response by corticosteroids while stress can play a part in initiating disorders such as malignant hyperpyrexia. There is also an increase in the blood coagulability which can lead to the production of microemboli and deep vein thrombosis (Ellis & Humphrey 1982).

ANAESTHETIC MANAGEMENT OF THE ENDOCRINE RESPONSE

There has been a widespread increase in interest in developing anaesthetic techniques which are believed to suppress the endocrine response to surgical trauma.

Psychological preparation

It has long been said that the pre-operative visit by the anaesthetist is as effective in relieving stress as premedication. Anxious patients have higher circulating catecholamine (Hjenidahl & Eliasion 1979). Recent studies have shown that acute and chronic psychogenic stress in animals are important factors in initiating the stress response (Armanio et al 1984). The effect of pre-surgical psychological counselling and preparation is currently being studied (Salmon 1986; personal communication).

High dose opioid anaesthesia

Morphine and nalorphine have been shown to suppress the cortisol response to surgery (McDonald 1959) and this has led to other studies of the use of large doses of intravenous morphine to modify the stress response (George et al 1974). The advent of more potent short acting analgesic agents has led to the use of fentanyl, alfentanil and sufentanyl.

Unfortunately the dose of intravenous opioids necessary to suppress the endocrine response was such that there was still profound respiratory depression at the end of operation, long after cessation of surgical stimulation. The use of high dose morphine implied that it could only be used in patients who were due to be ventilated post-operatively (e.g. after cardiac surgery). The technique was not suitable for patients who were expected to be returned to a normal surgical ward (Brandt et al 1978). Fentanyl, with a shorter plasma half life seemed more promising in inhibiting hypothalamic–pituitary stimulation but post-operative respiratory depression has also occurred. This has led to interest in even shorter acting drugs such as alfentanil and sufentanyl (Bovill et al 1983, Moller et al 1985).

It has been demonstrated that, as with morphine, the dose of fentanyl necessary to suppress the endocrine response is largely dependent upon the degree of surgical stimulation suffered by the patient. Thus a dose of 10 μg/kg body weight was sufficient to modify the rise in cortisol and hyperglycaemia seen during abdominal surgery (Haxholdt et al 1981) but that 50 μg/kg body weight was necessary during upper abdominal surgery and its modifying action lasted for only half as long (Cooper et al 1981). A similar effect could be seen in patients undergoing cardiac surgery using fentanyl in doses in excess of 50 μg/kg. Despite the use of high doses cardiopulmonary bypass, with the use of cold perfusate, and the attendant haemodynamic changes and surgical stimulation overcame the inhibitory effects of the opioid (Sebel et al 1981, Walsh et al 1981).

Maller (1985) has continued the study into the post-operative period after using alfentanil at a loading dose of 150 μg/kg body weight, followed by an infusion of 3 μg/kg/min in patients undergoing abdominal surgery. This dose suppressed the endocrine response during the surgical procedure but the effect had worn off after 2–3 hours into the post-operative period when the opioid levels had fallen. By this time the patients had returned to their wards and were receiving routine intramuscular opioid injections on demand for analgesia.

Bent et al (1984) have examined the effect of giving a known suppressive dose of fentanyl to patients in whom the endocrine response has already been established. Their patients received fentanyl 30 μg/kg body weight during pelvic surgery after 1 hour of surgery using inhalational anaesthesia alone. This resulted in a fall in blood lactate levels possibly due to a decrease in plasma catecholamines but the levels of cortisol and glucose which were already raised were unaffected, demonstrating the inability of high dose opioids to alter the established endocrine response. Hay et al (1985) confirmed this by demonstrating the inability of meptazinol or morphine to alter the established endocrine response when given at the end of surgery.

It has been suggested that endogenous opioids play an important role in the endocrine response, and that during high dose opioid anaesthesia there may be a negative feedback mechanism. Hall et al (1983) examined the role of endogenous opioids by using an infusion of naloxone, a potent opiate antagonist, in patients undergoing pelvic surgery with inhalational anaesthesia. He found no significant alteration in the levels of insulin, cortisol or growth hormone and concluded that endogenous opioids played only a minor role in the endocrine response.

Thus, in clinical practice it has become apparent over the last few years that high dose opioid anaesthesia is effective in suppressing the endocrine response only if it is initiated before surgical stimulation, and that it may be overcome if the stimulus is sufficiently strong. Suppression will only last for as long as adequate opioid levels are maintained. This implies that patients have to be ventilated in the post-operative period.

Afferent neuronal blockade

In clinical practice afferent neuronal blockade may be achieved either by local anaesthetics or by opioids given extradurally or intrathecally. The fact that afferent blockade may modify the endocrine response has been known for some years since animal studies showed that denervation of a limb prior to trauma to that limb prevented the rise in cortisol levels (Wise et al 1972). Recent advances have been achieved by employing a variety of different drugs and techniques to achieve this.

Bromage et al (1971) showed that, following epidural blockade with local anaesthetics for upper abdominal surgery, there was a suppression of the glucose response but plasma cortisol levels still rose with surgery despite adequate analgesia.

Engquvist et al (1977) confirmed this by demonstrating that for pelvic surgery a very extensive epidural blockade from T_4 to S_5 was required for complete suppression of the endocrine response: surgery in the upper abdomen, or incomplete blockade still caused a rise in cortisol levels.

Traynor et al (1982) combined thoracic epidural analgesia with the use of a vagal blockade at operation in patients undergoing upper abdominal surgery and although the hyperglycaemic response was suppressed there was still a rise in plasma cortisol levels. This was confirmed by Tsugi et al (1983) who used combined extradural analgesia with vagal blockade for gastrectomy, but in both these studies vagal blockade was performed after the onset of surgical stimulation. Epidural blockade and intravenous opioids are ineffectual in modifying the endocrine response if administered after surgical stimulation (Moller et al 1982). Tsuji (1983), however, noted that epidural blockade combined with splanchnic nerve blockade successfully suppressed the rise in cortisol levels.

Intrathecal local anaesthetics also suppressed the endocrine response but is only effective during the duration of the blockade (Moller et al 1984). Lush et al (1972) confirmed this finding with extradural local anaesthesia. In clinical practice this has led to epidural blockade being more widely used as the block may be maintained for longer into the post-operative period via an indwelling catheter, with the use of an infusion pump. In this way it is hoped that a more prolonged suppression of the endocrine response can be extended into the post-operative period. Kehlet et al (1980) and Rem et al (1980) have both shown that following epidural anaesthesia there is less immunodepression and nitrogen balance is maintained.

The use of opioids to provide analgesia when given extradurally and intrathecally has become more widespread since it was first described by Yaksh (1976). The apparent advantage for the patient over local anaesthetics is prolonged analgesia without the risk of sympathetic or motor blockade. After intrathecal diamorphine the average time for post-operative analgesia has been up to 16 hours (Child & Kaufman 1985) with 33% of the patients requiring no supplementary analgesia in the first 24 hours.

Christensen et al (1982) and Jorgensen et al (1982) showed that extradural morphine (4 mg) suppressed the adrenocortical and hyperglycaemic responses to surgery post-operatively but had little effect intra-operatively and that in fact a local anaesthetic, bupivacaine, produced greater suppression. This failure could be due to the need to use a higher dose when given extradurally as suggested by Bromage et al (1980), especially when a poorly lipid soluble drug such as morphine is used, which will only slowly cross the dura to its site of action on the opioid receptors of the dorsal columns of the spinal cord.

Child & Kaufman (1985) have shown that suppression of the endocrine response can be achieved intra-operatively by the use of intrathecal diamorphine which presents a high dose of opiate directly to the receptors before surgery starts. However, sympathetic and proprioceptive afferents are not effected by opioids (Kitihata & Collins 1981) and this may be the reason for incomplete suppression seen with spinal opiates. Again to reinforce the concept of multiple pathways transmitting afferent nociceptive responses, Piv et al (1983) have demonstrated in animals pathways, dependant upon γ-amino-butyric acid transmitter, which are unaffected by opioids but may be blocked by midazolam given intrathecally.

Efferent neuronal blockade

The interruption of the efferent limb of the sympathetic reflex, by α and β blockade has been studied to examine the role of the sympathetic system in initiating the endocrine response to surgery.

Giguene et al (1980) showed that α_1 blockade in rats reduced the release of ACTH from the pituitary, and α blockade by phentolamine reversed the insulin suppression that occurred during surgical stimulation (Walsh et al 1982). Labetalol, a combined α and β blocker was shown to reduce cortisol levels probably via an α mediated pituitary action (Child 1984).

β Blockade with propranolol 0.15 mg/kg in patients undergoing pelvic surgery was shown to have little effect on the hyperglycaemic response although insulin levels were decreased at 60 and 90 minutes into the operation as its β stimulated release was blocked (Cooper et al 1980). However, Lowidsen et al (1983) showed that β blockade prolonged hypoglycaemia by increasing tissue uptake of glucose directly and not by decreasing the hormones inhibiting insulin (epinephnine and growth hormone). These results show that adrenaline induced hepatic glycogenolysis is a major cause of the hyperglycaemia seen intra-operatively.

Hormone manipulation

Most of the anaesthetic techniques for modifying the endocrine response to surgery have concentrated on blocking the pathways transmitting stimuli which inhibit or stimulate the various hormones involved in the response. Recent work has studied manipulating the hormone levels directly.

The idea of being able to modify cortisol levels directly arose after it was discovered that patients being ventilated in certain intensive care units had lower cortisol levels than in groups previously treated and the only change in their overall management had been the introduction of the anaesthetic agent etomidate for their long term sedation (Fellows et al 1983).

Since then the exact nature of this suppression of cortisol has been elucidated, and it is apparent that adrenocortical function is altered by the inhibition of 11 β hydroxylation and 17 α hydroxylation. This is seen transiently after a bolus dose of etomidate (de Jong 1984) while a more prolonged suppression can be obtained with an etomidate infusion (Moore et al 1985) when aldosterone suppression is also seen.

It is unclear as yet whether this has any beneficial long term effect on the catabolic phase of negative nitrogen balance normally associated with raised cortisol levels. No patients showed any cardiovascular instability during this infusion despite low recorded cortisol levels of <100 mol/l.

Insulin appears to be the most important short term anabolic hormone effected during the stress response, and while many anaesthetic techniques have been used to prevent this inhibition, insulin levels can readily be maintained by intra-operative infusion. Hall et al (1983) have studied this by instituting a low dose insulin infusion regime on patients undergoing pelvic surgery. They showed that by increasing plasma levels, blood glucose fell during the surgical period, probably as a result of increased peripheral glucose uptake. The inhibition of substrate mobilization was continued with the reduction in fat metabolism, as plasma non-esterified fatty acids and later 3-hydroxybutyrate levels fell.

General patient care

Irrespective of the anaesthetic technique, the overall management of the patient has been shown to have some influence on the endocrine response. Mullin & Kilpatrick (1981) showed that good nutritional support both pre-operatively and post-operatively in surgical patients resulted in better immunocompetence with an increase in circulating lymphocytes, while there is evidence that adequate nutritional support post-operatively reduces the catabolic phase and decreases overall nitrogen losses. Many patients would benefit from a period of pre-operative feeding rather than undergoing surgery at the earliest convenient moment.

Although it is accepted practice to nurse neonates at an optimal temperature to minimize heat loss and reduce calorie demands, adult surgical patients have not been shown to benefit in the long term from being maintained in a thermoneutral environment during the post-operative period. Carli et al (1982) have shown that the prevention of intra-operative heat loss, however, has resulted in a decreased nitrogen loss post-operatively in elderly patients undergoing abdominal surgery. This may reflect the role of heat loss as an initiating factor in the stress response and the importance of preventing initiation rather than attempting to alter the established endocrine response.

CONCLUSIONS

Debate continues as to whether the endocrine response or its suppression is beneficial. However, the controversy has led to the introduction of many new drugs and techniques and has stimulated detailed research projects, one of which has served to improve patient care and wellbeing (see Update section).

REFERENCES

Akiyoshi T 1984 Immunodepression after surgery: impaired production of interleukin 2. Japanese Journal of Surgery 14: 384–386
Allison S P 1971 Changes in insulin secretion during open heart surgery. British Journal of Anaesthesia 43: 138–142

Armanio A 1984 The effect of acute and chronic psychogenic stress on corticoradrenal and pituitary-thyroid hormones in male rats. Hormone Research 20: 241–245

Bailey D R, Miller E D, Kaplan A J, Rogers P W 1975 The renin-angiotensin-aldosterone system during cardiac surgery with morphine-nitrous oxide anaesthesia. Anesthesiology 42: 538–541

Baracos V, Rodemann H P, Dinarello C A, Goldberg A 1983 Stimulation of muscle protein degradation and prostaglandin E_2 release by leukocytic pyrogen: a mechanism for the increased degradation of muscle proteins during fever. New England Journal of Medicine 308: 553–558

Bent J M, Paterson J L, Mashiter K, Hall G M 1984 Effects of high dose fentanyl anaesthesia on the established metabolic and endocrine responses to surgery. Anaesthesia 39: 19–24

Bessey P Q, Watters J W, Aoki T T, Wilmore P W 1984 Combined hormonal infusion simulates the metabolic response to injury. Annals of Surgery 200: 264–267

Bovill J G, Sebel P S, Tiolet J W, Tauber J L, Philbin D M 1983 The influence of sufentanil on endocrine and metabolic responses to cardiac surgery. Anesthesia and Analgesia 62: 391–395

Brandt M R, Harshin J, Prange Hausen A, Hummer L, Nistup Modsen S, Lugg I, Hellet H 1978 Influence of morphine anaesthesia on the endocrine-metabolic response to open heart surgery. Acta Anaesthetica Scandanavica 22: 400–405

Bromage P R, Camporesi E, Chestnut D 1980 Epidural narcotics for postoperative analgesia. Anaesthesia and Analgesia 59: 473–476

Bromage P R, Shibata H R, Willoughby H W 1971 The influence of prolonged epidural blockade on blood sugar and cortisol responses to operations upon the upper part of the abdomen and thorax. Surgery Gynecology and Obstetrics 132: 1051–1054

Buckingham J M, Palumbo P J, Woods J E 1976 Plasma insulin and glucose levels after renal transplantation. Urology 83: 210–212

Carli F, Clark M M, Woollen J W 1982 Investigation of the relationship between heat loss and nitrogen excretion in elderly patients undergoing major abdominal surgery under general anaesthesia. British Journal of Anaesthesia 54: 1023–1025

Child C S 1984 The effect of labetolol on the cardiovascular, hyperglycaemic and adrenocortical responses to surgery. Anaesthesia 39: 1192–1196

Child C S, Kaufman L 1985 Effect of intrathecal diamorphine on the adrenocortical hyperglycaemic and cardiovascular responses to major colonic surgery. British Journal of Anaesthesia 57: 389–393

Christensen P, Brandt M R, Pen J, Kellet H 1982 Influence of extradural morphine on the adrenocortical and hyperglycaemic response to surgery. British Journal of Anaesthesia 54: 23–27

Clarke R S J 1970 The hyperglycaemic response to different types of surgery. British Journal of Anaesthesia 42: 45–48

Clowes G H A Jr, George B C, Villee C A Jr, Saravis C A 1983 Muscle proteolysis induced by a circulating peptide in patients with sepsis or trauma. New England Journal of Medicine; 308: 545–459

Cooper C E, Nelson D H 1962 ACTH levels in plasma in pre-operative and surgically stressed patients. Journal of Clinical Investigation 41: 1599–1603

Cooper G M, Paterson J L, Mashiter K, Hall G M 1980 Beta adrenergic blockade and the metabolic response to surgery. British Journal of Anaesthesia 52: 1231–1236

Cooper G M, Paterson J L, Ward I D, Hall G M 1981 Fentanyl and the metabolic response to gastric surgery. Anaesthesia 36: 667–671

Cuthbertson D P 1942 Post-shock metabolism. Lancet 1: 433

Dinarello C A 1984 Interleukin-1. Reviews of Infectious Diseases 6: 51–55

Engquvist A, Brandt M R, Fernandes A, Kellet H 1977 The blocking effect of epidural analgesia on the adrenocortical and hyperglycaemic responses to surgery. Acta Anaesthesia Scandanavica 21: 330–335

Ellis F R, Humphrey D E 1982 Clinical aspect of endocrine and metabolic changes relating to anaesthesia and surgery. In: Watkins J, Salo M (Eds) Trauma, Stress and Immunity in Anaesthesia and Surgery. Butterworth Scientific, Guidford

Fellows I W, Bastaw M D, Byrne A J, Allison S P 1983 Adrenocorticol suppression in multiple injured patients: a complication of etomidate infusion. British Medical Journal 287: 1835–1837

Foster K J, Alberti K G M M, Binder C et al 1979 Lipid metabolites and nitrogen balance after abdominal surgery in man. British Journal of Surgery 66: 242–247

Fry D E, Griffiths H 1984 The inhibition by etomidate of the 11 beta hydroylation of cortisol. Clinical Endocrinology 20: 625–628

Fugiwara R 1984 Suppressive influence of surgical stress on the graft versus host reaction in mice. Acta Medica Okayama 38: 439–446

Gelfand R A, Matthews D E, Bier D M, Sherwin R S 1984 Role of counter regulatory hormones in the catabolic response to stress. Journal of Clinical Investigation 74: 2238–2240

George J M, Peier C E, Lavese R R, Rawer J M 1974 Morphine anaesthesia blocks cortisol and growth hormone response to surgical stress in humans. Journal Clinical Endocrinology and Metabolism 38: 736–739

Gignere V, Cate J, Labine F 1981 Characteristics of the alpha adrenergic stimulation of adrenocorticotrophin secretion in rat anterior pituitary cells. Endocrinology 109: 757–762

Grossman A, Besser G M 1982 Opiates control ACTH through nonadrenergic mechanism. Clinical Endocrinology 17: 287–290

Hall G M, Adrian T E, Bloom S R, Mashiter K 1983 The effects of naloxone on the circulating metabolites, gluco-regulatory hormones and gut peptides during pelvic surgery. Clinical Physiology 3: 49–58

Hall G M, Lucke J N 1982 Editorial. Brown fat — a thermogenic tissue of anaesthetic importance? British Journal of Anaesthesia 54: 907

Hall G M, Walsh E S, Paterson J L, Mashiter K 1982 Low dose insulin infusion and substrate mobilisation during surgery. British Journal of Anaesthesia 54: 517–521

Halter J B, Pflug A E, Porte D 1977 Mechanism of plasma catecholamine increases during sugical stress in man. Journal of Clinical Endocrinology and Metabolism 45: 936–939

Haxholdt O, Kehlet H, Dyrberg V 1981 The effect of fentanyl on the cortisol and hypoglycaemic response to abdominal surgery. Acta Anaesthetica Scandanavica 25: 434–436

Hoover-Plow J L, Clifford A J 1978 The effect of surgical trauma on protein turnover in rats. Biochemical Journal 176: 137–140

Hjeindahl P, Eliasson K 1978 Sympatho-adrenal and cardiovascular response to mental stress and orthostatic provocation in latent hypertension. Clinical Science 57: 293–296

Hume D M 1953 The neuroendocrine response to injury — the present status of the problem. Annal of Surgery 138: 548

Imura H, Kato Y, Ikeda M, Morimoto M, Yanvata M 1971 The effect of adrenergic blocking or stimulating agents on plasma growth hormone, immunoreactive insulin and blood free fatty acid levels in man. Journal of Clinical Investigation 50: 1069

Ishibashi M, Yamaji I 1984 Direct effects of catecholamines, thyrotropin releasing hormone and somatostatin on growth hormone and PRL secretion from adenomatous and non-adenomatous human pituitary cells in culture. Journal of Clinical Investigation 73: 66

Jaspan J B, Cohen D M, Karrison T, Tager H S, Rubenstein A H 1984 C-peptide and insulin secretion. Relationship between peripheral concentrations of C-peptide and insulin and their secretion rates in the dog. Journal of Clinical Investigation 74: 1821

De Jong F H, Mallias C 1984 Etomidate suppresses adrenocortical function by the inhibition of 11 beta hydroxylation. Journal of Clinical Endocrinology and Metabolism 59: 1143

Jorgensen B C, Andersen H B, Engquvist A 1982 Influence of epidural morphine on post-operative pain, endocrine-metabolic and renal responses to surgery. A controlled study. Acta Anaesthesica Scandanavica 26: 63

Kay N H, Allen M C, Bullingham R E S et al 1985 The influence of meptazinol on the metabolic and hormonal responses following major surgery; a comparison with morphine. Anaesthesia 40: 223–228

Kehlet H, Brandt M R, Pen J 1980 The role of neurogenic stimuli in mediating the endocrine-metabolic response to surgery. Journal of Parenteral and Enteral Nutrition 4: 152–156

Kitihata L M, Collins J G 1981 Spinal action of narcotic analgesics. Anesthesiology 54: 153

Kumar A, Pontopprdan H, Baraty R A, Laver M B 1974 Inappropriate response to increased plasma ADH during mechanical ventilation in acute respiratory failure. Anesthesiology 40: 215

Ledingham I M, Watt I 1983 Influence of sedation on mortality in critically ill multiple trauma patients. Lancet 1: 1270

Long C N H 1946 Relations of cholesterol and ascorbic acid to secretion of the adrenal corten. Recent Progress in Hormonal Research 1: 99

Lowidsen U, Christensen N, Lyngsoe J 1983 Effects of non-selective and beta selective blockade on glucose metabolism and hormone responses during insulin induced hypoglycaemia in normal man. Journl of Clinal Endocrinology and Metabolism 56: 876

Lush D, Thorpe J N, Richardson D J, Bowen D J 1972 The effect of epidural anaesthesia on the adrenocortical response to surgery. British Journal of Anaesthesia 44: 1169

McDonald R K, Evans F T, Weise V K, Patrick R W 1959 The effect of morphine and nalorphine on plasma hydrocortisone levels in man. Journal of Pharmacology and Experimental Therapy 125: 241–245

Mani F 1984 Depression of host defence mechanisms following cardiac surgery. Japanese Journal of Surgery 14: 377–383

Meyer E J, Lorenzi M, Bohannon N V 1979 Diabetic management by insulin infusion during major surgery. American Journal of Surgery 137: 323

Miller E D 1981 The role of the renin-angiotensin-aldosterone system in circulatory control and in hypertension. British Journal of Anaesthesia 53: 711

Moller I W, Rem J, Brandt M R, Kehlet H 1982 The effect of post-traumatic epidural analgesia on the cortisol and hyperglycaemic response to surgery. Acta Anaesthesica Scandanavica 26: 56–58

Moller I W, Hjouts E, Krantz T, Wardell E, Kehlet H 1984 The modifying effect of spinal anaesthesia on intra and postoperative advenocortical and hyperglycoenic responses to surgery. Acta Anaesthesica Scandinavica 28: 266–270

Moller I W, Krantz T, Wardell E, Kehlet H 1985 The effect of alfentanyl anaesthesia on the adrenocortical and hyperglycaemic response to abdominal surgery. British Journal of Anaesthesia 57: 591–594

Moore R W, Allen M C, Wood P J, Rees L S, Sear J W 1985 Perioperative endocrine effects of etomidate. Anaesthesia 40: 124–31

Moore F D 1953 Bodily changes in surgical convalescence I. The normal sequelae — observations and interpretations. Annals of Surgery 137: 289

Moran W H 1971 CPPB and vasopression secretion: editorial views. Anesthesiology 34: 501

Mullin T J, Kilpatrick J R 1981 The effect of nutritional support on immune competency in patients suffering from trauma, sepsis or malignant disease. Surgery 90: 610–615

Munro H N 1979 Editorial. Hormone and the metabolic response to injury. New England Journal of Medicine 300: 41

Newsome H H 1975 Editorial. Growth hormone in surgical stress. Surgery 77: 475

Nistrup Masden S, Fog-Moller F, Christiansen C, Vester-Andersen T, Engquvist A 1978 Cyclic AMP, adrenaline and noradrenaline in plasma during surgery. British Journal of Surgery 65: 191

Philbin D M, Baratz R A, Patterson R W 1970 The effect of carbon dioxide on plasma antidiuretic hormone levels during intermittent positive pressure breathing. Anesthesiology 33: 345

Philbin D M, Coggins C H 1978 Plasma antidiuretic hormone levels in cardiac surgical patients during morphine and halothane anaesthesia. Anesthesiology 49: 95

Piv D, Whitwam J G, Lok L 1983 Depression of nociceptive sympathetic reflexes by the intrathecal administration of midazolam. British Journal of Anaesthesia 55: 541

Rees L, Butler P W P, Gosling L, Besser G M 1970 Adrenergic blockade and the corticosteroid and growth hormone responses to methylamphetamine. Nature 22: 565

Rem J, Brandt M R, Kehlet H 1980 Prevention of postoperative lymphopenia and granulocytosis by epidural analgesia. Lancet 1: 283

Sambhi M P, Weil M H, Udhoji U N 1965 Acute pharmacodynamic effects of glucocorticoids. Cardia output and related haemodynamic changes in normal subjects and patients inshock. Circulation 31: 523

Sebel P S, Bovill J G, Shellenkus A P M, Hawker C D 1981 The hormonal responses to high dose fentanyl anaesthesia. A study in patients undergoing cardiac surgery. British Journal of Anaesthesia 53: 941–948

Simmons P S, Miles J M, Gerich J E, Haymond M W 1984 Increased proteolysis — an effect of increases in plasma cortisol within the physiological range. Journal of Clinical Investigation 73: 412

Stanley J C 1981a The glucose-fatty acid cycle. British Journal of Anaesthesia 53: 123

Stanley J C 1981b The glucose-fatty acid ketone body cycle. British Journal of Anaesthesia 53: 131

Stanley J C 1981c The regulation of glucose production. British Journal of Anaesthesia 53: 137

Thomasson B 1959 Studies on the content of 17 hydroxyortico steriods and its diurnal rhythm in the plasma of surgical patients. Scandinavian Journal of Clinicaland Laboratory Investigation 11: Supplement 42

Tomas F M, Munro H N, Young V R 1979 Effect of glucocorticoid administration on the rate of muscle protein breakdown in vivo in rats, as measured by urinary excretion of N-methylhistidine Biochemic Journal 178: 139

Traynor C, Paterson J L, Ward I D, Morgan M, Hall G M 1982 Effects of extradural analgesia and vagal blockade on the metabolic and endocrine responses to upper abdominal surgery. British Journal of Anaesthesia 54: 319–323

Tsugi H, Asoh T, Takenchi Y, Shirosaka C 1983 Attenuation of adrenocortical response to upper abdominal surgery with epidural blockade. British Journal of Surgery 70: 122–124

Tsugi H, Shirosaka C, Asoh T, Takenchi Y 1983 Influences of splanchnic nerve blockade on endocrine-metabolic responses to upper abdominal surgery. British Journal of Surgery 70: 437–479

Unger R H, Orci L 1976 Physiology and pathophysiology of glucagon. Physiological Reviews 56: 778

Walsh E S, Patterson J L, O'Riordon J B A, Hall G M 1981 the effect of high dose fentanyl anaesthesia on the metabolic and endocrine response to cardiac surgery. British Journal of Anaesthesia 53: 1153–1165

Walsh E S, Paterson J L, Mashiter K, Hall G M 1982 Effect of phentolamine on the metabolic response to gynaecological surgery. British Journal of Anaesthesia 54: 517

Weichselbaum T E, Margraf H W, Elman R 1953 Quantitative determination of 17 hydroxycorticosteroids and 21 hydroxy 20 ketosteroids in the peripheral blood of humans before and after operative trauma. Federation Proceedings 12: 287

Williamson D H, Farrell R, Kerr A 1977 Muscle protein catabolism after injury in man, as measured by urinary encretion of 3-methylhistidine. Clinical Science Molecular Medicine 52:527

Wilmore D W, Moyaln J A, Bristow B F, Mason A D, Pruitt B A 1974 Anabolic effects of human growth hormone and high caloric feelings following thermal injury. Surgery, Gynecology and Obstetrics 138: 875

Wise L, Mangraf H W, Ballinger W F 1972 Adrenocorticol function in severe burns. Archives of Surery 105: 213–220

Woolfson A M J, Heatley R V, Allison S P 1979 Insulin to inhibit protien catabolism after injury. New England Journal of Medicine 300:14

Yaksh T L, Rudy T A 1976 Analgesia mediated by a direct spiral action of narcotics. Science 192: 1357–1358

Paediatric anaesthesia and intensive care (1)

THE PRE-TERM INFANT

Over the last decade the prognosis for a prematurely born infant (less than 38 weeks' gestation) has continued to improve, whereas 20 years ago very few babies survived with birth weights below 1500 g. These infants survive because of the increased knowledge concerning immaturity of various organ systems. Many pre-term infants require surgery in the early post-natal period for conditions such as necrotizing enterocolitis, inguinal hernia, hydrocephalus and patent ductus arteriosus, and the expertise of the anaesthetist is required to supervise the care of this vulnerable group of patients.

Infant mortality in developed countries is now 7–12 per 1000 live births: since 1970 mortality has declined faster in the neonate compared with the post-neonatal group of infants. A further limited reduction in neonatal mortality is possible but can only be achieved at great expense (Boyle et al 1983, Bloom 1984). The cost of care for a 500–999 g baby may be up to 10 times greater than that for a baby over 1000 g, the costs increasing yearly. Walker et al (1984) analysed the cost-benefit for infants below 1000 g at birth and their subsequent development. 74% were normal or only minimally impaired, 10% were moderately handicapped while 16% were severely handicapped. The cost of care for a 900 g pre-term baby may be questioned on economic grounds. Saigal et al (1984) showed that prognosis is directly proportional to birth weight (Table 12.1).

Table 12.1 Progress of the pre-term infant in relation to birth weight

Birth weight (g)	Mortality (%)
501–600	97
601–700	69
701–800	51
801–900	38
901–1000	28

Neurological damage at birth or soon afterwards is probably the most important cause of handicap and may result, for example, in intraventricular cerebral haemorrhage. Care of pre-term babies involves minimizing the effects of such complications.

An accurate estimation of gestational age is important so that functional immaturity of physiological systems may be assessed in relation to identifying possible complications likely to arise during the course of anaesthesia.

Pulmonary maturity has been assessed by measuring the ratio of lecithin to sphingomyelin (L/S) in the amniotic fluid. An L/S ratio greater than 2 indicates pulmonary maturity with adequate surfactant and this is seen after the 35th week of gestation. Measurement of the surface active acid phospholipids, phosphatidyl inositol (PI) and phosphatidyl glycerol (PG) may predict more accurately the stage of lung development. PI reaches a peak at 35 weeks, but PG only appears at that stage and peaks at 40 weeks and is unaffected by, for example, maternal diabetes (Hallman et al 1977, O'Brien & Cefalo 1980, James et al 1984).

Ultrasonography for antenatal diagnosis of fetal defects is accurate even in the early stages of pregnancy (before 21 weeks) so that termination for anencephaly, for example, is both safe and legal (Gauderer et al 1984). Measurement of α-fetoprotein in maternal serum is used for screening for such neural tube defects (Persson et al 1983).

Congenital diaphragmatic hernia, oesophageal atresia, hydronephrosis, abdominal wall defects and hydrocephalus may also be diagnosed before birth with increasing accuracy as pregnancy proceeds. Such diagnosis allows the baby to be delivered at an optimal time for surgery in a unit in close proximity to, or even in, the regional paediatric centre. Surgery of the fetus is a possibility in some instances where the medical disorder may continue to damage the infant during the remainder of the pregnancy. For example, progressive hydrocephalus accentuates brain damage, while hydronephrosis leads to progressive renal damage.

Spielman et al (1984) described a technique for fetal surgery in utero under local analgesia. The mother was also given diazepam and morphine to reduce fetal movements. The incidence of induction of premature labour is high so β-methasone is given to accelerate fetal pulmonary development. Tocolytic agents such as ritodrine are not given as they are associated with maternal hypertension, dysrhythmia, pulmonary oedema and metabolic disturbances which may be transmitted to the fetus or the newborn if labour proceeds (Hermansen & Johnson 1984).

In fetal life breathing movements may be detected by ultrasonography. There is a diurnal rhythm as well as periods of breathing and periods of apnoea. Fetal breathing is reduced with hypoxia and hypoglycaemia and the pattern is modified with maternal barbiturates or general anaesthesia. It is possible that prolonged periods of fetal apnoea may result in apnoeic problems in neonatal life (Kaplan 1983). In premature labour, before 34 weeks' gestation, if no fetal breathing movements are detected, then delivery occurs within 48 hours, whereas if breathing movements are detected, the pregnancy will continue (Castle & Turnbull 1983).

After birth, clinical examination of the infant for assessment of gestational age is based on a percentile chart for weight at a particular gestational age (Lubchenco 1976) and clinical tests for passive tone as this changes with maturity (Smith & Smith 1982). The condition of a pre-term baby at birth is dependent on many factors such as maternal disease or drug ingestion and the type of delivery, with higher levels of fetal catecholamines after vaginal delivery (Hagnevik et al 1984). β-Blocking drugs taken during pregnancy may cause fetal growth retardation (Finnstrom et al 1984) and work on animals has shown that during maternal halothane anaesthesia fetal lambs suffer a marked fall in blood pressure (Biehl et al 1983).

Sykes et al (1983) have shown the need for continuous fetal heart monitoring together with frequent fetal blood sampling during delivery of pre-term babies. Clinical signs of fetal distress such as bradycardia are not as sensitive as biochemical determinations of fetal acidosis.

The pre-term baby has similar responses to the term baby to acute hypoxia by increasing perfusion to vital organs (Block et al 1984). The pre-term baby also responds to the stress of delivery with graded catecholamine release and though there is an increased adrenaline secretion in the pre-term baby, the catecholamine pattern is similar to the term baby (Newnham et al 1984). Bistoletti et al (1983) showed that surges of catecholamine release occur if the fetal pH falls below 7.25 causing abnormal fetal heart rate patterns. If bradycardia occurs the levels of noradrenaline are even higher, presumably putting the infant at an increased risk of intraventricular cerebral haemorrhage (see below).

PRE-TERM PHYSIOLOGY

The physiology of the newborn baby at term or prematurely born in relation to surgery and anaesthesia is reviewed (Hatch & Sumner 1985) but some aspects of the pre-term baby deserve closer attention. Immaturity of cardiorespiratory and metabolic functions complicates surgery in the peri-operative period, and respiratory function in ex-preterm babies may continue to be impaired during the first year of life, especially if the baby suffered from respiratory distress syndrome after birth (Steward 1982). The diaphragm of pre-term babies is particularly liable to fatigue, depending on the proportion of high oxidative type I muscle fibres present. Adults have 55% of these fibres whereas a full term infant has 25% and a pre-term baby before 30 weeks' gestation only 10% (Keens et al 1978).

Pre-term babies have difficulties in maintaining normal lung volumes, being likely to develop atelectasis with a marked increase in right–left intrapulmonary shunting as closing volume occurs during tidal breathing. The chest wall is very compliant and because the gas exchange units of the pre-term lung have a smaller diameter (75 μm) than the term (150 μm) or adult (250 μm), lung collapse will readily occur (Gregory & Stewart 1983). The high respiratory rate also helps to maintain the functional residual capacity (FRC) and during apnoea the FRC falls to very low levels as more gas has time to escape from the lungs (Gregory 1980). There is less efficient control of ventilation than for the term baby with a tendency to apnoea, together with the possibility of residual lung damage from mechanical ventilation earlier in life. Mild bronchopulmonary dysplasia is common and is manifested by mild tachypnoea with reduced lung compliance and oxygen dependency.

The pre-term cardiovascular system is immature and will not fully react to stress. The fetal myocardium develops less active tension during isometric contraction and only 30% of the myocardial muscle is contractile mass compared to 60% in the adult (Smith & Smith 1982) which allows these infants less capacity to increase stroke volume. For the cardiac output to increase there must be an increase in heart rate so that rates of up to 200/min are necessary to increase the cardiac output. Bradycardia, for example, in response to hypoxia, will cause a great fall in cardiac output. The

E

effect of the sympathetic nervous system, which develops earlier in fetal life than the vagus, predominates in pre-term babies. The carotid sinus baroreceptors respond to acute haemorrhage but the resulting tachycardia only partially compensates for this and the blood pressure will fall.

Patent ductus arteriosus

Patent ductus arteriosus (PDA) is a very frequent finding in pre-term babies as the mechanism for its closure is immature and factors such as a labile pulmonary vascular resistance with transitional circulation, hypoxia and fluid overload tend to maintain its patency. The patent ductus should be routinely screened by echocardiography as no murmur may be present (Fox & Duara 1983, Archer et al 1984).

Neutral thermal environment

The neutral thermal environment for a low birth weight baby may be as high as 36°C and is explained by very high evaporative heat loss in the first week of life. These babies, if subject to a thermal stress, may not react by increasing oxygen consumption but by a change in body temperature. Sauer et al (1984) suggest a better definition for neutral thermal environment — it is the ambient temperature at which the core temperature of the infant at rest is between 36.7°C and 37.3°C and the core and mean skin temperatures are changing less than 0.2°C and 0.3°C per hour respectively. In the first week of life the neutral temperature depends on the gestational age and post-natal age, but after the first week depends more on the increased body weight and the post-natal age. Heat loss from radiation may be reduced by a secondary heat shield within the incubator. Because of the difficulty in maintaining body temperature of very small newborns in the operating theatre, Besag et al (1984) have suggested that surgery for such conditions as necrotizing enterocolitis should take place in an open incubator in the intensive care unit rather than involve a move to the operating theatre. Other advantages of this are decreased handling of the baby, no interruption of vital monitoring and a reduced risk of displacing the tracheal tube and arterial or venous lines.

Retrolental fibroplasia

The failure to eradicate retrolental fibroplasia (RLF) is still unexplained as the causal relationship with supplemental oxygen was suggested as long ago as 1956 (Silverman 1982). The increasing survival of babies below 1500 g birth weight probably accounts for the increase in the number of infants with the cicatricial form. Although Betts et al (1977) suggested that RLF could occur from exposure to excess oxygen during anaesthesia, Merritt et al (1981) pointed out that the condition was likely to have several causative factors. Flynn (1984) has shown, by the use of logistic analysis, that major surgery with anaesthesia did not contribute to the risk of cicatricial RLF but he wondered why the incidence of the condition appeared to be rising despite technological sophistication in the delivery and monitoring of oxygen. Until recently it has proved impossible to use an animal model to induce scarring lesions by over-exposure to oxygen alone. However Bougle et al (1982) showed that in the retina of kittens, hyperoxia led to a fall in the enzyme superoxide dismutase which protected against the toxic effects of oxygen radicals, although this fall may be partially prevented by pretreatment with tocopheral (vitamin E). It was noticed by Phelps & Rosenbaum (1982) that the infants with scarring RLF had lower Pa_{O_2} than those with simple RLF and the proposal that relative hypoxia after hyperoxia leads to scarring may be borne out by animal work. Flower & Blake (1981) have produced cicatricial lesions experimentally in puppies by preventing oxygen induced vasoconstriction with aspirin which blocks the synthesis of prostaglandin so that retinal damage seemed to occur only where there was inadequate vasomotor tone to protect immature vessels. Thus scarring of the retina may occur without exposure to supplemental oxygen, possibly due to increased blood flow and raised transluminal pressure in the development of retinal vasculature.

INTENSIVE THERAPY

Respiratory support continues to be a key part of intensive care for the pre-term baby. Respiratory pathology is the major cause of morbidity and mortality in this age group. High frequency ventilation is established for use in pre-term babies and may have particular value in patients with air leaks — pneumothorax, pneumopericardium or pneumoperitoneum (Pokora et al 1983). Using high frequency jet ventilation the air leak was markedly reduced in 7 out of 9 patients, with reduction in Pa_{CO_2} and A-aDO_2. Gas mixing occurred throughout the tracheobronchial tree so there was no 'dead space' as such, but there was a continuous CO_2 gradient from the alveoli to the periphery. The jet stream with inadequate humidity may cause rapid mucosal damage and tracheal obstruction occurred in 3 out of 6 patients exposed to this type of ventilation for more than 20 hours. Vincent et al (1984) using high frequency ventilation at 6–7 Hz following cardiac surgery in infants, found that it allowed a reduction of peak ventilatory pressure at airway opening by 19% and peak tracheal pressure by 42% with favourable effects on the cardiovascular pressures. Franz et al (1983) showed that adequate gas exchange at low tracheal pressures was possible using high frequency ventilation in premature infants with primary lung disease. The effect on gas exchange may be greater on CO_2 reduction than increasing Pa_{O_2} and Truog et al (1984) found no change in the natural history of experimental hyaline membrane disease in premature primates. High frequency jet ventilation may not be suitable for all types of lung pathology. Mammel et al (1983) found that conventional ventilation was superior for experimental meconium aspiration. Jet ventilation caused an increased A-aDO_2, increased pulmonary vascular resistance and increased pulmonary artery pressure.

Complications of mechanical ventilation which may be life threatening are mediastinal air leaks and bronchopulmonary dysplasia. The use of pancuronium may prevent pneumothorax in very sick pre-term babies breathing against the ventilator (Greenough 1984). However useful relaxants are in this group of pre-term infants, they can produce complications. Runkle & Bancalari (1984) showed that overall oxygenation improved in a group of pre-term infants after the administration of pancuronium, the effect being most marked in those with meconium aspiration. In patients with hyaline membrane disease the Pa_{O_2} actually tended to fall. Pancuronium has cardiovascular side effects with a consistent rise in heart rate and blood pressure which lasted for 30–50 minutes after administration (Cabal et al 1985). Such sympathetic activity may predispose to intraventricular cerebral haemorrhage.

Suctioning is important to maintain the patency of very small endotracheal tubes. It should be used to remove secretions which become retained because of the absence of the normal expulsive mechanisms.

If enteral feeding is possible this should be done by the jejunal route as gastric contents are freely regurgitated in pre-term babies and will be aspirated around the non cuffed tracheal tube, even at the clinical levels of the distending pressure of 4–8 cm H_2O of CPAP (Goodwin et al 1984). If enteral feeding is impossible then total parenteral nutrition should be undertaken using peripheral veins on a rotational basis.

Total parenteral nutrition (TPN) for infants has been reviewed by Hatch & Sumner (1985). The regime is computerized and an aseptic technique is used for preparing the fluids in the hospital pharmacy. A central venous catheter of the Hickman type is implanted at open operation if TPN is required for prolonged periods.

Respiratory distress syndrome

Pulmonary function in respiratory distress syndrome (RDS) progressively deteriorates until a diuresis occurs and the functional residual capacity increases by up to 40% at the time of the maximum diuresis (Heaf et al 1982). The diuresis may represent the removal of excess lung fluid which is necessary for improvement in the respiratory distress syndrome.

Green et al (1983) have shown that if there is no spontaneous diuresis by the 2nd or 3rd day of life then frusemide therapy may improve the survival from RDS, possibly independent of its function as a diuretic. Frusemide does dilate the pulmonary vascular bed and in some cases may maintain the patency of the ductus arteriosus.

Trials of heterologous natural surfactants are encouraging (Gitling et al 1984) and the idea of replacing surfactant is an attractive one. However, the mortality is low with conventional supportive therapy and no further treatment seems necessary. There is evidence that a patent ductus arteriosus which requires closure may be a complication of artificial surfactant therapy.

Monitoring

New techniques for non-invasive monitoring of pre-term babies both in the operating theatre and intensive care unit continue to be developed. Direct monitoring, although invaluable, may be limited by difficulties with vascular access particularly over a long period of intensive care. Transcutaneous oxygen electrodes ($TcPO_2$) have been adopted by all workers in this field, although the use of carbon dioxide electrodes is limited by leakage of air which nullifies the readings. Modern pulse oximetry may prove to be a valuable alternative to $TcPO_2$ (Yelderman & New 1983).

A new non-invasive technique for the monitoring of cerebral oxygenation in pre-term babies has been described by Brazy et al (1985). The technique relies on transilluminating the anterior cerebral field with near infra-red light which undergoes differential absorption with changes in oxidation and reduction levels of cytochrome aa$_3$ in haemoglobin of tissue blood. The levels correlate well with transcutaneous PO$_2$ readings and show that even a brief bradycardia causes a precipitate fall in reduction of cytochrome aa$_3$.

REFERENCES
See end of chapter 13.

Paediatric anaesthesia and intensive care (2)

INTRAVENTRICULAR HAEMORRHAGE

Intraventricular haemorrhage (IVH) is the commonest serious neurological event of the neonatal period, affecting approximately 40–50% of all pre-term infants under 35 weeks' gestational age (Volpe 1981). This incidence increases as gestational age decreases and an incidence of 61% has been reported in infants born before 33 weeks (Ment et al 1984). The lesions are less severe in infants born in a perinatal unit (Clark et al 1981, Dolfin et al 1983) and it has been suggested that 'at-risk' infants should only be delivered in these special units. Pre-natal events, particularly the duration and course of labour may also be contributory factors (Meidell et al 1985): the incidence of IVH is apparently higher after vaginal delivery than Caesarean section (Greisen & Petersen 1983).

As may be expected from its prevalence in pre-term infants, there is a high association with the respiratory distress syndrome (RDS) and IVH is the commonest neuropathological finding in infants dying with this disorder (Leech et al 1979). Other associated factors include hypoxia, hypercarbia, acidosis, high ventilatory requirements and pneumothoraces (Lipscombe et al 1981, Levene et al 1982), although these must in part reflect the severity of the RDS. Severe acidosis appears to be an independently acting contributory factor. Sudden rises in blood pressure, particularly after a period of relatively low pressure, may also lead to IVH (Fujimura et al 1979).

The study of IVH, which usually occurs within 72 hours of birth, has been greatly helped in recent years by the introduction of computerized axial tomography and especially by portable real time ultrasonography. The initial lesion is bleeding into the highly vascular periventricular germinal matrix, typically over the head of the caudate nucleus (Leech & Kohnen 1974). In the pre-term infants of less than 28 weeks' gestational age, the vulnerable site is over the body of the caudate nucleus (Hambleton & Wigglesworth 1976). With increasing maturity the germinal matrix disappears and there is a more pronounced arterialization of the cerebral cortex and adjacent white matter. Haemorrhage in near term infants is thus more likely to occur from the choroid plexus (Donat et al 1978).

Although numerous theories have been proposed for the aetiology of IVH, several recent observations suggest an arterial mechanism. Pape & Wigglesworth (1979) have shown the periventricular capillary bed to be essentially an immature vascular rete.

These vessels cannot be categorized on light microscopy, and consist only of a layer of endothelium without smooth muscle, collagen or fibrin (Haruda & Blanc 1981). The metabolic activity of brain capillary endothelial cells suggests that they are particularly dependent on oxidative mechanisms (Goldstein 1979) and may easily be damaged by hypoxic events. Lou et al (1979) and Milligan (1980) have demonstrated a failure of autoregulation of cerebral blood flow in distressed pre-term infants and this may be of more importance. These infants are often hypotensive and hypoxic with maximal cerebral vasodilatation and it has been postulated that any rise in cerebral blood flow, particularly if sudden, may rupture the fragile, unsupported vascular matrix. This may occur with hypercarbia or acidosis, or following therapeutic measures such as the rapid infusion of blood volume expanders (Goldberg et al 1980) or the use of hyperosmolar solutions such as glucose or sodium bicarbonate (Papile et al 1978b). Events that raise systemic blood pressure may also be detrimental. Endotracheal suctioning for example has been shown to produce a marked systemic pressor response with a prominent and consistent rise in anterior cerebral artery blood flow (Perlman & Volpe 1983). This must raise doubts about the safety of awake intubation in sick pre-term infants.

Anterior cerebral blood flow patterns in pre-term infants ventilated in the treatment of severe RDS have been shown to be either fluctuating or stable (Perlman et al 1983a). 21 out of 23 infants with fluctuating cerebral flow developed IVH, while only seven out of 27 infants with a stable flow pattern did so. The flow pattern reflected the simultaneously recorded blood pressure and correlated with the severity of the RDS. Mean peak airway pressure was higher in the IVH group. Paralysing the infants resulted in the fluctuating flow pattern becoming stable. The presence of a patent ductus arteriousus has also been shown to produce abnormal cerebral blood flow patterns (Martin et al 1982) which must predispose to IVH, particularly as there is often an acute rise in blood pressure as the ductus is ligated (Marshall et al 1982).

It is uncertain how the presence of a pneumothorax predisposes to IVH. It may be due to a moderate rise in diastolic blood pressure due to an increase in peripheral resistance (Hill et al 1982). However, it has recently been shown in newborn dogs that the rapid evacuation of a pneumothorax leads to a precipitate rise in blood pressure and cerebral blood flow to supranormal levels (Batton et al 1984). Periventricular capillary pressure may also be raised when venous pressure is raised, for example, during positive pressure ventilation.

Abnormalities of coagulation have also been implicated in the pathogenesis of newborn IVH. McDonald et al (1984) have recently documented a significant association between hypocoagulability in the first few hours of life in pre-term infants with subsequent onset of progression of IVH. It is not yet known whether early and aggressive correction of this coagulopathy will reduce the morbidity and mortality of IVH.

There are numerous ways of classifying IVH, but that proposed by Papile et al (1978a) is usually used. In mild cases (grade 1), the bleeding is confined to the periventricular germinal matrix. Frequently however this periventricular haemorrhage (PHV) ruptures through the ependyma into the ventricles producing the more severe IVH (grade 2). In grade 3 IVH, there is associated ventricular dilatation, and in the severe lesions there appears to be extension into the cerebral parenchyma (grade 4). Recent studies would suggest however that this intraparenchymal lesion is a concomitant haemorrhagic infarction rather than a simple extension of the germinal matrix bleed (Flodmark et al 1980).

Treatment

Attempts to reduce the incidence of IVH have included the prophylactic use of drugs such as phenobarbitone which prevents sharp increases in blood pressure (Wimberley et al 1982). Ninan et al (1985) reported that intravenous phenobarbitone (10 mg/kg) reduced the increases in heart rate, blood pressure and intracranial pressure and the fall in oxygenation following endotracheal suctioning in pre-term infants. However although the lesions were less severe, the drug had no effect on the incidence of IVH (Bedard et al 1984).

A significant reduction in IVH following the use of tocopheral (vitamin E) has also been reported (Speer et al 1984). This agent, an antioxidant, may protect endothelial cell membranes thereby limiting the magnitude of haemorrhage in the germinal matrix. Unfortunately its use is associated with an increased incidence of necrotizing entercolitis (NEC) and generalized sepsis, presumably by interfering with oxygen dependent antimicrobial defences (Johnson et al 1985). This side effect does not occur if blood levels are kept below 3.5 mg/dl.

Ethamsylate, a capillary stablizing drug which has been used to limit bleeding in a number of surgical procedures, may also be useful (Cooke & Morgan 1984) but further studies are necessary.

Progressive haemorrhage increases morbidity and mortality and it is therefore important to prevent or limit the extend of IVH. Thorburn et al (1981) reported a mortality of over 50% in severe bleeding with a third of the survivors having a major neurodevelopmental handicap. Williamson et al (1983) found that 50% of the survivors at the age of 3 years had moderate or severe neurological abnormalities and 52% were intellectually impaired. The poor outcome would appear to be directly related to the magnitude of the underlying haemorrhage (Papile et al 1983) although there may be some contribution from the chronic ventricular dilatation that usually occurs after IVH. This has been attributed to an obliterative arachnoiditis (Larroche 1972) or less often to obstruction of the flow of cerebrospinal fluid (CSF) by necrotic debris within the ventricles at the aqueduct of Sylvius (Hill et al 1984). Ventricular dilatation is often managed conservatively and only rarely does hydrocephalus lead to increased intracranial pressure which requires a shunt (Allan et al 1984).

Intraventricular haemorrhage (IVH) is not the only problem associated with failure of autoregulation of cerebral blood flow in pre-term infants. Intracerebellar haemorrhage may occur in 15–22% of patients, although this is more difficult to detect (Perlman et al 1983b) and is presumably also due to hyperperfusion. Hypoperfusion of boundary zones between different arterial territories which may occur during hypotensive episodes, may cause infarction within the periventricular white matter — periventricular leukomalacia (PVL). This can lead to cystic degeneration (Levene et al 1983) often affecting the pyramidal tracts as they pass from the motor cortex through the internal capsule. The fibres affecting distribution to the leg, and which are closer to the ventricle, are more likely to be damaged resulting in spastic diplegia which is a significant handicap.

It is important therefore to ensure that meticulous attention is paid to these patients at risk, avoiding as far as possible fluctuations in blood pressure and cerebral blood flow. Hypotension should be avoided by early blood transfusion, care being taken to avoid the dangers of rapid transfusion.

APNOEA

Apnoeic episodes occur frequently in pre-term infants and, although incompletely understood, the underlying mechanisms are becoming clearer. Apnoea must be distinguished from the periodic respiration which is commonly seen in all newborn infants. This is associated with ventilatory pauses of less than 10 seconds, without bradycardia or cyanosis, and is of no pathological significance. True apnoea is usually defined as cessation of breathing for 20 seconds or more (Rigatto 1982) and is often accompanied by bradycardia and a fall in oxygenation. The incidence and severity of apnoea increase with decreasing gestational age, so prematurity is one of the main predisposing factors. In many cases the episodes can be attributed to insults such as infection, metabolic disturbances, or IVH, but in many cases the aetiology is unknown and is attributed to immaturity of the brain stem centres. Early studies suggest that apnoea was precipitated by hypoxia but this has been discounted by Gerhardt & Bancalari (1984) who found no difference in oxygenation and pulmonary mechanisms between pre-term infants with or without apnoea. They demonstrated significantly decreased inspiratory effort in the infants with apnoea, with alveolar hypoventilation and hypercapnia. This evidence of decreased respiratory centre activity was confirmed by a CO_2 response curve which is flattened and shifted to the right. CNS immaturity is also reflected by the decreased number of synaptic connections, incomplete dendritic arborization and the poor myelination seen in the pre-term brain (Schulte 1977). This results in decreased afferent traffic to, and a decreased state of excitation of, the reticular formation and respiratory centre. This is supported by the clinical observation that various external stimuli can increase respiratory activity and reduce the incidence of apnoea. Furthermore, auditory evoked brain stem conduction time is prolonged in pre-term infants and shortens with advancing gestational age, due to improving synaptic efficiency and myelination (Henderson-Smart et al 1983). This prolongation of auditory evoked brain stem conduction time correlates strongly with apnoea and confirms that brain stem immaturity underlies apnoea of prematurity.

Some of the repiratory reflexes are also immature in the pre-term infant. For example, the Hering-Breuer inflation reflex is poorly developed at 28–35 weeks' gestation and does not reach maximal strength until term (Gerhardt & Bancalari 1981). This reflects a decreased vagal input which would normally occur with each breath and contributes to the establishment of regular breathing. The intercostal phrenic inhibitory reflex, which may terminate inspiratory effort when distortions of the chest wall occur (Knill & Bryan 1976), is particularly active in pre-term infants (Thach et al 1978). Chest wall distortion in newborns is reportedly increased during REM sleep, possibly due to central inhibition of intercostal muscle tone (Schulte et al 1977). This may explain the increase of apnoeic episodes seen during REM sleep (Steinschneider & Weinstein 1983). The amount of REM sleep falls with increasing age as does the incidence of apnoea.

Pre-term infants are at a further disadvantage if subjected to increased respiratory loads, such as obstruction, as much of the extra effort is dissipated in deformation of their soft and compliant chest wall. Apnoea thus readily follows airway obstruction which is common in these infants. This is frequently a pharyngeal problem (Mathew et al 1982) and may be partly due to reduced genioglossus muscle tone (Dransfield et al 1983). CPAP is well established treatment for apnoea but a recent paper suggests that it selectively reduces apnoea with an obstructive component, possibly by splinting the pharyngeal airway (Miller et al 1985). It may also act by stabilizing the chest wall thereby eliminating the intercostal phrenic inhibitory reflex. CPAP does not appreciably alter the ventilatory response to CO_2 (Durand et al 1983) and since hypoxia is not a cause of apnoea it presumably does not act by improving oxygenation.

Aminophylline is widely used in the treatment of idiopathic apnoea and is more effective than CPAP (Jones 1982). The mechanism is probably through inhibition of phosphodiesterase with increased levels of cyclic AMP. Cyclic AMP plays an important role in the action of a variety of neurotransmitters, and is also increased by catecholamines. Aminophylline may therefore compensate for the deficiency of catecholamines seen in pre-term infants with apnoea (Kattwinkel et al 1976). There is an increase in the sensitivity of the respiratory centre to CO_2 (the slope is increased and shifted to the left) with increased minute ventilation and decreased arterial CO_2 (Gerhardt et al 1979). Aminophylline also increases the strength of the Hering-Breuer reflex indicating an increase in vagal afferents from pulmonary stretch receptors. It is not clear whether this increased central stimulation is due to increased sensitivity of peripheral receptors or a facilitation of afferent signals or both. The ability to compensate for increased loads in pre-term infants is also increased (Gerhardt et al 1983).

Pre-term infants with a history of apnoea are more likely than full term infants to have apnoeic episodes following anaesthesia (Steward 1982). Indeed in one study, all prematurely born infants with a history of apnoea operated on before 41 weeks' gestational age required post-operative respiratory support (Liu et al 1983). This was irrespective of surgical procedure or anaesthetic technique. All infants with a history of apnoea, particularly those of less than 46 weeks' gestational age, should therefore be closely monitored for 24 hours post-operatively.

SUDDEN INFANT DEATH SYNDROME

The sudden infant death syndrome (SIDS) occurs more frequently in premature infants. The condition is a distinct clinical entity and is diagnosed by unexpected death from which the post-mortem examination fails to reveal an adequate cause for death and accounts for approximately 1500 deaths per year in the United Kingdom. In addition there is a group of infants who suffer a severe episode of apnoea and cyanosis, the 'near miss' SIDS, some of whom may die subsequently. This group of infants has been the subject of considerable physiological study as it was assumed that the 'near miss' is related to SIDS, because its peak age distribution was slightly earlier than that of SIDS and because of the sudden subsequent death of many 'near miss' patients.

Although the syndrome may be multifactorial, current research focuses on possible defective control of ventilation in these infants (Leading Article 1984) and abnormal ventilatory responses have been detected in the parents of SIDS babies (Schiffman et al 1980). There is an increased frequency of apnoeic pauses in a group of infants some of whom were SIDS victims later (Steinschneider et al 1982). Jeffery et al (1983) investigated 58 infants with 'near miss' SIDS and 6 twins of SIDS victims and found that some had idiopathic apnoea, minimal tracheomalacia, aberrant innominate arteries or temporal lobe seizures, but 55 had gastro-oesophageal reflux.

It has been shown that reflux precedes apnoea and that this could be induced by instilling dilute acid (not water or milk) on to the lower oesophagus (Herbst et al 1979). Simultaneous tracings of oesophageal pH, heart rate, impedence pneumography and nasal airflow showed that apnoea did not settle until surgical or medical treatment for reflux was undertaken. Reflux appeared to be a common factor in very many 'near miss' SIDS infants. Reflux may not be diagnosed on barium swallow, but requires the use of a milk scan (radioisotope gastroesophageal scintingraphy) (MacFayden et al 1983).

The phase of active sleep may be associated with depressed reflexes and therefore a vulnerability to asphyxia in infants (Jeffery et al 1980). The response to mild respiratory obstruction or liquid in the pharynx may be reflex apnoea but no arousal occurs because the ventilatory responses to CO_2 retention and hypoxia are also depressed. This forms the basis of the sleep apnoea theory for SIDS, but there may be also a further defect in the neurochemical control of respiration which allows non-arousal to occur during the sleep apnoea.

Recent investigation has concerned chemoreceptor activity in the carotid body which triggers acute responses to hypoxia. Perrin et al (1984) found that the carotid bodies of infants with SIDS had increased catecholamine levels including a ten-fold increase in dopamine when compared with infants dying from other defined causes.

Olson et al (1982) found that in adults, dopamine infusion may impair the ventilatory response to hypercapnia and hypoxia. It is thus possible that the high concentration of dopamine within the carotid bodies hampers the response of the infant to hypoxia and it is thus possible that there is a neurochemical cause of SIDS. An 'at-risk' infant could become severely asphyxiated during active sleep from very many minor problems such as upper respiratory tract infection, nasal obstruction or gastro-oesophageal reflux.

The prevention of SIDS is a controversial topic. Carpenter et al (1983) have shown that keeping a close watch at home on babies at risk will reduce mortality. They developed a scoring system which depends on maternal age, previous pregnancies, length of labour and gestational age. There is, however, a continuing mortality from SIDS. Home monitoring of apnoea and/or ECG, has been undertaken by Duffty et al (1982) in babies who have suffered 'near miss' SIDS and in siblings of SIDS victims. They used apnoea alarms of the transducer pad or chest impedence type or heart rate monitors for at least 2 months. The monitors were set to alarm after 20 seconds' apnoea or at a heart rate below 80. Many episodes of apnoea and bradycardia were detected, some being multiple and some serious. At least one patient later succumbed to SIDS. Simpson (1983) has reviewed the difficulties of home monitoring. The presence of the monitor at home may induce parental neurosis, although in practice the machines are well accepted. The selection of the patients may be difficult and although siblings of SIDS victims and those with 'near miss' are included, those with low birth weight from special care baby units, with bronchopulmonary dysplasia and those from narcotic dependent mothers should also be included. This would make for a large group requiring expensive machines. There is also concern about the efficacy of the monitors. Apnoea alarms are unreliable and many only detect central but not obstructive apnoea and breathing movements may continue until the moment of death.

There is a great deal of knowledge concerning the aetiology of this tragic cause of infant death, but its prevention still remains to be elucidated.

REFERENCES

Allan W C, Dransfield D A, Tito A M 1984 Ventricular dilation following periventricular-intraventricular hemorrhage: outcome at age 1 year. Pediatrics 73: 158–162

Archer L N J, Glass E J, Godman M J 1984 Silent ductus arteriosus in idiopathic respiratory distress syndrome. Acta Paediatrica Scandinavica 73: 652–656

Ariagno R L, Guilleminault C, Korobkin R, Owenboeddiker M, Baldwin R 1983 Near miss for sudden infant death syndrome infants — a clinical problem. Pediatrics 71: 726–736

Batton D G, Hellman J, Nardis E E 1984 Effect of pneumothorax-induced systemic blood pressure alterations on the cerebral circulation in newborn dogs. Pediatrics 74: 350–353

Bedard M P, Shankaran S, Slovis T L, Pantoja A, Dayal B, Poland R 1984 Effect of prophylactic phenobarbital on intraventricular hemorrhage in high-risk infants. Pediatrics 73: 435–439

Besag F M C, Singh M P, Whitelaw A G L 1984 Surgery of the ill, extremely low birth weight infant: should transfer to the operating theatre be avoided? Acta Paediatrica Scandinavica 73: 594–595

Betts E K, Downes J J, Schaffer D B, Johns R 1977 Retrolental fibroplasia and oxygen administration during general anesthesia. Anesthesiology 47: 518–520

Biehl D R, Côté J, Wade J G, Gregory G A, Sitar D 1983 Uptake of halothane by the fetal lamb in utero. Canadian Anaesthetists' Society Journal 30: 24–27

Bistoletti P, Nylund L, Lagercrantz H, Hjemdahl P, Ström H 1983 Fetal scalp catecholamines during labor. American Journal of Obstetrics and Gynecology 147: 785–788

Block B S B, Llanos A J, Creasy R K 1984 Responses of the growth retarded fetus to acute hypoxemia. American Journal of Obstetrics and Gynecolgoy 148: 878–881

Bloom B S 1984 Changing infant mortality: the need to spend more while getting less. Pediatrics 73: 862–866

Bougle D, Vert P, Reichart E, Hartemann D Heng E L 1982 Retinal superoxide dismutase activity in new born kittens exposed to normobaric hyperoxia: effect of vitamin E. Pediatric Research 16: 400–402

Boyle M H, Torrance G W, Sinclair J C, Horwood S P 1983 Economic evaluation of neonatal intensive care of very low birth weight infants. New England Journal of Medicine 308: 1330–1337

Brazy J E, Lewis D V, Mitnick M H, Van der Vliet F J 1985 Non-invasive monitoring of cerebral oxygenation in preterm infants: preliminary observations. Pediatrics 75: 217–225

Cabal L A, Siassi B, Artal R, Gonzalez F, Hodgman J, Plajsteck C 1985 Cardiovascular and catecholamine changes after administration of pancuronium in distressed neonates. Pediatrics 75: 284–287

Carpenter R G, Gardner A, Jepson M, Taylor E M, Salvin A, Sunderland R, Emery J L et al 1983 Prevention of unexpected infant death. Lancet i: 723–727

Castle B M, Turnbull A C 1983 The presence or absence of fetal breathing movements predicts the outcome of preterm labour. Lancet 2: 471–472

Clark C E, Clyman R I, Roth R S, Sniderman S H, Lane B, Ballard R A 1981 Risk factor analysis of intraventricular hemorrhage in low-birth-weight infants. Journal of Pediatrics 99: 625–628

Cooke R W I, Morgan M E I 1984 Prophylactic ethamsylate for periventricular haemorrhage. Archives of Disease in Childhood 59: 82–83

Dolfin T, Skidmore M B, Fong K W, Hoskins E M, Shennan A J 1983 Incidence, severity, and timing of subependymal and intraventricular hemorrhages in preterm infants born in a perinatal unit as detected by serial real-time ultrasound. Pediatrics 71: 541–546

Donat J F, Okazaki H, Kleinburg H, Reagan T J 1978 Intraventricular hemorrhages in full-term and premature infants. Mayo Clinic Proceedings 53: 437–441

Dransfield D A, Spitzer A R, Fox W W 1983 Episodic airway obstruction in premature infants. American Journal of Diseases of Children 137: 441–443

Duffty P, Bryan M H 1982 Home apnoea monitoring in 'near miss' sudden infant death syndrome (SIDS) and in siblings of SIDS victims. Pediatrics 70: 69–74

Durand M, McCann E, Brady J P 1983 Effect of continuous positive airway pressure on the ventilatory response to CO_2 in preterm infants. Pediatrics 71: 634–638

Finnstrom O, Ezitis J, Ryden G, Wichman K 1984 Neonatal effects of beta blocking drugs in pregnancy. Acta Obstetrica et Gynecologica Scandinavica (suppl.) 118: 91–93

Flodmark O, Becker L E, Harwood-Nash D C, Fitzhardinge P M, Fitz C R, Chuang S H 1980 Correlation between computed tomography and autopsy in premature and full-term neonates that have suffered perinatal asphyxia. Radiology 137: 93–103

Flower R W, Blake D A 1981 Retrolental fibroplasia: evidence for a role of the prostaglandin cascade in the pathogenesis of oxygen-induced retinopathy in the newborn beagle. Pediatric Research 15: 1293–1302

Flynn J T 1984 Oxygen and retrolental fibroplasia: update and challenge. Anesthesiology 60: 397–399

Fox W W, Duara S 1983 Persistent pulmonary hypertension in the neonate: diagnosis and management. Journal of Pediatrics 103: 505–514

Frantz I D, Werthammer J, Stark A R 1983 High frequency ventilation in premature infants with lung disease: adequate gas exchange at low tracheal pressure. Pediatrics 71: 483–488

Fujimura M, Salisbury D M, Robinson R O, Howat P, Emerson P M, Keeling J W, Tizard J P M 1979 Clinical events relating to intraventricular haemorrhage in the newborn. Archives of Disease in Childhoo 54: 409–414

Gauderer M W L, Jassani M N, Izant R J 1984 Ultrasonographic antenatal diagnosis: will it change the spectrum of neonatal surgery? Journal of Pediatric Surgery 19: 404–407

Gerhardt T, Bancalari E 1981 Maturational changes of reflexes influencing inspiratory timing in newborns. Journal of Applied Physiology 50: 1282–1285

Gerhardt T, Bancalari E 1984 Apnea of prematurity: 1. Lung function and regulation of breathing. Pediatrics 74: 58–62

Gerhardt T, McCarthy J, Bancalari E 1979 Effects of aminophylline on respiratory center activity and metabolic rate in premature infants with idiopathic apnea. Pediatrics 63: 537–542

Gerhardt T, MCarthy J, Bancalari E 1983 Effects of aminophylline on respiratory center and reflex activity in premature infants with apnea. Pediatric Research 17: 188–191

Gitlin J D, Parad R, Taeusch H W 1984 Exogenous surfactant therapy in hyaline membrane disease. Seminars in Perinatology 8: 272–282

Goldberg R N, Chung D, Goldman S L, Bancalari E 1980 The association of rapid volume expansion and intraventricular hemorrhage in the preterm infant. Journal of Pediatrics 96: 1060–1063

Goldstein G W 1979 Pathogenesis of brain edema and hemorrhage: role of the brain capillary. Pediatrics 64: 357–360

Goodwin S R, Graves S A, Haberkern C M 1984 Aspiration in intubated premature infants. Pediatrics 75: 85–88

Green T P, Thompson T R, Johnson D E, Lock J E 1983 Diuresis and pulmonary function in premature infants with respiratory distress syndrome. Journal of Pediatrics 103: 618–623

Greenough A, Wood S, Morley C J, Davis J A 1984 Pancuronium prevents pneumothoraces in ventilated premature babies who actively expire against positive pressure inflation. Lancet 1: 1–3

Gregory G A 1980 Respiratory care of the child. Critical Care Medicine 8: 582–587

Gregory G A, Steward D J 1983 Life threatening perioperative apnea in the ex-'premie'. Anesthesiology 59: 495–498

Greisen G, Petersen M B 1983 Intraventricular hemorrhage and method of delivery of very low birth weight infants. Journal of Perinatal Medicine 11: 495–498 67–73

Hagnevik K, Faxelius G, Irestadt L, Lagercrantz H, Lundell B, Persson B 1984 Catecholamine surge and metabolic adaptation in the newborn after vaginal delivery and caesarian section. Acta Paediatrica Scandinavica 73: 602–609

Hallman M, Feldman B H, Kirkpatrick E, Gluck L 1977 Absence of phosphatidylglycerol (PG) in respiratory distress syndrome in the newborn. Pediatric Research 11: 714–720

Hambleton G, Wigglesworth J S 1976 Origin of intraventricular haemorrhage in the preterm infant. Archives of Disease in Childhood 51: 651–659

Haruda F, Blanc W A 1981 The structure of intracerebral arteries in premature infants and the autoregulation of cerebral blood flow. Annals of Neurology 10: 303

Hatch D J, Sumner E 1985 Neonatal anaesthesia and perioperative care. Arnold, London

Heaf D P, Belik J, Spitzer A R, Gewitz M H, Fox W W 1982 Changes in pulmonary function during diuretic phase of respiratory distress syndrome. Journal of Pediatrics 101: 103–107

Henderson-Smart D J, Pettigrew A G, Campbell D J 1983 Clinical apnea and brain stem neural function in preterm infants. New England Journal of Medicine 308: 353–357

Herbst J J, Minton S D, Book L S 1979 Gastro esophageal reflux causing respiratory arrest and apnea in newborn infants. Journal of Pediatrics 95: 763–768

Hermansen M C, Johnson G L 1984 Neonatal supraventricular tachycardia following prolonged maternal ritodrine adminstration. American Journal of Obstetrics and Gynecology 149: 798–799

Hill A, Perlman J M, Volpe J J 1982 Relationship of pneumothorax to occurrence of intraventricular hemorrhage in the premature newborn. Pediatrics 69: 144–149

Hill A, Shackleford G D, Volpe J J 1984 A potential mechanism of pathogenesis for early posthemorrhagic hydrocephalus in the premature newborn. Pediatrics 73: 19–21

James D K, Chiswick M L, Harkes A, William S M, Tindall V R 1984 Maternal diabetes and neonatal respiratory distress. Maturation of fetal surfactant. British Journal of Obstetrics and Gynaecology 91: 316–324

Jeffery H E, Reid I, Rahilly P, Read D J C 1980 Gastro-oesophageal reflux in 'near miss' sudden infant death infants in active but not quiet sleep. Sleep 3: 393–399

Jeffery H E, Rahilly P, Read D J C 1983 Multiple cause of asphyxia in infants at high risk of sudden infant death. Archives of Disease in Childhood 58: 92–100

Johnson L, Bowen F W, Abbasi S, Herrmann N, Weston M, Sacks L, et al 1985 Relationship of prolonged pharmacologic serum levels of vitamin E to incidence of sepsis and necrotising enterocolitis in infants with birthweight 1500 grams or less. Pediatrics 75: 619–638

Jones R A K 1982 Apnoea of prematurity. A controlled trial of theophylline and facemask continuous positive airways pressure. Archives of Disease in Childhood 57: 761–765

Kaplan M 1983 Fetal breathing movements. American Journal of Diseases of Children 137: 177–181

Kattwinkel J, Mars T T, Fanaroff A A, Klaus M H 1976 Urinary biogenic amines in idiopathic apnea of prematurity. Journal of Pediatrics 88: 1003–1006

Keens T G, Bryan A L, Levison H, Ianuzzo D C 1978 Development pattern of muscle fiber types in human ventilatory muscles. Journal of Applied Physiology 44: 909–913

Kiely E 1984 One hundred consecutive central venous catheters in children. Zeitschrift Für Kinderchirurgie 39: 332–336

Knill R, Bryan A C 1976 An intercostal-phrenic inhibitory reflex in human newborn infants. Journal of Applied Physiology 40: 352–356

Larroche J C 1972 Post-haemorrhagic hydrocephalus in infancy: anatomical study. Biology of the Neonate 20: 287–299

Leader 1984 Ventilatory dysfunction and sudden death syndrome. Lancet ii: 558–559

Leech R W, Kohnen P 1974 Subependymal and intraventricular hemorrhages in the newborn. American Journal of Pathology 77: 465–475

Leech R W, Olsson M I, Alvord E C 1979 Neuropathologic features of idiopathic respiratory distress syndrome. Archives of Pathology and Laboratory Medicine 103: 341–343

Levene M I, Fawer C-L, Lamont R F 1982 Risk factors in the development of intraventricular haemorrhage in the preterm neonate. Archives of Disease in Childhood 57: 410–417

Levene M I, Wigglesworth J, Dubowitz V 1983 Hemorrhagic periventricular leucomalacia in the neonate: a real-time ultrasound study. Pediatrics 71: 794–797

Lipscombe A P, Thorburn R J, Reynolds E O R, Stewart A L, Blackwell R J, Cusick G, Whitehead M D 1981 Pneumothorax and cerebral haemorrhage in preterm infants. Lancet i: 414–416

Liu L M P, Cote C J, Goudsouzian N G, Ryan J F, Firestone S, Dedrick D F et al 1983 Life threatening apnea in infants recovering from anesthesia. Anesthesiology 59: 506–510

Lou H C, Lassen N A, Friis-Hansen B 1979 Impaired autoregulation of cerebral blood flow in the distressed newborn. Journal of Pediatrics 90: 119–121

Lubchenco L O 1976 The high risk infant. Saunders, Philadelphia

MacFadyen U M, Hendry G M A, Simpson H 1983 Gastro-oesophageal reflux in near miss sudden infant death syndrome or suspected recurrent aspiration. Archives of Disease in Childhood 58: 87–89

Mammel M C, Gordon M J, Connett J E, Boros S J 1983 Comparison of high frequency jet ventilation and conventional mechanical ventilation in a meconium aspiration model. Journal of Pediatrics 103: 630–634

Marshall T A, Marshall F, Reddy P P 1982 Physiologic changes associated with ligation of the PDA in preterm infants. Journal of Pediatrics 101: 749–753

Martin C G, Snider R, Katz S M, Peabody J L, Brody J P 1982 Abnormal cerebral blood flow patterns in preterm infants with a large PDA. Journal of Pediatrics 101: 587–593

Mathew O P, Roberts J L, Thach B T 1982 Pharyngeal airway obstruction in preterm infants during mixed and obstructive apnea. Journal of Pediatrics 100: 964–968

McDonald M M, Johnson M L, Rumack C M, Koops B L, Guggenheim M A, Babb C, Hathaway W E 1984 Role of coagulopathy in newborn intracranial hemorrhage. Pediatrics 74: 26–31

Meidell R, Marinelli P, Pettet G 1985 Perinatal factors associated with early-onset intracranial hemorrhage in premature infants. American Journal of Diseases of Children 139: 160–163

Ment L R, Duncan C C, Ehrenkranz R A et al 1984 Intraventricular hemorrhage in the preterm neonate: timing and cerebral blood flow changes. Journal of Pediatrics 104: 419–425

Merritt J C, Sprague D N, Merritt W E, Ellis R A 1981 RLF : A multifactorial disease. Anesthesia and Analgesia 60: 109–111

Miller M J, Carlo W A, Martin R J 1985 Continuous positive airway pressure selectively reduces obstructive apnea in preterm infants. Journal of Pediatrics 106: 91–94

Milligan D W A 1980 Failure of autoregulation and intraventricular haemorrhage in preterm infants. Lancet i: 896–898

Newnham J P, Marshall C L, Padbury J F, Lam R W, Hobel C J, Fisher D A 1984 Fetal catecholamine releases with preterm delivery. American Journal of Obstetrics and Gynecology 149: 888–893

Ninan A, O'Donnell M, Hamilton K, Sankaran K 1985 Physiological changes induced by endotracheal instillation and suctioning in critically ill preterm infants with and without sedation. Pediatric Research 19: 355A (No. 1468)

O'Brien W F and Cafalo R C 1980 Clinical applicability of amniotic fluid tests for fetal pulmonic maturity. American Journal of Obstetrics and Gynecology 136: 135–144

Olson L G, Hensley M J, Saunders N A 1982 Ventilatory responsiveness to hypercapnic hypoxia during dopamine infusion in humans. American Review of Respiratory Disease 126: 783–787

Pape K W, Wigglesworth J S 1979 Haemorrhage, ischaemia and the perinatal brain. In: Clinics in Developmental Medicine, Nos. 69/70. Spastics International Medical Publications, William Heinemann, London

Papile L A, Burstein J, Burstein R, Koffler H 1978a Incidence and evolution of subependymal and intraventricular hemorrhage: a study of infants with birthweights less than 1500 gms. Journal of Pediatrics 92: 529–534

Papile L A, Burstein J, Burstein R, Koffler H, Koops B 1978b Relationship of intravenous sodium bicarbonate infusions and cerebral intraventricular hemorrhage. Journal of Pediatrics 93: 834–836

Papile L A, Munsick-Bruno G, Schaefer A 1983 Relationship of cerebral intraventricular hemorrhage and early childhood neurologic handicaps. Journal of Pediatrics 103: 273–277

Perlman J M, Volpe J J 1983 Suctioning in the preterm infant: effects on cerebral blood flow velocity, intracranial pressure and arterial blood pressure. Pediatrics 72: 329–334

Perlman J M, McMenamin J B, Volpe J J 1983a Fluctuating cerebral blood flow velocity in respiratory distress syndrome. Relation to the development of intraventricular hemorrhage. New England Journal of Medicine 309: 204–209

Perlman J M, Nelson J S, McAlister W H, Volpe J J 1983b Intracerebellar hemorrhage in a premature newborn: diagnosis by real-time ultrasound and correlation with autopsy findings. Pediatrics 71: 159–162

Perrin D G, Cutz E, Becker L E, Bryan A C, Madapallimatum A, Sole M J 1984 Sudden infant death syndrome: increased carotid-body dopamine and noradrenaline content. Lancet ii: 535–537

Persson P H, Kullander S, Gennser G, Grennert L, Laurell C B 1983 Screening for fetal malformations using ultrasound and measurements of alpha-fetoprotein in maternal serum. British Medical Journal 286: 747–749

Phelps D L, Rosenbaum A L 1982 The effect of marginal hypoxemia during the recovery period in oxygen-induced retinopathy in the kitten. Clinical Research 30: 146A

Pokora T, Bing D, Mammel M C, Boros S J 1983 Neonatal high-frequency jet ventilation. Pediatrics 72: 27–32

Rigatto H 1982 Apnea. Pediatric Clinics of North America 29: 1105–1116

Runkle B, Bancalari E 1984 Acute cardiopulmonary effects of pancuronium bromide in mechanically ventilated newborn infants. Journal of Pediatrics 104: 614–617

Saigal S, Rosenbaum P, Stoskopf B, Sinclair J C 1984 Outcome in infants 501–1000 g birth weight delivered to residents of the McMaster Health Region. Journal of Pediatrics 105: 969–976

Sauer P J J, Dane H J, Visser H K A 1984 New standards for neutral thermal environment of healthy very low birthweight infants in week one of life. Archives of Disease in Childhood 59: 18–22

Schiffman P L, Westlake R E, Santiago T V, Edelman N H 1980 Ventilatory control in parents of victims of sudden infant death syndrome. New England Journal of Medicine 302: 486–491

Schulte F J 1977 Apnea. Clinics in Perinatology 4: 64–76

Schulte F J, Busse C, Eichhorn W 1977 Rapid eye movement sleep, motoneurone inhibition and apneic spells in preterm infants. Pediatric Research 11: 709–713

Silverman W A 1982 Retinopathy of prematurity: oxygen dogma challenged. Archives of Disease in Childhood 57: 731–733

Simpson H 1983 Sudden unexpected infant death. Home monitoring. Archives of Disease in Childhood 58: 469–471

Smith P C, Smith N T 1982 The special considerations of the premature infant. In: Steward D J (ed) Some Aspects of Paediatric Anaesthesia. Elsevier, Amsterdam

Speer M E, Blifeld C, Rudolph A S, Chadda P, Holbein M E B, Hittner H M 1984 Intraventricular hemorrhage and vitamin E in the very low-birth-weight infant: evidence for efficacy of early intramuscular vitamin E administration. Pediatrics 74: 1107–1112

Spielman F T, Seeds J W, Corke B C 1984 Anaesthesia for fetal surgery. Anaesthesia 39: 756–759

Steinschneider A, Weinstein S 1983 Sleep respiratory instability in term neonates under hyperthermic conditions — age, sex, type of feeding, and rapid eye movements. Pediatric Research 17: 35–41

Steinschneider A, Weinstein S L, Diamond E 1982 The sudden infant death syndrome and apnea/obstruction during neonatal sleep and feeding. Pediatrics 70: 858–863

Steward D J 1982 Preterm infants are more prone to complications following minor surgery than are term infants. Anesthesiology 56: 304–306

Sykes G S, Molloy P M, Johnson P, Stirral G M, Turnbull A C 1983 Fetal distress and the condition of newborn infants. British Medical Journal 287: 943–945

Thach B T, Frantz I D, Adler A M, Taeusch H W 1978 Maturation of reflexes influencing inspiratory duration in human infants. Journal of Applied Physiology 45: 203–211

Thorburn R J, Lipscombe A P, Stewart A L, Reynolds E O R, Hope P L, Pape K E 1981 Prediction of death and major handicap in very preterm infants by brain ultrasound. Lancet i: 1119–1121

Truog W E, Standaert T A, Murphy J H, Woodrum D E, Hodson W A 1984 Effects of prolonged high-frequency oscillatory ventilation in premature primates with experimental hyaline membrane disease. American Review of Respiratory Disease 130: 76–80

Vincent R N, Stark A R, Lang P, Close R H, Norwood W I, Casteneda A R et al 1984 Hemodynamic response to high-frequency ventilation in infants following cardiac surgery. Pediatrics 73: 426–430

Volpe J J 1981 Neonatal intraventricular hemorrhage. New England Journal of Medicine 304: 886–891

Walker D-J B, Feldman A, Vohr B R, Oh W 1984 Cost-benefit analysis of neonatal intensive care for infants weighing less than 1000 g at birth. Pediatrics 74: 20–25

Williamson W D, Desmond M M, Wilson G S, Murphy M A, Rozelle J, Carcia-Prats J A 1983 Survival of low-birth-weight infants with neonatal intraventricular hemorrhage. American Journal of Diseases of Children 137: 1181–1184

Wimberley P D, Lou H C, Pedersen J, Hejl M, Lassen N A, Friis-Hansen B 1982 Hypertensive peaks in the pathogenesis of intraventricular hemorrhage in the newborn. Abolition by phenobarbitone sedation. Acta Paediatrica Scandinavica 71: 537–542

Yelderman M, New W 1983 Evaluation of pulse oximetry. Anesthesiology 59: 349–352

Advances in trauma emergency care

INTRODUCTION

Trauma kills more people under 14 years of age in the USA compared with the combined total who die as a result of heart disease, cancer, pneumonia, intestinal disease and meningitis (Cowley & Trump 1971). It is the leading cause of death in those under 38 years of age and it is the fourth commonest cause of death for the entire population (Chicago National Safety Council 1982). Trauma is responsible in economic costs (1982 dollars) for 6.9% of health care expenditures and 2.3% of the United States gross national product. Over 61 billion dollars was spent on trauma in fiscal 1982 of which 19.3 billion dollars were treatment related expenses and 41.7 billion dollars were forgone earnings (Munoz 1984). Fatalities cost over twice the amount of non fatalities. Advances in trauma emergency care are important, therefore, both from the medical and economic standpoint (Norman & Moles 1977). This review describes the concept of regional trauma care and outcome evaluation and prediction. Systematic approaches to the management of patients with massive blunt trauma are discussed and advances in emergency care that increase the likelihood of survival are identified.

REGIONAL TRAUMA CARE

Data from several trauma centres in the USA suggest that fatalities can be prevented by implementation of regional trauma systems (Boyd 1973, Cowley 1975, Gill et al 1976, West et al 1979, Lowe et al 1983, Cales 1984, Wright et al 1984). Potentially salvageable patients are more likely to die in centres where facilities for the treatment of trauma are inadequate (West et al 1979, Lowe et al 1983, Cales 1984, Wright et al 1984). In one study, 25% of the fatalities and 16% of the results in non-trauma centres were considered inappropriate for the severity of injury incurred (Lowe et al 1983). In addition, small hospitals (less than 200 beds) had a higher percentage of early deaths from trauma than large hospitals. The age of those dying and the injury severity scores were higher after implemtation of regional trauma centres (Cales 1984). There was an improvement in trauma care and death rate from motor vehicle accidents as demonstrated when a 24% drop in mortality was reported after an emergency care system was started in Jacksonville, Florida (Waters & Wells 1973). In several states and regions in the USA, graded echelon trauma care systems have been developed. The first of these was the Maryland emergency medical services graded echelon system established in 1973. The major elements of such trauma care systems are shown in Table 14.1. These requirements are described in full by the American College of Surgeons Committee on Trauma Care (American College of Surgeons 1976, 1979, 1980). The commitment of an area wide emergency medical service can best be described by one such as the Maryland Emergency Medical Services System which is outlined in Table 14.2 The expertise, facilities and cost of a level 1 trauma centre is summarized in Table 14.3. Further details of the patient population,

Table 14.1   Major elements of a graded-echelon trauma care system

Central emergency telephone number
Centralized dispatch of ambulances and helicopters
Rapid field stabilization and resuscitation
Paramedic triage with field hospital communication
Rapid transport with ongoing resuscitation to a designated hospital
Designation and grading of standard of specified receiving hospitals
Ongoing training programs for medical personnel and the public

Table 14.2   Maryland emergency medical services (EMS) system

Population	4.25 million (3.1 million > 18 years)
	3 054 000 registered motor vehicles
Roads	26 631 miles of State roads including 362 miles of motorway
Traffic accidents	130 009 592 were disabling and 650 people died (1984)
Trauma centers	Level 1 (Shock Trauma) + four Level 2 in Baltimore City; in each of four other EMS regions in State, one Level 2 regional trauma center
Telecommunications	Posts in Baltimore City and each of 23 counties
Transport	Helicopters: 5 State Police, 2 State Park, 1 Army
	Ambulances: Local, State and City

Table 14.3   Shock trauma — Level 1 Trauma Centre, Baltimore City (73 beds attached to a 785 bed University Hospital; 24 hour consultant and junior staff in-house coverage)

	Consultants	Fellows
Surgeons	11	10
Anesthesia/critical care	9	5
Neurosurgeons	3	4
Medical	2	
Nurses	207	
Annual operating costs and salaries	$22.4 million	
Field operations	$2.7 million	

mortality (Mackenzie et al 1979, 1982, 1983, 1984) and respiratory management (Mackenzie et al 1981) are described elsewhere.

OUTCOME, EVALUATION AND PREDICTION

Assessment of injury is essential for the appropriate allocation of therapeutic resources, for evaluation of the patient's status, prediction of outcome and for comparison of trauma care delivery. Blunt non-penetrating trauma, resulting from high speed motor vehicle injuries, rapid deceleration, high velocity missiles or blast injuries may cause massive multisystem tissue disruption and has a less favourable outcome than penetrating trauma. Penetrating trauma, occuring from low velocity missiles such as knife and hand gun injuries, or falls resulting in isolated bony injuries, presents a specific, easily defined problem. Single organ injury frequently occurs and the comparatively little systemic disturbance that ensues often allows hospital discharge within a few days of injury. Several indices of injury are currently used to allow comparisons of the extent of, and outcome from, injury. These include the Glasgow Coma Scale (Teasdale & Jennett 1974), Injury Severity Score (Baker & O'Neill 1976), Anatomic Index (Champion et al 1980), Trauma Score (Champion et al 1983) and Acute Physiology and Chronic Health Evaluation Index (APACHE) (Knaus et al 1985). These indices make numerical descriptions of the overall severity of injury possible in patients who have received multiple injuries and physiological derangements. The Glasgow Coma Scale (GCS) is incorporated into many of the other indices and, although it has limitations, it is generally considered a good predictor of outcome from head injury. The Trauma Score, based on systolic blood pressure, respiratory rate, respiratory effort, capillary refill and the GCS was shown to be a good predictor of survival from blunt trauma. It is thought to be a useful mechanism for assessment of patient triage (decision on urgency of treatment) and to identify anomalous outcomes in the management of the patients with multiple sites of trauma. The Injury Severity Score excludes those patients dead on arrival and has successfully been used to compare outcome from trauma in different states and countries. Patients with injury severity scores in excess of 30 should receive specialized trauma care. However, the objections to Injury Severity Score are that ranking of severity is based on subjective impressions and it is correlated with outcome only in patients with low mortality. Combinations of moderately severe injuries may result in a higher score than a fatal head injury. The Anatomic Index was developed to eliminate these problems and correlated with outcome when mortality was in excess of 20% (Champion et al 1980). The Anatomic Index is most useful when assessing patients with blunt trauma and head injury such as may occur following motor vehicle injuries.

Twelve variables make up the APACHE 2 scoring system. The score and the acute risk of death rises on the basis of initial physiological abnormality, with chronic health problems at advanced age. Since trauma mortality and morbidity are functions of both injury severity and quality of medical care, survival rates and the quality of care in two facilities, systems or countries can only be compared when patient injury severity is controlled. These indices are, therefore, a major development in emergency trauma care that enable clinicians, epidemiologists and health service researchers to obtain objective evidence for allocation and evaluation of emergency medical services.

SYSTEMATIC APPROACH TO THE MANAGEMENT OF MASSIVE BLUNT TRAUMA

The patient subjected to massive trauma is difficult to assess when admitted to hospital or even at the scene of injury for often no history is obtainable and the patient may be unconscious, unresponsive or under the influence of alcohol or other drugs. No past medical history is available and since trauma is unique to each individual, the extent and the site of injuries in unknown when the patient is first seen. The traditional management approach of history taking, physical examination and establishment of a diagnosis before starting therapy are clearly inappropriate in the management of massive trauma. Systematic approaches with use of treatment protocols and alogarithms to determine management options are familiar to the military forces and are advantageous in the management of the severely injured patient. As a result of treatment protocols it is thought that time is not wasted by the trauma team, which consists of anaesthetists, surgeons and nurses in determining possible management plans. Communication between the medical and nursing personnel is simplified, because each member of the resuscitation team knows their role, and can function independently during the period immediately after injury or reception of the injured patient. The therapeutic and diagnostic approach to the patient must be coordinated if survival is to occur. Protocols provide a simple coordinated approach to patient management, greatly simplify training of staff, facilitate research by establishing a standardized data base, and allow for evaluation of the changes in the protocols by comparison with previous practices (Cowley & Dunham 1982a).

The major importance of these systematic approaches are that (1) priorities of management are identified and may be executed in an orderly fashion, (2) the likelihood of omission in examination, investigation or therapy, is reduced, and (3) treatment is not delayed until the diagnosis is complete but investigation and therapy occur simultaneously guided by protocol priorities. The following guidelines were developed at the Maryland Institute of Emergency Medicine (Shock Trauma) and have been annually updated and revised. Similar management protocols are used and advocated at other trauma centres (Hassett et al 1984) and they illustrate a protocol approach to management of massive blunt trauma that has reduced mortality from 25.4% in 1972 to 11.7% of admissions in 1984 (Fig. 14.1). Protocols may readily be developed into alogarithms with treatment/diagnosis options (Gill & Long 1979) that may be useful in less well equipped or staffed medical facilities and may also be computerized.

First priorities

1. Establish an airway and provide adequate ventilation with 100% oxygen. Auscultate the chest and confirm bilateral and equal breath sounds.
2. Control external haemorrhage with pressure. Order blood from blood bank, establish intravenous lines and begin volume infusion.

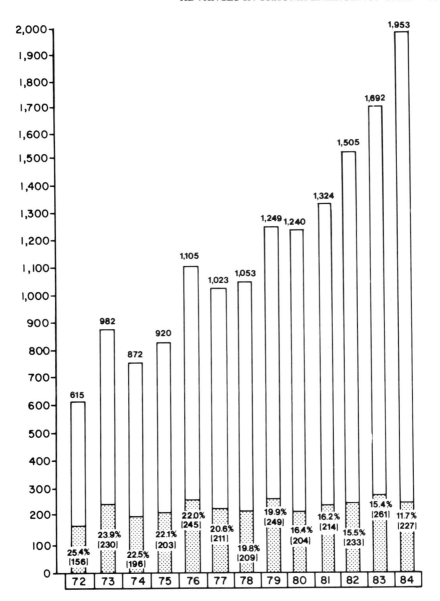

Fig 14.1 Number of patients admitted to the Maryland Institute for Emergency Medical Services (Shock Trauma) for fiscal years 1972–84. Mortality is shown as the dotted portion of the histogram. The percentage mortality varies from 25.4% to 11.7%. The number of patients this mortality represents is shown in brackets.

3.   Two minute evaluation in which the central nervous system is examined (follows commands, speaks, moves all extremities, note extremity response to noxious stimuli, pupil size — equality and light reaction). Pulse and blood pressure are measured and an inspection of the naked patient (front and back) is rapidly carried out to identify the visible extent of trauma. The head, chest, abdomen, pelvis and extremities are briefly examined.

4. Baseline blood samples to include haemoglobin, haematocrit, white blood cell and platelet count, urea, electrolytes, glucose, creatinine, osmolality, lactate and a coagulation profile (PT, PTT, fibrinogen) are made on 70 ml of blood drawn on insertion of a central venous monitoring and infusion catheter. An arterial blood gas is obtained by percutaneous femoral artery sampling. Results are sent by computer and are available within 5–20 min depending on the investigation.
5. Monitor patient with ECG.
6. Pass nasogastric tube.

Second priorites

1. Emergency investigations.
   X-ray. (a) All motor vehicle accident and major trauma victims receive a cervical spine X-ray to include C7. (b) Once spinal injury and fractures are excluded a true erect chest X-ray is taken with the patient sitting up. This identifies a haemo/pneumothorax; tracheal tube, nasogastric tube and central venous catheter placement; ruptured diaphragm, bony fractures and aortic rupture. Failure to identify the aortic arch or widening of the mediastinum on a true erect chest X-ray is an indication for an aortogram. (c) A pelvic X-ray is taken.
   Peritoneal lavage. 1 litre of normal saline is infused by peritoneal dialysis catheter through a periumbilical incision while the patient is tilted head up and then head down (to detect a bleeding ruptured diaphragm or pelvic haematoma). The fluid is siphoned off, and the finding of frank blood or inability to read newsprint through the lavage tubing or bottle is a positive result, making exploratory laparotomy mandatory. More quantitative analysis of the lavage fluids, showing an erythrocyte count $> 100\,000$ mm^3 or leucocyte count greater > 500 mm^3, presence of gastro-intestinal contents or an elevated amylase are also positive results (Cowley & Dunham 1982b).
   Pass a urinary catheter and send urine sample for analysis of urine electrolytes, creatinine and test for blood, protein and glucose. Haematuria requires intravenous pyelography if the blood pressure is at least 70 mmHg systolic. A cystogram is performed when haematuria occurs with a major pelvic fracture.
2. Splint extremity fractures. Air inflatable or rigid splints may be used.
3. A brief and pertinent history is obtained when available.
4. An arterial catheter is inserted if indicated for blood pressure and sample monitoring.

Third priorities

1. Systematic examination includes the face and scalp, the extremities, the back as well as rectal, pelvic and perineal examinations.
2. Additonal systematic investigations may include specific X-rays, computerized tomographic (CT) scans and contrast studies, if indicated.
3. Speciality consultation is obtained from neurosurgeons, orthopaedic surgeons, urologists, etc.
4. A multidisciplinary discussion then takes place with the anaesthetist, surgeon, nurse and consultants' participation and subsequent management is determined.

Despite this approach to the management of trauma victims, mortality still ranges from 24–11.7% in trauma centres (Fischer et al 1984, Geritsen et al 1984) (see Fig. 14.1). In Maryland, State laws are very supportive of the Medical Examiners or Coroner's office and a high autopsy rate is mandatory by law. Of 2320 motor vehicle accident victims who were admitted to the Maryland Institute of Emergency Medicine in a four year period, 1703 came directly by helicopter from the scene of injury and 383 (22.4%) of the direct admissions died. Autopsies were performed on 366 (95.6%) of these deaths (Mackenzie et al 1982). If only the deaths occurring within 24 hours of a motor vehicle accident are considered, there was a 98.3% (300:305) autopsy rate (Mackenzie et al 1984). From these data the commonest cause of death within 24 hours of injury is craniocerebral trauma which accounted for 40.3% (123:305) of all deaths. Haemorrhage was the second commonest cause of death, accounting for 30.2% (92:305) of all deaths; and cervical spine injury was the third commonest cause of death, accounting for 12.5% (38:305) of all deaths. Unconsciousness on admission was the most sensitive indicator of a fatal prognosis. Management of head injured patients included tracheal intubation, mechanical hyperventilation, routine intracranial pressure monitoring, corticosteroids and for one of the four years, barbiturate therapy for persistent intracranial pressure greater than 25 mmHg (Mackenzie et al 1982). This latter barbiturate intervention did not alter death from head injury compared to prior years.

Despite multiple long bone, pelvic and other bony fractures fat embolus was extremely rare, the syndrome being detected in only one patient in 5 years of admissions. This may be related to early reduction of fractures with internal fixation of femurs (Riska et al 1980, Browner et al 1984) and use of external Hoffman type fixation of the tibial, humeral and pelvic fractures. These techniques allow early mobilization of the trauma victim and are thought to reduce post-operative pulmonary complications (Mackenzie et al 1981, Browner et al 1984).

Surgical intervention is carried out after resuscitation on admission, since operating conditions, because of oedema, haematoma and sepsis, are frequently less optimal 1 week or more later. Once again priorities are established, but all surgery is carried out in one session. This may mean that several teams of surgeons may operate simultaneously on the same patient. All surgical procedures are completed before the patient is transferred to the critical care unit having had all the general surgical, orthopaedic, neurosurgical and plastic surgery necessary for recovery. While this approach results in operating and anaesthesia times of 8–24 hours, it saves repeated subsequent visits for surgery and provides for optimum management on admission and may prevent the development of pulmonary complications, fat emboli and sepsis.

ADVANCES IN VENTILATION, INFUSION AND PUMPING

Advances in emergency trauma care that increase the chance of survival are discussed in the context of Weil & Shubin's (1969) VIP approach (ventilation, infusion, pumping) to the management of shock.

Airway ventilation

Advances in airway ventilation management are outlined in Table 14.4. The oesophageal obturator has been modified since its introduction including the design

Table 14.4   Advances in airway management

Esophageal obturator
Pharyngeo-tracheal lumen airway
Cricothyroidotomy
Jet ventilation

of an oesophageal gastric tube airway. However, it is still reported to be associated with oesophageal rupture, inadvertent tracheal intubation and inadequate CO_2 removal (Bryson et al 1978, Taryle et al 1979, Smith et al 1983). However, the obturator does protect the airway from aspiration of gastric contents and oxygenation is usually adequate. The obturator is, therefore, an improvement on simple oropharyngeal airways, although non anaesthesia personnel have difficulty with the mask fit. New mask designs include a 'jellyfish' mask (Respironics Corporation) which is reported to provide maximum tidal volume at two different lung compliances with minimum leak in comparison to two alternative mask designs, one having an inflatable cuff rim (Stewart et al 1984). The pharyngotracheal lumen (PTL) airway has a two tube, two cuff, system that is inserted blind and it functions as a tracheal tube if the trachea is entered or as an oesophageal obturator with oesophageal intubation. It protects against aspiration, gastric dilatation and inflation and provides similar oxygenation and CO_2 removal to mechanical ventilation through a conventional orotracheal tube (Niemann et al 1984). There are limited clinical data on the use of the PTL airway and conventional tracheal intubation is still the management of choice when the necessary expertise is available.

Cricothyroidotomy is an emergency surgical procedure that may be life saving and indicated with acute upper airway obstruction or when orotracheal intubation cannot be accomplished. A minitracheostomy kit facilitates the performance of a cricothyroidotomy (Portex). However, the long term morbidity from laryngeal granuloma is disturbing when these stoma are used for prolonged tracheal intubation. For this reason a cricothyroidotomy should be converted to a translaryngeal or tracheostomy intubation as soon as possible once the initial airway emergency is overcome. Cannulae passed through the cricothyroid membrane may be used for O_2 insufflation or for ventilation by means of a jet device that interrupts the 3.3 kPa (50 psi) oxygen pressure source intermittently (Sanders 1967). Oxygen insufflation with as little as 1 l/min may maintain O_2 levels for 10–30 min whereas the jet ventilator also removes CO_2 (similar to the technique of ventilation during shared airway endoscopy procedures). The jet ventilator can provide adequate oxygenation and ventilation in animals in shock with an open thorax and aortic cross clamp (Jorden et al 1984) and in humans in the presence of large airway leaks (Turnbull et al 1981). The cricothyroid membrane puncture may also be used to facilitate orotracheal intubation and avoid tracheostomy by passage of a guide wire in a retrograde manner through the larynx and out of the mouth. An orotracheal tube is then directed into the larynx down the guide wire which is removed when the tube enters the trachea. A long Intracath substitutes for a guide wire if this is not available.

Infusion

Advances in infusion techniques relate to the rapidity and type of fluid replacement. Two new techniques can greatly increase the infusion of intravenous (i.v.) fluids based on Poiseuilles law. Large bore intravenous tubing (5.0 mm ID compared to standard 3.2 mm ID) in conjunction with a 10 French (Fr) gauge short i.v. catheter or 8 Fr pulmonary artery catheter introducer enabled infusions of 1200–1400 ml/min of crystalloid or whole blood to be made (Millikan et al 1984). The ability to replace fluids rapidly is frequently the limiting factor in survival of trauma patients. Multiple i.v. sites may hinder movement around the patient, although multiple i.v. sites above and below the diaphragm on the right and left sides of the patient are also advocated. Multiple i.v. sites ensure circulatory volume replacement and avoid expansion of an undetected extravascular haematoma that may have occurred as a result of vessel rupture. Iserson & Reeter (1984) developed a reservoir that allows up to five individual infusion bags to be run simultaneously. They feed through a single large diameter tube to a 4.5 mm ID catheter which provides a maximum of 650 ml/min gravity flow and a pressurized flow of more than 1600 ml/min.

The delay while awaiting for cross matching of blood in the bleeding patient who cannot be adequately resuscitated with crystalloid or colloid infusion may result in exsanguination. Type O blood can be transfused without fear of major incompatability reactions; however the red cell does contain antibodies than cause minor transfusion reactions (Gervin & Fischer 1984). To avoid Rh sensitization in Rh negative patients low titre O blood with no Rh antigen may be used. This blood is extemely scarce and when O Rh positive packed cells were transfused no haemolytic transfusion reactions occurred. Absence of reactions was found even though 170 of the 343 patients transfused with 1945 units of group O Rh positive blood during a 22 month period were of blood groups other than O. However, of the 47 Rh negative patients 36 developed anti-Rh (D) antibody (Sohmer et al 1979). Resuscitation of trauma patients with type specific uncrossmatched blood also resulted in lack of transfusion reactions. Subsequent major crossmatch failed to identify either blood incompatibility or significant antibodies in 236 patients who received 1141 units of blood in a 3 year period (Gervin & Fischer 1984).

Pumping

Whether or not pneumatic trousers are beneficial in the prehospital management of hypovolaemic shock remains controversial. Several authors have found no increase in cardiac filling pressures with application of military anti-shock trousers (MAST) and in one controlled study of their field use there was no significant improvement in blood pressure, mortality or trauma score in patients who had MAST (Bukell et al 1984). There was no demonstrable benefit of MAST in patients with cardiac arrest (Mackersie et al 1984) although others report increased survival (Mahoney & Murick 1983). Further prospective randomized studies are required to establish the usefulness of external counterpressure.

After remaining unchanged for 20 years, cardiopulmonary resuscitation (CPR) techniques have undergone recent major changes. The new era of research into CPR began with the suggestion that alterations in pleural pressure and not direct cardiac compression are the cause of blood flow during CPR (Criley et al 1976, Chandra et al 1981, Weisfeldt 1983). Elevations in pleural pressure are transmitted to all the chambers of the heart and vascular structures in the chest. Because of the unique valvular arrangements in the heart that only permits forward flow and because there are competent venous valves at the thoracic inlet and the pulmonary valve remains closed, changes in pleural pressure are transmitted to the extra thoracic arteries resulting in forward flow of blood. In most adults during CPR, chest compression results in a pressure gradient between arteries and veins outside the thorax, due to the pleural pressure elevations transmitted to the heart. Children with compliant chest walls or thin adults may be exceptions (Rogers 1985).

Pleural pressure may be elevated to obtain maximum cardiac output by synchronizing chest compression and ventilation during CPR (Chandra et al 1981, Weisfeldt 1983). However, simultaneous compression and ventilation or 'new' CPR results in higher intracranial pressures than seen with conventional CPR. Although carotid blood flow increases, the benefit, in terms of cerebral perfusion pressure (mean arterial – mean intracranial pressure) with simultaneous chest compression and ventilation (SCV–CPR) is not as great as was initially hoped. However, cerebral perfusion pressure is still increased in comparison to conventional CPR. Use of an abdominal binder or simultaneous abdominal compression further augments the pleural pressure changes and carotid blood flow during SCV–CPR (Chandra et al 1981).

Pharmacological intervention following cardiac arrest has also changed. Cerebral perfusion may be improved by increasing resistance in extracerebral vessels with α agonist drugs such as adrenaline in high doses (Michael et al 1984). Since there are not thought to be α adrenergic receptors active in the cerebral vessels, blood flow is diverted to the cerebral circulation. Other α agonists may also be useful and are presently being investigated (Rogers 1985). Calcium entry blocking drugs such as nimodipine (Steen et al 1985), lidoflazine and verapamil (Vaagnes et al 1984) improve outcome following cardiac arrest, possibly by preventing vascular smooth muscle vasospasm resulting in improved post-ischaemic cerebral blood flow. Recent data implicate the calcium ion as a triggering element in a number of adverse reactions in a wide variety of tissues after shock, sepsis, trauma and anoxia. Reports have also linked ionic calcium shifts to abnormalities in cellular metabolism, intracellular release of free fatty acids, and production of oxidative free radicals. Calcium and the prostaglandins are also involved in the non-reflow phenomenon that occurs with cerebral vascular constriction after cardiac arrest. All of these factors are implicated in neuronal injury after ischaemic anoxia and the use of calcium blockers and other pharmacological agents in post-resuscitation brain injury clearly requires further investigation (White et al 1983, Shapiro 1985). While outcome from ventricular fibrillation cardiac arrest is quite favourable, much improvement is required in asystolic cardiac arrest. The ability to improve myocardial perfusion during resuscitation may be crucial in determining a successful outcome. It seems likely in the future that not one but several CPR techniques will be used to resuscitate patients depending on the cause and type of cardiac arrest. To prevent neurological disability calcium channel blockers may be used alone or in combination with other agents such as indomethacin, barbiturates, free radial scavengers (White et al 1983) and α agonists (Michael et al 1984).

REFERENCES

American College of Surgeons 1976 Optimal hospital resources for care of the seriously injured. Bulletin ACS 61(9): 15–22

American College of Surgeons Committee on Trauma. 1979 Hospital resources for optimal care of the injured patient. Bulletin ACS 64(8): 43–48

American College of Surgeons Committee on Trauma Care 1980 Appendix A, C1, C2, D, E, F. Bulletin ACS 65(2): 9–35

Baker S P, O'Neill B 1976 The injury severity score: an update. Journal of Trauma 16: 882–885

Boyd D R 1973 A symposium on the Illinois trauma program. A systems approach to the care of the critically injured. Journal of Trauma 12: 275–276

Browner B D, Burgess A R, Robertson R J et al 1984 Immediate closed antegrade Ender nailing of femoral fractures in polytrauma patients. Journal of Trauma 24: 921–927

Bryson T K, Benumof J F, Ward C F 1978 The esophageal obturator airway: a clinical comparison to ventilation with a mask and oropharyngeal airway. Chest 74: 537–539

Bukell W H, Pepe P L, Applebaum D J et al 1984 Effect of antishock trousers on the trauma score. A prospective analysis (abstract). Annals of Emergency Medicine 12: 402

Cales R H 1984 Trauma mortality in Orange County: the effect of implementation of a regional trauma system. Annals of Emergency Medicine 13: 1–10

Champion H R, Sacco W J, Lepper R C et al 1980 An anatomic index of injury severity. Journal of Trauma 20: 197–202

Champion H R, Sacco W J, Hunt T R 1983 Trauma severity scoring to predict mortality. World Journal of Surgery 7: 4–11

Chandra N, Weisfeldt M L, Tschik J et al 1981 Agumentation of carotid flow during CPR by ventilation at high airway pressure simultaneous with chest compression. American Journal of Cardiology 48: 1053–1063

Chicago National Safety Council 1982 Accident facts

Cowley R A 1975 Total emergency medical system for State of Maryland. Maryland State Medical Journal 24: 37–45

Cowley R A, Dunham C M (eds) 1982a Shock trauma/ critical care manual. University Park Press, Baltimore, p XII

Cowley R A, Dunham C M (eds) 1982b Shock trauma/critical care manual. University Park Press, Baltimore, p146

Cowley R A, Trump B 1971 Todays neglected disease — trauma. Bull. University Maryland School of Medicine 56: 19–25

Criley J M, Blarfuss A H, Kissel G L 1976 Cough-induced cardiac compression. Journal of the American Medical Association 236: 1246–1250

Fischer R P, Flynn T C, Miller P W 1984 Urban helicopter response to the scene of injury. Journal of Trauma 24: 946–951

Geritsen S M, Loenhoul T Van, Gunbrere J S F 1984 Prognostic signs and mortality in multiply injured patients. Injury 14: 89–92

Gervin A S, Fischer R P 1984 Resuscitation of trauma patients with type specific uncrossmatched blood. Journal of Trauma 24: 327–331

Gill W, Champion H R, Long W B et al 1976 A clinical experience of multiple trauma in Maryland. Maryland State Medical Journal 25: 55–58

Gill W B, Long W B 1979 Shock trauma manual. William and Wilkins, Baltimore

Hassett J, LaDuca J, Seibel R, Border J R 1984 Priorities in multiple injuries: a brief review. Injury 14: 12–16

Iserson K V, Reeter A K 1984 Rapid fluid replacement. A new methodology. Annals of Emergency Medicine 12: 97–100

Jorden R C, Moore E E, Marx J A et al 1984 Percutaneous transtracheal ventilation in a canine shock model with an open thorax. Annals of Emergency Medicine 13: 22–25

Knaus W A, Draper E A, Wagner D P et al 1985 APACHE II — A severity of disease classification system. Critical Care Medicine 13: 818–829

Lowe D K, Gatley H L, Goss J R et al 1983 Patterns of death, complication and error in the management of motor vehicle accident victims. Implication for a regional system of trauma care. Journal of Trauma 23: 503–509

Mackenzie C F 1981 History and literature review of chest physiotherapy, chest physiotherapy program, patient population and respiratory care at MEIMSS. In: Mackenzie C F (ed) Chest physiotherapy in the intensive care unit. Williams and Wilkins, Baltimore, ch 1, p 1–26

Mackenzie C F, Shin B, Fisher R et al 1979 Two year mortality in 760 helicopter transported patients direct from the road accident scene. American Surgeon 45: 101–108

Mackenzie C F, Shin B, Fisher R et al 1982 Four year mortality of trauma victims admitted directly from the accident by helicopter. Anesthesiology 57: A96

Mackenzie C F, Shin B, Fisher R et al 1983 Transports des accidentes de la voie publique (vehicules a moteur): mortalite pour 1703 primaries en 4 ans. Convergences Medicales 2: 427–431

Mackenzie C F, Shin B, Sodestrom C et al 1984 Would field deployment of physicians reduce deaths within 24 hours of road traffic accidents. Anesthesiology 61: A93

Mackersie R C, Christensen J M, Lewis F R 1984 The pre hospital use of external counter pressure. Does MAST make a difference? Journal of Trauma 24: 882–888

Mahoney B D, Murick M J 1983 Pneumatic trousers in refractory prehospital cardiopulmonary arrest. Annals of Emergency Medicine 12: 8–12

Michael J R, Gueria A D, Koehler R C et al 1984 Mechanisms whereby epinephrine augments cerebral and myocardial perfusion during cardiopulmonary resuscitation in dogs. Circulation 69: 822–835

Millikan J S, Cam T L, Hansborough J 1984 Rapid volume replacement for hypovolemic shock. A comparison of techniques and equipment. Journal of Trauma 24: 428–431

Munoz E 1984 Economic cost of trauma, United States, 1982. Journal of Trauma 24: 237–244

Neimann J T, Rosborough J P, Myers R A M et al 1984 The pharyngeotracheal lumen airway: preliminary investigation of a new adjunct. Annals of Emergency Medicine 13: 591–596

Norman J, Moles M 1977 Trauma and immediate care: editorial. British Journal of Anaesthesia 49: 641–642

Riska E B, Bonsdorff H von, Hakkinen S et al 1980 Prevention of fat embolism by early internal fixation of fractures in patients with multiple injuries. Injury 8: 110–116

Rogers M C 1985 Cardiopulmonary resuscitation — present and future. IARS Review Course Lectures. Presented at the 159th International Anesthesia Research Society Congress, Houston Texas, 10–13 March, p 124–126

Sanders R D 1967 Two ventilating attachments for bronchoscopes. Delaware Medical Journal 39: 170–175

Shapiro H M 1985 Post cardiac arrest therapy: calcium entry blockade and brain resuscitation. Anesthesiology 62: 384–387

Smith J P, Bodai B I, Seifkin A et al 1983 The esophageal obturator airway. A review. Journal of the American Medical Association 250: 1081–1084

Sohmer P R, Etter M, Shin B et al 1979 Rh sensitization subsequent to transfusion with uncrossmatched Rh positive packed red cells (abstract). Critical Care Medicine 7: 144

Steen P A, Giswold S E, Milde J H et al 1985 Nimodipine improves outcome when given after complete cerebral ischemia in primates. Anesthesiology 62: 406–414

Stewart R D, Kaplan R, Thompson F et al 1984 Influence of mask design on bag-mask ventilation (abstract). Annals of Emergency Medicine 13: 404

Taryle D A, Chandler J E, Good J T et al 1979 Emergency room intubations. Complication and survival. Chest 75: 541–543

Teasdale G, Jennett B 1974 Assessment of coma and impaired consciousness. Lancet ii: 81–84

Turnbull A D, Carlon G, Howland W S et al 1981 High frequency jet ventilation in major airway or pulmonary disruption. Annals of Thoracic Surgery 32: 468–474

Vaagnes P, Cantadore R, Safar P et al 1984 The effect of lidoflazine and verapamil on neurologic outcome after 10 minutes ventricular fibrillation cardiac arrest in dogs (abstract). Critical Care Medicine 12: 228

Waters J M, Wells C H 1973 The effects of a modern emergency medical care system in reducing automobile crash deaths. Journal of Trauma 13: 645–647

Weil M H, Shubin H 1969 The 'VIP' approach to the bedside management of shock. Journal of the American Medical Association 207: 337–340

Weisfeldt M L 1983 Augmentation of cerebral perfusion by simultaneous chest compression and lung inflation with abdominal binding following cardiac arrest in dogs. Circulation 67: 266–275

West J G, Trunkey D D, Lim R C 1979 System of trauma care. A study of two counties. Archives of Surgery 114: 455–460

White B C, Windegar C D, Wilson R F et al 1983 Possible role of calcium blockers in cerebral resuscitation: a review of the literature and synthesis for future studies. Critical Care Medicine 11: 202–207

Wright C S, McMurty R Y, Pickard J 1984 A postmortem review of trauma mortalities — a comparative study. Journal of Trauma 24: 67–68

Medico-legal aspects of anaesthesia

INTRODUCTION

When the author began in anaesthetic practice over three decades ago, an unexpected death or misadventure which occurred during anaesthesia and surgery was viewed by the patient's family and by the public at large as a tragic event which was often unavoidable and one of the accepted risks of surgery. 'Doctors are only human and are doing their best', was a common reaction to unfortunate events.

At the present time the attitude of the general public has altered in that not only do they expect a first class medical service but also in the event of a complication arising, they are entitled to compensation. There is no shortage of skilled lawyers available to facilitate the process of litigation for negligence while doctors are often called upon to testify against their colleagues. It is not suggested that gross carelessness or lack of knowledge does not occasionally lead to damage, nor that under such circumstances the patient should not receive adequate compensation, but the majority of anaesthetic mishaps although often avoidable are due to normal human error or lack of judgement rather than negligence. Considering the calculated risks anaesthetists have to take every day, the number of serious mishaps which do occur are remarkably few. Even so, every mishap is one too many and continuing efforts must be made to reduce the incidence of such events by improving the supervision and training of anaesthetists and the quality of equipment.

The changing public attitude to medical mishaps is clearly unpleasant and may eventually have some damaging effect on medical care, in that the doctor is forced into defensive medicine. There is evidence that this has already happened in some countries, but in the UK there has been little development in that direction so far. It is believed that some benefit has accrued from the increase in public scrutiny. We seem to be more aware of the weaknesses in our professional standards and more willing to learn from our own and our colleagues mistakes, thus improving our overall standard of care.

This chapter presents a study of the anaesthetic mishaps which have occured over the last 10 years and which have been reported to The Medical Protection Society (MPS). This study should highlight the nature and magnitude of the problem and suggest possible ways of reducing the incidence of mishaps and of mitigating the difficulties involved in dealing with claims for damages when they occur.

It is probably fair to say that the anaesthetist more than any other medical practitioner is, during his active professional life, in a position where a lapse of concentration, error of technique or lack of knowledge can have sudden and devastating consequences. It is these more serious consequences of error to which attention will be paid but it must be remembered that anaesthetists are often assessed by their ability to avoid minor injuries such as broken teeth, sore throat, muscle pains and bruising at injection sites. The patient judges our skills by our ability to prevent minor insults rather than our professional knowledge and expertise. Similarly the surgeon is often judged by the neatness of his scar rather than his technical dexterity within the abdomen.

STUDIES OF ANAESTHETIC MISHAPS

Studies of anaesthetic mishaps are often difficult and incomplete, as collecting information depends upon the cooperation of so many professional people. Since 1949 the Association of Anaesthetists has attempted to acquire accurate information on the causes of mortality for anaesthesia on a number of occasions. The first of these was published by Edwards et al (1956). This study surveyed 1000 deaths and found that aspiration of vomit was the major cause of death. Dinnick (1964) reported on a further 600 cases and observed that the major cause of death was hypovolaemia. Following the publication of the first report and the widespread attention drawn to it the incidence of aspiration of vomit has declined. In 1982 the results of a further study on mortality related to anaesthesia were published and it was noted that aspiration of vomit and hypotension resulted in 30% of the deaths (Lunn & Mushin 1982).

A further study by the Association of Anaesthetists and the Association of Surgeons have recently been set up to assess morbidity and mortality in relation to surgical procedures. Bearing in mind the potential seriousness of anaesthetic mishaps it is fortunate that such events are relatively infrequent. In a study set up in New South Wales (1959) and covering a period of 20 years, a mortality rate of 1:30 000 anaesthetics administered was reported. In the 1982 study edited by Lunn & Mushin, a death rate of 1:10 000 directly attributable to anaesthesia was reported but this study included deaths up to 6 days after operation. Both studies indicated the expected mortality rate from anaesthesia but did not attempt to give any indication of anaesthetic morbidity. Serious complications such as brain damage and major peripheral nerve damage were excluded as were the minor hazards associated with anaesthesia. Information on the incidence of major anaesthetic mishaps is difficult to collect, including those which might have resulted in death (near miss). It is relatively easy, however, to acquire data on those cases in which a suit for damages has been filed. As a member of the Council of the Medical Protection Society the author has been able to review such cases which have been considered by the Society during the last 10 years. It is from this source that the problems involved will be discussed. The figures presented do not in any way represent the overall incidence of anaesthetic complications but they do give some insight into the number of claims made against anaesthetists who are members of the Medical Protection Society and outlines the nature of the mishaps which are encountered.

Figure 15.1 shows a steady increase in the number of claims made against anaesthetists from the period 1974–83. The reason for this increase is complex but for the most part must represent a trend towards more litigation rather than a deterioration in anaesthetic standards or an increase in the number of anaesthetists who are members of the Society. Whatever the cause, it is certain that these mishaps lead to intense suffering for patients and their relatives while the anaesthetist also suffers considerable psychological trauma.

In a recent study in the USA, Sollazzi & Ward (1984) recorded a ratio of suicide or attempted suicides of one in every 45 anaesthetists who were sued for negligence. The claims also represent a greatly increasing financial burden on our medical protection societies. It is thus of paramount importance that an attempt should be made to reduce the incidence of errors by improvements in technique and monitoring as well as being aware of possible hazards that have already been alluded to in the anaesthetic literature.

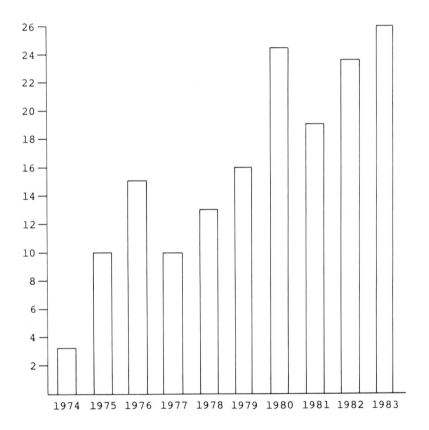

Fig. 15.1 Number of claims made against anaesthetists 1974–83.

MAJOR ANAESTHETIC MISHAPS

Table 15.1 represents the primary cause of mishaps during anaesthesia or in the recovery period and which leads to death or serious neurological damage.

Table 15.1 Primary causes of mishaps resulting from anaesthesia leading to death or serious neurological damage

Cause of mishap	Number of cases	Percentage of major mishaps
Respiratory obstruction	19	18.6
8 Recovery room		
11 Theatre		
Respiratory insufficiency	12	11.9
9 Recovery room		
Oesophageal intubation	12	11.9
Aspiration of vomit	8	7.8
Disconnection of apparatus	5	4.9
Difficult intubation	3	2.9
Endobronchial intubation	1	1.0
Transposition of gases	1	1.0
Hypotension	9	8.8
4 Hypotensive anaesthesia		
3 Regional anaesthesia		
2 Anterior spinal artery syndrome		
Hypovolaemia	5	4.9
Halothane jaundice	4	3.9
Malignant hyperpyrexia	3	2.9
Equipment failure	4	3.9
Myocardial infarct	5	4.9
Anaphylaxis	1	1.0
Dental surgery anaesthesia	7	6.8
5 Operator anaesthetist		
2 Independent anaesthetist		
Unknown cause	3	
Total	102	

Respiratory obstruction

Respiratory obstruction was the commonest cause of a major catastrophe during the 10 year period under review and accounted for 18.6% of all major claims. Slightly less than 50% of these episodes occurred in the recovery period when the patients were being supervised by nursing staff (see below). The remainder occured immediately after induction or during the course of the operation. Two cases, which resulted in severe brain damage, were due to severe bronchospasm, and may be said to have been unavoidable but all the others were almost certainly avoidable and could have been prevented if constant vigilance had been exercised by the anaesthetist. In four of these cases the patient had been left unattended by the anaesthetist when the incident occurred, a situation which is certainly negligent and may even be regarded as criminally negligent by some courts. Recently, a judge pronounced that such cases might in the future carry a verdict of manslaughter and many anaesthetists would accept this view.

The tragedies resulting from patients being left unattended poses the problem of delegation of patient care during the course of operation. Is there any justification for leaving a nurse or an ODA in charge of an anaesthetized patient while the anaesthetist is called upon to attend to other possible problems outside the operating theatre. The answer must be an emphatic negative and there is little doubt that the courts would take the same view unless the anaesthetist could prove convincingly that his absence from the theatre was necessary to save another life. Often the anaesthetist is summoned from the operating theatre to take or make an 'urgent' telephone call. Anaesthetists often work singlehanded, are frequently called upon to assist with emergencies elsewhere and may be under considerable pressure to leave the patient in the care of untrained staff. The author recommends that all anaesthetic trolleys should be equipped with a telephone or that during prolonged surgery the anaesthetist should have an assistant trained in anaesthesia.

Respiratory depression

It is often difficult from the reports available to distinguish between events which are due to unrecognized and untreated respiratory depression or obstruction. Where there was clear evidence of untreated drug excess or failure to adequately reverse muscle relaxants, the cases were classified as due to respiratory insufficiency (see Table 15.1).

Twelve such cases leading to severe brain damage were reported to the Society. In nine of these, cardiac arrest occurred during the recovery period, whilst the patient was in the charge of nursing staff. In every case death or brain damage was clearly avoidable. In many instances there was clear evidence that the anaesthetist delegated the care of a patient who was still in an unsafe condition to nursing staff who were inadequately trained to recognize respiratory problems. In addition there was a failure in some of the cases to give specific instructions as to the degree of supervision required when the patient was still under the influence of respiratory depressant drugs.

Disconnection of apparatus

Unfortunately, by modern technical standards, much of the anaesthetic equipment in constant use in our hospitals falls below acceptable safety standards. For example, methods used for connecting patients to ventilating circuits are unstable and frequently fall apart. Constant vigilance by the anaesthetist in charge should of course avoid a tragedy from this cause but such tragedies do still occur. Five such cases were reported to the Society in the 10 year period under review. In one case the anaesthetist was known to have been absent from the theatre at the time of the event while in another the anaesthetist, who had no trained assistant, was setting up a CVP line in an emergency when the disconnection occurred and was undetected.

There is little excuse for failing to fit airway pressure alarms on all ventilator circuits, but this is still not the routine practice in the UK. These devices are readily available and reasonably reliable and should therefore be considered standard equipment. Their use would certainly reduce the incidence of unrecognized disconnections but cannot and should not replace constant vigilance by the anaesthetist.

Intubation problems

Problems in performing tracheal intubation was the second commonest cause of mortality or serious morbidity during anaesthesia in this series, accounting for 15% of all major claims. Although intubation problems will occur from time to time during

F

anaesthesia and failure to intubate successfully may be unavoidable, it is seldom possible to defend any case where a patient dies or suffers brain damage as a direct result of unsuccessful tracheal intubation. It is only in the most exceptional case that a competent anaesthetist will be unable to maintain adequate oxygenation with a mask during attempts at intubation.

Oesophageal intubation

A discussion of the hazards of oesophageal intubation with most anaesthetists produces a fairly standard answer — 'it commonly occurs but it is easy to recognize and should never be missed'. Is this always the case? Howells & Reithmuller (1982) described a case in which great difficulty was experienced in recognizing the mistake and the author has on record from MPS files 12 cases of oesophageal intubation in which cardiac arrest or brain damage occurred before the error was recognized. Five of these cases were in dark skinned patients in whom cyanosis was less readily detectable. In three of these cases more than one experienced anaesthetist confirmed the presence of tracheal intubation before cardiac arrest intervened and the tube was subsequently found to be in the oesophagus. In all cases clear respiratory sounds were said to be present on auscultation. The implication of these reports is that even a conscientious and careful anaesthetist may be unable to distinguish between a tracheal and oesophageal intubation by the commonly employed methods. It is possible that monitoring the end expired carbon dioxide is the only certain way of providing a clear differential diagnosis.

  We can learn three things from these observations.
1. We cannot rely upon auscultation of the chest and epigastrium to make the correct diagnosis.
2. If there is a failure to maintain oxygenation immediately after attempted tracheal intubation, then oesophageal intubation must be suspected, the tube withdrawn and the patient ventilated with a mask. Only in the event of a firm diagnosis of some other cause for cyanosis should artificial respiration be continued through the tube. If intubation has been difficult, then the tube may be kept in situ and lung inflation achieved with a mask placed over the tube.
3. Special care must be taken with dark skinned patients.

Unsuccessful intubation

Difficult or impossible intubation will occur from time to time and no anaesthetist should be criticized for failing to intubate a patient but failure to maintain adequate oxygenation of the patient during attempts at intubation cannot be defended. In this series there were three such cases leading to cardiac arrest or severe brain damage. In none of them was difficulty anticipated as a result of the pre-operative history. In each case many attempts at intubation were made using repeated doses of suxamethonium accompanied by periods of hypoxaemia.

  These rare tragedies could have been avoided if the anaesthetist had abandoned his endeavours after the second or third attempt. There are few operations which cannot be carried out without intubation or postponed until alternative procedures have been considered. Even emergency caesarian section or laparotomy can be performed more safely under general anaesthetia using a face mask and oesophageal tube rather than exposing the patient to considerably greater risk of repeated attempts at intubation.

Aspiration of vomit

Regurgitation with the inhalation of gastric contents can occur at any time during anaesthesia but death or brain damage from this cause is commonest during late pregnancy. Eight cases were reported to the Society during the period 1973–84, seven of which occurred during the induction of anaesthesia for operative obstetrics. In four of these cases both cricoid pressure and an adequate antacid therapy were reported as having been used. In the other four cases either cricoid pressure was not used or it was not reported as having been used. Since the introduction of cricoid pressure by Sellick (1961) the incidence of this complication has undoubtedly decreased, but despite these precautions, regurgitation and inhalation does still occur. This may well be due to the unwelcome fact that obstetric anaesthesia is still being administered by junior anaesthetists without adequate supervision. It would seem that an improvement in our anaesthetic technique for these vulnerable patients is still necessary (Green 1978).

Hypotension

Uncontrolled drug induced hypotension accounted for nearly 9% of the cases reported to the Society. Four were associated with controlled hypotensive anaesthesia. These cases have caused particular concern as all the patients were undergoing cosmetic surgery in which the degree of hypotension used was not considered to be essential for the success of the operation. The other four cases followed epidural or spinal anaesthesia. In two of these cases the regional anaesthesia resulted in permanent paralysis from anterior spinal artery thrombosis and two were maternity cases which suffered cardiac arrest following epidural anaesthesia for delivery (one vaginal delivery and the other for caesarean section).

OBSTETRIC ANAESTHESIA

It is not the intention in this chapter to single out any particular subspeciality of anaesthesia for discussion, but obstetric anaesthesia does merit our special attention for two reasons. The first is that it represents 12% of all anaesthetic related claims reported and secondly because the anaesthetist has dual responsibility for mother and child. Errors in technique and adverse reactions involving the mother can adversely affect the baby.

Five claims were for minor injury such as retained epidural catheters, post-spinal headache and awareness. The remaining 15 were either death or severe brain damage. Two of the major tragedies involved epidural anaesthesia, but the remaining 13 resulted from difficult intubation, oesophageal intubation or chemical pneumonitis as the result of inhaled gastric contents. The significance of these hazards during general anaesthesia in the obstetric patient has been emphasized in the confidential report into maternal deaths in the UK (1979). During 1976–79 there were 30 maternal deaths due directly to the administration of anaesthesia in pregnancy (17.2

deaths per million pregnancies). Eleven of these patients died as the direct result of inhalation of gastric contents and 16 from the complications associated with intubation. These figures are disturbing and have led to a considerable increase in the number of operative deliveries performed under epidural anaesthesia in the last 5 years. It remains to be seen whether this move towards regional rather than general anaesthesia reduces the hazards of operative obstetrics (Green & Taylor 1982).

Experience of claims against anaesthetists for episodes which occurred during general anaesthesia for operative obstetrics reinforce the importance of ensuring that a senior anaesthetist, experienced in obstetric anaesthesia, with a competent assistant, is always available for these patients. The recent trend to resort to epidural or spinal anaesthesia for operative obstetrics may reduce some of the hazards inherent in general anaesthesia but epidural anaesthesia for caesarean section may also introduce hazards such as hypotension and is not a technique to be delegated solely to junior anaesthetic staff.

MINOR ANAESTHETIC MISHAPS

The esteem of individual anaesthetists and the speciality in general is seldom judged by the occurrence of errors leading to a major catastrophe which are fortunately rare. Reputation is judged by the frequency of minor damage, such as bruises, cut lips, sore throats, corneal abrasions and muscle pains. Only a very small number of these minor injuries lead to claims for damages but nevertheless they result in considerable discomfort to patients. These comparatively minor mishaps can be avoided by the use of meticulous care and technique.

Dental damage

Damage to teeth and crowns represent by far the commonest minor injury leading to claims against the anaesthetist. Twenty-six claims for dental damage were reported (Table 15.2). This is a gross underestimate of the numbers that have occurred as many of these claims are considered without reference to the medical committee of the MPS and are not included in this series. The majority of these claims cannot be defended as it appears that the anaesthetists were unaware of the presence of crowns

Table 15.2   Minor claims against anaesthetists 1974–83

Cause of claim	Number of cases	Percentage of all claims
Dental damage	26	16.6
Peripheral nerve injury	9	5.8
Awareness	6	3.8
Extra vascular injection	4	2.6
Thrombophlebitis	4	2.6
Broken catheter in epidural space	2	1.3
Headache (spinal tap)	2	1.3
Cut cheek	1	0.6
Total	54	

in the upper jaw, had inserted an oral airway when a nasopharyngeal airway would have sufficed or had failed to use a crown guard during intubation.

Awareness

Awareness during anaesthesia is causing considerable concern as the incidence would seem to be increasing. In this series one of the five cases occurred during caesarean section and was successfully defended in court — the others could not be defended. Three cases arose during major abdominal surgery when the anaesthetic technique involved the use of a muscle relaxant, fentanyl, nitrous oxide and oxygen while one occurred during a difficult intubation. It is not always fully appreciated how close to awareness some patients are, when relaxant-narcotic techniques are employed and the extent of psychological trauma that results even after a short period of awareness. A recent award of £13 000 made in an English court, to a mother who experienced awareness during caesarean section might well be a forerunner to further claims.

MONITORING AND RECORD KEEPING

Monitoring

Despite the development during the last decade of many excellent monitors available for the anesthetized patient there is only one monitoring system that should be relied upon for the ultimate safety of the patient and that is the three senses of the well trained anaesthetist — sight, sound and touch. Any damage to a patient which is accompanied by a failure to apply these senses continuously during the course of an anaesthetic is unlikely to be defendable in a court of law. Fortunately, major anaesthetic tragedies are rare, but it may be this very infrequency of human error leading to major patient damage which lulls many anaesthetists into a sense of false security and allows him to relax the standard of vigilance that is needed to avoid the tragedies that do occur. If we were to extrapolate from the mortality figures reported by Lunn & Mushin (1982), in which they recorded 227 deaths related to anaesthesia in 1 147 000 operations, half of which were considered avoidable, we can expect to have one avoidable anaesthetic related death in every 10 000 anaesthetics administered. This is, on average, about one in every 10 years of an anaesthetist's professional career.

If individual human vigilance is not to be relied upon, then other supplementary monitoring and alarm systems must be employed. Each device has its own range of usefulness and no instrument on its own will cover all respiratory or circulatory situations. At the present stage of our knowledge of monitoring techniques and the ever increasing liability of being involved in litigation, some if not all the following 'fail safe' non-invasive monitoring devices should be considered mandatory in all but very short minor cases:

Automatic sphygmomanometer	(with high–low level alarm)
Digital pulse monitor	(with high–low rate alarm)
Electrocardiogram	(with high–low rate alarm)

Capnometer	(with high–low CO_2 and respiratory rate alarm)
Disconnection alarm	(on every ventilating system)
Oxygen analyser	(with high–low O_2 alarm)

Unfortunately the most readily available monitoring devices do not give a sufficiently rapid indication of the commonly occurring ventilation problems. For example, the ECG, pulse monitors and automatic sphygmomanometer will not alert the anaesthetist to oesophageal intubation, disconnection, kinked tube or anoxic mixtures until long after the observant anaesthetist has located the problem or the patient has already been irreparably damaged. The oesophageal intubation requires a capnograph for rapid diagnosis of the problem: the disconnection — an incircuit airway pressure alarm and the kinked tube or anoxic mixture — an O_2 analyser, in order to alert the anaesthetist in time.

Mishaps may occur when the anaesthetist has become fatigued during a prolonged or protracted operating list and cannot be deemed to be negligent. It is to protect the patient in this situation that monitoring devices with alarms are essential. It is however imprudent to impose minimal requirements for monitoring as any firm recommendation will tend to become mandatory when judgements in law are involved. The type of operation may dictate the nature of the monitoring which undoubtedly would be more extensive for complex cardiac surgery than that for minor operative procedures. In some countries the use of routine temperature monitoring would be considered essential for the early detection of malignant hyperpyrexia. The introduction of numerous alarm systems into the operating theatre may make the noise emitted by these warning devices difficult to interpret (Kerr 1985). Nevertheless the use of a blood pressure recorder, pulse monitor and perhaps an electrocardiogram on all patients requiring anaesthesia for more than a few minutes should be considered essential.

Record keeping

One of the most common difficulties encountered by those trying to defend the anaesthetists against negligence claims made by their patients, is the lack of a proper record of the events which took place. Contemporaneous record keeping is often difficult or impossible during anaesthesia especially during a crisis, but there is absolutely no merit in distorting the account of events to ameliorate the situation. Attempts at transforming an indefensible case to a defensible one can readily be detected by a discerning prosecuting counsel, adding to the torment and anguish of the anaesthetist.

PATIENT RECOVERY

The recovery room is now an integral part of the operating theatre suite and no anaesthetist would question the value of the facilities they offer. He must always be aware however that patient safety during the recovery period can only be judged by the efficiency of the staff who run them. In this survey of claims 20% of the major injuries resulting from anaesthesia (death or brain damage) occurred during the recovery period, and most of these in a properly designed recovery room.

In a survey of events which occurred in the recovery room Eltringham (1979) noted that cardiovascular complications were the most common cause for concern. It is therefore interesting to observe that in this series where claims for damages were involved, 15 out of the 21 events which occurred in the recovery period were respiratory problems in which the nursing staff failed to respond to the situation. In eight cases the respiratory problem was due to uncorrected respiratory obstruction. In the remaining nine cases there was clear evidence of an unacceptable degree of respiratory depression when patient care had been transferred to the recovery ward staff. In a significant number of these cases the anaesthetist had already left the theatre suite when cardiac arrest and/or subsequent brain damage occurred.

These rare but tragic events inevitably lead to a claim for damages which can seldom be defended. This poses the question of liability after patient care had been delegated to nursing staff. At present in the UK, the courts are likely to take the view that the anaesthetist must accept the major responsibility for the care of his patient during the recovery period until that patient is conscious and able to control his own vital functions. Should the evidence show that the nurse in charge has fallen short of her expected skills, then the hospital authorities may have to accept some of the responsibility, but the major liability will always remain with the anaesthetist. As improved recovery facilities are developed and higher standards for training of recovery ward staff are promoted, this legal attitude may change. In Canada and the USA where recovery room services have been established for over 25 years a different view is often held and incidents leading to patient damage which occur in the recovery rooms are considered to be the primary responsibility of the hospital authority nursing staff, and not the anesthesiologist (Farman 1979).

STATUS OF ANAESTHETIST

In the UK the National Health Service aims to provide a trained anaesthetist (Consultant or Senior Registrar) for the majority of routine surgical procedures performed under general anaesthesia whereas in the private sector fully trained anaesthetists are practically always involved in providing anaesthesia. Despite these overall principles 60% of the claims against anaesthetists which reached the Society were against junior anaesthetists working without supervision. One of the most disturbing features arising out of the analysis of these claims is that in many cases the anaesthetist was clearly insufficiently trained to manage the particular case by himself. An inadequate staff structure was often the reason quoted for persuading junior staff to undertake work beyond their capabilities. More often it was the failure of the consultant in charge of administration to ensure that suitably trained staff were available and assigned to operating lists appropriate to the stage of their training.

CONCLUSION

It is hoped that this analysis of cases will be helpful in clarifying the problems of morbidity and mortality in relation to anaesthesia since most of the cases were uncomplicated without coexisting or life threatening disease. Claims against such doctors are not usually made when serious complications arise in patients with pre-existing disability.

It is also hoped that a review of a few of the mishaps which have occurred during the last decade will alert anaesthetists to the problem and help to prevent some of the mishaps occurring in the future.

'Where the consequence of error is certain to be disastrous it is impossible to be too careful.'

REFERENCES

Dinnick O P 1964 Deaths associated with anaesthesia: a report of 600 cases. Anaesthesia 19: 536–556
Edwards G, Morton H J V, Pask E A, Wylie W D 1956 Deaths associated with anaesthesia: a report of 1000 cases. Anaesthesia 11: 194–220
Eltringham R J 1979 Complications in the recovery room. Journal of the Royal Society of Medicine 72: 278–280
Farman J V 1979 Do we need recovery rooms? Journal of the Royal Society of Medicine 72: 270–272
Fisher J L 1970 Recovery from anaesthesia. Canadian Medical Association Journal 109: 348–350
Green R A 1978 Anaesthesia for caesarian section. Anaesthesia 33: 70–71
Green R A, Taylor T H 1984 Anaesthesia mishaps in the UK. International Anesthesiology Clinics 22: 73–89
Holland R 1984 Anaesthesia related mortality in Australia. International Anesthesiology Clinics 22: 61–71
Howells T H, Reithmuller R J 1980 Signs of endotracheal intubation. anaesthesia 10: 984–986
Kerr J H 1985 Warning devices. British Journal of Anaesthesia 57: 696–708
Lunn J N, Mushin W W 1982 Mortality associated with anaesthesia. Nuffield Provincial Hospital Trust, London
Report on Confidential Enquiries into Maternal Deaths in England and Wales 1970–72, 1975 and 1979. H M Stationary Office, London.
Sellick B A 1961 Cricoid pressure to control regurgitation of stomach contents during the induction of anaesthesia. Lancet ii: 404–406
Solazzi R W, Ward R J 1982 The spectrum of medical liability cases. International Anesthesiology Clinics 22: 43–59

Anaesthesia and laser surgery

The problem created for the anesthetist by the introduction of the surgical laser is the risk not only of the combustion of volatile anaesthetic agents but also the ignition of equipment such as plastic endotracheal tubes which readily ignite in the presence of oxygen. Laser light, unlike ordinary light, retains a high intensity of energy at great distances from its source. It is said to be coherent in that the light is not only of a single wavelength but also oscillating in time. Thus the essentials of laser light are that it is monochromatic and appears as a non-divergent pencil beam of light. In contrast ordinary light is a mixture of many wavelengths and is emitted in all directions. Lasers are named after the materials from which the light is emitted. The material determines the laser's wavelength, which in turn governs the absorption of the light in tissues and hence the use of that particular laser. At present the most widely used lasers in medicine are the carbon dioxide, argon and Nd-Yag (neodymium-yttrium aluminium garnet) lasers.

CARBON DIOXIDE LASER

The wavelength of light from the carbon dioxide laser is 10.6 μ, which is in the far infra-red. The light from the carbon dioxide laser is absorbed by water and glass. In the tissues light energy is strongly absorbed by cellular water, causing it to boil, and this bursts the cell. Thus the cell temperature does not exceed 100°C, causing minimal heating of adjacent cells. This in turn produces very little oedema so that the carbon dioxide laser is now widely used in laryngeal surgery. It is in the larynx in particular that the laser beam can come into contact with the endotracheal tube.

All plastics and rubbers are heated by the carbon dioxide laser at a rate of approximately 5000°C per second, thus causing conflagration in an atmosphere enriched by oxygen or nitrous oxide. It should not be forgotten that nitrous oxide supports combustion more vigorously than 100% oxygen (Leonard 1975, Hirshman & Smith 1980, Chilcoat et al 1983). The ignition of the endotracheal tube is instantaneous, very vigorous and occurs without warning. Combustion is similar in appearance to a magnesium flare; toxic fumes are also emitted. If this happens at operation it is highly unlikely that the patient will survive. Any flammable anaesthetic agents such as diethyl ether or cyclopropane are absolutely contra-indicated when lasers are used. Earlier work (Leonard 1975) suggested that halothane, enflurane and isoflurane were not flammable in the concentrations used in anaesthesia.

There are a number of methods for avoiding ignition of the endotracheal tube although none is perfect (Hermens et al 1983, Gupta et al 1984). It should be stressed again that ignition of the endotracheal tube is to be avoided at all costs, otherwise results are likely to be catastrophic. However, with due care and attention accidents can be prevented.

Adults

There are three methods of preventing ignition of the endotracheal tube:

1. By protecting the standard tube.
2. By the use of a specially designed laser proof tube.
3. By avoiding the use of an endotracheal tube.

Protected standard endotracheal tube.

Three methods have been suggested for protecting standard tubes. The most widely used is half inch (1.25 cm) wide adhesive aluminium tape which is wound spirally around the tube so that it just overlaps, from the upper edge of the cuff for a distance of approximately 4 in (10 cm). The winding starts at the cuff in order to prevent the beam from getting under the edge of the tape (Fig 16.1). The optimum size of tube is a 5 mm microlaryngeal tube which allows a tolerably good view of the vocal cords and also protects the airway. Artificial ventilation is effective with this tube but a disadvantage is that, although the wrapped portion is laser proof, it is virtually rigid, can kink easily and the edges are traumatic to the vocal cords. This is the reason for limiting the protected area of the tube to 4 inches.

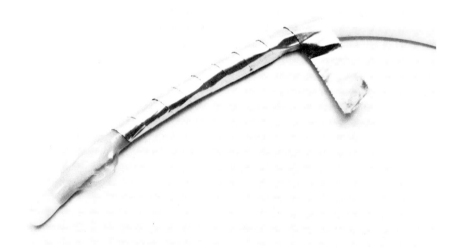

Fig. 16.1 Aluminium protected standard endotracheal tube.

During laser surgery it is essential to protect the cuff by inserting a pledget of wet cotton wool between the vocal cords and the cuff. As this pledget readily dries out it is frequently replaced by the surgeon to keep the area moist. It is also advantageous to fill the cuff with saline instead of air: this allows a leak in the cuff to be readily identified and at the same time reduces the likelihood of ignition. It is inadvisable to attach the tube firmly to the patient because in the event of ignition the tube should be removed as quickly as possible. Incandescent particles have been reported to have caused ignition of a tube on one occasion (Hirshman & Smith 1980).

The second method of protecting the tube involves the use of wrappings of thin aluminium foil. The aluminium foil consists of the leader portion of recording tapes which is not only thin and difficult to apply but is laminated with a thin sheet of plastic which is also flammable.

The third method suggested is the use of wet gauze but this is unwieldy as it has to be wrapped evenly around the tube. It also has to be kept in position and must be wet but not moist, otherwise it will be just as inflammable. A tube prepared in this way is also very bulky. These alternative methods cannot be recommended as they are less effective than the use of aluminium tape in protecting the tube.

There has been considerable debate as to whether tubes made of plastic are safer than those made of rubber. The use of either material can be dangerous. Although the flammability and penetration of rubber are slightly less, the vapours emitted from rubber are more toxic than from plastic. There is a minor advantage in the use of PVC in that combustion is self extinguishing if there is no more oxygen (Russell 1983).

Pashayan & Gravenstein (1985) have recently shown that a mixture of 40% oxygen and 60% helium increases the mean time to ignition to 42.6s, or 25.3s with 2% halothane, at a laser power of 10 W. It has therefore been suggested that the duration of each burst of laser activity should be restricted to less than 10s.

Specially designed endotracheal tubes

Norton & De Vos (1978) have produced a laser proof endotracheal tube which so far has proved most successful. It is made of stainless steel, is flexible and is relatively non-traumatic to the vocal cords. It is completely laser proof but unfortunately is thick walled and has no cuff. A small tube has to be used to obtain a good view of the vocal cords but the resistance to expiration is high. It can be used in conjunction with an injector technique, especially in patients with chronic pulmonary disease or obesity (see below).

An attempt has been made to produce a laser proof cuffed tube that is flexible and soft with thin walls. Tubes were coated with aluminium flakes, except for the cuff, and then covered with a thin reflective layer of aluminium, silver or gold. This was applied by vacuum extraction of metal vapour and resulted in a soft and flexible standard tube with a high degree of reflective surface. On inspection with the naked eye the tube appeared to be satisfactory but unfortunately laser resistance was present on only 90% of the surface. Reluctantly, the project had to be abandoned (Russell 1983).

A further attempt to produce a laser proof tube has resulted in the manufacture of a silicone tube which is largely covered by aluminium paste. In an atmosphere of less than 25% oxygen in nitrogen it will not fully ignite. When the laser beam impinges on its surface it temporarily ignites producing silicone dioxide which forms a flame-shield (Gabba et al 1984). Thus the tube has the advantage that in air it is laser resistant but not laser proof, provided it is not hit repeatedly by the laser beam. The silicone tube is soft and flexible with a cuff like an ordinary endotracheal tube. Unfortunately it must only be used in the presence of air as it ignites like PVC if the oxygen concentration is greater than 25%. Nitrous oxide is also not permissible as it supports combustion even better than oxygen. Once this tube does ignite it is not self extinguishing in air. Very few patients who require laser surgery to the larynx are fit and healthy, and since the anaesthetic techniques employ supplementary oxygen the use of the silicone tube is limited. The tube may also be used with pulsed lasers and at a maximum energy level of 3900 W/cm².

The conditions under which silicone tubes can be used for laser surgery are clearly stated on the packaging and these must be strictly adhered to otherwise unnecessary fatalities may arise.

General anaesthesia without an endotracheal tube

The high pressure Venturi injector, which has been widely used for many years for bronchoscopy, offers an attractive alternative to the use of an endotracheal tube. The normal surgical laryngoscope is tubular and acts as an attachment for the injector (Fig 16.2). Ventilation of the patient is adequate in the majority of patients provided the tip of the laryngoscope is near the cords. The laryngoscope is completely laser proof, it gives a clear view of the larynx, it is not traumatic and during the procedure a high percentage of oxygen may be given. As with bronchoscopy, general anaesthesia is

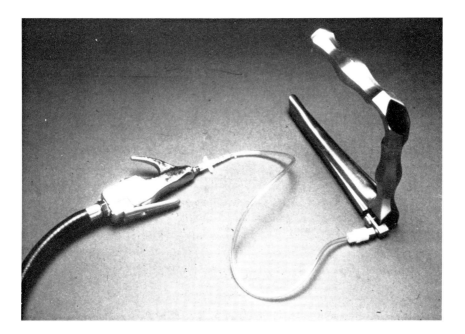

Fig. 16.2 Normal surgical laryngoscope acting as attachment for injector.

maintained intravenously. The author can only recall one obese patient who was difficult to ventilate adequately and who had two strictures of the trachea, one of which was subglotic and the other further down the trachea. In such circumstances a smaller size of Norton tube may be used as an extension to the injector. Suggestions that the injector might be used with other plastic endotracheal extensions such as the Bennett tube are untenable as any plastic or rubber which comes within the operative field must be laser proof.

The possible disadvantages of the injector are twofold. The slight movement of the vocal cords between inspiration and expiration produced by the change in gas flow may result in small lesions of the cords moving out of view of the operating microscope. In practice this can be overcome by keeping the respiration regular so that the surgeon can predict the movement of the cords. The other theoretical disadvantage is that where a lesion is on the medial edge of the cords, the trachea below this is unprotected. In practice this does not seem relevant.

For the majority of adult patients it is the author's preference to use the injector technique. Laryngoscopy is performed with either a wrapped tube or an unwrapped tube in situ, and anaesthesia is maintained with a combination of inhalational agents and muscle relaxants. After the lesion has been assessed by the surgeon and laser therapy is indicated, the tube is removed and the injector is applied. Inhalational agents are withdrawn and anaesthesia is maintained by intravenous drugs. Close cooperation between the surgeon and anaesthetist is particularly important when carbon dioxide lasers are applied to the larynx (Conacher et al 1985). The use of high frequency ventilation may have a place in laser surgery.

Children

In children where the larynx is so much smaller, a technique without an endotracheal tube becomes even more attractive. Much has been written about the advantages and disadvantages of using injectors in children, especially the dangers of causing pneumothorax, but despite this the injector has its advocates (Scamman & McCabe 1984).

The carbon dioxide laser is the treatment of choice for juvenile viral papillomata as it does not stimulate further profusion of the growth. This unpleasant condition may present in young children of 1 or 2 years of age in whom the small larynx becomes invisible and impossible to intubate. The author's preference is to anaesthetize these children with deep inhalational anaesthesia with spontaneous respiration throughout (Rita et al 1983). Anaesthesia is induced by either intravenous or inhalational methods and maintained with oxygen and halothane. Halothane is preferred to enflurane as the end-tidal carbon dioxide is less under deep anaesthesia. When the level of anaesthesia is adequate for laryngeal surgery to be tolerated an 8 or 10 French gauge suction catheter is passed through the nose into the nasopharanx, with a marker 3–4 cm from the tip to ensure that it does not protrude from the nasopharanx into the operating field. High flows of 6–8 l/min of oxygen are insufflated down the tube, 3% halothane being required to maintain the correct depth of anaesthesia. A high flow is also necessary to prevent incandescant particles from becoming lodged in the catheter during expiration (Hirshman & Smith 1980). This technique results in low arterial PCO_2 levels and may even cause apnoea (Van Hasselt et al 1985).

Tracheostomy tubes

Tracheostomy tubes in either adults or children need to be protected in the same way as laryngeal tubes. This is best done by changing to a silver Negus tracheostomy tube or some such similar metal tube. As this tube is metal it is laser proof. If the patient already has a fenestrated Negus tube in position it should be changed for a non-fenestrated one in order to facilitate ventilation. As a stricture or stenosis is lasered there may be problems of excessive leakage causing difficulties with ventilation, which may be overcome with the Venturi injector as above.

It is important to have metal connectors joining the tracheotomy tube to the anaesthetic circuit. There is a danger too that if a fenestrated tube is inadvertently used plastic or rubber connections are a risk. A metal connector suitably wedged forms a better and safer connection. Anaesthesia is maintained by inhalational agents but in adults muscle relaxants are also used to reduce movement of the vocal cords.

Oral lesions

These are also increasingly treated with the carbon dioxide laser. Although there remains a risk that the endotracheal tube may come into the field of the laser, protection is easier. An ordinary full size endotracheal tube with a metal connection may be wrapped with aluminium tape, leaving 3 cm clear above the cuff so that the larynx is not damaged. If the lesion concerned extends anywhere near the larynx, it should be regarded as a laryngeal lesion and treated as such. Wrapping of the tube must start at the lowest point and work upwards and over the connector so that no plastic is showing in the surgical field. It is advisable to move the tube to the opposite side of the mouth from the lesion. With the airway thus secured any suitable general anaesthetic may be administered.

THE ND-YAG LASER (Neodymium-yttrium aluminium garnet)

The laser works in the near infra-red range and therefore has the same theoretical risks as the carbon dioxide laser. The principle difference is that the light can be transmitted down fibre-optic pathways and can therefore be applied more directly to the lesions. There should be no flammable materials between the end of the fibre-optic instrument and the area to which it is directed. If there is any doubt, similar precautions should be taken as detailed for the carbon dioxide laser. It is fairly easy to arrange in vitro tests to assess flammability if there is further doubt.

ARGON LASER

This laser does not generally produce any problems for the anaesthetist. It produces visible light in the blue green spectrum and is used where it may be absorbed by red or black pigments. It is advocated for treatment of port wine stains of the face, where it may possibly come in contact with the endotracheal tube and cause ignition of a red rubber tube. Although in the majority of cases this is unlikely, it is prudent to use a plastic disposable tube wrapped in aluminium tape as outlined above.

REFERENCES

Chilcoat R T, Byles P H, Kellamn R M 1983 The hazard of nitrous oxide during laser endoscopic surgery. Anesthesiology 59: 258

Conacher I D, Paes M C, Morritt G N 1985 Anesthesia for carbon dioxide laser surgery on the trachea. British Journal of Anaesthesia 57: 448–450

Gabba D M, Hayes D M, Goode R L 1984 Incendiary characteristics of a new laser-resistant endotracheal tube. Anesthesiology 61: A147

Gupta B, Lingham R P, McDonald J S, Gage F 1984 Hazards of laser surgery. Anesthesiology 61: A146

Hermens J M, Bennett M J, Hirshman C A 1983 Anesthesia for laser surgery. Anesthesia and Analgesia 62: 218–229

Hirshman C A, Smith J 1980 Indirect ignition of the endotracheal tube during carbon dioxide laser surgery. Archives of Otolaryngology 106: 639–641

Kroll D A, Morris M D, Norton M L 1984 Hazards of laser degradation of methylmethacrylate. Anesthesiology 61: 115–116

Kumar A, Frose E 1981 Prevention of fire hazard during laser microsurgery. Anesthesiology 54: 350

Leonard P F 1975 Lower limits of flammability of halothane, enflurane and isoflurane. Anesthesia and Analgesia 54: 238–240

Norton M L, De Vos P 1978 New endotracheal tube for laser surgery of the larynx. Annals of Otology, Rhinology and Laryngology 87: 554–557

Pashayan A G, Gravenstein J S 1985 Helium retards endotracheal tube fires from carbon dioxide lasers. Anesthesiology 62: 274–277

Rita L, Seleny F, Holinger L D 1983 Anesthetic management and gas scavenging for laser surgery of infant subglottic stenosis

Rogers R C, Gibbons J, Cosgrove J, Copple D L 1985 High-frequency jet ventilation for tracheal surgery. Anaesthesia 40: 32–36

Russell C A 1984 Tracheal tubes for laser surgery. Anaesthesia 39: 293–294

Scammen F L, McCabe B F 1984 Evaluation of supraglottic jet ventilation for laser surgery of the larynx. Anesthesiology 61: A447

Van Hasselt G, Wainwright A C, Carruth J A S 1985 Personal Communication

Some aspects of the use of narcotics in general anaesthesia

INTRODUCTION

An opiate or narcotic analgesic is a drug (synthetic or naturally occurring) having, to a greater or lesser extent, morphine like properties. The term opioid was coined by Acheson in 1957 and defined as a synthetic morphine substitute. The meaning has now widened to include any substances which bind to opioid receptors and may therefore include compounds having no analgesic action.

Analgesia

Analgesia is defined as a painless state and this is reflected in the demotic expression 'pain killer'. In common medical usage an analgesic drug is one that reduces the sensation of pain. There is a difficulty here because, if the subject feels any pain at all, he cannot be in the anaesthetic state which implies the absence of awareness. However, as will be described below, it has been shown that analgesic drugs can significantly decrease the concentration of inhalational agents required to produce anaesthesia and therefore the anaesthetic state can be regarded as having an analgesic component. At least that is one interpretation of the findings.

Narcotics

Narcotics were used before the introduction of general anaesthesia and continue to be used as adjuncts either as premedicant agents or to allay post-operative pain. Smith (1908), in a paper on morphine-scopolamine anaesthesia, cited 157 references on the subject. Scrutiny of the paper showed that morphine and scopolamine alone was used in only four cases and having been found to be unsatisfactory the two drugs were thereafter used as supplements to chloroform.

Crile and Lundy, who wrote classical papers on the anaesthetic state, considered that the desirable analgesic component of that state was best achieved with the use of local anaesthetic agents and, indeed, recent work has confirmed their value in suppressing the metabolic response to surgery. Abolition of the autonomic reflex response to surgical stimulus was not considered an integral part of general anaesthesia until the publication of a paper entitled 'Changing concepts concerning depth of anesthesia' by Woodbridge (1957). But before this date it had become apparent that nitrous oxide alone was not always able to provide adequate anaesthesia in the paralysed patient, and in 1947 Neff et al published the first paper describing the use of a narcotic analgesic —pethidine — as a supplement to nitrous oxide analgesia. This technique was introduced to the UK following a visit to Dr Neff at Stanford by Drs Mushin and Randell-Baker (1949) who advocated a method using 66% nitrous oxide with gallamine or d-tubocurarine and incremental doses of 25 mg of pethidine at 25–30 minute intervals. Techniques of this type immediately became very popular and have continued in widespread use ever since. Early work on the subject is reviewed by Siker (1956) and of interest is a paper describing laryngeal intubation under pethidine (Ruben & Andreassen 1951). Also of interest is a double blind comparison in 113 patients (Holmes 1976) of equianalgesic doses of pethidine, morphine, phenoperidine and fentanyl as a supplement to nitrous oxide anaesthesia. Very little difference was found between the four opiates and Bouill et al (1984) have commented that it is a characteristic of opiates that when they are combined with anaesthetic agents differences between them tend to be obscured.

Reduction of the amount of anaesthetics required

Claude Bernard, in a lecture published in 1869, stated that morphine injected into dogs before anaesthesia reduced by half the amount of chloroform needed for surgical anaesthesia (Schrier 1982). A similar view was expressed on the effect of morphine on cyclopropane anaesthesia by Waters & Schmidt (1934). Experimental work in support of this is reviewed by Robbins (1958) in his monograph on cyclopropane. Morphine was found to produce marked reductions in the blood cyclopropane concentrations required for various levels of anaesthesia. Two groups of workers (Seevers et al 1934, Robbins et al 1939) found that although morphine reduced the blood concentration of cyclopropane necessary for anaesthesia, the concentration at which respiratory arrest occurred was not altered. It was therefore considered that morphine increased the safety margin in the use of cyclopropane.

In the human subject 10 mg of morphine given intravenously 10 minutes before anaesthesia was shown to reduce by 15% the blood ether concentration required for a given depth of anaesthesia as defined by electroencephalographic changes (Taylor et al 1957).

Later work has approached this subject by demonstrating reductions in maximum alveolar concentration (MAC) for various anaesthetic agents produced by narcotic analgesics. Reductions in MAC for human subjects given pre-anaesthetic doses of morphine in the range 8–15 mg have been shown for halothane to be —7% (Saidman & Eger 1964) and for fluroxene —20% (Munson et al 1965).

More recently experiments to measure the anaesthetic sparing effects of various doses of different narcotic analgesics have given interesting results. Initial work showed that the strong narcotic analgesics, morphine and pethidine (μ agonists) reduced the cyclopropane requirement for anaesthesia in rats by an amount directly proportional to the log dose of the analgesic, but the agonist/antagonist drug pentazocine showed a ceiling effect, in that at doses above 20 mg/kg no further reduction in anaesthetic requirement occurred (Hoffman & Difazio 1970). This finding for narcotic analgesics of the agonist/antagonist class has been confirmed in dogs for butorphanol with which maximum reduction of enflurane MAC was found to be 11% and nalbuphine, for which the maximum reduction in enflurane MAC was found to be 8% (Murphy & Hug 1982a). Of great interest is the finding by the same authors that if a sufficiently high dose is used a ceiling effect is found with morphine and also with fentanyl (Murphy & Hug 1982b). However large the dose of morphine or fentanyl the MAC could not be reduced by more than 65%.

Naloxone

The introduction of naloxone, an effective narcotic antagonist active at the μ receptor, promised a more flexible approach to effective analgesia. However, it soon became apparent that it was best reserved for emergency use. The duration of effective action of one bolus dose in the clinical range (0.1–0. 4 mg) does not usually exceed 30 minutes so that most patients who have received a relative overdose of any one of the commonly used narcotic analgesics require either a continuous slow infusion or repeated doses; both methods needing close supervision. An alternative method of supplemental intramuscular doses of naloxone has been advocated (Longnecker et al 1973) but there must always be a degree of uncertainty about the rate of absorbtion by this route.

Naloxone reverses analgesia to the same or similar degree that it reverses respiratory depression, but it is possible by judicious titration to reverse respiratory depression sufficiently while retaining a degree of analgesia (Krycke et al 1976, Drummond et al 1977). Because of the considerable differences between the kinetics of the drugs involved, such a titration technique requires very close supervision.

There are further objections to the routine use of naloxone; it may cause vomiting even with small doses (Longnecker et al 1973), restlessness and shivering (Tammisto & Tigerstedt 1979), pruritis and copious diuresis (Smith 1984), serious or indeed fatal ventricular dysrhymthias (Michaelis et al 1974, Andree 1980), acute left ventricular failure with severe pulmonary oedema (Flache et al 1977) and severe hypertension (Azar & Turndorf 1979, Estillo & Cottrell 1981). Most of these side-effects have been described in patients who were already seriously ill but Prough et al (1984) have recorded two cases of acute pulmonary oedema in healthy teenagers following small doses of naloxone given post-operatively to reverse doses of 0.5 mg of fentanyl given during the course of minor surgery. These authors consider that naloxone may cause neurogenic pulmonary oedema.

The severe cardiovascular effects sometimes seen when naloxone is used to reverse narcotics resemble those of massive catecholamine release (Tanaka 1974). Patschke et al (1977) observed increases in heart rate, cardiac output and blood pressure in anaesthetized dogs on administering naloxone after large doses of morphine: they considered that these effects might be due to an acute abstinence syndrome. These effects could not be due to stress and pain because the dogs remained anaesthetized while the effects were seen.

Finally, the work of Kruynack & Gintautas (1982) should be noted: they proposed that naloxone may govern the duration of narcosis through the actuation of an opposing arousal system in the CNS unrelated to pharmalogical competition for opiate receptors.

SEQUENTIAL ANALGESIA

In a classic paper Lasagna & Beecher (1954) showed that nalorphine, which had hitherto been regarded solely as an antagonist of morphine, did itself have analgesic properties in man, and the possibility that mixtures of morphine and nalorphine might be clinically useful was investigated by a number of workers. The object of these investigations was to see if it were possible by the use of such mixtures to separate, or at least diminish, the undesirable side-effects associated with narcotic analgesia, particularly respiratory depression. Nalorphine does oppose the respiratory depression produced by morphine (Bellville & Fleischli 1968), but the relationship in action between the two drugs proved to be of a complex biphasic nature (Houde & Wallenstein 1956) and it was these observations that led to the concept of 'receptor dualism' (Martin 1967).

Nalorphine

Nalorphine, in doses sufficient to have an analgesic action, also has a high incidence of psychotomimetic effects (dysphoria, delirium and hallucinations) but the introduction of similar compounds with a lower incidence of these σ-agonist actions makes possible a clinical use of the drug interactions described above.

Fentanyl

In 1968 de Castro & Viars described a technique in which high doses of fentanyl were used as the sole anaesthetic agent and the respiratory depressant action reversed by a large dose of pentazocine which left the patients breathing adequately but having analgesia for the immediate post-operative period from the pentazocine. They named this technique anésthesie analgésie sequentielle (AAS), a name which unfortunately embodies some confusion between analgesia and anaesthesia. Rifat (1972) described a series of elderly patients undergoing hip surgery who, after intravenous induction and intubation (using propranidid and suxamethonium), received fentanyl 25 μg/kg and alcuronium and were ventilated with 50/50 N_2O/O_2. Additional doses of fentanyl 125 μg were given if required. At the end of the operation, after reversal of the relaxant and withdrawal of the nitrous oxide, 48 of the 52 patients studied remained unconscious and were apnoeic or ventilation was inadequate. The were given pentazocine 1 mg/kg and all resumed adequate ventilation and regained consciousness. After extubation all maintained adequate blood gases, blood pressure and central venous pressure and in only three patients was any further analgesic needed in the first five post-operative hours.

Pentazocine

Pentazocine, although more useful than nalorphine, is nevertheless an unsatisfactory drug in many ways. It has a weak analgesic action reaching a ceiling at a dose of about 1 mg/kg and at or above this level the incidence of psychotomimetic effects becomes unacceptably high, reaching 50% is some series (Wood et al 1974). Pentazocine also has undesirable haemodynamic effects and may increase heart rate and systemic and pulmonary blood pressure (Alderman et al 1972) due to, or at least associated with, catecholamine release not connected with hypercarbia (Tammisto et al 1979).

Nalbuphine

Nalbuphine appears considerably superior to pentazocine for use in sequential analgesia. It is considerably more potent than pentazocine as a μ antagonist, being about one fourth as potent as nalorphine in this respect (Jasinski & Mansky 1972). In common with other drugs in this class, both respiratory depression (Romagnoli & Keats 1980, Julien 1982) and analgesia (Gal et al 1982) appear to be limited, i.e. there is a ceiling effect, at about the same dosage level of approximately 0.15 mg/kg. At this dose the analgesic potency is equal to or slightly less than the same dose of morphine and has a similar duration of action (Beaver & Feise 1978). The incidence of psychotomimetic effects seen with nalbuphine is very much less than with pentazocine (Jasinski 1979, Stambaugh 1982).

Magruder et al (1982) reported a series of 15 patients having a wide variety of operations under anaesthesia by a nitrous oxide, narcotic and relaxant technique after thiopentone induction. The narcotic analgesic used was oxymorphone or hydro-morphone and the dose given (0.05 mg/kg plus 0.05 mg) was equivalent to 30 mg of morphine in a 70 kg patient. At the end of the surgery, after reversal of the muscle relaxant and extubation, all the patients had inadequate ventilation confirmed by blood gas estimation and were given nalbuphine 0.1 mg/kg. The response was dramatic and in all patients blood gas estimation showed that ventilation had returned to normal. There were no episodes of renarcotization and analgesia remained adequate for an average period of 7 hours. Following this the authors adopted a routine of 0.1 mg/kg of nalbuphine intravenously and 0.2 mg/kg intramuscularly given post-operatively in 150 patients. They found that this provided protection against respiratory depression even after total morphine doses of 2–3 mg/kg. The only adverse effect observed was a transient pressor response to nalbuphine producing an increase in heart rate and mean arterial pressure of about 20% above control value.

Latasch et al (1984 described a series of 60 patients undergoing elective general surgery who were anaesthetized by a fentanyl–relaxant–nitrous oxide technique. The dose of fentanyl used was 7 μg/kg, plus a standardized dose of 0.2 mg every 15 minutes. At the end of the operation, after reversal of the relaxant, a further 0.2 mg of fentanyl was given to ensure apnoea during transfer to the recovery room where 20 mg of nalbuphine was given intravenously to all the patients. As in the previous series there was a dramatic and rapid response. In no patient was more than 1.5 minutes required from apnoea to ventilation adequate to maintain normal blood gas, and after 2 minutes all patients were so alert that they required immediate extubation. Pain was measured by a visual analogue scale and pain relief was found to be excellent throughout the following 3 hours without further nalbuphine. No renarcotization was seen in this series and the authors note that no significant changes in heart rate or blood pressure occurred. Side-effects included vomiting in 5 patients and shivering in 12 patients immediately after the nalbuphine. As already mentioned shivering is commonly seen when naloxone is administered.

Finally, Moldenhauer et al (1985) reported on 21 patients undergoing aorto-coronary bypass who were anaesthetized by a high dose fentanyl technique. In the first group of patients, who received a mean dose of 120 μg/kg of fentanyl, reversal by nalbuphine was carried out on the morning after the day of surgery. In the second group, who received a mean dose of 97 μg/kg of fentanyl, nalbuphine was given within 8 hours of the end of surgery. In this series nalbuphine was administered with great caution in 15 μg/kg increments (i.e. 1.05 mg for a 70 kg man) every 30–60 minutes until the Pa_{CO_2} decreased to <48 mmHg or until a total dose of 150 μg/kg had been given. The total dose of nalbuphine required for each patients was in the range 1–10 mg. Three patients in the first group experienced an immediate increase in pain on receiving nalbuphine which was relieved by giving additional nalbuphine. Four patients in the series experienced transient increases of more than 25% in mean arterial pressure after nalbuphine administration. Renarcotization (Pa_{CO_2} >55 mmHg) occurred in three patients of the second group and further nalbuphine was required 2–3 hours after the last dose. The object of treatment by nalbuphine in this series was to shorten the time to extubation which was carried out about 20 minutes after the effective or final dose of nalbuphine.

The successful use of nalbuphine in sequential analgesia has been demonstrated but clearly further study is required to determine the optimum dosage and timing. The transient rise in blood pressure is also a cause for concern.

Buprenorphine

Buprenorphine is the other clinically available drug which has similar actions to pentazocine and nalbuphine, i.e. antagonism at the μ receptor and partial agonist effects at the κ receptor. Its use in the technique of sequential analgesia was suggested by de Castro (1979). Results have not yet been reported but the interactions between fentanyl and buprenorphine on opioid receptors isolated from rat brain have been described by Boas & Villiger (1985). Buprenorphine was able to displace fenanyl from receptors in a concentration and time dependent manner. Buprenorphine was very slow in its action compared to fentanyl and appeared less selective than fentanyl. From the data in this paper it appears that buprenorphine may be of value in the technique of sequential analgesia, but its time of onset and offset is markedly different from nalbuphine and pentazocine and it is hardly at all antagonized by naloxone.

There is a further problem with buprenorphine. The dose response curve for analgesia, at least in rats and mice, has been clearly shown to be both biphasic and bell shaped. After attaining a peak of analgesia, increasing the dose results in the analgesia decreasing towards zero (Dum & Herz 1981).

The complexities of this subject are discussed by Hull (1985) and emphasized by a recent report (Schmidt et al 1985) which described a series of 12 patients given either 30 or 40 μ/kg of buprenorphine as a component of balanced anaesthesia for cholecystectomy. Six of these patients were in pain post-operatively and were given naloxone 0.04 mg/min (the maximum dose given was 0.44 mg). This dose of naloxone was found to give a prolonged period of pain relief. In these patients the buprenorphine produced respiratory depression when first given but the recorded respiratory rates post-operatively suggested that that respiratory depression was not present at that time and these rates were not affected by the naloxone. It is tempting to speculate that the naloxone pushed the patient's analgesic state back up the far side of the bell shaped curve.

The technique of sequential analgesia has the potential of overcoming one of the major disadvantages of using high doses of narcotic analgesics during anaesthesia, i.e. a prolonged post-operative respiratory depression. However, it is not yet clear whether, when an agonist/antagonist analgesic is used in this manner, it completely displaces the powerful agonist substituting its own more limited degree of analgesia or whether it only partially reverses the action of the more potent drug. As has already been described, by using naloxone, partial reversal of fentanyl is possible, but clearly the kinetics of these drug interactions are complex and they may prove to be unsuitable for routine clinical use. Further research may bring us closer to being able to produce profound analgesia without respiratory depression.

NARCOTICS AND THE STRESS RESPONSE

The stress response is the name given to the complex of hormonally regulated metabolic changes which occur in response to surgery and, to a much lesser extent, to anaesthesia. The subject is reviewed in this volume by Bailey & Child (see Ch. 11). There are reasons for thinking that the abolition or at least the reduction of some of the components of the stress response may be beneficial to the patient. For instance a prominent feature of the stress response is the release of catecholamines from the adrenal medulla with consequent stimulation of α and β receptors. This results in large increases in heart rate and blood pressure which may produce cardiac ischaemia in susceptible patients. It is not surprising therefore to find that it is the blocking of this response which has been chiefly studied. However, it is not at all clear whether it is desirable to supress many of the the components of the stress response. John Hunter thought not. In 1794 he wrote 'there is a circumstance attending accidental injury which does not belong to disease, namely that the injury done, has in all cases a tendency to produce, both the disposition and the means of cure' (Cuthbertson 1976). Other more recent writers have observed 'teleologically, sympathetic activity has relevance in enabling a stressed individual to survive a traumatic event such as injury after attack by animals. It is not clear whether the surgeon should be placed in this category' (Ellis & Humphrey 1982). Certainly some of the components of the response such as the adrenal cortical response and the liberation of growth hormone are concerned with maintenance of intravascular volume and substrate mobilization necessary to promote healing processes and clearly these should be preserved.

Abdominal surgery

In 1959 it was shown that morphine suppressed release of adrenocorticotropic hormone (ACTH) from the anterior pituitary (McDonald 1959). The effect of large doses of this drug on cortisol and growth hormone release were investigated by George et al (1974). They found that 1 mg/kg of morphine reduced, but did not abolish, the hormonal changes during abdominal surgery and that 4 mg/kg for cardiac surgery abolished responses only until the start of bypass. This finding has been confirmed by other workers. Large doses of fentanyl (50 μg/kg) were found by Hall et al (1978) to abolish the endocrine response to tubal microsurgery under halothane anaesthesia. But McQuay et al (1979) who studied patients undergoing more stressful major pelvic surgery found that the plasma cortisol response was not blocked by fentanyl. In major gastric surgery the same dose of fentanyl was found to reduce but not to abolish the stress response inter-operatively. All patients in this series required respiratory support post-operatively (Cooper et al 1981).

In another series of upper abdominal cases 50 μg/kg of fentanyl abolished the cortisol response but in this series no effect on the hyperglycaemic response was found (Haxholdt et al 1981). Alfentanil (150 μg/kg followed by infusion of 3 μg/kg/min) was found to inhibit increases in plasma cortisol and glucose during abdominal hysterectomy but the effect did not extend more than 3 hours into the post-operative period (Moller et al 1985).

The circulatory response to laryngoscopy and intubation has been found to be more easily blocked by narcotics. Martin et al (1982) found that 8 μ/kg of fentanyl greatly reduced the response, and Kautto (1982) observed that 6 μg/kg of this drug completely abolished changes in the blood pressure and pulse rate observed during laryngoscopy and intubation.

To conclude this aspect of the use of narcotics, Roisen et al (1981) determined the anaesthetic dose blocking the adrenergic cardiovascular responses to skin incision for various anaesthetic agents and for morphine. This dose was termed the MAC BAR. In a group of unpremedicated patients the dose of morphine required was found to be 1.13 mg/kg and this compared with a mean alveolar concentration of 1.1% of halothane.

Cardiac surgery

The era of narcotic anaesthesia for cardiac surgery was ushered in by the classic paper of Lowenstein et al (1969) which showed how slight were the effects of morphine (1 mg/kg) on cardiovascular function. It became apparent that very large doses of morphine, up to 3 mg/kg body weight, could be safely given to patients with cardiac disease and that these doses resulted in modification of the cardiovascular stress response.

The use of narcotic analgesics in connection with anaesthesia for cardiac surgery has generated a great deal of research and the subject has recently been comprehensively reviewed (Bovill et al 1984). A necessarily brief summary will be given. The frequent requirement of post-operative mechanical ventilation after cardiac surgery which necessitated retention of the endotracheal tube was, in practice, the major factor resulting in the introduction of large doses of morphine as part of the aneasthetic technique for cardiac surgery (Lowenstein & Philbin 1983). As attempts were made to achieve better results by using larger doses of morphine various difficulties became apparent. A fall in peripheral vascular resistance had been a feature noted in a pioneering paper of Lowenstein et al (1969) and indeed it had been considered a desirable feature in patients with congestive failure as it constituted a form of internal phlebotomy. But Stanley et al (1973), in a series of cardiac cases in which the average dose of morphine was 11 mg/kg, found that the requirement of blood to be excessively high both during and after surgery. It has also been shown that there was a requirement for a significant rate of transfusion volume during cardiac surgery when a more moderate dose of 3 mg/kg of morphine was used compared to the fluid requirements under halothane anaesthesia (Stanley et al 1974).

It seems that a large part, if not all, of this undesirable effect of high dose morphine is due to histamine release.

In a series of coronary artery bypass graft (CABG) patients, Rosow et al (1982) found that a dose of 1 mg/kg of morphine produced an average 750% peak increase in plasma histamine and that this was associated with a fall in systemic vascular resistance (SVR) and blood pressure. In contrast patients given fentanyl (50 μg/kg) showed no increase in plasma histamine. Evidence suggesting that the cardiovascular changes were entirely due to histamine was provided by Philbin et al (1981) who found that the administration of a combination of an H_1 antagonist (diphenhydramine) and an H_2 antagonist (cimetidine) completely blocked the fall in SVR and diastolic blood pressure in CABG patients given 1 mg/kg of morphine.

As mentioned above, the synthetic narcotic analgesic fentanyl does not cause histamine release and for this and other reasons is now routinely use in anaesthesia for cardiac surgery in doses of 50–100 μ/kg. Two analogues of fentanyl, sufentanil and alfentanil, have also been used in high dosage for cardiac surgery. Their actions resemble fentanyl closely, but they differ in potency and duration of action. Sufentanil appears to be between five and seven times as potent as fentanyl and to have a similar duration of action (Rolly et al 1979, de Langa et al 1982) and alfentanil to have a potency and duration of action of approximately one-third that of fentanyl. Some of the problems associated with the use of high doses of narcotic analgesics will be briefly discussed below.

Awareness

Lowenstein et al (1969) noted that 1 mg/kg morphine often did not cause unconsciousness if adequate oxygenation and CO_2 excretion were maintained, and since then a number of cases of awareness during high dose fentanyl/oxygen anaesthesia have been reported. In one case (Mummaneni et al 1980) a patient given 75.8 μg/kg of fentanyl and 10 mg of diazapem for a cardiac operation had no awareness but for a further operation shortly afterwards the same patient was given 72 μg/kg of fentanyl and no diazepam and awareness and recall were reported. In another case (Hilgenberg 1981) in which a total of 90 μg/kg of fentanyl was given there was no sign of lightness of anaesthesia during the operation but there was undoubted recall, although this was only of hearing. In a third case, despite the patient being given a total dose of 96 μg/kg of fentanyl, diazepam 10 mg and 50% nitrous oxide for 25 minutes the stimulation of the incision, entering the chest and diathermy of the edges of the pericardium resulted in a hypertensive crisis requiring thiopentone, enflurane and other drugs, including 50% nitrous oxide for the rest of the operation (Mark & Greenberg 1983). The patient recalled the incident afterwards. These and other occurrences confirm that narcotic analgesics alone cannot be relied upon to produce anaesthesia and this view is expressed by Wong (1983) in an editorial robustly entitled, 'Narcotics are not expected to produce unconsciousness and amnesia'. Certainly general anaesthetic agents appear to depress the central nervous system in a fairly non-specific manner related to their lipid solubility whereas narcotic analgesics have a critical structure and bind specifically to localized receptors producing analgesia, respiratory depression and other specific effects. These receptors are currently being subdivided into various subgroups, none of which appear to subserve the functions of unconsciousness or amnesia. The failure of narcotic analgesics at what ever dose to reduce MAC by more than 65% (referred to previously) is a further indication that these drugs do not produce anaesthesia. The use of even more potent narcotic analgesics may be beneficial in that, since less of the drug is required, there is less likelihood of side-effects (Janssen 1982). But greater specificity is even less likely to include lack of awareness and absence of recall among the spectrum of activities. The contrary view has been expressed by Stanley (1983), who considers that the higher the dose the less likely is incomplete amnesia.

The subdivision of receptors referred to above is of great interest as it contains the possibility of a drug producing effective analgesia without respiratory depression or addictive liability.

Not least of the problems of preventing awareness with fentanyl/O_2 anaesthesia, is the finding that the addition of nitrous oxide in the presence of high concentrations of fentanyl and other narcotic analgesics produces significnt myocardial depression with increased systemic vascular resistance and lowered cardiac output and blood pressure (Wong et al 1973). This view has been challenged by Balasaraswathi et al (1981) and Michaels et al (1984). The direct action of nitrous oxide on the myocardium is depressant but this is usually compensated for by the sympa-thomimetic effects of the gas (Eisele 1985). Blocking of these effects by the narcotic may uncover the direct depressant action.

Cardiovascular depression is also observed when diazepam is added to high dose fentanyl (Reves et al 1984), and depressant interactions between morphine and halothane (Stoelting et al 1974) and fentanyl and enflurane (Bennet and Stanley 1979) have also been described. Despite these findings benzodiazepine and inhalational anaesthetic agents, particularly isoflurane, are widely used to ensure a lack of awareness when high dose narcotics analgesics are used. It seems likely that the extent of the depressant interactions on the myocardium between narcotic analgesics and other sedative or anaesthetic drugs depends very much on the state of the coronary circulation (Barash & Kopriva 1982).

Muscular rigidity

A peculiar feature of very large doses of narcotic analgesics is the production of muscular rigidity confined to the thorax and abdomen becoming apparent during induction. This effect, which appears supraspinal in origin but whose cause is unknown, was first noted by Hamilton & Cullen (1953) and with morphine by Freund et al (1973). With morphine the rigidity is relatively mild and manifests itself by expiratory grunting and restrictive expiration. If vigorously exhorted, conscious patients are able to overcome this by spontaneous effort (Lowenstein & Philbin 1983).

With fentanyl and the newer narcotic analgesics the rigidity may be much more severe and may prevent inflation of the lungs by intermittent positive pressure. It has been pointed out by Andrews & Prys-Roberts (1983) that there is no reason why abdominal or thoracic rigidity should prevent inflation of the lungs and that this must be due to closure of the glottis. Whatever may be the truth of this, the phenonemon requires the routine use of muscle relaxants to overcome it. In one large series (Grell et al 1970) muscle rigidity sufficient to impair intermittent positive pressure ventilation was seen when the dose of fentanyl exceeded 0.25 mg. It has been claimed that rigidity is not seen with slow infusion rates (Stanley & Webster 1978). An infusion rate of 1 mg/min of fentanyl to a total dose of 50 μg/kg produced truncal rigidity in 88% of patients which was reduced by pretreatment with small doses of pancuronium (12.2 μg/kg) or metocurine (50 μg/kg) (Jaffe & Ramsey 1983). Hill et al (1981), who found a high incidence of rigidity at the relatively slow fentanyl infusion rate of 3 μg/kg/min occurring after a mean dose of 14.6 μg/kg, considered that this could be prevented by simultaneous infusion of pancuronium at the rate of 12 μg/kg/min without the patient becoming aware of paralysis. Finally, it has been reported (Christian et al 1983) that three patients developed truncal rigidity between 5 and 8 hours after induction of anaesthesia. The dose of fentanyl in these cases was 75 or 115 μg/kg and the rigidity was immediately abolished by naloxone. The authors consider that acid-base changes had effected brain fentanyl concentrations. This effect has been shown to occur in dogs. Ainslie et al (1979) found marked increases in brain concentration with decreasing Pa_{CO_2}.

Despite, or because of, the great amount of work on high dose narcotic analgesics for cardiac surgery very little has been done to demonstrate that this technique is actually superior in results to any other. A recent paper (Fischerstrom et al 1985) describes results in a series of 20 unselected patients with severe angina undergoing coronary artery bypass surgery. They were randomly allocated to two groups, one group receiving nitrous oxide/halothane and the other nitrous oxide/fentanyl/diazepam. A benign and almost identical haemodynamic response to surgery and bypass was observed in the two groups. The dose of fentanyl used in this series was very moderate.

A careful review of the question of narcotic versus inhalation anaesthesia for cardiovascular surgery (Roizen 1984) concludes that no available data justify the belief that any narcotic or inhalational agent is superior in its effect on outcome when used in conjunction with cardiac or vascular surgery.

REFERENCES

Ainslie S G, Eisele J H, Corkill G 1979 Fentanyl concentrations in brain and serum during respiratory acid base changes in the dog. Anesthesiology 51: 293–297
Alderman E L, Barry W H, Graham A F, Harrison D C 1972 Haemodynamic effects of morphine and pentazocine differ in cardiac patients. New England Journal of Medicine 287: 623–627
Andree R 1980 Sudden death following naloxone administration. Anaesthesia and Analgesia 59: 782–784
Andrews C J H, Prys-Robert C 1983 Fentanyl — a review. Clinics in Anaesthesiology 1: 97–121
Azar I, Turndorf H 1979 Severe hypertension and multiple atrial premature contractions following naloxone administration. Anaesthesia and Analgesia 58: 524–525
Bailey P, Gerbode F, Garlington L 1958 An anesthetic technique for cardiac surgery which utilizes 100% oxygen as the only inhaland. Archives of Surgery 76: 437–440
Balasarasuathi K, Kumar P, Rao T, El Etr A A 1981 Left ventricular and diastolic pressure as an index for nitrous oxide use during coronary artery surery. Anesthesiology 55: 708–709
Barash P G, Koprina C J 1982 Narcotics and the circulation. In: Kitahata, Collins (eds) Narcotic Analgesics in Anesthesiology. Williams and Wilkins, Baltimore
Beaver W T, Feise G A 1978 A comparison of the analgesic effect of intramuscular nalbuphine and morphine in patients with postoperative pain. Journal of Pharmacology and Experimental Therapeutics 204: 487–496
Bennett G M, Stanley T H 1979 Cardiovascular effects of fentanyl during enflurane anesthesia in man. Anaesthesia and Analgesia 58: 179–182

Bellville J W, Fleischli G 1968 The interaction of morphine and nalorphine on respiration. Clinical Pharmacology and Therapeutics 9: 152-162

Boas R A, Villiger J W 1985 Clinical actions of fentanyl and buprenorphine. The significance of receptor binding. British Journal of Anaesthesia 57: 192-196

Bovill J G, Sebel P S, Stanley T H 1984 Opioid analgesics in anesthesia with special reference to their use in cardiovascular anesthesia. Anesthesiology 61: 731-755

Christian C C, Waller J L, Moldenhauer C C 1983 Postoperative rigidity following fentanyl. Anesthesia and Anesthesiology 58: 275-277

Cooper G M, Paterson J L, Ward I D, Hall G M 1981 Fentanyl and the metabolic response to gastric surgery. Anaesthesia 36: 667-671

Cuthbertson D P 1976 Surgical metabolism: historical and evolutionary aspects In: Wilkinson, Cuthbertson (eds) Metabolism and the Response to Injury. Tunbridge Wells

De Castro J 1979 British Journal of Clinical Pharmacology 7: 319S

De Castro J, Viars P 1968 Archives of Medicine 23: 170-171

De Langa S, Boscoe M J, Stanley T H, Pace N 1982 Comparison of sufentanil-O_2 and fentanyl-O_2 for coronary artery surgery. Anesthesiology 56: 112-118

Drummond G B, Davie I T, Scott D B 1977 Naloxone: dose dependant antagonism of respiratory depression by fentanyl in anesthetised patients. British Journal of Anaesthetics 49: 151-154

Dum J, Herz A 1981 In vivo receptor binding of opiate partial agonist buprenorphine correlated with its agonistic and antagonistic effects. Journal of Pharmacology 74: 627

Eiselle J H 1985 Cardiovascular effects of nitrous oxide. In: Eger E I (ed) Nitrous Oxide. Edward Arnold, London

Ellis R F, Humphrey D E 1982 Clinical aspects of endocrine and metabolic changes relating to anaesthesia surgery. In: Watkins (ed) Trauma, Stress and Immunity in Anaesthesia and Surgery. Butterworth, London

Estillo A E, Cottrell J E 1981 Naloxone, hypertension and ruptured cerebral aneurysm. Anesthesiology 54: 352

Fischerstrom A, Oliquist, Settergren G 1985 Comparison of fentanyl and halothane as supplement to nitrous-oxide-oxygen anaesthesia for coronary artery surgery Acta Anaesthesiologica Scandinavica 29: 16-21

Flache J W, Flache W E, Williams G D 1977 Acute pulmonary oedema following naloxone reversal of high dose morphine anaesthesia. Anesthesiology 47: 376-378

Freund F G, Martin W E, Wong K, Hornbein T F 1973 Abdominal muscle rigidity induced by morphine and nitrous oxide. Anesthesiology 38: 358-362

Gal T J, Difazio C A 1982 Analgesic and respiratory depressant activity of nalbuphine: a comparison with morphine. Anesthesiology 57: 367-374

George J M, Reier C E, Lanese R R, Rower J M 1974 Morphine anaesthesia blocks cortisol and growth hormone response to surgical stress in humans. Journal of Clinical Endocrinology and Metabolism 38: 736-739

Grell F L, Koom R A, Nedson J S 1970 Fentanyl in anesthesia. A report of 500 cases. Anesthesia and Analgesia 49: 523-532

Hall G M, Young C, Holdcroft A, Labgband-Zadeh J 1978 Substrate mobilisation during surgery. A comparison between halothane and fentanyl anaesthesia. Anaesthesia 33: 924-930

Hamilton W K, Cullen S C 1953 Effect of levallorphan tartrate upon opiate induced respiratory depression. Anesthesiology 14: 550-554

Haxoldt O S, Kehlet H, Dryberg V 1981 Effect of fentanyl on the cortisol and hyperglycaemic response to abdominal surgery. Acta Anaesthesiologica Scandavica 25: 434-436

Hilgenberg J C 1981 Intraoperative awareness during high dose fentanyl-oxygen anesthesia. Anesthesiology 54: 341-343

Hill A B, Nahrwold M L, Rosayro M De, Knight P R, Jones R M, Bottes R E 1981 Prevention of rigidity during fentanyl oxygen induction of anesthesia. Anesthesiology 55: 452-454

Hoffman J C, Difazio C A 1970 The anesthesia sparing effecting of pentazocine, meperidine and morphine. Archives Internationales de Pharmacodynamie et de Therapie 186: 261-268

Holmes C McK 1976 Supplementation of general anaesthesia with narcotic analgesics. British Journal of Anaesthesia 48: 907-913

Houde R W, Wallenstein S L 1956 Clinical studies of morphine-nalorphine combinations. Federation Proceedings 15: 440-441

Hull C J 1985 Receptor binding and its significance. British Journal of Anaesthesia 57: 131-133

Jaffe T B, Ramsey F M 1983 Attenuation of fentanyl-induced truncal rigidity. Anesthesiology 58: 562-564

Janssen P A S 1982 Potent new analgesics tailor made for different purposes. Acta Anaesthesiologica Scandanavica 26: 262-268

Jasinski D R, Mansky P A 1972 Evaluation of nalbuphine for abuse potential. Pharmacology and Therapeutics 13: 78–90
Jasinski D R 1979 Human pharmacology of narcotic antagonists. British Journal of Clinical Pharmacology 7: 287S
Julien R M 1982 Effects of nalbuphine on normal and onymorphone-depressed ventilatory responses to carbon-dioxide challenge. Anesthesiology 97: A320
Kaufman L 1982 Anaesthesia and the endocrine response. In: Kaufman L (ed) Anaesthesia Review 1, Churchill Livingstone, Edinburgh
Kaufman L 1984 Update. Anaesthesia and the endocrine response. In: Kaufman L (ed) Anaesthesia Review 2. Churchill Livingstone. Edinburgh
Kautto U M 1982 Attenuation of the circulatory resonse to laryngoscopy and intubation of fentanyl. Acta Anaesthesiologica Scandanavica 26: 217–221
Kraynack B J, Gintautas J G 1982 Nalaxone: analeptic action unrelated to opiate receptor antagonism. Anesthiology 56: 251–253
Krycke B J, Finck A J, Shah N K, Snow J C 1979 Naloxone antagonism after narcotic supplemented anaesthesia. Anaesthesia and Analgesia 55: 800–805
Lasagna L, Beecher H K 1954 The analgesic effectiveness of nalorphine and nalorphine-morphine combinations in man. Journal of Pharmacology and Experimental Therapeutics 122: 356–363
Latasch L, Probst S, Dudziak R 1984 Reversal by nalbuphine of respiratory depression caused by fentanyl. Anaesthesia and Analgesia 63: 814–816
Longnecker D, Grazis P, Eggers G 1973 Naloxone for antagonism of morphine induced respiratory depression. Anaesthesia and Analgesia 52: 447–452
Lowenstein E, Hallowell P, Levina F H, Duggett W M, Austin G, Laver M B 1969 Cardiovascular response to large doses of morphine in man. New England Journal of Medicine 281: 1389–1393
Lowenstein E, Philbin D M 1983 Narcotic anaesthesia clinics. Anesthesiology 1: 5–15
McDonald R K, Evans F T, Weise V K, Patrick R W 1959 Effect of morphine and nalorphine on plasma hydrocortisone levels in man. Journal of Pharmacology and Experimental Therapeutics 125: 241–245
McQuay H J, Moore R E, Paterson G M C, Adams A P, 1979 Plasma fentanyl concentrations and clinical observations during and after operatioń. British Journal of Anaesthesia 515: 543–550
Magruder M R, Delaney R D, Difazio C A 1982 Reversal of narcotic-induced respiratory depression with nalbuphine hydrochloride. Anesthesiology 9: 34–37
Mark J B, Greenberg L M 1983 Intraoperative awareness and hypertensic crisis during high dose fentanyl-diazepam-oxygen anaesthesia. Anaesthesia and Analgesia 62: 698–700
Martin D E, Rosenberg H, Aukburg S F 1982 Low dose fentanyl blunts circulatory responses to tracheal intubation. Anaesthesia and Analgesia 61: 680–684
Martin W R 1967 Opioid antagonists. Pharmacology Review 10: 463–521
Michaels I, Trout J R, Barash P G 1984 In: Estafanous F G (ed) Opiods in Anesthesia. Boston, 256–260
Michaelis L L, Hickey P R, Clark T A 1974 Ventricular irritability associated with the use of naloxone hydrochloride. Journal of Thoracic Surgery 18: 608–614
Moldenhauer C C, Roach G W, Finlayson D C 1985 Nalbuphine antagonism of ventilatory depression following high-dose fentanyl anesthesia. Anesthesiology 62: 647–650
Moller I W, Krantz T, Wandall E, Kehlet H 1985 Effect of alfentanil anaesthesia on the adrenocortical and hyperglycaemic response to abdominal surgery. British Journal of Anaesthesiology 57: 591–594
Munson E S, Saidman L J, Eger E I 1965 Effect of nitrous oxide and morphine on the minimum anaesthetic concentration fluroxene. Anesthesiology 26: 134–139
Mummaneni N, Rao T L, Montoya A 1980 Awareness and recall with high dose fentanyl-oxygen anesthesia. Anaesthesia and Analgesia 59: 948–949
Murphy M R, Hug C C 1982a The enflurane sparing effect of morphine, butorphanol and nalbuphine. Anesthesiology 57: 489–492
Murphy M R, Hug C C 1982b The anesthetic potency of fentanyl in terms of its reduction of enflurane MAC. Anesthesiology 57: 485–488
Muskin W W, Rendell-Baker L 1949 Pethidine as a supplement to nitrous oxide anaesthesia. British Medical Journal 2: 472
Neff W, Mayer E C, De La Luz Perales M 1947 Nitrous oxide and oxygen anaesthesia with curare relaxation. California Medicine 66: 67–69
Patschke D, Eberlein H J, Tarnour J, Simmerman G 1977 Antagonism of morphine with naloxone in dogs: cardiomuscular effects with special reference to the coronary circulation. British Journal of Anaesthesia 49: 525–533
Philbin D M, Moss J, Akins C W 1981 The use of H_1 and H_2 histamine antagonists with morphine anesthesia. Anesthesiology 55: 292–296

Prough D S, Roy R, Bumgarner J, Shannon G 1984 Acute pulmonary oedema in healthy teenagers following conservative doses of IV naloxone. Anesthesiology 60: 485–486

Rifat K 1972 Pentazocine in sequential analgesic anaesthesia. British Journal of Anaesthesia 44: 175–182

Robbins B H, 1958 Cyclopropane anaesthesia, 2nd ed. Williams and Wilkins, Baltimore

Robbins B H, Baxter J H, Fitzhugh D G 1939 Studies of cyclopropane V. The effect of morphine, barbital and amytal upon the concentration of cyclopropane in the blood required for anesthesia and respiratory arrest. Journal of Pharmacology and Experimental Therapeutics

Roison M F 1984 Does choice of anesthetic (narcotic versus inhalational) significantly affect cardiovascular outcome after cardiovascular surgery? In: Estafanous F G (ed) Opioids in Anesthesia. p 180–189

Roison M F, Horrigan R W, Frazer B M 1981 Anesthetic doses blocking adrenorgic (stress) and cardiovascular responses to incision-MAC-BAR. Anesthesiology 54: 390–398

Romagnoli A, Keats A S 1980 Ceiling effect for respiratory depression by nalbuphine. Clinical Pharmacology and Therapeutics 27: 478–485

Rolly G, Kay B, Cock X 1979 A double blind comparison of high doses of fentanyl and sufentanil in man. Acta Anaesthesiologica Belgica 30: 247–254

Rosow C E, Moss J Philbin D M, Saverese J J 1982 Histamine release during morphine and fentanyl anaesthesia. Anesthesiology 56: 93–96

Ruben H, Andreassen 1951 Pharmacological effects of pethidine on the larynx seen during intubation. British Journal of Anaesthesia 23: 33–38

Saidman L J, Eger E I 1964 Effect of nitrous oxide and of narcotic premedication on the alveolar concentration of halothane required for anesthesia. Anesthesiology 25: 302–306

Schmidt J F, Chraemmer J, Orgensen B, Pederson J E, Risbo A 1985 Postoperative pain relief with naloxone. Severe respiratory depression and pain after high dose buprenorphine. Anaesthesia 40: 583–586

Schrier R I 1982 Narcotic analgesics for preoperative and postoperative medication. In: Kitahata L M (ed) Narcotic Analgesics in Anesthesiology. Williams and Wilkins, Baltimore, p 177–187

Seevers M H, Meek W J, Rovenstine E A, Stiles J A 1934 A study of cyclopropane anesthesia with especial reference to gas concentrations, respiratory and electrocardiographic changes. Journal of Pharmacology and Experimental Therapeutics 51: 1–17

Siker E S 1956 Analgesic supplements to nitrous oxide anaesthesia. British Medical Journal 2: 1326–1331

Smith R R 1908 Scopolamine-morphine anaesthesia. Surgeries in Gynaecology and Obstetrics 7: 414–420

Smith T C 1984 Opioid reversal and postoperative narcotic analgesia. In: Estafanous F G (ed) Opioids in Anaesthesia. Butterworths, Boston, p 297–301

Stambaugh J E 1982 Evaluation of nalbuphine. Current Therapeutic Research 31: 393–401

Stanley T H 1983 High dose fentanyl. Mt Sinai Journal of Medicine 50: 308–311

Stanley T H, Gray: N H, Stanford W, Armstrong R 1973 The effects of high-dose morphine on fluid and blood requirements in open heart procedures. Anesthesiology 38: 536–541

Stanley T H, Isern-Amaral J, Gray N H, Patton C P 1974 Comparison of blood requirements during morphine and halothane anaesthesia for open-heart surgery. Anesthesiology 41: 34–38

Stanley T H, Webster L R 1978 Anesthetic requirements and cardiovascular effects of fentanyl-oxygen and fentanyl-diazepam-oxygen anesthesia in man. Anesthesia and Analgesia 57: 411–416

Stoelting R K, Creasser C E, Gibbs P S 1974 Circulatory effects of halothane added to morphine anaesthesia in patients with coronary artery disease. Anaesthesia and analgesia 53: 445–449

Tammisto T, Tigersted I 1979 Restlessness and shivering after naloxone reversal of fentanyl-supplemented anaesthesia. Acta Anaesthesiologica Scandinavica 23: 51–56

Tanaka G Y 1974 Hypertensive reaction to naloxone. Journal of American Medical Association 228: 25–26

Taylor H E, Doerr J C, Gharib A 1957 Effect of preanesthetic medication on ether content of arterial blood required for surgical anaesthesia. Anesthesiology 18: 849

Waters R M, Schmidt E R 1934 Cyclopropane anaesthesia. Journal of the American Medical Association 103: 975–983

Wong K C 1983 Narcotics are not expected to produce unconsciouness and amnesia. Anaesthesia and Analgesia 62: 625–626

Wong K C, Martin W E, Hornbein T F 1973 The cardiovascular effects of morphine sulfate with oxygen and with nitrous oxide in man. Anesthesiology 38: 542–549

Wood A J T, Moir D C, Campbell C 1974 Medicines, evaluation and monitoring group: central nervous system effects of pentazocine. British Medical Journal 1: 305–307

Woodbridge P D 1957 Changing concepts concerning depth of anaesthesia. Anesthesiology 18: 536–550

Recovery from neuromuscular blockade

INTRODUCTION

Many anaesthetists administer an anticholinesterase at the end of an operation to reverse the effects of non-depolarising neuromuscular blockade as otherwise the relaxant might still be occupying a variable but possibly significant number of receptor sites. In the absence of neuromuscular monitoring, the anaesthetist often equates neuromuscular function with tests of muscle strength such as head raising, hand grip strength and eye opening, or with respiratory measurements such as tidal volume, respiratory rate or inspiratory and expiratory airway pressures. Unfortunately tidal volume and respiratory rate are influenced by a number of factors unrelated to neuromuscular function, and the presence of spontaneous breathing is in itself no guarantee of adequate recovery from neuromuscular blockade (Ali et al 1975). Competitive neuromuscular blocking agents have been shown to have a sparing effect on the muscles of respiration compared with peripheral muscles (Foldes et al 1961, Johansen et al 1964, Gal & Smith 1976, Wymore & Eisele 1978) so that spontaneous breathing can occur in the presence of marked peripheral neuromuscular blockade. Although gas exchange may be satisfactory in a patient whose airway is safeguarded by an endotracheal tube, this degree of recovery may not leave enough margin of safety for an extubated patient who subsequently develops an obstructed airway. Two studies have questioned the adequacy of anaesthetists' clinical judgement. Harrison (1978) carried out a retrospective survey of anaesthetic deaths in his hospital in South Africa and found that 10 out of 53 were due to inadequate respiration following myoneural blockade. More recently Viby Mogensen (1979) used a nerve stimulator to assess patients post-operatively. These patients, who had been returned to the ward, had all been judged by the anaesthetist on clinical grounds to be adequately 'reversed'. He found that 42% of patients had a train of four ratio less than 0.7, and concluded that residual paralysis remained a problem in patients who had not been quantitatively assessed.

Neuromuscular monitoring

The need has evolved for precise monitoring during routine anaesthesia to reduce the risk of residual paralysis. The simplest method is to stimulate a peripheral nerve and

---

*'Reversal' — more correctly reversal of the neuromuscular blocking action of the drug but 'reversed' or 'reversal' appear to be in *common* usage.

quantify the response of the muscle supplied by that nerve before and after the administration of a neuromuscular blocking agent. The ulnar nerve at the wrist is often chosen because of ease of access, and following supramaximal stimulation the response of the adductor pollicis muscle is measured. The twitch response is a relatively insensitive test as 75–80% of receptors have to be occupied before there is any diminution of the control height (Paton & Waud 1967). The sensitivity can be increased by using tetanic stimulation at 50 Hz (Gissen & Katz 1969) or a train of four at 2 Hz (Ali et al 1970). A few studies have correlated tests of neuromusclar function with adequacy of ventilation and clinical signs of recovery. Walts and colleagues (1970) concluded that sustained muscular contraction in response to tetanic stimulation at 30 Hz for 5 seconds was useful because it correlated with 90% recovery of vital capacity in human volunteers. They concluded that head raising was unreliable as it did not return to control but perhaps this is because head raising is a more sensitive test. Johansen et al (1964) found that head lift and hand grip strength were 38% and 48% of control when both inspiratory and expiratory flow rates were more than 90% of control. Ali & Kitz (1973) found that when the vital capacity was more than 90%, tetanus at 50 Hz was fully sustained and the mean train of four ratio was 75% following block with tubocurarine and this ratio was found to be lower in conscious volunteers who fulfilled similar criteria (60%) (Ali et al 1975). Thus the train of four may be a more sensitive test of neuromuscular function than the tetanus performed at frequencies up to 50 Hz and the ratio need not recover to unity to be judged adequate. Miller (1976) collated data from several studies and suggested that the sensitivities of the tests were head lift > hand grip > inspiratory force, inspiratory and expiratory flow and normal train of four > sustained tetanus at 30 Hz, and normal vital capacity > normal tidal volume and twitch height. He concluded that even when the test was normal it could not be assumed that all receptors were free of relaxant. This underlines the importance of achieving the maximum possible recovery of neuromuscular function at the end of surgery.

Observation of the single twitch is not particularly useful as observer memory of the control twitch response may be unreliable. The train of four is independent of the control and therefore might be more useful. However, evaluation of the ratio of the train of four using a simple peripheral nerve stimulator even in expert hands is poor (Savarese & Ali 1977, Viby Mogensen et al 1983). A more precise evaluation can be obtained using a force transducer with recording apparatus (Ali & Savarese 1976). This provides both a record for the control single twitch and allows detailed analyses of fade of the tetanus or train of four. Alternatively the EMG can be recorded and simple EMG devices are now available which provide an analogue reading of the compound electromyogram with a meter display of the assessment of block (Lam et al 1981). Close correlation between this EMG measurement and that of the force transducer has recently been found for single twitch and train of four modes (Windsor et al 1985). These workers confirmed the increased accuracy of this method of monitoring over the simple nerve stimulator. It should be remembered that for equivalent degrees of depression of the single twitch response, the degree of fade of the train of four (or tetanus) will vary with different drugs as each drug has its own individual profile (Williams et al 1980). Hence the place of the ratio of the train of four as a measure of neuromuscular function has yet to be defined.

An assessment of the degree of neuromuscular block present during anaesthesia is important since the reversal agent is more effective if there is already some degree of spontaneous recovery (Katz 1967a, 1971, Kopman 1979). Thus Katz (1971) noted that if the twitch height was less than 20% of control following neuromuscular blockade with pancuronium, reversal with neostigmine could take 30 minutes. However, if the twitch height was at least 20–25% of control then reversal could be accomplished in under 10 minutes. The recovery time following neuromuscular blockade with atracurium and vecuronium has been shown to be considerably reduced (Baird et al 1982, Baird & Kerr 1983).

Neuromuscular function

During neuromuscular transmission, acetylcholine released from the nerve ending diffuses across the neuromuscular junction to the motor endplate where it brings about depolarization. It is then hydrolysed to choline and acetate by the action of acetylcholinesterase which is probably not a structural component of either the pre- or post-junctional membrane but rather coats them both. This reaction is completed within milliseconds and enables the endplate to become rapidly repolarized and responsive to subsequent nerve impulses at physiological rates up to 50 Hz. It has been estimated that at least 10 active acetylcholinesterase sites are available for each molecule of acetylcholine released. This explains the efficiency of destruction.

ANTICHOLINESTERASES

A single molecule of acetylcholinesterase appears to contain six binding and hydrolytic sites for acetylcholine (Wright & Plummer 1973). Each active centre consists of two subsites: one is known as the anionic site and bears an overall negative charge which binds the positively charged end of the acetylcholine molecule and orientates it properly for hydrolysis; the other is known as the esteratic site and is responsible for the actual cleavage of the molecule. Drugs that inhibit or inactivate acetylcholinesterase are called anticholinesterases and those currently available for their anti-relaxant action at the neuromuscular junction are neostigmine, pyriodostigmine and edrophonium. They act either by reversible occupation of the active site of the enzyme, e.g. edrophonium, or by acylation of the esteratic site, e.g. carbamates (physostigmine, pyridostigmine and neostigmine) and organophophates. Whereas edrophonium has a brief duration of action presumably owing to its reversible attachment and subsequent rapid renal clearance, deacylation of carbamylated and phosphorylated acetylcholinesterase takes longer and until this has taken place the enzyme is inhibited. Both neostigmine and pyridostigmine form the same carbamylated enzyme: thus the duration of enzyme inhibition produced by these drugs is the same and exceeds that produced by edrophonium.

There is still uncertainty as to whether the effect on neuromuscular transmission is due to inhibition of acetylcholinesterase. Hobbiger (1976), in a comprehensive review of the subject, classified only the carbamates as anticholinesterases as it was more difficult to show that the in vivo action of the 3- and 4-hydroxyanilinium ions, e.g. edrophonium, was the result of acetylcholinesterase inhibition. They are classed as anticholinesterases for the purpose of this review to distinguish them from other drugs with facilitatory actions at the neuromuscular junction whose mechanisms of action do not necessarily involve acetylcholinesterase inhibition. These drugs include guanidine, tetraethylammonium and 4-aminopyridine, all of which increase the prejunctional output of acetylcholine by blocking ion channels: they have been tried with varying degrees of success in disorders in which the output is absolutely or relatively low.

Actions of anticholinesterases

Much of the present controversy about the mode and site of action of anticholinesterases is attributable to the widely held view that the role of acetylcholinesterase is confined to terminating the action of acetylcholine at the postsynaptic membrane. Experimental evidence suggests that this is not the case and that when acetylcholinesterase is inhibited, acetylcholine gains access to nerve terminals and exerts actions there. The classic facilitatory effects of anticholinesterases were originally described by Brown and colleagues (1936). They showed that after physostigmine administration a single indirect stimulus produced repetitive firing in the muscle. Masland & Wigton (1940) reproduced these results with neostigmine. This repetitive firing increases the twitch tension (Feng & Li 1941). Not only does the muscle action potential become repetitive but so does the nerve action potential and this can be recorded antidromically. These nerve action potentials are associated with fasciculations which are a characteristic finding after anticholinesterase treatment in resting muscles (Masland & Wigton 1940, Feng & Li 1941, Eccles 1942, Riker & Wescoe 1946, Boyd & Martin 1956, Meer & Meeter 1956, Blaber & Goode 1968). The period of transmitter release is prolonged by anticholinesterases (Hubbard & Schmidt 1961). Probably the most satisfactory explanation for this repetitive firing is the existence of acetylcholine receptors at the prejunctional membrane (Werner 1960, Hubbard & Schmidt 1961). It seems that after acetylcholinesterase inhibition, acetylcholine can reach and excite presynaptic receptors on nerve endings not normally affected. As the most terminal part of the nerve ending is resistant to depolarization (Hubbard et al 1965, Blaber & Goode 1968) the cholinoceptors in question are probably located more centrally at the first node of Ranvier as illustrated in Figure 18.1. One of the physiological roles of junctional acetylcholinesterase may be to protect these receptors from effects of acetylcholine.

Repetitive firing following stimulus evoked action potentials disappears at stimulation frequencies more than 2 Hz and is not important in voluntary movements. The main effect of anticholinesterases on normal unblocked muscle is to cause paralysis partly by producing depolarizing block from a postsynaptic action of acetylcholine, and perhaps partly by a reduction in acetylcholine release which probably accounts for tetanic fade (Barnes & Duff 1953, Blaber & Bowman 1963a, Bowman 1980). It is only when neuromuscular transmission is impaired by drugs such as tubocurarine or in myasthenia gravis that anticholinesterases lead to facilitation.

Anticurare action

Whilst it is widely accepted that the anticurare action of anticholinesterase drugs is due to inhibition of acetylcholinesterase which allows the accumulation of

G

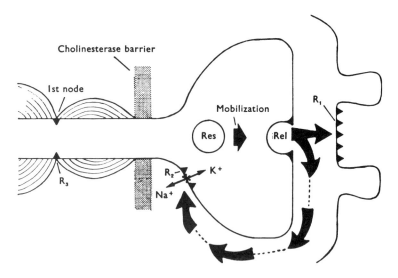

Fig 18.1 Postulated sites of cholinoceptors at the neuromuscular junction (W. C. Bowman). R_1 represents the post-junctional cholinoceptors that mediate the production of the epp when acetylcholine combines with them. R_2 represents hypothetical cholinoceptors on the nerve terminal membrane. Stimulation of these cholinoceptors by transmitter acetylcholine hastens mobilization of reserve acetylcholine (Res) into the readily releasable store (Rel), thereby maintaining transmitter output during high frequencies of stimulation. Block of the R_2 receptors results in tetanic fade. Different antagonists have different relative affinities for the R_1 and R_2 cholinoceptors. R_3 represents hypothetical cholinoceptors at the first node of Ranvier. These are normally protected from transmitter acetylcholine by a cholinesterase barrier, but they may be stimulated after cholinesterase inhibition producing localized depolarization that generates propogating action potentials which travel in both directions from the site of initiation.

transmitter in the synaptic gap, there is disagreement about the site or sites at which this additional acetylcholine exerts its anticurare effect (MacLagen 1976). Most workers subscribe to the view that anticurare action is a postsynaptic effect due to the increased concentration of acetylcholine competing with the antagonist for cholinoceptors. The other possible explanation involves a presynaptic mechanism resulting in an increase in the acetylcholine output from the nerve terminals, which would have exactly the same anticurare effect. This argument rests on whether it can be shown conclusively that anticholinesterase drugs are capable of increasing the quantal content of the end plate potential either by a direct presynaptic action or via a presynaptic action of the excess acetylcholine itself (Riker et al 1957, 1959, Blaber & Bowman 1959, Blaber & Christ 1967, Hobbiger 1976). Blaber (1970, 1972, 1973) showed that tubocurarine reduced the rate of refilling of the acetylcholine store in nerve terminals and demonstrated that edrophonium, in the cat, could reverse this tubocurarine induced reduction in the rate of refilling of the acetylcholine store.

Physostigmine

Physostigmine was originally obtained from the Calaber bean in West Africa and brought to England in 1840. In 1895 Jolly noted the resemblance between the signs of myasthenia and those of tubocurarine poisoning in animals and suggested that physostigmine (already known to antagonise tubocurarine) might be of therapeutic value. However it was nearly 40 years later before Mary Walker reported its successful use in myasthenia gravis in 1934, and then in 1935 reported that neostigmine was more effective. Physostigmine has been shown to antagonize non-depolarizing block in animals but the result was disappointing in humans (Baraka 1978). Its lipid solubility and consequent ability to cross the blood–brain barrier causing excitant effects is the reason for its application in the reversal of general anaesthesia. It is not used in an anticurare agent.

Neostigmine

This is the agent widely used in the UK for reversal of neuromuscular blockade. It is used in a dose range 2.5–5 mg. A number of isolated cases of cardiac arrest occurred following its use (Clutton Brock 1949, MacIntosh 1949) and these were attributed to the bradycardia associated with its muscarinic action. These reports led to the practice of always giving atropine prior to neostigmine, but Kemp & Morton (1962) demonstrated that in healthy patients a mixture of atropine 1.2 mg and neostigmine 2.5 mg did not cause bradycardia. This was attributed to the fact that atropine acted within 25 to 45 seconds whereas neostigmine took 100 seconds. Kjellberg & Tammisto (1970) confirmed these findings.

Pyridostigmine

Pyridostigmine is an analogue of neostigmine with one-quarter of its potency so that 10 mg pyridostigmine is equivalent to 2.5 mg neostigmine. Katz (1967b) concluded that it exhibited fewer muscarinic effects on the bowel and myocardium than neostigmine although Fogdall & Miller (1973) found there was no difference. It has been shown to have a slower onset and a longer duration of action than neostigmine as well as a greater therapeutic ratio (Miller et al 1974).

Edrophonium

Edrophonium has not been widely used as an anticurare agent in this country because of its apparently short duration of action, unreliable antagonism and the possibility of recurarization (Artusio et al 1950, MacFarlane et al 1950, Doughty & Wyllie 1952, Hunter 1952, Nastuk & Alexander 1954, Katz 1967, Sugai & Payne 1975). Some of these reports of unsustained and unreliable reversal may have been the result of inadequate dosage (10–20 mg) or imprecise/lack of monitoring of neuromuscular function. Recent studies measuring the twitch response suggest that in higher dosage (0.5–1.0 mg/kg) the duration of action of edrophonium is almost the same as neostigmine. In equiantagonistic doses the duration of antagonism during steady state 90% block of the twitch response was 66 minutes for edrophonium (0.5 mg/kg) and 76 minutes for neostigmine (0.04 mg/kg) (Cronelly et al 1982), while the pharmacokinetic profile of the two drugs has been shown to be very similar (Morris et

---

*'Curarization' — this implies paralysis with the non-depolarizing drug, D tubocurarine but again the term has crept into usage even when applied to drugs with a similar action.

al 1981). There has therefore been a revival of interest in edrophonium as an anticurare agent.

The importance of cholinesterase inhibition in the mechanism of action of edrophonium is unclear. The anticholinesterase activity of edrophonium has been shown to be less than that of neostigmine (Randall & Lehman 1950, Smith et al 1957) and muscarinic side-effects of edrophonium are less obvious which supports this finding. However, inhibition of acetylcholinesterase by a reversible inhibitor is probably not comparable to that produced by acylating anticholinesterases. As stated above, part of the anticurare action of edrophonium may involve an increase in the quantal release of acetylcholine. In the clinical situation Donati et al (1983) examined the ratio of the train of four when the first twitch was less than 70% of control during the antagonism of pancuronium blockade using neostigmine and edrophonium. They found the highest ratio after edrophonium, and concluded that it was a more effective antagonist at the presynaptic receptor than the other two drugs. If the train of four ratio reflects only presynaptic effects, then these results would support a presynaptic predominance for edrophonium. However, the results of dose response studies are confusing. Dose response curves for edrophonium during antagonism of a steady state tubocurarine induced neuromuscular block are not parallel to those obtained after neostigmine or pyridostigmine, suggesting different mechanisms of action (Cronelly et al 1982), but are parallel to those obtained with neostigmine after a bolus dose of pancuronium (Breen et al 1985). If these drugs have different mechanisms of action, then combining edrophonium and pyridostigmine or neostigmine might be expected to potentiate reversal. The possibility that combinations of anticholinesterases may work synergistically has been studied but so far results have been disappointing (Bevan et al 1984, Cronelly & Miller 1984).

Efficacy as anticurare agents

Large doses of edrophonium (0.5–1.4 mg/kg) produce adequate and sustained reversal of neuromuscular block provided that some spontaneous recovery of neuromuscular function has occurred. Bevan (1979), Kopman (1979), Ferguson et al (1980), Morris et al (1981) and Cronelly et al (1982) have demonstrated satisfactory reversal of pancuronium and tubocurarine blockade once the twitch response had recovered to 5–10% of control. This has also been noted following vecuronium (Baird et al 1982) and atracurium (Baird & Kerr 1983, Jones et al 1984).

As an anticurare agent edrophonium has a quicker onset of action than neostigmine and pyridostigmine (Ferguson et al 1980, Cronelly et al 1982, Jones et al 1984, Astley et al 1986). This may be due to the initial rapid release of acetylcholine from the motor nerve terminal or it may be related to the size of the molecule. Ramzan (1982) postulated that the difference in speed of onset of any two drugs was proportional to the rate of diffusion of the cation through the plasma, and since the molecular weight of edrophonium is 166 compared to 223 for neostigmine, it may diffuse more rapidly to its site of action. Bevan (1979) proposed that atropine should be administered 30–60 s before the edrophonium to block the undesirable muscarinic side-effects. However, these effects are less pronounced following the use of edrophonium than with neostigmine (Randall 1950, Baird et al 1982, Cronelly et al 1982). Whether this is due to the fact that edrophonium is not constantly in combination with the acetylcholinesterase enzyme or that its anticholinesterase properties are weak is not clear.

When attempts are made to reverse patients with no apparent recovery of neuromuscular function the outcome is often unsatisfactory and protracted. Neostigmine may be preferable to edrophonium under these circumstances. Katz (1967b) noticed that whereas neostigmine could reverse patients with no twitch response this was not always possible after edrophonium. He had used a small dose of edrophonium (10 mg) which may have accounted for his results, but more recent experiments (Rupp et al 1984) have confirmed this disappointing result with edrophonium when compared to the action of neostigmine. In our experiments we have monitored twitch and tetanic responses simultaneously during antagonism with edrophonium of atracurium induced neuromuscular blockade. We found that recovery of the twitch response after reversal with edrophonium, although longer than that with neostigmine, was still facilitated but that recovery of the tetanic response was unsustained (Fig 18.2). The recovery of the tetanic response of patients reversed in this way was not significantly different from control patients recovering spontaneously from the same dose of atracurium. The recovery of the tetanic response of patients reversed with neostigmine was significantly faster than the control group. One explanation of this failure may be that edrophonium depleted the acetylcholine stores as a result of increased prejunctional output and at high frequencies of stimulation mobilization could not be maintained.

In anaesthetized patients renal excretion accounts for 75% of the clearance of pyridostigmine, 70% of the clearance of edrophonium and 50% of the clearance of neostigmine (Conelly et al 1979, Morris et al 1981). Thus, in patients with impaired renal function excretion of these drugs is decreased and, in fact, this decrease has been shown to be greater than the corresponding decrease for pancuronium and tubocurarine (Miller et al 1977, Somogyii et al 1977). Reports of 'recurarization' in anephric patients are therefore probably not due to the muscle relaxant outlasting the antagonist.

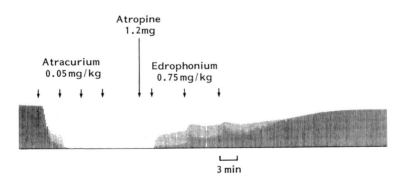

Fig 18.2 Tetanic response following neuromuscular blockade with incremental dose atracurium. The patient was reversed with edrophonium 0.75 mg/kg from 100% block of the tetanic response. Note the initial improvement in tetanic height and tetanic fade, followed immediately by worsening of both. The recovery of the tetanic response of patients reversed in this way did not differ significantly from control patients.

A certain degree of caution is necessary when using reversal agents. Perhaps the term 'recurarization' is not always correct as it is possible under certain circumstances to reintroduce neuromuscular block with anticholinesterases. The neuromuscular blocking properties of neostigmine have been known for many years (Briscoe 1936, Goodman & Gilman 1956, Blaber & Bowman 1963b, Payne et al 1980). If in the absence of neuromuscular monitoring neuromuscular function has recovered it is possible to reintroduce neuromuscular block with both neostigmine and edrophonium. The block so introduced following neostigmine can last for up to 20 minutes (Fig. 18.3). This is long enough to be clinically significant and may be the explanation for some reports of neostigmine resistant curarization. Payne et al (1980) found that although a single dose of 5 mg of neostigmine given to antagonize competitive neuromuscular block could result in neuromuscular block, it was more pronounced after two doses of 2.5 mg of neostigmine. In contrast, neuromuscular block reintroduced under the same conditions following edrophonium (Fig 18.4) is not as constant and is much shorter in duration (of the order of 3 minutes).

The reintroduction of neuromuscular block appears to depend upon the dose of anticholinesterase, the dose of muscle relaxant (more obvious with small doses), the individual relaxant (more likely after the shorter acting drugs) and the degree of spontaneous recovery already present when the anticholinesterase is administered. This neuromuscular block becomes more obvious at faster rates of stimulation (Briscoe 1937) and in practical terms means that it is only evident after tetanic

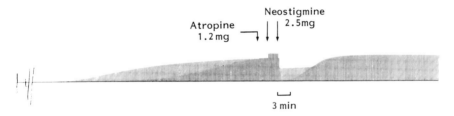

Fig 18.3 Tetanic response following neuromuscular blockade with 0.08 mg/kg of vecuronium. The patient was given neostigmine twice at a time when recovery was almost complete (note absence of tetanic fade). There was improvement of neuromuscular function after the first dose. After the second dose however tetanic height fell and tetanic fade was reintroduced which in this patient took longer than 15 min to recover.

Fig 18.4 Tetanic response following neuromuscular block with incremental atracurium. This patient received edrophonium 0.5 mg/kg twice when recovery was almost complete. Note the initial and immediate facilitation after the first dose followed quickly by a fall in tetanic height associated with the reintroduction of fade. This deterioration is deeper and more persistant after the second dose, but still lasts less than 3 min.

stimulation so that if the twitch response is examined this effect is missed. Payne et al (1980) concluded that if recovery of the tetanic response had reached 50%, then a single dose of reversal agent should be sufficient to provide adequate reversal and further doses could result in block. The neuromuscular block seen following edrophonium is probably not clinically significant. We have failed to demonstrate neuromuscular block after pyridostigmine when given to reverse competitive neuromuscular blockade. The fact that fade is demonstrable after edrophonium suggests that in fact its mode of action may be similar to that of neostigmine.

The facilitatory drug 4-aminopyridine potentiates the effects of neostigmine and pyridostigmine in animals and man in doses smaller than those producing antagonism of neuromuscular blockade (Booij et al 1978, Miller et al 1979). Booij et al (1980) have reported the successful use of 4-aminopyridine alone in the reversal of vecuronium induced blockade. However, the central nervous stimulation that occurs may limit its use as a sole antagonist but if used in combination with an anticholinesterase it does reduce the atropine requirement (Miller et al 1979).

CONCLUSION

Following the recent introduction of two competitive neuromuscular blocking agents of medium duration, atracurium and vecuronium, antagonism of neuromuscular block can be achieved more quickly and reliably than by following block after drugs with a longer duration of action. However, the duration of action of three times the ED95 dose of both drugs has been shown to be approximately 1 hour (Agoston 1980, Payne & Hughes 1981, Basta 1982, Fragen 1982, Katz 1982), and therefore after short surgical procedures reversal is advisable. This is especially important if there is no recovery area, a situation that regrettably still exists in many hospitals (Lunn & Mushin 1982). Reversal is also advised after the use of longer acting agents. However, if neuromuscular monitoring indicates good recovery of function and there are trained staff to supervise the patient in the immediate post-operative period, then reversal after neuromuscular block with either atracurium or vecuronium is not obligatory as the offset of these two drugs in comparison with other competitive neuromuscular blocking agents is so rapid. Edrophonium acts more quickly than neostigmine, requires less atropine, and is associated with a smaller risk of reintroduction of block of any clinical significance, but is less effective when there is no apparent recovery of neuromuscular function. Neostigmine is more effective under these circumstances but its administration may result in neuromuscular block if repeated doses are given to a patient who has already recovered. Our preliminary results with pyridostigmine suggest that its efficacy as an anticurare agent lies between edrophonium and neostigmine, with less propensity to produce neuromuscular block in its own right.

Clinical assessment of the adequacy of neuromuscular function is notoriously inaccurate (Harrison 1978, Viby Mogensen et al 1979), but neuromuscular monitoring now offers the opportunity for a proper assessment of neuromuscular function. It is to be hoped that further advances in this field will enable the anaesthetist to reverse neuromuscular blockade quickly, without risk to the patient.

REFERENCES

Agoston S, Salt P, Newton D, Bencini A, Boomsma P, Erdmann W 1980 The neuromuscular blocking action of Org NC45 a new pancuronium derivative in anaesthetised patients. British Journal of Anaesthesia 52:53S
Ali H H, Kitz R J 1973 Evaluation of recovery from non-depolarizing neuromuscular block using a digital neuromuscular transmission analyser. Anaesthesia and Analgesia 52: 740–743
Ali H H, Savarese J J 1976 Monitoring of neuromuscular function. Anesthesiology 45: 216–242
Ali H H, Utting J E, Gray C 1970 Stimulus frequency in the detection of neuromuscular block in humans. British Journal of Anaesthesia 42: 967–977
Ali H H, Wilson R S, Savarese J J, Kitz R J 1975 The effect of tubocurarine on the indirectly elicited train of four muscle response and respiratory measurement in humans. British Journal of Anaesthesia 47: 570–573
Artusio J F, Riker W F, Wescoe W C 1950 Studies on the interrelationship of certain cholinergic compounds. Journal of Pharmacology and Experimental Therapeutics 100: 227–236
Astley B A, Hughes R, Payne J P 1986 Antagonism of atracurium induced neuromuscular blockade by neostigmine or edrophonium. British Journal of Anaesthesia 58: 1290–1295
Baird W L M, Bowman W C, Kerr W J 1982 Some actions of OrgNC45 and of edrophonium in the anaesthetised cat and in man. British Journal of Anaesthesia 54: 375–384
Baird W L M, Kerr W J 1983 Reversal of atracurium with edrophonium. A pilot study in man. British Journal of Anaesthesia 55: 63S
Baraka A 1978 Antagonism of neuromuscular block by physostigmine in man. British Journal of Anaesthesia 50: 1075–1076
Barnes J M, Duff J I 1953 The role of cholinesterase at the myoneural junction. British Journal of Pharmacology 8: 334–339
Basta S J et al 1982 Clinical pharmacology of atracurium besylate. A new nondepolarising muscle relaxant. Anesthesia and Analgesia 61: 723–729
Bevan D R 1979 Reversal of Pancuronium with edrophonium. Anaesthesia 34: 614–619
Bevan D R, Doherty W G, Breen P J, Donati F 1984 Mixtures of neostigmine and edrophonium are not synergistic. Anesthesiology 61: A298
Blaber L C 1970 The effect of facilitatory concentrations of decamethonium on the storage and release of transmitter at the neuromuscular junction of the cat. Journal of Pharmacology and Experimental Therapeutics 175: 664–672
Blaber L C 1972 The mechanism of the facilitatory action of edrophonium in the cat. British Journal of Pharmacology 46: 498–507
Blaber L C 1973 The prejunctional actions of some nondepolarising blocking drugs. British Journal of Pharmacology 47: 109–116
Blaber L C, Bowman W C 1959 A comparison between the effects of edrophonium and choline in the skeletal muscles of the cat. British Journal of Pharmacology 14: 456–466
Blaber L C, Bowman W C 1963a The effects of some drugs on the repetitive discharges produced in nerve and muscle by anticholinesterases. International Journal of Neuropharmacology 2: 1–16
Blaber L C, Bowman W C 1963b Studies on the repetitive discharges evoked in motor nerve and skeletal muscle after injections of anticholinesterase drugs. British Journal of Pharmacology 20: 326–344
Blaber L C, Christ D D 1967 The action of facilitatory drugs on the isolated tenuissimus muscles of the cat. International Journal of Neuropharmacology 6: 473–484
Blaber L C, Goode J W 1968 A comparison of the action of facilitatory and depolarising drugs at the mammalian motor nerve terminal. International Journal of Neuropharmacology 7: 429–440
Booij L H D J, Miller R D, Crul J F 1978 Neostigmine and 4-aminopyridine antagonism of lincomycin and pancuronium neuromuscular blockade. Anesthesia and Analgesia 57: 316–321
Booij L H D J, Van Der Pol F, Crul J F, Miller R D 1980 Antagonism of Org NC45 neuromuscular blockade by neostigmine, pyridostigmine and 4-aminopyridine. Anaesthesia and Analgesia 59: 31–34
Bowman W C 1980 Pharmacology of neuromuscular function. Wright, Bristol, ch 3 p 18
Boyd I A, Martin A R 1956 The endplate potential in mammalian muscle. Journal of Physiology 132: 74–91
Breen P J, Doherty W G, Donati F, Bevan D R 1985 The potencies of edrophonium and neostigmine as antagonists of pancuronium. Anaesthesia 40: 844
Briscoe G 1936 Shift in optimum rate of stimulation due to prostigmin. Journal of Physiology 86: 48P
Briscoe G 1937 Optimum stimulation rates for red and white skeletal mammalian muscles, and shift in rates produced by the eserine group. Journal of Physiology 90: 10P–11P
Brown G L, Dale H H, Feldberg W J 1936 Reactions of the normal mammalian muscle to acetylcholine and to eserine. Journal of Physiology 87: 394–425

Clutton Brock J 1949 Death following neostigmine. British Medical Journal 1: 1007

Cronelly R, Miller R D 1984 Onset and duration of edrophonium-pyridostigmine mixtures. Anesthesiology 61: A301

Cronelly R, Morris R B, Miller R D 1982 Edrophonium: duration of action and atropine requirement in humans during halothane anaesthesia. Anesthesiology 57: 261–266

Cronelly R, Stanski D R, Miller R D, Sheiner L B, Sohn Y J 1979 Renal function and the pharmacokinetics of neostigmine in anaesthetised man. Anesthesiology 51: 222–226

Donati F, Ferguson A, Bevan D R 1983 Twitch depression and train of four ratio after antagonism of pancuronium with edrophonium, pyridostigmine or pyridostigmine. Anesthesia and Analgesia 62: 314–316

Doughty A G, Wylie W D 1952 Antidotes to true curarising agents. British Journal of Anaesthesia 24: 66–80

Eccles J C, Katz B, Kuffler S W 1942 Effects of eserine on neuromuscular transmission. Journal of Neurophysiology 5: 211–230

Feng T P, Li T H 1941 Studies on the neuromuscular junction; a new aspect on the phenomena of eserine potentiation and post tetanic facilitation in mammalian muscles. Chinese Journal of Physiology 16: 37–56

Ferguson A, Egerszegi P, Bevan D R 1980 Neostigmine, pyridostigmine and edrophonium as antagonists of pancuronium. Anesthesiology 53: 390–394

Fogdall R P, Miller R D 1973 Antagonism of tubocurarine and pancuronium induced neuromuscular blockades by pyridostigmine in man. Anesthesiology 39: 504–509

Foldes F F, Monte A P, Brunn H M, Wolfson B 1961 Studies with muscle relaxants in unanaesthetised subjects. Anesthesiology 22: 230–236

Fragen R J, Robertson E N, Booij L H D J, Crul J F 1982 A comparison of vecuronium and atracurium in man. Anesthesiology 57: A253

Gal T J, Smith T C 1976 Partial paralysis with tubocurarine and the ventilatory response to carbon dioxide. Anaesthesiology 45: 22–28

Gissen A J, Katz R L 1969 Twitch, tetanus and post tetanic potentiation as indices of nerve muscle block in man. Anesthesiology 30: 481–487

Harrison G G 1978 Death attributable to anaesthesia. A 10 year survey (1967–1976). British Journal of Anaesthesia 50: 1041–1046

Hobbiger F 1976 Pharmacology of anticholinesterase drugs. In: Zaimis (ed) Handbook of Experimental Pharmacology. Springer-Verlag, Berlin vol 42: ch 4c, p 487

Hubbard J I, Schmidt R F 1961 Stimulation of the motor nerve terminals. Nature 191: 1103–1104

Hubbard J I, Schmidt R F, Yokota T 1965 The effect of acetylcholine upon mammalian motor nerve terminals. Journal of Physiology 181: 810–829

Hunter A 1952 Tensilon: a new anticurare agent. British Journal of Anaesthesia 24: 175–186

Johansen S H, Jorgensen M, Molbech S 1964 Effect of tubocurarine on respiratory and non respiratory muscle power in man. Journal of Applied Physiology 19: 990–994

Jolly F 1895 Pseudoparalysis myasthenica. Neurol. Zbl 14: 34

Jones R M, Pearce A C Williams J P 1984 Recovery characteristics following antagonism of atracurium with neostigmine or edrophonium. British Journal of Anaesthesia 56: 453–456

Katz R L 1967a Neuromuscular effects of tubocucarine, edrophonium and neostigmine in man. Anaesthesiology 28: 327–336

Katz R L 1967b Pyridostigmine (Mestinon) as an antagonist of tubocurarine. Anesthesiology 28: 528–534

Katz R L 1971 Clinical neuromuscular pharmacology of pancuronium. Anaesthesiology 34: 550–556

Katz R L, Stirt J, Murray A L, Lee C 1982 Neuromuscular effects of atracurium in man. Anaesthesia and Analgesia 61: 730–734

Kemp S W, Morton H J V 1962 The effect of atropine and neostigmine on the pulse rate of anaesthetised patients. Anaesthesia 17: 170–175

Kjellberg M, Tammisto T 1970 Heart rate changes after atropine and neostigmine given for the reversion of muscle paralysis. Acta Anaesthesiologies Scandinavica 14: 203

Kopman A 1979 Edrophonium antagonism of pancuronium induced neuromuscular blockade in man. Anesthesiology 51: 139–142

Lam H S, Cass N M, NG KC 1981 Electromyographic monitoring of neuromuscular block. British Journal of Anaesthesia 53: 1351–1356

Lunn J, Mushin W 1982 Mortality associated with anaesthesia. Nuffield Provincial Hospitals Trust

Macfarlane D W, Pelikan E W, Unna K R 1950 Evaluation of curarising drugs in man. Journal of Pharmacology and Experimental Therapeutics 100: 382–392

Macintosh R R 1949 Death following injection of neostigmine. British Medical Journal 1: 852

Maclagen J 1976 Competitive Neuromuscular Blocking Agents. In: Zaimis (ed) Handbook of Experimental Pharmacology. Springer Verlag, Berlin, 42. Ch 4b, p 421

Masland R L, Wigton R S 1940 Nerve activity accompanying fasciculation produced by prostigmin. Journal of Neurophysiology 3: 269–275

Meer C, Meeter E 1956 The mechanism of action of anticholinesterases II. The effect of DFP on the isolated rat phrenic nerve diaphragm preparation. A. Irreversible effects. Acta Physiologica et Pharmalogica Neerlandica 4: 454–471

Miller R D 1976 Antagonism of neuromuscular blockade. Anesthesiology 44: 318–330

Miller R D, Van Nyhuis L S, Eger E I Vitez T S, Way W L 1974 Comparative times to peak effect and durations of action of neostigmine and pyridostigmine. Anesthesiology 41: 27–33

Miller R D, Matteo R S, Benet L Z, Sohn Y J 1977 The pharmacokinetics of tubocurarine in man with and without renal failure. Journal of Pharmacology and Experimental Therapeutics 202: 1–7

Miller R D, Booij L H D J, Agoston S, Crul J F 1979 4 Aminopyridine potentiates neostigmine and pyridostigmine in man. Anesthesiology 50: 416–420

Morris R B, Cronelly R, Miller R D, Stanski D R, Fahey M R 1981 Pharmacokinetics of edrophonium and neostigmine when antagonising tubocurarine neuromuscular blockade in man. Anesthesiology 54: 399–402

Morris R B, Cronelly R, Miller R D, Stanski D R, Fahey M R 1981 Pharmacokinetics of edrophonium in anephric and renal transplant patients. British Journal of Anaesthesia 53: 1311–1313

Nastuk W L, Alexander J T 1954 The actions of 3 hydroxy phenyldimethylethylammonium (tensilon) on neuromuscular transmission in the frog. Journal of Pharmacology and Experimental Therapeutics 111: 302–328

Paton W D M, Waud D R 1967 The margin of safety of neuromuscular transmission. Journal of Physiology 191: 59–90

Payne J P, Hughes R 1981 Evaluation of atracurium in anaesthetised man. British Journal of Anaesthesia 53: 45–54

Payne J P, Hughes R, Al Azawi S 1980 Neuromuscular blockade by neostigmine in anaesthetised man. British Journal of Anaesthesia 52: 69–76

Ramzan I M 1982 Molecular weight of cation as a determinant of speed of onset of neuromuscular blockade. Anesthesiology 57: 247

Randall L O 1950 Anticurare action of phenolic quaternary ammonium salts. Journal of Pharmacology and Experimental Therapeutics 100: 83–93

Randall L O, Lehman G 1950 Pharmacological properties of some neostigmine analogues. Journal of Pharmacology and Experimental Therapeutics 99: 16–32

Riker W F, Roberts J, Standaert F G, Fujimori H 1957 The motor nerve terminal as the primary focus for drug induced facilitation of neuromuscular transmission. Journal of Pharmacology and Experimental Therapeutics 121: 286–312

Riker W F, Werner G, Roberts J, Kuperman A 1959 Pharmacologic evidence for the existence of a presynaptic event in neuromuscular transmission. Journal of Pharmacology and Experimental Therapeutics 125: 150–158

Riker W F, Wescoe W C 1946 The direct action of prostigmine on skeletal muscle. Its relationship to the choline esters. Journal of Pharmacology and Experimental Therapeutics 88: 58–66

Rupp S M, McChristian J W, Miller R D 1984 Neostigmine antagonises a profound neuromuscular block more rapidly than edrophonium. Anesthesiology 61: A297

Savarese J J, Ali H H 1977 Accurate prediction of individual metocurine dosage by clinical observation of the threshold of fade on the train of four stimulation. Meeting of the American Society of Anesthetists. New Orleans (Abstract)

Smith C M, Mead J C, Unna K R 1957 Antagonism of tubocurarine III. Time course of action of pyridostigmine, neostigmine and edrophonium in vivo and in vitro. Journal of Pharmacology and Experimental Therapeutics 120: 215–228

Somogyi A A, Shanks C A, Triggs E J 1977 The effect of renal failure on the disposition and neuromuscular blocking action of pancuronium bromide. European Journal of Clinical Pharmacology 12: 23–29

Sugai N, Payne J P 1975 The skeletal muscle response to edrophonium during neuromuscular blockade by tubocurarine in anaesthetised man. British Journal of Anaesthesia 47: 1087–1092

Viby Mogensen J, Engbaek J, Jensen N H, Jorgensen B C, Ording H 1983 New developments in clinical monitoring of neuromuscular transmission: monitoring without equipment. Clinical experiences with Norcuron, Symposium. Geneva, p 66–71

Viby Mogensen J, Jorgensen B C, Ording H 1979 Residual curarisation in the recovery room. Anesthesiology 50: 539–541

Walker M B 1934 Treatment of myasthenia gravis with physostigmine. Lancet 1: 1200–1201

Walker M B 1935 Case showing effect of prostigmine on myasthenia gravis. Proceedings of the Royal Society of Medicine 28: 759–761

Walts L F, Levin N, Dillon J B 1970 Assessment of recovery from tubocurarine. Journal of the American Medical Association 213: 1894–1896

Werner G 1960 Neuromuscular facilitation and antidromic discharges in motor nerves and their relation to activity in motor nerve terminals. Journal Neurophysiology 23: 171–187

Williams N E, Webb S N, Calvey T N 1980 Differential effects of myoneural blocking drugs on neuromuscular transmission. British Journal of Anaesthesia 52: 1111–1114

Windsor J P W, Sebel P S, Flynn P J 1985 The neuromuscular transmission monitor. Anaesthesia 40: 146–151

Wright D L, Plummer D T 1973 Multiple forms of acetylcholinesterase from human erythrocytes. Journal of Biochemistry 133: 521–527

Wymore M L, Eisele J H 1978 Differential effects of tubocurarine on inspiratory muscles and two peripheral muscle groups in anaesthetised man. Anesthesiology 48: 360–362

Pain

INTRODUCTION

The discovery by Kosterlitz and Hughes in 1975 of endogenous peptides that bind to opiate receptors has produced a wealth of literature and started a new era of research accompanied by new terminology (Hughes 1983). It soon became clear that the original concept of one endogenous peptide, endorphin (endogenous morphine), was an over simplification and that many peptides and receptor sites were involved. The abundance of published literature, at times conflicting and intensely stimulating, may have overwhelmed clinicians wandering through the maze of detail. The concepts set out in this synopsis may already have been overtaken by more recent research studies.

Anatomical and physiological aspects of pain perception were reviewed by Jordan (1984), noting the importance of the periaqueductal grey area in the brain and the substantia gelatinosa, a semitranslucent area of grey matter that caps the posterior horn (Kitahata & Collins 1981, Yaksh 1981). Small diameter afferent fibres terminate in lamina one and two of the dorsal horn and this would include C fibres. Laminae one to five appear to accept afferents from A δ and C fibres. The substantia gelatinosa (lamina 2) has a high concentration of opioid receptors, endorphins and substance P and is the region where sensory impulses are modulated. Substance P (not P for pain) is released following stimulation of small primary afferent fibres and this is opposed by the action of endorphins. Intrathecal opiates inhibit the release of substance P. In contrast to local anaesthetics, spinal opioids do not produce motor blockade or hypotension as the sympathetic nerves also appear to be unaffected.

Yaksh & Rudy (1976) demonstrated that intrathecal opioids produced naloxone reversible analgesia in animals. The technique was applied to man for the relief of pain for inoperable carcinoma by Wang et al (1979). The opioid acts at the C and A δ afferent inputs to the 'wide dynamic range' (WDR) cells in the spinal cord which are associated with the transmission of pain. The spinal opioid appears to be particularly effective in suppressing noxious stimuli associated with C fibre stimulation and less effective in blocking A δ afferents which are readily affected by local anaesthetics. This appears to be confirmed clinically when spinal opioids reduce the pain associated with deep sensation but have a reduced effect on the pain of the skin incision which is mediated by A δ fibres. Homma et al (1983) demonstrated that the suppression of WDR neurones was dose dependent and that adrenaline was also capable of exerting a similar effect (Collins et al 1984). Adrenaline in a dose of 10 μg (a fifth of the dose used in previous experiments) applied to the cord had little effect on noxious stimulation but produced significant suppression of WDR neurones following fentanyl administered spinally 30 minutes previously. It was concluded that there was an interaction between α adrenergic drugs and spinal opiates, suggesting that the prolongation of action of the opioid was not due to the vasoconstrictor action of the adrenaline. The neural mechanisms of pain are outlined by Casey (1982) and by Mense (1983).

CLASSIFICATION OF RECEPTORS

Detailed reviews of opioid receptors and endorphins appeared in Anaesthesia Review 3 (Jordan 1985, Pinnock 1985). Martin et al (1976) suggested that there were three receptor sites which would explain some of the actions of opioids. The μ receptor was associated with analgesia, respiratory depression and small pupils. The μ receptor provided analgesia and sedation while the σ receptor was responsible for the psychological effects seen after drugs such as phencyclidine. Other studies suggested that enkephalins acted at δ receptor sites in the brain but the significance of this receptor at the moment appears to be speculative. Other receptors proposed included ϵ (epsilon) and τ, and other unidentified sites (Yaksh 1984a, Yaksh 1984b).

Morphine acts selectively at μ receptor sites which are are present in the cerebral cortex, in the periaquaductal grey region (PGR), and also in the dorsal horn. In addition to morphine, diamorphine, pethidine and fentanyl are agonists at μ receptors, as is the agonist/antagonist buprenorphine. Wood et al (1982) supported the concept of μ_1 receptors which mediated analgesia and regulation of cholinergic neurones, while μ_2 receptors were responsible for respiratory depression and dopaminergic neurones.

There are a high proportion of κ receptors in the spinal cord at which the peptide dynorphin appears to be active. The agonist/antagonist drugs such as pentazocine, nalbuphine and butorphanol have antagonist properties at μ sites and agonist properties at κ sites. For a guide to the structure and terminology of endogenous opioid peptoids see Cox (1982).

Yaksh (1983) demonstrated the presence of μ and δ receptor sites in the spinal cord of primates. He also showed that the onset and duration of action of opioids applied to the spinal cord was closely related to the lipid partition coefficient of the drug. Of the agents employed he showed that the order of potency was greater with beta endorphin, followed by morphine, L-methadone, pethidine and D-methadone. It was also noted that daily intrathecal morphine resulted in a loss of response over a 5–9 day period and sensitivity recurred after an opiate free interval of 7 days. Drugs with δ activity may be of value in patients who have become intolerant to intrathecal morphine. In further in vivo studies, Schmauss & Yaksh (1984) suggested that there were μ, δ and κ receptor sites in the spinal cord and proposed that there are μ/δ receptors for cutaneous thermal transmission and μ/κ receptors for visceral chemical transmission.

Tyers (1980) and Skingle & Tyers (1980) found that at least two opioid receptors were involved in the mediation of antinociception. Tests employing heat were specific for μ agonists, while tests employing pressure, chemical writhing or tooth pulp stimulation responded to both μ and κ agonists. Bryant et al (1983) suggested that buprenorphine and morphine produce analgesia by interacting with different opioid receptors and concluded that the predominant site of analgesia for buprenorphine was supraspinal. In man the epidural dose of buprenorphine was similar to that given intramuscularly, while in animal studies the intrathecal dose exceeded the parenteral dose. Hayes & Tyers (1983) found that μ receptor agonists produced their antinociceptive effects and side-effects by acting at the μ receptor. κ Receptor agonists appeared to produce little analgesia and the opioid side-effects are probably due to interaction with μ or other receptors.

Further detailed studies are still required to unravel the complexity of actions of opioids at receptor sites. Recent reviews include electrophysiology of opioids (Duggan & North 1984) and pharmacology of opioids (Martin 1984).

ENDOGENOUS OPIOIDS AND THE CARDIOVASCULAR SYSTEM

The cardiovascular effects of endogenous opioids have been reviewed extensively by Holaday (1983). The results are difficult to evaluate as there may be more than one type of opioid receptor involved and there may be secondary factors such as the peripheral action of catecholamines, histamine release or even other central transmitters. The state of consciousness of the animals may determine the outcome of experiments. In addition selectivity for specific receptors by opioids may not be absolute.

The intraventricular injection of enkaphalin analogues produce hypotension and bradycardia. Selective μ, δ and κ agonists injected into the anterior hypothalamus produced hypotension and bradycardia in low doses but had no effect on heart rate at higher doses (Pfeiffer et al 1983). Studies in man with intravenous infusions of DAMME (Met-enkephalin analogue) showed that it caused a fall in blood pressure by attenuating barorceptor function, an action medicated by μ receptors (Rubin et al 1983).

Interest in endogenous opioids and blood pressure suggested the possibility that they might be involved in circulatory failure associated with endotoxic shock. There are studies to suggest that pre-treatment with naloxone prevents the aftermath of endotoxic shock and that naloxone improves oxygenation, partly by improving haemodynamics and reducing shunting of the cardiac output, and partly by reversing the depressant effects of endogenous opioids (Mamazza et al 1984). The role of opioid antagonists in treatment remains speculative. Small doses of naloxone are ineffectual, while the results of larger doses may be influenced by the pH of naloxone which is highly acid (pH = 3). Other suggestions are that improvements in the circulation are due to the associated improvement in blood volume with intravenous fluids (see naloxone, p. 222).

INTRATHECAL MORPHINE

Intrathecal morphine has been studied in patients undergoing thoracotomy when 0.25 mg or 0.5 mg was injected at the end of operation. The duration of analgesia varied from 1 to 20 hours for the smaller dose and from 1 to 40 hours for the larger dose. The concentration of morphine in the CSF was high and dose dependant whereas the plasma concentration was low. The CSF concentration was higher than that seen after epidural administration of 6 mg of morphine (Nordberg et al 1984a).

Nordberg and colleagues also commented on the delayed onset of analgesia with spinal morphine. Post-operative pain relief could be prolonged using intrathecal cinchocaine and morphine compared with cinchocaine alone (Moore et al 1984a).

Moore et al (1984b) also measured the concentration in the CSF of morphine and heroin and found that after either 2 mg of intrathecal diamorphine or 2.5 mg of intrathecal morphine the concentration in the CSF was 4000 times greater than after the injection of 1 mg/kg of intravenous morphine. As diamorphine was more rapidly removed from the CSF than morphine because of its lipophilic properties they recommended that diamorphine was likely to be safer because of its shorter half-life.

Opioids are known to effect motility of the gastrointestinal tract. Animal studies have shown that intrathecal use of the drug appears to be species specific, affecting mice but not rats (Vaught et al 1983).

EPIDURAL MORPHINE

Pharmacokinetic aspects of epidural morphine analgesia have been studied by Nordberg et al (1983) who found that the technique produced dose dependent analgesia with concentrations of morphine much higher in the CSF than in the plasma. Although the elimination half-life for morphine was similar in the CSF and plasma, the long duration of morphine given epidurally seems to be dependent on a high concentration being present locally. In a further study Nordberg et al (1984) found that plasma morphine concentrations after epidural injection was similar to that found after intramuscular injections. Morphine reached a peak in the CSF after 135 ± 40 minutes and despite the high concentrations reached in the CSF it was calculated that only 2% of the drug administered epidurally was available to the CSF compartment. It is of interest that after 100 mg of intramuscular pethidine the drug can be detected within 18 minutes in the CSF reaching a maximum after 90 minutes. The transfer of pethidine from plasma in the CSF appears to occur very rapidly (Boreus et al 1983).

Other studies have compared plasma morphine concentrations of morphine following epidural morphine or diamorphine, assuming that diamorphine is a lipophilic form of morphine (Phillips et al 1984) (see p. 198). Plasma morphine concentrations were greater after epidural diamorphine than after epidural morphine, the mean peaks occurring at 5 and 10 minutes respectively. Similar results were reported with larger doses of the drugs by Watson et al (1984) who also calculated that the effects of epidural diacetyl morphine crossing the dura was 55% of that of morphine, presumably due to absorption of the heroin by epidural veins. Respiratory depression, particularly the rate, was noted only after epidural diacetyl morphine.

Respiration

Camporesi et al (1983) compared the respiratory response to CO_2 following 10 mg of morphine given either intravenously or epidurally. The maximum respiratory depression following intravenous injection occurred after 30 minutes which coincided with the peak level of morphine in the plasma. After epidural morphine, measurements of plasma levels, recorded 30 and 60 minutes later were significantly

powered pump controlling a reservoir of morphine. Pain relief was effective in 100 patients without respiratory depression as assessed by blood gas analyses. 20% of the patients developed urinary retention (Chrubasik 1984).

Respiratory depression

The mechanism for this is still unclear. There have been suggestions that the morphine is absorbed via the epidural veins and then exerts a central action or it diffuses into the CSF and is carried rostrally. The breakdown products of morphine may also be active centrally. It may be dose related while the profound analgesia may reduce respiratory drive (Bullingham et al 1982, Cousins & Mather 1984). Jensen et al (1984) concluded that positioning the patient in 45° of head up elevation did not protect against respiratory depression following epidural morphine. Hammond (1984) reported on the successful use of nalbuphine to increase respiratory activity following the use of epidural diamorphine.

Respiratory function

It might have been expected that adequate post-operative pain relief would have resulted in an improvement in post-operative pulmonary function. Although pain relief was better with epidural morphine than those receiving intramuscular analgesia, there was no significant improvement in the epidural group in lung function tests as assessed by vital capacity (VC), forced expiratory volume over 1 second (FEV1) or functional residual capacity (FRC) (Bonnet et al 1984). In orthopaedic operations of the lower limb, pulmonary function was more affected following spinal bupivacaine and morphine compared with patients who had general anaesthesia with halothane. The FVC, the FRC and closing capacity (CC) were reduced in both groups. In the spinal group there was impaired gas distribution and V/Q mismatch, possibly due to hypoventilation (Hedenstierna & Lofstrom 1985).

Urinary function

There have been numerous reports of urinary retention following epidural morphine (see above). It has been suggested that there is an endogenous role for opioids in the lower urinary tract by decreasing detrusor pressure and increasing sphincteric tone. However, Husted et al (1985) agreed that epidural morphine did cause urinary retention, the effect occurring mainly in patients with abnormal detrusor function. Naloxone remains the treatment of choice.

Intrathecal diamorphine (heroin)

In a personal series of 250 cases given 2.5–5 mg intrathecal diamorphine in 4– 10 ml saline, without barbotage, the average duration of pain relief following major abdominal surgical procedure was 16 hours: 46 patients required no further analgesia in the first 24 hours, while a further 70 were pain-free beyond this time. Two patients had no post-operative narcotic analgesia whatsoever. In one control group when general anaesthesia was supplemented by intravenous fentanyl the first dose of analgesia was administered 1 hour post-operatively whereas in another control group given 10 mg of intravenous diamorphine, the first dose was given 4 hours post-operatively. There appears to be no direct relationship between the duration of analgesia and the patients' weight and height, dose of diacetyl-morphine or the volume injected. Age appears to be the only factor involved and is an indication for the use of smaller doses in the elderly.

Respiratory depression in the post-operative period was noted in elderly patients in the initial studies when patients were monitored with an impedance plethysmograph. Respiratory depression has seldom occurred since the technique of anaesthesia was modified to include minimal premedication, etomidate instead of thiopentone and atracurium (without reversal) instead of d-tubocurarine. Monitoring the end-tidal CO_2 confirms that apnoea at the end of operation is usually due to a low PCO_2. When intrathecal diamorphine was used for diagnostic or therapeutic purposes no instances of respiratory depression were noted.

In addition, the use of intrathecal diamorphine following induction of anaesthesia suppressed some of the endocrine responses to surgical stimulation, produced cardiovascular stability during operation and possibly contributed to improved patient well-being in the post-operative period (Child & Kaufman 1985). ADH levels rose with surgical stimulation, this being less in the intrathecal diamorphine group compared with the controls but the levels could be reduced in both groups by the use of intravenous bumetanide given at the beginning of surgery (Kaufman & Bailey 1986).

Barron & Strong (1984) used intrathecal diamorphine for total hip replacement and for spinal operations (laminectomy: spinal fusion — when the drug was injected by a surgeon under direct vision). The dose administered was in the range of 0.005–0.015 mg/kg but the duration of analgesia was not dose dependent. The total dose did not exceed 1 mg. There were no episodes of post-operative respiratory depression and the incidence of pruritus was low. In a series of 250 patients many had analgesia for more than 24 hours. (In an addendum, the series by Barron & Strong (1984) refers to 1000 patients).

Intravenous morphine

Marshall et al (1985) compared the use of continuous intravenous morphine with intramuscular morphine for the relief of post-operative pain following cholecystectomy. The patients receiving the intravenous infusion received as much supplementary morphine as the control group in the first 24 hours and considerably more in the 24 hour period when the intravenous infusion was discontinued. Side-effects, such as nausea and vomiting, were more common in the intravenous group.

Subcutaneous morphine

The introduction of small and reliable syringe pumps has led to a renewed interest in the use of subcutaneous infusions of morphine. Waldmann et al (1984) compared the blood levels of morphine sulphate given intravenously with that given subcutaneously via a 21-gauge butterfly needle inserted into the deltoid muscle. Using a method of assay which only measured morphine and not the metabolities they found at doses of

25 μg/kg per hour there was no significant difference in the serum levels made in the first 24 hours. The use of additional analgesics was the same for both groups of patients. The technique has been extended to provide pain relief following major abdominal operations and appears to be effective in a dose of 2–6 mg per hour without producing respiratory complications (B. Varley, personal communication).

CANCER PAIN RELIEF

The principles of management and relief of chronic pain are extensively reviewed by Brena & Chapman (1985). The review includes not only chapters on theories of pain and the philosophy of management but also discusses the use of nerve stimulation, nerve block, psychotherapy and drug therapy (Wall & Melzack 1984, Foley 1985). The aim of pain relief in cancer patients should be the continuous suppression of pain, the removal of the fear and memory of pain, while at the same time having an alert patient (Tuttle 1985). Doses of drugs should be adequate and administered before the effects of previous injections have worn off, while at the same time an attempt should be made to increase the interval between injections. Tuttle (1985) reviewed the place of the non-narcotic analgesics including acetylsalicylic acid, the use of antidepressants, as well as drugs used to prevent nausea and vomiting.

Oral morphine in cancer patients

There is a marked individual variation in the effectiveness of oral morphine administered to cancer patients. Sawe et al (1985) have studied the kinetics of morphine and the main metabolite, morphine-3-glucuronide and found the availability after oral administration varied between 30% and 69% and the plasma clearance rate ranged from 18.6 to 34 ml/kg per minute. In vivo studies were matched with in vitro studies from microsomes extracted from liver biopsy. The ratio of morphine-3-glucuronide to morphine reflects the extent of hepatic metabolism, but further studies are necessary to evaluate the usefulness of this ratio in the urine or plasma to predict the efficacy of hepatic metabolism of morphine.

Buccal morphine has been advocated by Bell et al (1985) in preference to intramuscular injection. Tablets containing 10 mg of morphine base given buccally were just as effective as intramuscular injection. The plasma levels were slightly lower following the buccal route but the decline in plasma level was slower compared with intramuscular injection.

Intravenous morphine

The safety and efficacy of continuously administered intravenous morphine for severe cancer pain has been studied by Citron et al (1984). The patients were given an initial bolus injection of 2–5 mg of morphine which was repeated at 10 minute intervals until the pain was relieved. Thereafter continuous intravenous morphine was administered with the hourly dose equal to the sum of the bolus doses. In some of the studies the mean dose was 20 mg per hour and this was increased by 33% if pain relief was inadequate and there were no signs of respiratory depression. In a small number of patients there was a reduction in PaO_2 and an increase in $PaCO_2$ which did not warrant alteration in the drug regime. Blood gas values tended to return to baseline levels despite an increase in the dose of morphine. The technique was considered to be safe and effective although somnolence and slowing of the respiratory rate were occasionally noted. The use of a respiratory monitor, such as an inductance or impedance plethysmograph, would be an additional safeguard. There have also been studies on patient controlled analgesia by continuous infusion; the literature has been reviewed by Graves et al (1983).

Epidural and spinal morphine

The differential effects of epidural morphine in the treatment of 55 patients with pain associated with cancer were reported by Arner & Arner (1985). The epidural catheter was tunnelled subcutaneously to the anterior abdominal wall and injections made through a micropore filter. The daily dose of morphine varied from 4–480 mg. Twenty-one patients became completely free from pain, the best results being obtained when the pain was continuous and deep seated in origin. Cutaneous pain appeared to be unaffected. This would confirm the view that opioids act on receptors mediated by C-fibres and not on A δ fibres which supply the skin.

The use of reservoirs have been described by Coombs et al (1982) where a reservoir is implanted under the skin, either on the interior abdominal wall or under the clavicle. The catheter may be silastic or stainless steel coated with fluoropolymers (Racz et al 1982). Penn et al (1984) have reported the use of epidural or intrathecal morphine using implanted pumps which were either of the constant rate of infusion type or were programmable. Wang (1985) found that long-term use of intrathecal morphine administered by catheter or implanted pump effectively relieved pain in 74% of the patients treated. There was no serious respiratory depression but somnolence, itching and sphincter disturbances were common. With long-term use there was mechanical failure of the pumps and tolerance to morphine developed. With intraspinal morphine, using implanted reservoirs, Coombs et al (1983) noted that patients were able to reduce their oral requirements of analgesics, although it was also noted that tolerance occurred as a significant increase in the dose of infused morphine was required.

REFERENCES

Arner S, Arner B 1985 Differential effects of epidural morphine in the treatment of cancer-related pain. Acta Anaesthesiologica Scandinavica 29: 32–36
Bailey P M, & Sangwan S 1985 Anaesthesia (in press)
Barron D W, Strong J E 1984 The safety and efficacy of intrathecal diamorphine. Pain 18: 279–285
Bell M D D, Murray G R, Mishra P, Calvey T N, Weldon G R, Williams N E 1985 Buccal morphine – the new route for analgesia? Lancet 1: 71–73
Bonnet F, Blery C, Satan M, Simonet O, Brage D, Gaudy J 1984 Effect of epidural morphine on post-operative pulmonary dysfunction. Acta Anaesthesiologica Scandinavica 28: 147–151
Boreus L O, Skoldefors E, Ehrnebo M 1983 Appearance of pethidine and norpethidine in cerebrospinal fluid of man following intramuscular injection of pethidine. Acta Anaesthesiologia Scandinavica 27: 222–225
Brena S F, Chapman S L 1985 Chronic pain: management principles. In: Clinics in Anaesthesiology. W B Saunders, London
Bromage P R, Camporesi E M, Durant P A, Neilsen C H 1983 Influence of epinephrine as an adjuvant to epidural morphine. Anaesthesiology 58: 257–262

Bryant R M, Olley J E, Tyers M B 1983 Antinociceptive action of morphine and buprenorphine given intrathecally in the conscious rat. British Journal of Pharmacology 78: 659–663

Bullingham R E S 1983 Opiate analgesic. In: Clinics in Anaesthesiology. W B Saunders, London

Bullingham R E S, McQuay H J, Moore R A 1982 Extradural and intrathecal narcotics. In: Atkinson R S, Langton-Hewer C (eds) Recent Advances in Anaesthesia and Analgesia No. 14 Churchill Livingstone, Edinburgh Ch 10, p 141–156

Camporesi E M, Nielsen C H, Bromage P R, Durant P A C 1983 Ventilatory CO_2 sensitivity after intravenous and epidural morphine in volunteers. Anesthesia and Analgesia 62: 633–640

Casey K L 1982 Neural mechanisms of pain: an overview. Acta Anaesthesiologica Scandinavica Suppl 74: 13–20

Child C, Kaufman L 1985 Effect of intrathecal diamorphine on the adrenocortical, hyperglycaemic and cardiovascular responses to major colonic surgery. British Journal of Anaesthesia 57: 389–393

Chrubasik J 1984 Epidural, on-demand, low-dose morphine infusion for postoperative pain. Lancet 1: 107

Citron M L, Johnston-Early A, Fossieck B E, Krasnow S H, Franklin R, Spagnolo S V, Cohen M H 1984 Safety and efficacy in continuous intravenous morphine for severe cancer pain. American Journal of Medicine 77: 199–204

Collins J G, Kitahata L M, Matsumoto M, Homma E, Suzukawa M 1984 Spinally administered epinephrine suppresses noxiously evoked activity of WDR neurons in the dorsal horn of the spinal cord. Anesthesiology 60: 269–275

Coombs D W, Saunders R L, Gaylor M, Pageau M G 1982 Epidural narcotic infusion reservoir: implantation technique and efficacy. Anesthesiology 56: 469–473

Coombs D W, Saunders R L, Gaylor M S et al 1983 Relief of continuous chronic pain by intraspinal narcotics infusion via an implanted reservoir. Journal of the American Medical Association 250: 2336–2339

Cousins M J, Mather L E 1984 Intrathecal and epidural administration of opioids. Anesthesiology 61: 276–310

Cox B M 1982 Endogenous opioid peptides: a guide to structures and terminology. Life Sciences 31: 1645–1658

Duggan A W, North R A 1984 Electrophysiology of opioids. Pharmacological Reviews 35: 219–281

Foley K M 1985 The treatment of cancer pain. New England Journal of Medicine 313: 84–95

Glenski J A, Warner M A, Dawson B, Kaufman B 1984 Postoperative use of epidurally administered morphine in children and adolescents. Mayo Clinic Proceedings 59: 531

Graves D A, Foster T S, Batenhorst R L, Bennett R L, Baumann T J 1983 Patient-controlled analgesia. Annals of Internal Medicine 99: 360–366

Hammond J E 1984 Reversal of opioid-associated late-onset respiratory depression by nalbuphine hydrochloride. Lancet 2: 1208

Hayes A G, Tyers M B 1983 Determination of receptors that mediate opiate side effects in the mouse. British Journal of Pharmacology 79: 731–736

Hedenstierna G, Lofstrom J 1985 Effect of anaesthesia on respiratory function after major lower extremity surgery. Acta Anaesthesiologica Scandinavica 29: 55–60

Holaday J W 1983 Cardiovascular effects of endogenous opiates systems. Annual Review of Toxicology 23: 541–594

Hughes J (ed) 1983 Opioid peptides. British Medical Bulletin 39: 1–106

Homma E, Collins J G, Kitahata L M, Matsumoto M, Kawahara M 1983 Suppression of noxiously evoked WDR dorsal horn neuronal activity by spinally administered morphine. Anesthesiology 58: 232–236

Husted S, Djurhuus J C, Husegaard H C, Jepsen J, Mortensen 1985 Effect of postoperative extradural morphine on lower urinary tract function. Acta Anaesthesiologica Scandinavica 29: 183–185

Inturrisi C E, Max B M, Foley K M, Schultz M, Shin S U, Houde R W 1984 The pharmacokinetics of heroin in patients with chronic pain. New England Journal of Medicine 310: 1213–1217

Jordan C C 1984 Anatomical and physiological aspects of pain perception. In: Kaufman L (ed) Anaesthesia Review 2. Churchill Livingstone, Edinburgh

Jordan C C 1985 Opioid receptors. In: Kaufman L (ed) Anaesthesia Review 3. Churchill Livingstone, Edinburgh

Kafer E R, Brown J T, Scott D, Findlay J W A, Butz R F, Teeple E, Ghia J N 1983 Biphasic depression of ventilatory responses to CO_2 following epidural morphine. Anesthesiology 58: 418–427

Kaufman L, Bailey P M 1986 (in press)

Kitahata L M, Collins J G 1981 Spinal action of narcotic analgesics. Anesthesiology 54: 153–163

Kuo R J 1984 Epidural morphine for post-hemorrhoidectomy analgesia. Diseases of the Colon and Rectum 27: 529–530

Mamazza J, Hinchey E J, Chiu R C J 1984 The pulmonary effects of opiate blockade in septic shock. Journal of Surgical Research 36: 625–630

Marshall H, Porteous C, McMillan I, Macpherson S G, Nimmo W S 1985 Relief of pain by infusion of morphine after operation: does tolerance develop? British Medical Journal 291: 19–21

Martin W R 1984 Pharmacology of opioids. Pharmacological Reviews 35: 285–323

Martin W R, Eades C G, Thompson J A et al 1976 The effects of morphine and nalorphine-like drugs in the non-dependent and morphine-dependent chronic spinal dog. Journal of Pharmacology and Experimental Therapeutics 197: 517–532

Mense S 1983 Basic neurobiologic mechanisms of pain and analgesia. American Journal of Medicine 75(5A) 4–14

Mehnert J H, Dupont T J, Rose D H 1983 Intermittent epidural morphine instillation for control of postoperative pain. American Journal of Surgery 146: 145

Molke Jensen F, Madsen J B, Guldager H, Christensen A A, Eriksen H O 1984 Respiratory depression after epidural morphine in the postoperative period. Influence of posture. Acta Anaesthesiologica Scandinavica 28: 600–602

Moore A, Bullingham R, McQuay H, Allen M, Baldwin D, Cole A 1984a Spinal fluid kinetics of morphine and heroin. Clinical Pharmacology and Therapeutics 35: 40–45

Moore R A, Paterson G M C, Bullingham R E S, Allen M C, Baldwin D, McQuay H J 1984b Controlled comparison of intrathecal cinchocaine with intrathecal cinchocaine and morphine. British Journal of Anaesthesia 56: 837–841

Nordberg G, Hedner T, Mellstrand T, Dahlstrom B 1984a Pharmacokinetic aspects of intrathecal morphine analgesia. Anesthesiology 58: 345–551

Nordberg G E T, Hedner T, Mellstrand T, Dahlstrom B 1984a Pharmacokinetic aspects of intrathecal morphine analgesia. Anesthesiology 60: 448–454

Nordberg G, Hedner T, Mellstrand, Borg L 1984b Pharmacokinetics of epidural morphine in man. European Journal of Clinical Pharmacology 26: 233–237

Owen J A, Nakatsu K 1983 Diacetylmorphine (heroin) hydrolases in human blood. Canadian Journal of Physiology and Pharmacology 61: 870–875

Penn R D, Paice J A, Gottschalk W, Ivankovich A D 1984 Cancer pain relief using chronic morphine infusion. Journal of Neurosurgery 61: 302–306

Pfeiffer A, Feuerstein G, Kopin I J, Faden A I 1983 Cardiovascular and respiratory effects of mu-, delta- and kappa-opiate agonists microinjected into the anterior hypothalamic area of awake rats. Journal of Pharmacology and Experimental Therapeutics 225: 735–741

Phillips D M, Moore R A, Bullingham R E S, Allen M C, Baldwin A, Fisher A, Lloyd J W, McQuay H J 1984 Plasma morphine concentrations and clinical effects after thoracic extradural morphine or diamorphine. British Journal of Anaesthesia 56: 829–836

Pinnock C A 1985 Endorphins. In: Kaufman L (ed) Anaesthesia Review 3. Churchill Livingstone, Edinburgh

Racz G B, Sabonghy M, Gintautas J, Kline W 1982 Intractable pain therapy using a new epidural catheter. Journal of the American Medical Association 248: 579–581

Rechtine G R, Cleveland C M R, Bohlman H H 1984 The use of epidural morphine to decrease postoperative pain in patients undergoing lumbar laminectomy. Journal of Bone and Joint Surgery 66: 113–116

Rubin P C, Howden C W, McLean K, Reid J L 1983 Endogenous opioids and baroreflex control in humans. Hypotension 5: 535–538

Sawe J, Kager L, et al 1985 Oral morphine in cancer patients: in vivo kinetics and in vitro hepatic glucuronidation. British Journal of Clinical Pharmacology 19: 495–501

Schmauss C, Yaksh T L 1984 In vivo studies on spinal opiate receptor systems mediating antinociception. II. Pharmacological profiles suggesting a differential association of mu, delta and kappa receptors with visceralchemical and cutaneous thermal stimuli in the rat. Journal of Pharmacology and Experimental Therapeutics 228: 1–12

Shulman M S, Sandler A, Brebner J 1984 The reversal of epidural morphine induced somnolence with physostigmine. Canadian Anaesthetic Society Journal 31: 678–680

Skingle M, Tyers M B 1980 Further studies on opiate receptors that mediate antinociception: tooth pulp stimulation in the dog. British Journal of Pharmacology 70: 323–327

Snir-mor I, Weinstock M, Davidson J T, Bahar M 1983 Physostigmine antagonizes morphine-induced respiratory depression in human subjects. Anesthesiology 59: 6–9

Stenseth R, Sellevold O, Breivik H 1985 Epidural morphine for postoperative pain: experience with 1085 patients. Acta Anaesthesiologica Scandinavica 29: 148–156

Tuttle C B 1985 Drug management of pain in cancer patients. Canadian Medical Association Journal 132: 121–134

Tyers M B 1980 A classification of opiate receptors that mediate antinociception in animals. British Journal of Pharmacology 69: 503–512

Vaught J L, Cowan A, Gmerek D E 1983 A species difference in the slowing effect of intrathecal morphine on gastrointestinal transit. European Journal of Pharmacology 94: 181–184

Wall P D, Melzack R 1984 Textbook of Pain. Churchill Livingstone, Edinburgh

Waldmann C S, Eason J R, Rambohul E, Hanson G C 1984 Serum morphine levels. Anaesthesia 39: 768–771

Wang J K 1985 Intrathecal morphine for intractable pain secondary to cancer of pelvic organs. Pain 21: 99–102

Wang J K, Nauss L A, Thomas J E 1979 Pain relief by intrathecally applied morphine in man. Anesthesiology 50: 149–151

Watson J, Moore A, McQuay H, Teddy P, Baldwin D, Allen M, Bullingham R 1984 Plasma morphine concentrations and analgesic effects of lumbar extradural morphine and heroin. Anesthesia and Analgesia 63: 629–634

Wood P L, Richard J W, Thakur M 1982 Mu opiate isoreceptors: differentiation with kappa agonists. Life Sciences 31: 2313–2317

Yaksh T L 1981 Spinal opiate analgesia: characteristics and principles of action. Pain 11: 293–346

Yaksh T L 1983 In vivo studies on spinal opiate receptor systems mediating antinociception. I. Mu and delta receptor profiles in the primate. Journal of Pharmacology and Experimental Therapeutics 226: 303–316

Yaksh T L 1984a Multiple opioid receptor systems in brain and spinal chord: Part I. European Journal of Anaesthesiology 1: 171–199

Yaksh T L 1984b Multiple opioid receptor systems in brain and spinal chord: Part II. European Journal of Anaesthesiology 1: 201–243

Yaksh T L, Rudy T A 1976 Analgesia mediated by a direct spinal action of narcotics. Science 192: 1357–1358

Yaksh T L, Muller H (ed) 1982 Spinal Opiate Analgesia — Experimental and Clinical Studies. Springer Verlag, Heidelberg

Update (1)

ABDOMINAL ANAESTHESIA

The influence of anaesthetic agents on the outcome of intestinal anastamoses has been reported in previous reviews (see Anaesthesia Review 1, 2, 3). The consensus of opinion is that the blood supply to the bowel, in particular the mucosa, is the critical factor and even in diverticular disease where the bowel is said to be made more irritable by drugs such as morphine and neostigmine the breakdown is minimal (Sangwan 1986). The use of atracurium has largely displaced the necessity for the use of atropine and neostigmine.

Recent animal studies by Tverskoy et al (1985) have confirmed earlier reports that halothane decreased vascular resistance and increased mesenteric blood flow. Enflurane produced little change but isoflurane increased splanchnic vascular resistance and decreased blood flow. The vasoconstriction was abolished by phentolamine, suggesting that the effect had been produced by circulating catecholamines.

Clonidine

Clonidine has been used successfully to treat five patients with pseudo intestinal obstruction following prolonged use or abuse of narcotic analgesics which inhibit gastrointestinal activity. Clonidine alleviates the symptoms of opiate withdrawal presumably by its central α_2 agonistic effects (Sandgren et al 1984).

APPARATUS

Oxygen masks

Hill et al (1984) found that at low inspiratory flow rates high volume Ventimasks and low volume Inspiron Accurox masks acted as fixed performance devices. Some

REFERENCES

Sandgren J E, McPhee M S, Greenberger N L 1984 Narcotic bowel syndrome treated with clonidine. Annals of Internal Medicine 101: 331–334
Sangwan S 1985 (in press)
Tverskoy M, Gelman S, Fowler K C, Bradley E L 1985 Intestinal circulation during inhalation anaesthesia. Anesthesiology 62: 462–469

variation occurred when peak inspiratory rate increased. They recommended that more consistent results were obtained with low volume masks and high oxygen flow rates.

Neonatal endotracheal tubes

The accurate placing of endotracheal tubes in the neonate requires skill as well as ingenuity in preventing them from becoming misplaced. Heller & Cotton (1985) have advocated the use of an endotracheal tube in which a fibre optic strand had been incorporated in the wall, allowing the tube to be visualized externally at will. The technique avoids the use of X-rays which are often used to check the position of the endotracheal tube.

Nebulizers

Douglas et al (1985) studied the flow rate of jet nebulizers which were being used to provide bronchodilator treatment with β-adrenergic agents (Rimiterol). It had been accepted that flow rates below 6 l/min produced droplets which were too large to be deposited in the tracheobronchial tract. However, the present study in chronic stable asthmatics showed that by reducing the flow rate from 8 l/min to 4 l/min the bronchodilator response was similar as assessed by FEV_1 and FVC.

Automatic blood pressure monitors

Johnson & Kerr (1985) have evaluated the accuracy of 5 non-invasive automatic blood pressure monitors. These included the Dinamap, the Narco and the Sentron which sense by oscillotonometry, the Copal which is semi-automatic and senses by auscultation and the Vitastat which can employ either method. Compared with intra-arterial recordings, the Copal and Sentron produced the best results. The Copal, being semi-automatic, is much cheaper than the fully automatic machines.

Pereira et al (1985) were critical of non-invasive methods of measurements, finding that oscillation tended to underestimate the intra-arterial systolic pressure while the diastolic pressure was inaccurate. They suggested that non-invasive methods of measurement of blood pressure were unsuitable for critically ill patients and for hypotensive anaesthesia. Arnold & McDevitt (1985), in studies of the indirect blood pressure measurement during intravenous isoprenaline infusions, found that the Dinamap recorder underestimated systolic pressure but overestimated diastolic pressure. The errors did not remain linear.

Van Egmond et al (1985) concluded that the Accutorr was a reasonably accurate oscillotonometer which is fully automated with an inflation phase of only 15–20 s and with a built in printer. A later model (2100 Datascope) incorporates the e.c.g. warning alarms which can be pre-set and can store data which can be displayed or printed at will. The Finapres is more accurate (derived from FINger Arterial PRESsure). Instead of an arm cuff there is a finger cuff which incorporates a photoplethysmograph measuring the absorption of the wavelength (infra-red) which is specific for arterial blood. Pressure readings were lower with the Finapres because of the distal position of the finger cuff. The intra-arterial method was considered the most accurate method of measurement.

REFERENCES

Arnold J M O, McDevitt D G 1985 Indirect blood pressure measurement during intravenous isoprenaline infusions. British Journal of Clinical Pharmacology 19: 114–116
Douglas J G, Leslie M J, Crompton G K, Grant I W B 1985 Is the flow rate used to drive a jet nebuliser clinically important? British Medical Journal 290: 29
Heller R M, Cotton R B 1985 Early experience with illuminated endotracheal tubes in premature and term infants. Pediatrics 75: 664–666
Hill S L, Barnes P K, Hollway T, Tennant R 1984 Fexed performance oxygen masks: an evaluation. British Medical Journal 288: 1261–1263
Johnson C J H, Kerr J H 1985 Automatic blood pressure monitors. A clinical evaluation of five models in adults. Anaesthesia 40: 471–478
Pereira E, Pry-Roberts C, Dagnino J, Anger C, Cooper G M, Hutton P 1985 Auscultatory measurement of arterial pressure during anaesthesia: a reassessment of the Korotkoff sounds. European Journal of Anaesthesiology 2: 11–20
Van Egmond J, Hasenbros M, Crul J F 1985 Invasive v. non-invasive measurement of arterial pressure. British Journal of Anaesthesia 57: 434–444

BLOOD TRANSFUSION

The increased use of blood components has led to more units of blood being supplied in concentrated form. Calkins et al (1982) have shown that the rapid administration of plasma reduced cells resulted in a 270% increase in plasma free haemoglobin. There was no correlation between flow rate and haemolysis. In addition to viscosity and flow rate, haemolysis may be influenced by pressure applied to the transfusion bag. They recommended that packed red cells should always be diluted with normal saline.

The National Blood Transfusion Service has recently introduced CPD red cell concentrates to which 100 ml of saline, adenine, glucose and mannitol (SAG-M) has been added. The saline dilutes the red cells, the adenine extends red cell storage, the glucose is for red cell metabolism and the mannitol reduces haemolysis. The shelf-life of the stored blood is up to 6 weeks. If more than 4 units of SAG-M blood are given, a bottle of plasma protein fraction (PPF) should be given simultaneously with the 5th unit which will also cover the 6th unit. Further PPF should be given to cover the 7th and 8th and a bottle for the 9th and 10th units. This will ensure there is not only adequate red cell replacement but also protein replenishment. In liver disease or in patients with a low plasma albumin, PPF should be administered with every 2 units of blood. (With the older type of CPD adenine suspended red cells, the routine use of PPF is unnecessary.)*

The use of blood substitutes has been outlined by Singer & Goldstone (1985). Kemner et al (1984) have drawn attention to the interaction between fluids used for

---

*   These recommendations are taken from the National Blood Transfusion Service leaflet dated December 1984.

resuscitation with morphine. The half-life of morphine was prolonged when stroma free haemoglobin was administered, the effect being pH dependent and specific to morphine but not to codeine. The volume of distribution increased. There is also decreased binding of morphine to stroma free haemoglobin solutions compared with plasma. Fluosol DA 20% solution also prolonged the half-life of morphine by altering total body clearance. The dose of morphine administered to patients resuscitated with these solutions will need to be adjusted and possibly regulated by measuring serum concentrations.

There are possible problems associated with administering sodium lactate solutions (Hartmann's solution) through the same blood transfusion set as that used for whole blood. The calcium in the Hartmann's solution may inactivate the anticoagulant in the blood, leading to clot formation especially if the blood is warmed (Blagdon & Gibson 1985).

Derrington (1985) has commented on the disadvantages of microfilters. Apart from expense and the reduction in transfusion rate, haemolysis may occur with depth filters when pressure is applied to the transfusion bag. Proteins and white cells, present in solution in plasma, can pass through the filter and cause microaggregates in the lung. Derrington (1985) concluded that their routine use was not justified. On the other hand, microparticulate induced phlebitis was reduced by approximately two-thirds using a filter (0.22 μm) (Falchuk et al 1985).

REFERENCES

Blagdon J, Gibson T 1985 Potential hazard of clotting during blood transfusion using a blood warming pack. British Medical Journal 290: 1475–1476
Calkins J M, Vaughan R W, Cork R C, Barberii J, Eskelson C 1982 Effects of dilution, pressure, and apparatus on hemolysis and flow rate in transfusion, of packed erythrocytes. Anesthesia and Analgesia 61: 776–780
Derrington M C 1985 The present status of blood filtration. Anaesthesia 40: 334–347
Falchuk K H, Peterson L, McNeil B J 1985 Microparticulate-induced phlebitis: its prevention by in-line filtration. New England Journal of Medicine 312: 78–80
Kemner J M, Snodgrass W R, Worley S E, Hodges G R, Clark G M, Hignite C E 1984 Interaction of oxygen-carrying resuscitation fluids with morphine. Journal of Laboratory Clinical Medicine 104: 433–444
Singer C R J, Goldstone A H 1985 Recent advances in blood transfusion and blood products (1 & 2). In: Kaufman L (ed) Anaesthesia Review 3. Churchill Livingstone, Edinburgh, p 156–182

CARDIAC ARREST

Keenan & Boyan (1985) reported on the incidence of cardiac arrest over a 15 year period. Of the 27 cardiac arrests (incidence of 1.7 per 10 000), 14 patients subsequently died. A detailed analysis of the deaths show that the incidence was three times higher for children compared with adults and six times higher for emergencies compared with elective surgery. Inadequate ventilation including oesophageal intubation and displaced endotracheal tube were implicated in half the cardiac arrests while one-third resulted from overdose of inhalational agents. Progressive bradycardia was observed in 26 cases prior to the cardiac arrest. Eighteen of the arrests occurred during induction.

REFERENCES

Keenan R L, Boyan P 1985 Cardiac arrest due to anaesthesia. A study of incidence and causes. Journal of the American Medical Association 253: 2373-2377

OBSTETRICS

The physiology of the myometrium and uterine cervix have been reviewed in detail by Huszar & Naftolin (1984). The biochemical events associated with initiation of labour are also discussed, including factors affecting the myometrium and the cervix. The inhibition of premature labour is affected by β-adrenergic drugs and it has been suggested that calcium channel blocking agents might be equally effective. There is also the possibility of developing a prostaglandin synthesis inhibitor which does not cross the placental barrier.

Oxytocin release was inhibited by opioids in animal experiments and this was related to the intensity of analgesia. The order of potency was buprenorphine followed by morphine, pethidine, meptazinol and finally pentazocine. The close relationship between potency of analgesia and suppression of oxytocin release might explain the inhibition of labour during opioid analgesia (Clarke & Wright 1984).

Ritodrine

Marks & De Chazal (1984) reported on the successful use of epidural analgesia for caesarean section in a patient who developed ritodrine induced pulmonary oedema. Early in the pregnancy two episodes of supra-ventricular tachycardia were treated with verapamil. The pulmonary oedema appears to have been precipitated by propranolol given to control the heart rate.

Cano et al (1985) have confirmed the hypokalaemic action of ritrodrine. This effect has been ascribed to β-adrenergic stimulation (see Anaesthesia Review 2), but Cano et al (1985) found that the use of intravenous ritrodrine resulted in an increase in plasma insulin levels. There was also a significant fall in the serum calcium.

Curet et al (1984) studied the association between ruptured membranes, tocolytic therapy and respiratory distress syndrome (RDS) and found that although premature rupture of the membranes and isoxuprine individually lowered the incidence of RDS, there was an increased incidence of the complication when these factors were present together. They advised that β-adrenergic stimulating agents should not be administered to patients unless the membranes were intact.

Antacid therapy

The last confidential enquiry into maternal deaths (1982) still considered the skill of the anaesthetist to be the most important factor in the prevention of inhalation of gastric contents. H_2-receptor antagonists and antacids must still be considered as adjuncts and should not engender a false sense of security. There have been many studies on the use of drugs but little on improving technique. If antacids leave the stomach rapidly the results are ineffectual whereas if they leave slowly the gastric volume increases (O'Sullivan & Bullingham 1983). The duration of action of sodium citrate is short and may not last the duration of the surgical procedure (O'Sullivan et al 1984). Sodium citrate is no longer on the limited prescribing list as recommended by the DHSS. The most reliable regime of raising intragastric pH is the oral administration of ranitidine (150 mg) on the evening before operation and then again 2-4 hours prior to induction of anaesthesia. The intramuscular injection of metoclopramide (10 mg) 1-2 hours before induction decreases gastric fluid volume (O'Sullivan et al 1985).

In animal studies, Mizus et al (1985) showed that pre-treatment with aminophylline or isoprenaline prevented the increase in pulmonary artery pressure produced following the instillation of hydrochloric acid into the lungs. Isoprenaline was effective even after the acid was instilled. Isoprenaline and aminophylline also reduced the pulmonary vascular permeability following acid injury to the lung.

Anaesthesia

Bernstein et al (1985) recommended the use of ketamine as an alternative to thiopentone as an induction agent for caesarean section. Fetal PO_2, acid-base values and Apgar scores were comparable. The advantage of ketamine is that it produces amnesia and analgesia and maintains the blood pressure if there is severe bleeding.

Milsom et al (1985) compared the effects of epidural analgesia and general anaesthesia on maternal haemodynamics during caesarean section in fit patients. Many of the changes during epidural analgesia are associated with sympathetic blockade resulting in vasodilatation and a decrease in peripheral resistance, whereas those occurring during general anaesthesia may be ascribed to sympathetic stimulation occurring during intubation or due to the response to surgical manipulations under light anaesthesia. There was no difference in Apgar scores at 1 and 5 minutes or in umbilical vein pH with either technique. The study suggested that epidural analgesia might be preferred in patients with pre-eclamptic toxaemia.

Epidural analgesia may be associated with neonatal jaundice. Clark & Landaw (1985) were able to show that bupivacaine compared with mepivacaine and lignocaine was the most soluble, had the highest binding to protein and was the most potent in altering red blood cell viscosity. Bupivacaine was bound twice as much to neonatal red cells than to adult red cells. The reduced survival time of fetal red cells may lead to neonatal jaundice.

Pre-eclampsia

Plasma colloid osmotic pressure was lower in pre-eclamptic patients and was reduced during labour, reaching the lowest level between 16 and 18 hours following delivery (Zinaman et al 1985). They felt it was prudent to curtail the use of crystalloid solutions and resort to a colloid infusion instead.

Collins et al (1985) found that diuretics appeared to exert a favourable influence on blood pressure in pre-eclampsia but despite this there was very little improvement in post-natal survival.

A more encouraging approach has been suggested by Beaufils et al (1985). In order to prevent intravascular coagulation which often precedes pre-eclampsia, they administered low doses of aspirin (150 mg/d) and dipyridamole (300 mg/d) at the end of the first trimister. Although the patients treated were particularly prone to the possibility of pre-eclampsia, treatment appeared to prevent its occurrence by inhibiting platelet aggregation. (This was not measured during the trial. There were no complications to the mother or the fetus.)

Epilepsy in pregnancy

There are possible hazards during pregnancy of administering drugs for the control of epilepsy. Drugs may be cleared more rapidly during pregnancy and they may also cross the placental barrier. There is also the possibility of congenital malformations especially with valproic acid. A regime for the management of the pregnant epileptic has been outlined by Dalessio (1985), advocating the use of phenytoin or carbamazepine. The plasma levels should be monitored at regular intervals.

Vitamin K deficiency may develop in patients on phenytoin, phenobarbitone or primidone and may result in increased bleeding during labour. Even if vitamin K levels are normal in the mother they may be reduced in the newborn, leading to haemorrhagic disease. Treatment is with vitamin K or even freshly frozen plasma (Dalessio 1985).

Asthma

The alterations in pulmonary physiology during pregnancy have been summarized by Greenberger & Patterson (1985). There are reductions in residual volume and functional residual capacity and the alveolar-arterial oxygen gradient increases. During an asthmatic attack the minute volume has to increase and greater negative inspiratory pressures are required to ventilate obstructed bronchi. Maternal hypoxaemia would have a major effect on fetal oxygenation. Despite the possibility of teratogenetic effects of drugs (based on animal experiments), Greenberger & Patterson (1985) recommended the use of β-agonists such as terbutaline, metaproterenol and albuterol. These drugs act primarily at β_2 receptors which also produce uterine relaxation. Steroids may be necessary and it is suggested these are preferably given by inhalation. Hydrocortisone may be necessary during labour.

Phaeochromocytoma

Phaeochromocytoma during pregnancy presents risks to the mother and fetus. The diagnosis may be confused with pre-eclampsia. Venuto et al (1984) reported a patient who developed a phaeochromocytoma at 32 weeks. The blood pressure was controlled with prazosin and the baby delivered by elective caesarean section. The phaeochromocytoma was removed at a further operation 18 days later. Schenker & Granat (1982) reviewed the literature and concluded that before 20–24 weeks the tumour should be excised whereas after that period the pregnancy should be allowed to continue with the blood pressure controlled by adrenergic blocking agents. They recommended delivery by caesarean section and preferably removal of the tumour at the same operation.

Pain relief

Use of spinal opioids was referred to in Anaesthesia Review 3 (Ch. 12, 13). Animal experiments reported by Craft et al (1984) showed that epidural fentanyl had no effects on maternal or fetal arterial blood pressure, acid base status or on maternal central venous pressure, systemic and pulmonary vascular resistance, cardiac output and intrauterine pressure. Following the injection of 50 μg of fentanyl the maternal level reached a maximum at 60 minutes and the fetal at 45 minutes. After 100 μg of fentanyl the maximum maternal level was at 45 minutes and the fetal at 15 minutes. Measurements of β-endorphin levels during labour have shown these are significantly decreased following 1 mg of intrathecal morphine which also provided excellent pain relief (Abboud et al 1984a). In a further study Abboud et al (1984b) demonstrated that intrathecal morphine, 0.5 mg or 1 mg in 7.5% dextrose, provided excellent pain relief during labour. The onset of analgesia was from 15–60 minutes after injection and it lasted for 6–8 hours. Analgesia was less satisfactory when there was distension of the perineum either by the application of forceps or during the crowning of the fetal head. There was a high incidence of complications including pruritus, nausea and vomiting, urinary retention and drowsiness but these side-effects could be alleviated by naloxone which did not appear to affect analgesia. There was no significant maternal respiratory depression. In a personal series using intrathecal diamorphine, the onset of analgesia was rapid, there was no maternal or fetal respiratory depression and, although perineal distension required additional analgesia such as nitrous oxide and oxygen, patients experienced no perineal pain for up to 24 hours following episiotomy.

Intrathecal administration of small doses of morphine revealed only small amounts in the maternal blood. Following epidural morphine, however, there was a rapid uptake via the epidural venous plexus and blood concentrations approached that following intramuscular administration, producing levels likely to depress fetal respiration (Camporesi & Redick 1983).

REFERENCES

Abboud T K, Goebelsmann U, Raya J et al 1984a Effect of intrathecal morphine during labor on maternal plasma beta-endorphin levels. American Journal of Obstetrics and Gynecology 149: 709–710

Abboud T K, Shnider S M, Dailey P A et al 1984b Intrathecal administration of hyperbaric morphine for the relief of pain in labour. British Journal of Anaesthesia 56: 1351–1360

Beaufils M, Donsimoni R, Uzan S, Colau J C 1985 Prevention of pre-eclampsia by early antiplatelet therapy. Lancet 1: 840–842

Bernstein K, Gisselsson L, Jacobsson L, Ohrlander S 1985 Influence of two different anaesthetic agents on the newborn and the correlation between foetal oxygenation and induction-delivery time in elective caesarean section. Acta Anaesthesiologica Scandinavica 29: 157–160

Camporesi E M, Redick L F 1983 Clinical aspects of spinal narcotics: postoperative management and obstetric pain. In: R E S Bullingham (ed) Clinics in Anesthesiology. W B Saunders, London, Vol 1, p 57–70

Cano A, Tovar I, Parrilla J J, Abad L 1985 Metabolic disturbances during intravenous use of ritodrine: increased insulin levels and hypokalemia. Obstetrics and Gynecology 65: 356–360

Clark D A, Landaw S A 1985 Bupivacaine alters red blood cell properties: a possible explanation for neonatal jaundice associated with maternal anesthesia. Pediatric Research 19: 341–343

Clarke G, Wright D M 1984 A comparison of analgesia and suppression of oxytocin release by opiates. British Journal of Pharmacology 83: 799–806

Collins R, Yusuf S, Peto R 1985 Overview of randomised trials of diuretics in pregnancy. British Medical Journal 290: 17–23

Craft J B, Robichaux A G, Kim H-S, Thorpe D H, Mazel P, Woolf W A, Stolte A 1984 The maternal and fetal cardiovascular effects of epidural fentanyl in the sheep model. American Journal of Obstetrics and Gynecology 148: 1098–1104

Curet L B, Vijaya Rao A, Zachman R D, Morrison J C, Burkett G, Poole W K, Bauer C, Collaborative Group on Antenatal Steroid Therapy 1984 Association between ruptured membranes, tocolytic therapy, and respiratory distress syndrome. American Journal of Obstetrics and Gynecology 148: 263–368

Dalessio D J 1985 Current concepts: seizure disorders and pregnancy. New England Journal of Medicine 312: 559–563

Greenberger P A, Patterson R 1985 Current concepts: management of asthma during pregnancy. New England Journal of Medicine 312: 897–902

Huszar G, Naftolin F 1984 The myometrium and uterine cervix in normal and preterm labor. New England Journal of Medicine 311: 571–581

Marks R J, De Chazal R C S 1984 Ritodrine-induced pulmonary oedema in labour. Anaesthesia 39: 1012–1014

Milsom I, Forssman L, Biber B, Dottori O, Rydgren B, Sivertsson R 1985 Maternal haemodynamic changes during caesarean section: a comparison of epidural and general anaesthesia. Acta Anaesthesiologica Scandinavica 29: 161–167

Mizus I, Summer W, Farrukh I, Michael J R, Gurtner G H 1985 Isoproterenol or aminophylline attenuate pulmonary edema after acid lung injury. American Review of Respiratory Disease 131: 256–259

O'Sullivan G, Harrison B J, Bullingham R E S 1984 The use of radiotelemetry techniques for the in vivo assessment of antacids. Anaesthesia 39: 987–995

O'Sullivan G, Bullingham R E S 1983 Antacid prophylaxis in the pregnant patient. Anaesthesia 38: 998–999

O'Sullivan G, Sear J W, Bullingham R E S, Carrie L E S 1985 The effect of magnesium trisilicate mixture, metoclopramide and ranitidine on gastric pH, volume and serum gastrin. Anaesthesia 40: 246–253

Schenker J G, Granat M 1982 Phaeochromocytoma and pregnancy — an updated appraisal. Australian and New Zealand Journal of Obstetrics and Gynaecology 22: 1–10

Venuto R, Burstein P, Schnider R 1984 Pheochromocytoma: antepartum diagnosis and management with tumor resection in the puerperium. American Journal of Obstetrics and Gynecology 150: 431–432

Zinaman M, Rubin J, Lindheimer M D 1985 Serial plasma oncotic pressure levels and echoencephalography during and after delivery in severe pre-eclampsia. Lancet 1: 1245–1247

PHAEOCHROMOCYTOMA

In a recent review Bravo & Gifford (1984) considered the diagnosis, localization and management of phaeochromocytoma. They referred to the classic triad of sweating, tachycardia and headache which seemed to confirm the diagnosis in the majority of patients. However the presentation may vary (see Anaesthesia Review 1, 2, 3). Ganguly et al (1984) described a patient with adrenal phaeochromocytoma who initially was thought to have sick sinus syndrome requiring a pacemaker and who exhibited cyclic fluctuations of blood pressure. The hypertension was accompanied by bradycardia and the hypotension by tachycardia. The fluctuations in blood pressure were not controlled by phentolamine. As the patient had been previously treated with a diuretic, saline was infused which controlled the fluctuations of blood pressure.

A high proportion of Bravo & Gifford's patients (based on a study of 43 cases) had false negative results for urinary vanilmandelic acid and for metanephrines. Blood levels of catecholamines were raised but they are also raised in patients with hypertension: 4 of their patients with phaeochromocytonma had levels which fell within the range of values for patients with essential hypertension but none within the range for normotensives. Plasma catecholamines may be raised in patients with chronic renal failure (Darwish et al 1984). If the tests are not completely confirmatory, provocative tests include intravenous glucagon or suppression with clonidine. Bravo & Gifford (1984) considered that the concentration of total plasma catecholamines over 2000 pg/ml (11.82 mmol/l) to be diagnostic. The operative management consisted of the prior administration of phenoxybenzamine or prazosin. During operation phentolamine and sodium nitroprusside were used to control hypertension and propranolol to abolish arrhythmias.

Lenders et al (1985) reported on the use of nifedipine, a calcium channel blocker, in the treatment of phaeochromocytoma of the bladder. Nifedipine prevented the rise in blood pressure as well as the symptoms of headache and nausea during micturition despite a sharp increase in plasma catecholamine levels.

Catecholamines

Plasma potassium rises during muscular exercise and falls rapidly when exertion is stopped. Williams et al (1985) have confirmed that during exercise there are high plasma levels of noradrenaline and that the β-adrenergic receptors prevent the acute rises of plasma potassium during exercise, while α-adrenergic receptor stimulation augments hypokalaemia and prevents the fall in serum potassium when exercise stops.

The use of adrenaline in lignocaine anaesthetic solutions for dental analgesia causes a significant fall in serum potassium levels in young adults (Meechan & Rawlins 1985).

β-Adrenergic receptor stimulation increases cellular levels of potassium whereas α-adrenergic stimulation has the opposite effect (Williams et al 1984).

Plasma noradrenaline has been studied in patients with chronic congestive cardiac failure and it has been found that a single resting venous blood sample was the most satisfactory guide to prognosis. Heart rate, plasma renin, serum sodium and stroke volume index were of less value than the level of noradrenaline in determining prognosis. The level is also higher in patients who died from progressive heart failure than in those who died suddenly (Cohn et al 1984).

Studies on leucocytes from normal volunteers have shown that noradrenaline depresses membrane sodium pump activity. This effect is suppressed by propranolol implying that the effect is mediated by β receptors. There may be implications in this study in patients with hypertension (Riozzi et al 1984).

β-Adrenergic blocking drugs

Biochemical and physiological aspects of adrenergic receptors are considered by Lefkowitz et al (1984), while the biochemical mechanisms of β-adrenergic receptor regulation are reviewed by Stiles et al (1984). Age appears to affect adrenoceptor function and there is evidence that there is diminished responsiveness to β receptor stimulation although this effect is less marked at the α-adrenergic receptor (Kelly & O'Malley 1984). Metoprolol alters the cardiorespiratory response to exercise with greater utilization of oxygen and creatine phosphate stores (Hughson 1984).

Paradoxical hypertension may develop following repair of coarctation of the aorta and this can be prevented by the administration of propranolol for 2 weeks prior to surgery and for 1 week in the post-operative period. Following resection of the aneurysm there are increased concentrations of plasma noradrenaline and plasma renin. Compared with patients treated for coarctation without β-blockers, the patients given propranolol had not only reduced blood pressure but also decreased plasma renin. There were no differences in the level of plasma noradrenaline (Gidding et al 1985).

Drug interactions with β-blockers have been summarized by Beeley (1984). Enzyme inducing drugs increase the clearance of β-blockers such as propranolol and metoprolol which are lipid soluble and readily metabolized. β-Blockers may affect other drugs by reducing hepatic blood flow. Although propranolol lowers the clearance of lignocaine by decreasing hepatic blood flow by 25%, the main cause of reduced clearance is due to the inhibition of hepatic enzymes which metabolize lignocaine (Bax et al 1985). There are also interactions with antiarrhythmic drugs, while recovery from hypoglycaemia may be prolonged in diabetics on treatment.

REFERENCES

Bax N D S, Tucker G T, Lennard M S, Woods H F 1985 The impairment of lignocaine clearance by propranolol — major contribution from enzyme inhibition. British Journal of Clinical Pharmacology 19: 597–603
Beeley L 1984 Drug interactions and β-blockers. British Medical Journal 289: 1330–1331
Bravo E L, Gifford R W 1984 Current concepts: pheochromocytoma: diagnosis, localization and management. New England Journal of Medicine 311: 1298–1303
Cohn L N, Levine T B, Olivari M T et al 1984 Plasma norephinephrine as a guide to prognosis in patients with chronic congestive heart failure. New England Journal of Medicine 311: 819–823
Darwish R, Elias A N, Vaziri N D, Pahl M, Powers D, Stokes J D 1984 Plasma and urinary catecholamines and their metabolites in chronic renal failure. Archives of Internal Medicine 144: 69
Ganguly A, Grim C E, Weinberger M H, Henry D P 1984 Rapid cyclic fluctuations of blood pressure associated with an adrenal pheochromocytoma. Hypertension 6: 281–284
Gidding S S, Rocchini A P, Beekman R et al 1985 Therapeutic effect of propranolol on paradoxical hypertension after repair of coarctation of the aorta. New England Journal of Medicine 312: 1224–1228
Hughson R L 1984 Alterations in the oxygen deficit-oxygen debt relationships with β-adrenergic receptor blockade in man. Journal of Physiology 349: 375–387
Kelly J, O'Malley K 1984 Adrenoceptor function and ageing. Clinical Science 66: 509–515
Lenders J W M, Sluiter H E, Thien T 1985 Treatment of a phaeochromocytoma of the urinary bladder with nifedipine. British Medical Journal 290: 1624–1625
Lefkowitz R J, Caron M G, Stiles G L 1984 Mechanisms of membrane-receptor regulation: biochemical, physiological and clinical insights derived from studies of the adrenergic receptors. New England Journal of Medicine 310: 1570–1579
Meechan J G, Rawlins M D 1985 The effect of adrenaline in lignocaine anaesthetic solutions on serum potassium levels. Proceedings of the British Pharmacological Society, p C39
Riozzi A, Heagerty A M, Bing R F, Thurston H, Swales J D 1984 Noradrenaline: a circulating inhibitor of sodium transport. British Medical Journal 289: 1025–1027
Stiles G L, Caron M G, Lefkowitz R J 1984 β-adrenergic receptors: biochemical mechanisms of physiological regulation. Physiological Reviews 64: 661–743
Williams M E, Gervino E V, Rosa R M et al 1985 Catecholamine modulation of rapid potassium shifts during exercise. New England Journal of Medicine 312: 823–827
Williams M E, Rosa R M, Silva P, Brown R S, Epstein F H 1984 Impairment of extrarenal potassium disposal by α-adrenergic stimulation. New England Journal of Medicine 311: 145–149

Update (2)

ALLERGY

Induction agents may be associated with allergic type responses. Recently thiopentone has been implicated in immune haemolytic anaemia and renal failure. A patient acquired an antibody against thiopentone and the I antigen on the red cell membrane appeared to be the receptor site for the thiopentone antibody complex (Habibi et al 1985). The patient had undergone two previous operations with thiopentone as the induction agent without mishap but on this occasion haemoglobinuria resulted post-operatively. Later acute intravascular haemolysis and renal failure occurred. Detailed blood studies confirmed the presence of a specific antibody against thiopentone.

Harle et al (1984), using choline-sepharose assay, were able to detect IgE antibodies to suxamethonium in patients who recently had severe anaphylactoid reactions after the drug was administered. There is likely to be cross sensitivity to other drugs such as d-tubocurarine and alcuronium (Fisher & Baldo 1983).

REFERENCES

Fisher M, Baldo B 1983 Adverse reactions to alcuronium. An Australian disease? Medical Journal of Australia 1: 630–632
Habibi B, Basty R, Chodez S, Prunat A 1985 Thiopental-related immune hemolytic anemia and renal failure: specific involvement of red-cell antigen I. New England Journal of Medicine 312: 353–355
Harle D G, Baldo B A, Fisher M M 1984 Detection of IgE antibodies to suxamethonium after anaphylactoid reactions during anaesthesia. Lancet 1: 930–932

ANAESTHETIC AGENTS

Midazolam

Midazolam (Hypnovel), a recently introduced benzodiazepine, seemed a reasonable alternative to diazepam but Byatt et al (1984) reported accumulation of the drug after repeated doses leading to prolonged respiratory depression and prolonged unconsciousness. Although midazolam was used with other drugs such as papaverine or chloral hydrate, the prolonged period of recovery may be misinterpreted and a neurological cause sought for the failure to regain consciousness.

Etomidate

Duthie et al (1985) found that a single dose of etomidate produced no clinically significant adrenocortical suppression in patients undergoing minor surgical procedures. However there was some inhibition of 11-β-hydroxylase activity. During the first hour of anaesthesia (nitrous oxide, oxygen and halothane), the plasma cortisol fell transiently but this also occurred in the control group of patients who were given thiopentone. In a further study Moore et al (1985) investigated the peri-operative endocrine effects of etomidate infusion during hysterectomy: they too concluded that etomidate suppressed only adrenocortical function without producing any obvious clinical adverse effects. They even suggested that etomidate might be a novel means of producing stress free anaesthesia (misnomer: the stress of surgery is still present, the response to surgical stimuli is suppressed). Bailey (1986) has confirmed the drop in plasma cortisol levels, reaching a maximum of 20% 30 minutes after induction with etomidate.

Atropine

The effect of low doses of atropine on gastrointestinal secretions has shown that acid and pepsin secretion are readily inhibited. Patients with duodenal ulcer require 2–4 times as much to produce the same effect. The maximum increase in heart rate following atropine was much less in patients with duodenal ulcer (Hirschowitz et al 1984). Hyoscine has a greater affinity apparently for the muscarinic receptor compared with atropine (Cohen & Haberman 1984). Atropine also inhibited secretion of growth hormone (Taylor et al 1985).

REFERENCES

Bailey P 1986 (in press)
Byatt C M, Lewis L D, Dawling S, Cochrane G M 1984 Accumulation of midazolam after repeated dosage in patients receiving mechanical ventilation in an intensive care unit. British Medical Journal 289: 799–800
Cohen S, Haberman F 1984 The molal volumes of atropine and hyoscine in relation to their respective potencies. British Journal of Pharmacology 83: 807–811
Duthie D J R, Fraser R, Nimmo W S 1985 Effect of induction of anaesthesia with etomidate on corticosteroid synthesis in man. British Journal of Anaesthesia 57: 156
Hirschowitz B I, Molina E, Ou Tim L, Helman C 1984 Effects of very low doses of atropine on basal acid and pepsin secretion, gastrin and heart rate in normals and DU. Digestive Diseases and Science 29: 790–796
Moore R A, Allen M C, Wood P J, Rees L H, Sear J W 1985 Peri-operative endocrine effects of etomidate. Anaesthesia 40: 124–130
Taylor B J, Smith P J, Brook C G D 1985 Inhibition of physiological growth hormone secretion by atropine. Clinical Endocrinology 22: 497–501

ANALGESIC AGENTS

Fentanyl

Fentanyl is frequently used to provide analgesia during the course of surgery. It had been claimed that it had little effect on cardiovascular stability, but recent reports (Lowenstein & Philbin 1983) have shown that this only appears to be so in the presence of pancuronium which predisposes to tachycardia. Caution must be exerted when using other muscle relaxants such as d-tubocurarine and atracurium. During cardiopulmonary bypass, serum concentrations of fentanyl fall but rise when lung ventilation and perfusion are restarted. This would indicate that fentanyl is sequestered in the lungs during bypass surgery (Bentley et al 1983).

Bidwai et al (1976) reported that large doses of fentanyl (1 ml/kg) decreased urinary output, presumably as a result of ADH release. This effect could be antagonized by nitrous oxide. However Ecoffey et al (1984) noted that fentanyl caused a marked fall in blood pressure, a factor known to release ADH, but despite this fentanyl suppressed the ADH response to osmotic and hypotensive stimuli.

Alfentanil

The advantages of alfentanil have been discussed by Stanski & Hug (1982). Alfentanil has a smaller volume of distribution, clearance and terminal half-life (Bower & Hull 1982). It is less lipid soluble than fentanyl.

In elderly patients over the age of 65 years plasma clearance was reduced and the plasma terminal half-life prolonged compared with a group of young adults, following the administration of 50 μg/kg (Helmers et al 1984). These changes are probably due to a reduction in hepatic metabolism. Two cases of unexpected respiratory depression have been reported by Sebel et al (1984) following alfentanil infusion and which responded to intravenous naloxone.

In animal studies high dose alfentanil resulted in an increase in left ventricular end diastolic pressure (D'Aubioul et al 1984). They also noted that there was a marked increase in heart rate and A-V dissociation and despite this they concluded that alfentanil was safe in animals provided they were artificially ventilated.

Alfentanil, in an initial dose of 150 μg/kg followed by a continuous infusion of 3 μg/kg per minute, suppressed intra-operative plasma cortisol and glucose concentrations in only the first 3 hours of the post-operative period (Moller et al 1985). Although there was cardiovascular stability during operation, the high incidence of nausea and vomiting and the frequency of respiratory depression made the technique unacceptable. In addition Moller et al (1985) concluded that the inhibition of the endocrine response to surgery was too limited in duration to contribute to a reduction in post-operative morbidity.

Morphine

Care is necessary in the use of intravenous morphine for sedation in patients with decreased renal function. Ball et al (1985) reported on reduced morphine clearance in patients with renal impairment and warned of the dangers of mistaking cerebral damage for overdose of morphine especially in patients in intensive care situations. However, despite this potential hazard, sleep time in patients in intensive care units following major abdominal surgery was less than 2 hours per night following high dose morphine anaesthesia. Sleep time was measured by continuously recording the electroencephalogram, electrooculogram and electromyogram (Aurell & Elmqvist 1985).

In addition Rodriguez et al (1983) favoured the use of intravenous morphine (initial dose of 1 or 4 mg/kg followed by an infusion of 0.2 or 0.5 mg/kg per hour) to reduce post-operative metabolic rate. This was in a study of patients who became hypothermic during surgical procedures and the use of morphine, although significantly increasing the rewarming period, produced a significant reduction in shivering, heat loss and myocardial work while at the same time maintained cardiovascular function.

Laska et al (1984) noted that caffeine potentiated the action of mild analgesics and in animal studies this had been confirmed for morphine (Misra et al 1985). The mode of action of caffeine is unknown and may involve pharmacokinetic or dispositional factors. Naloxone still reverses the analgesia. The clinical implications of this are still to be evaluated.

REFERENCES

D'Aubioul J, Van Gerven W, Van de Water A, Xhonneux R, Reneman R S 1984 Cardiovascular and some respiratory effects of high doses of alfentanil in dogs. European Journal of Pharmacology 100: 79–84

Aurell J, Elmqvist D 1985 Sleep in the surgical intensive care unit: continuous polygraphic recording of sleep in nine patients receiving postoperative care. British Medical Journal 290: 1029–1032

Ball M, Moore R A, Fisher F, McQuay H J, Allen M C, Sear J 1985 Renal failure and the use of morphine in intensive care. Lancet 1: 784–786

Bentley J B, Conahan T J, Cork R C, 1983 Fentanyl sequestration in lungs during cardiopulmonary bypass. Clinical Pharmacology and Therapeutics 34: 703–706

Bidwai A V, Liu W S, Stanley T H, Bidwal V, Loeser E A, Shaw C L 1976 Canadian Anaesthetists Society Journal 23: 296–302

Bower S, Hull C J 1982 Comparative pharacokinetics of fentanyl and alfentanil. British Journal of Anaesthesia 54: 871–877

Ecoffey C, Simon D, Samil K, Diraison P M, Poggi J, Noviant Y, Ardaillou R 1984 Antidiuretic hormone response to osmotic stimulus under fentanyl anaesthesia. Acta Anaesthesiologica Scandinavica 28: 245–248

Helmers H, Van Per A, Woestenborghs R, Noorduin H, Heykants J 1984 Alfentanil kinetics in the elderly. Clinical Pharmacology and Therapeutics 36: 239–243

Laska E M, Sunshine A, Mueller F, Elvers W B, Siegel C, Rubin A 1984 Enhancement of analgesic efficacy of acetaminophen, aspirin and their combination with caffeine: caffeine as an analgesic adjuvant. Journal of the American Medical Association 251: 1711–1718

Lowenstein E, Philbin M 1983 Narcotic anaesthesia. In: Bullingham R E S (ed) Clinics in Anaesthesiology. WB Saunders, London. p 5–15

Misra A L, Pontani R B, Vadlamani N L 1985 Potentiation of morphine analgesia by caffeine. British Journal of Pharmacology 84: 789–791

Moller I W, Krantz T et al 1985 Effect of alfentanil anaesthesia on the adrenocortical and hyperglycaemic response to abdominal surgery. British Journal of Anaesthesia 54: 591–594

Rodriguez J L, Weissman C et al 1983 Morphine and postoperative rewarming in critically ill patients. Circulation 68: 1238–1246

Sebel P S, Lalor J M, Flynn P J, Simpson B A 1984 Respiratory depression after alfentanil infusion. British Medical Journal 289: 1581–1582

Stanski D R, Hug C C 1982 Alfentanil — a kinetically predictable narcotic analgesic. Anaesthesiology 57: 435–437

CIMETIDINE

Cimetidine may influence the action of other drugs such as diazepam, propranolol and local anaesthetic agents by impairing microsomal oxidation in the liver. However in a controlled study by Greenblatt et al (1984), in patients who were given cimetidine and diazepam, plasma concentrations of diazepam and desmethyldiazepam were elevated compared to controls but the increase was of minimal clinical importance. Greenblatt et al (1984) added a note of caution on their findings as the patients that they studied were under 50 years of age. The use of cimetidine and diazepam in the elderly and in patients with medical disorders such as liver disease or chronic respiratory disease may result in increased drowsiness and sedation.

The kinetics of cimetidine has been studied by Ziemniak et al (1984) in patients suffering from shock from burns. Cimetidine is often administered to burns patients to prevent stress gastric ulcers. During the early period of resuscitation with intravenous fluids there was decreased renal clearance of cimetidine but an increase in non-renal clearance. The latter may be due to increased oxidative metabolism in the liver associated with the hypermetabolic state following major burns. The increased clearance of cimetidine was related to the extent of the burn (Martyn et al 1985).

REFERENCES

Greenblatt D J, Abernethy D R, Morse D S, Harmatz J S, Shader R I 1984 Clinical importance of the interaction of diazepam and cimetidine. New England Journal of Medicine 310: 1639–1643
Martyn J A J, Greenblatt D J, Abernethy D R 1985 Increased cimetidine clearance in burn patients. Journal of the American Medical Association 253: 1288–1291
Ziemniak J A, Watson W A, Saffle J R, Smith I L, Russo J, Warden G D, Schentag J J 1984 Cimetidine kinetics during resuscitation from burn shock. Clinical Pharmacology and Therapeutics 36: 228–233

DOMPERIDONE

Four cases of cardiac arrest have been reported following intravenous administration of domperidone. In three of the patients hypokalaemia was present. It was felt that domperidone may precipitate fatal cardiac arrhythmias when given in dose to suppress the emetic effects of cytotoxic drugs (Roussak et al 1984).

REFERENCES

Roussak J B, Carey P, Parry H 1984 Cardiac arrest after treatment with intravenous domperidone. British Medical Journal 289: 1579

ENDOCRINE RESPONSE TO STRESS

Naloxone and experimental shock

Evans et al (1984a) devised a model in animals to assess endotoxic shock. They found that naloxone increased the survival rate of experimental shock in the animal. It caused only limited and transient increases in mean arterial pressure with little effect on the cardiac index. Peripheral resistance rose but substantial increases in the mean arterial pressure were produced if there was volume replacement. The findings suggested that naloxone had a myocardial action (Evans et al 1984a). However DeMaria et al (1985) found that 0.4–1.2 mg of naloxone given intravenously was ineffectual in maintaining the blood pressure in septic shock in man.

The use of opioid antagonists, not only in septic and cardiogenic shock but also in chronic obstructive pulmonary disease and mental illness, has been reviewed by McNicholas & Martin (1984). Although endogenous opioids do not appear to be involved in the regulation of respiration in man they did have effects on respiratory function tests in chronic asthmatic patients. Al-Damluji et al (1983) administered 8 mg of naloxone at midnight (i.e. 24:00) and this was followed by a continuous infusion of 5.6 mg per hour until 10:00 on two successive nights. The peak expiratory flow rate (PEFR), FEV_1 and FVC which were measured at 06:00 were unchanged but between 08:00 and 20:00 on the day following the first infusion there was a significant improvement in PEFR and FEV_1; the explanation for this was unclear. There is a possibility that this could be due to the increase in ACTH and cortisol that occurs following the use of naloxone.

Naloxone failed to antagonize the respiratory depressant effects of midazolam (Forster et al 1983) but it did appear to interact with halothane when assessing the effect of morphine on the heart rate response to noxious stimuli (Kissin et al 1984).

Steroids

Steroids, including dexamethesone or methylprednisolone, improved shock reversal if administered within 4 hours of shock and thereafter it did not improve the survival rate (Sprung et al 1984).

Endocrine response

The endocrine and metabolic responses to surgery have been reviewed by Kehlet (1984) and Hall (1985). Christenson et al (1982) reported that even elective surgery was followed by a pronounced feeling of fatigue lasting for at least a month in about a third of the patients. Christenson & Kehlet (1984) found that the fatigued patients had a pronounced loss in weight in the triceps and upper arm muscle circumference and had reduced levels of serum transferrin. The pre-operative measurement of various parameters including electrolytes, lymphocytes, minerals and albumin were unhelpful in determining which patients would become fatigued. They have suggested that it might be possible to increase pre-operative nutrition and advocated regional analgesia to inhibit the afferent impulses suppressing the endocrine metabolic response to surgery.

Plasma β-endorphin levels have been measured in patients undergoing elective major surgery under anaesthesia with nitrous oxide, oxygen and halothane (the muscle relaxant was not identified). McIntosh et al (1985) found that there was a true circadian rhythm pre-operatively of β-endorphin levels and this was altered or even abolished for up to 72 hours. The plasma level or β-endorphin was significantly elevated following surgery.

Moore (1984) while not commenting on muscle fatigue noticed that there was less depression in lymphocytes following epidural analgesia compared with general anaesthesia. Hall (1985), however, stressed the limitations of current techniques of suppressing the endocrine response to surgical stimulation by inhibition of afferent stimulation, the use of high dose opioid analgesia or by the infusion of hormones such as insulin or adrenoceptor blocking agents. Hall (1985) reported excellent analgesia and suppression of plasma cortisol following epidural diamorphine but no effect on circulating glucose, lactate, pyruvate or non-esterified fatty acid (NEFA). However, Child & Kaufman (1985) found that the intrathecal administration of diamorphine was far superior than intravenous fentanyl in that the hyperglycaemic response to stimulation was delayed and the adrenocortical response was significantly depressed both during and following major abdominal surgery. In the intrathecal group of patients there were no significant cardiovascular responses to the incision. Analgesia following operation was required at 0.97 ± 0.2 hours in the fentanyl group of patients but only after 17.27 ± 3.4 hours in the patients who had intrathecal diamorphine. Although intrathecal opiates do not appear to affect autonomic afferent stimulation, the cardiovascular and hormonal responses appear to be suppressed.

Hjortso et al (1985) compared patients undergoing major abdominal surgery receiving fentanyl during operation and morphine post-operatively with a group of patients receiving extradural analgesia initially with etidocaine and later with morphine. Although pain relief was significantly better in the extradural group, the regime failed to prevent the increase in urine excretion of cortisol, adrenaline, noradrenaline and nitrogen. They concluded that the relief of pain had little influence on the endocrine response to stimulation during abdominal surgery. Hakanson et al (1985) concluded that thoracic extradural morphine was less effective than thoracic extradural bupivacaine in suppressing metabolic responses to upper abdominal surgery. In an earlier study Moller et al (1984) concluded that spinal anaesthesia with tetracaine had only a transient inhibitory effect on the endocrine response to surgery and it was related to the level of sensory loss produced by the local anaesthetic solution. Epidural analgesia with etidocaine also failed to modify post-operative depression of delayed hypersensitivity (Hjortso et al 1984).

Korinek et al (1985) compared extradural morphine with extradural bupivacaine and found in the latter group that plasma ADH was unchanged in the post-operative period whereas in those patients receiving extradural morphine there was the delayed but step-wise increase in plasma ADH, indicating that the morphine has reached the brainstem. A recent study by Kaufman & Bailey (1986), on the other hand, showed that the rise in ADH following surgical stimulation was less in patients given intrathecal diamorphine compared with controls. Bumetanide (0.0075 mg/kg) appeared to prevent the release of ADH. They also confirmed that the urinary output did not appear to be related to ADH levels.

A recent study of the endocrine response to surgery in patients undergoing hysterectomy has shown that isoflurane failed to suppress the rise in blood glucose levels during surgery and in the post-operative period. Adrenaline and noradrenaline levels increased significantly. There was an initial fall in cortisol levels which may have been due to the etomidate, but 1 hour following the end of surgery the levels had tended to rise again (Rolly et al 1984).

Nitrogen balance

Burns et al (1981) claimed that naftidrofuryl reduced the urinary excretion of nitrogen following surgical procedures in a study limited to the first 2 days following operation. Jackson et al (1984), who studied the negative nitrogen loss for a period of 5 days, were unable to substantiate these results. The blood lactate concentration was lower than in the control groups. This may have been due to the selection of the groups or associated with the intravenous therapy.

The post-operative negative nitrogen balance following elective surgery may be reduced by the pre-operative infusion of 5 mg/kg per minute of glucose. This regime reduced the excretion of urea and 3-methylhistidine, suggesting not only a reduction in protein breakdown but a possible improvement in protein synthesis for the first 24 hours following operation (Crowe et al 1984).

Bessey et al (1984) simulated the metabolic response to injury by infusing volunteers with cortisol, glucagon and epinephrine and were able to produce negative nitrogen balance, potassium loss, hyperglycaemia, hyperinsulinaemia, insulin resistance, sodium retention and leukocytosis. None of the hormones given individually produced the same effect as the administration of all three hormones together. Glucose and insulin concentrations rose, suggesting some degree of insulin resistance.

There is still debate as to the choice of fluid for resuscitation following severe shock or even to reduce the negative nitrogen loss following surgery. Smith & Norman (1982) found that in the animal model a mixture of crystalloid and colloid produced better results than either given alone. Plasma renin and plasma aldersterone reached high levels in burn patients treated with low volume colloids during resuscitation. The highest levels were reached 5 days following the burn. Clinical deterioration occurred in two patients who had secondary peaks of these hormone levels 7–14 days after the accident (Griffiths et al 1983).

Catecholamines

Catecholamine levels have been measured in man in response to maximal stress. In intensive care situations there are moderate elevations of plasma noradrenaline up to 1.37 ng/ml, whereas acute stress such as occurs during cardiac arrest may result in levels reaching 35.9 ng/ml. Patients in intensive care units developed plasma levels of noradrenaline which were twice as great as normal resting controls but the level rose 32 times following cardiac arrest. With close monitoring it is possible to survive plasma adrenaline concentrations as high as 273 ng/ml (Wortsman et al 1984). From animal studies Tilders et al (1985) found that argine vasopressin (AVP) was involved in activating the pituitary–adrenal system, possibly by potentiating the effects of corticotrophin releasing hormone (CRF).

Opioids

Pfeiffer & Herz (1984) have reviewed the endocrine actions of opioids, drawing attention to their widespread effects and variation according to species. In rats opioids stimulate ACTH and corticosterone secretion whereas in man there is inhibition. In both species naloxone stimulates the release of ACTH. Oxytocin and vasopressin release are inhibited by opioids at posterior pituitary level in animals, the effects being reversed by naloxone. There is an action at pancreatic level increasing levels of insulin and glucagon, an effect which may not be reversed by naloxone in man. Bouloux et al (1985) have shown that the enkephalin (DAMME) depressed noradrenaline and adrenaline levels, reducing their release in response to hypoglycaemia. Naloxone had no effect on basal plasma catecholamine levels but caused a significant increase in adrenaline response to hypoglycaemia. DAMME and naloxone had no effect on the blood glucose response to hypoglycaemia induced by insulin (see Ch. 19).

Thus it can be seen that the hormonal and metabolic responses to stress and surgical stimulation are complex. The results may be influenced by the extent and duration of surgery, laboratory measurements and the timing of sampling as well as the technique of anaesthesia. Etomidate caused a 20% reduction in cortisol levels, a factor which may effect the subsequent rise. The various hormones may interact and it is speculative to extrapolate results from animal experiments. The philosophical debate as to whether suppression of the endocrine response to trauma and surgical stimulation is beneficial remains unresolved.

REFERENCES

Al-Damluji S, Thompson P J, Citrone K M, Turner-Warwick M 1983 Effect of naloxone on circadian rhythms in lung function. Thorax 38: 914–918
Bessey P Q, Watters J M, Aoki T T, Wilmore D W 1984 Combined hormonal infusion simulates the metabolic response to injury. Annals of Surgery 200: 264–281
Blandford R L, Burden A C 1984 Abnormalities of cardiac conduction in diabetics. British Medical Journal 289: 1659
Bouloux P-M G, Grossman A, Lytras N, Besser G M 1985 Evidence for the participation of endogenous opioids in the sympathoadrenal response to hypoglycaemia in man. Clinical Endocrinology 22: 49–56
Burns H J G, Galloway D J, Ledingham I McA 1981 Effect of naftidrofuryl on the metabolic response to surgery. British Medical Journal 283: 7–8
Child C S, Kaufman L 1985 Effects of intrathecal diamorphine on the adrenocortical, hyperglycaemic and cardiovascular responses to major colonic surgery. British Journal of Anaesthesia 57: 389–393
Christensen T, Kehlet H 1984 Postoperative fatigue and changes in nutritional status. British Journal of Surgery 71: 473–476
Christensen T, Bendix T, Kehlet H 1982 Fatigue and cardiorespiratory function following abdominal surgery. British Journal of Surgery 69: 417–419
Crowe P J, Dennison A, Royle G T 1984 The effect of pre-operative glucose loading on postoperative nitrogen metabolism. British Journal of Surgery 71: 635–637
DeMaria A, Effernan J J, Grindlinger G A, Craven D E, McIntosh T K, McCabe W R 1985 Naloxone versus placebo in treatment of septic shock. Lancet 1: 1363–1365
Evans S F, Hinds C J, Varley J G 1984a A new canine model of endotoxin shock. British Journal of Pharmacology 83: 433–442
Evans S F, Hinds C J, Varley J G 1984b Effects of intravascular volume expansion on the cardiovascular response to naloxone in a canine model of severe endotoxin shock. British Journal of Pharmacology 83: 443–448
Forster A, Morel D, Bachmann M, Gemperle M 1983 Respiratory depressant effects of different doses of midazolam and lack of reversal with naloxone — a double-blind randomized study. Anesthesia and Analgesia 62: 920–924
Griffiths R W, Millar J G B, Albano J, Shakespeare P G, 1983 Observations on the activity of the renin-angiotensin-aldosterone (RAA) system after low volume colloid resuscitation for burn injury. Annals of the Royal College of Surgeons of England 65: 212–215
Hakanson E, Rutberg H, Jorfeldt K L, Martensson J 1985 Effects of the extradural administration of morphine or bupivacaine, on the metabolic response to upper abdominal surgery.British Journal of Anaesthesia 57: 394–399
Hall G M 1985 The anaesthetic modification of the endocrine and metabolic response to surgery. Annals of the Royal College of Surgeons of England 67: 25–29
Hjortso N-C, Andersen T, Frosig F, Neumann P, Rogon E, Kehlet H 1984 Failure of epidural analgesia to modify postoperative depression of delayed hypersensitivity. Acta Anaesthesiologica Scandinavica 28: 128–131
Hjortson N-C, Christensen N J, Kehlet H 1985 Effects of the extradural administration of local

anaesthetic agents and morphine on the urinary excretion of cortisol, catecholamines and nitrogen following abdominal surgery. British Journal of Anaesthesia 57: 400–406

Jackson J M, Khawaja H T, Weaver P C, Talbot S T, Lee H A 1984 Naftidrofuryl and the nitrogen, carbohydrate, and lipid responses to moderate surgery. British Medical Journal 289: 581–586

Kaufman L, Bailey P M 1986 (in press)

Kehlet H 1984 Epidural analgesia and the endocrine-metabolic response to surgery. Update and perspectives. Acta Anaesthesiologica Scandinavica 28: 125–127

Kissin I, Kerr C R, Smith L R 1984 Effect of morphine on the heart rate response to noxious stimulation: interaction with halothane and naloxone. Pain 18: 351–358

Korinek A M, Languille M, Bonnet F, Thibonnier M, Sasano P, Lienhart, A, Viars P 1985 Effects of postoperative extradural morphine on ADH secretion. British Journal of Anaesthesia 57: 407–411

McIntosh T K, Bush H L, Palter M, Hay J R, Aun F, Yeston N S, Egdahl R H 1985 Prolonged disruption of beta-endorphin dynamics following surgery. Journal of Surgical Research 38: 210–215

McNicholas L F, Martin W R 1984 New and experimental therapeutic roles for naloxone and related opioid antagonists. Drugs 27: 81–93

Moller I W, Hjortso E, Krantz T, Wandall E, Kehlet H 1984 The modifying effect of spinal anaesthesia on intra- and postoperative adrenocortical and hyperglycaemic response to surgery. Acta Anaesthesiologica Scandanivica 28: 266–269

Moore T C 1984 Anesthesia associated with depression in lymphocyte traffic: less with regional anesthesia than with general anesthesia in sheep. American Journal of Surgery 148: 71–76

Pfeiffer A, Herz A 1984 Endocrine actions of opioids. Hormone Metabolic Research 16: 286–397

Rolly G, Versichelen L, Moerman E 1984 Cardiovascular, metabolic and hormonal changes during isoflurane N₂O anaesthesia. European Journal of Anaesthesiology 1: 327–334

Smith J A R, Norman J N 1982 The fluid of choice for resuscitation of severe shock. British Journal of Surgery 69: 702–705

Sprung C L, Caralis P V, Marcial E H et al 1984 The effects of high-dose corticosteroids in patients with septic shock. The New England Journal of Medicine 311: 1137–1143

Tilders F J H, Berkenbosch F, Vermes I, Linton E A, Smelik P G 1985 Rise of epinephrine and vasopressin in the control of the pituitary-adrenal response to stress. Federation Proceedings 44: 155–160

Wortsman J, Frank S, Cryer P E 1984 Adrenomedullary response to maximal stress in humans. American Journal of Medicine 77: 779–784

MUSCLE RELAXANTS

Atracurium

Bell et al (1984) reported on the use of atracurium in five patients with myasthenia gravis who were having thymectomy. An initial dose of 0.1 mg/kg was administered followed by incremental doses which were reversed at the end of surgery with atropine and neostigmine.

A patient with myasthenia gravis being treated with pyridostigmine and neostigmine orally had a laparotomy for resection of colon for diverticular disease (personal series). Only 10 mg of atracurium were administered for an operation lasting 2.5 hours. The relaxant was not reversed, there was no post-operative respiratory insufficiency and the oral anticholinesterase drugs were not required post-operatively.

REFERENCES

Bell C F, Florence A M, Hunter J M, Jones R S, Utting J E 1984 Atracurium in the myasthenic patient. Anaesthesia 39: 961–968

SLEEP APNEA SYNDROME (SAS)

This subject has been reviewed by Apps(1983) and Partridge (1984). The paraphysiology of sleep apnea and its causes has also been reviewed by Cherniack

(1984). In a further study by Guilleminault et al (1983) of 400 patients with SAS, 48% were found to have cardiac arrhythmias during sleep. These included ventricular tachycardia, sinus arrest for up to 13 seconds and second degree atrioventricular block. Nearly 20% had premature ventricular contractions. The arrhythmias usually occurred during a fit of obstructive apnea. Tracheotomy was performed in 50 patients after which SAS disappeared as well as the significant cardiac arrhythmias (ventricular ectopic beats still occurred).

The frequency of ventricular arrhythmias is reduced during sleep but this increases in patients with neurological disorders associated particularly with cerebrovascular disease (Rosenberg et al 1983). SAS was present in 30% of patients with essential hypertension (Lavie et al 1984).

SAS is often associated with obesity but Onal et al (1985) have shown that there are significant correlations between daytime measurements of pulmonary volume and inspiratory airway conductance and the severity of sleep induced apnea. Airway resistance appears to contribute to SAS but there are still central neurological factors.

The apnea of SAS may be considered to be central, obstructive or have a mixture of these components. In the post-operative period use of drugs in the course of anaesthesia may give rise to a situation similar to SAS in that they can cause respiratory depression and, unless the airway is protected, respiratory obstruction may ensue. There is also the possibility of patients with unrecognized SAS presenting for operation.

REFERENCES

Apps M C P 1983 Sleep-disordered breathing. British Journal of Hospital Medicine 30: 339–347
Cherniack N S 1984 Sleep apnea and its causes. Journal of Clinical Investigation 73: 1501–1506
Guilleminault C, Connolly S J, Winkle R A 1983 Cardiac arrhythmia and conduction disturbances during sleep in 400 patients with sleep apnea syndrome. American Journal of Cardiology 52: 490–494
Lavie P, Ben-Yosef R, Rubin A-H E 1984 Prevalence of sleep apnea syndrome among patients with essential hypertension. American Heart Journal 108: 373–376
Onal E, Leech J A, Lopata M 1985 Relationship between pulmonary function and sleep-induced respiratory abnormalities. Chest 87: 437–441
Partridge M R 1984 Sleep apnea syndromes. In: Kaufman L (ed) Anaesthesia Review 2 Churchill Livingstone, Edinburgh, p 10–20
Rosenberg M J, Uretz E, Denes P 1983 Sleep and ventricular arrhythmias. American Heart Journal 106: 703–709

Index